# Zinc in Plants

Current Knowledge and Recent Advances

# Zinc in Plants

## Current Knowledge and Recent Advances


Edited by

**Durgesh Kumar Tripathi**
Crop Nanobiology and Molecular Stress Physiology Laboratory,
Amity Institute of Organic Agriculture, Amity University Uttar Pradesh,
Noida, Uttar Pradesh, India

**Vijay Pratap Singh**
Department of Botany, C.M.P. Degree College, University of Allahabad,
Prayagraj, Uttar Pradesh, India

**Sangeeta Pandey**
Amity Institute of Organic Agriculture, Amity University Uttar Pradesh,
Noida, Uttar Pradesh, India

**Shivesh Sharma**
Department of Biotechnology, Motilal Nehru National Institute of
Technology, Allahabad, Prayagraj, Uttar Pradesh, India

**D.K. Chauhan**
University of Allahabad, Prayagraj, Uttar Pradesh, India

Academic Press is an imprint of Elsevier
125 London Wall, London EC2Y 5AS, United Kingdom
525 B Street, Suite 1650, San Diego, CA 92101, United States
50 Hampshire Street, 5th Floor, Cambridge, MA 02139, United States



ISBN: 978-0-323-91314-0

For Information on all Academic Press publications
visit our website at https://www.elsevier.com/books-and-journals

*Publisher:* Mara Conner
*Acquisitions Editor:* Nancy Maragioglio
*Editorial Project Manager:* Hilary Carr
*Production Project Manager:* Omer Mukthar
*Cover Designer:* Vicky Pearson Esser

Typeset by MPS Limited, Chennai, India

# Contents

# List of contributors

**Hamada Abdelgawad** Department of Botany and Microbiology, Faculty of Science, Beni-Suef University, Beni-Suef, Egypt11 Department of Biology, University of Antwerp, Antwerp, Belgium

**Nimisha Amist** Department of Botany, Ewing Christian College, Prayagraj, Uttar Pradesh, India

**Wardah Azhar** Institute of Crop Science, Department of Agronomy, Ministry of Agriculture and Rural Affairs Key Lab of Spectroscopy Sensing, Zhejiang University, Hangzhou, P.R. China

**Md. Abdul Azim** Biotechnology Division, Bangladesh Sugarcrop Research Institute, Pabna, Bangladesh

**Zeba Azim** Plant Physiology Laboratory, Department of Botany, University of Allahabad, Prayagraj, Uttar Pradesh, India

**Palak Bakshi** Department of Botanical and Environmental Sciences, Guru Nanak Dev University, Amritsar, Punjab, India

**Garima Balyan** Department of Biotechnology, Faculty of Biosciences and Biotechnology, Invertis University, Bareilly, Uttar Pradesh, India; Crop Nano Biology and Molecular Stress Physiology Lab, Amity Institute of Organic Agriculture, Amity University Uttar Pradesh, Noida, Uttar Pradesh, India

**Renu Bhardwaj** Department of Botanical and Environmental Sciences, Guru Nanak Dev University, Amritsar, Punjab, India

**Tamanna Bhardwaj** Department of Botanical and Environmental Sciences, Guru Nanak Dev University, Amritsar, Punjab, India

**Oscar Cruz-Alvarez** Universidad Autónoma de Chihuahua, Facultad de Ciencias Agrotecnológicas, Chihuahua, Mexico

**Kamini Devi** Department of Botanical and Environmental Sciences, Guru Nanak Dev University, Amritsar, Punjab, India

**Shalini Dhiman** Department of Botanical and Environmental Sciences, Guru Nanak Dev University, Amritsar, Punjab, India

**Neeraj Kumar Dubey** Department of Botany, Rashtriya Snatkottar Mahavidyalaya, Jaunpur, Uttar Pradesh, India

**Ahmed El-Sawah** Department of Agricultural Microbiology, Faculty of Agriculture, Mansoura University, Mansoura, Egypt

**Dalila Jacqueline Escudero-Almanza** Universidad Autónoma de Chihuahua, Facultad de Ciencias Agrotecnológicas, Chihuahua, Mexico

**Arti Gautam** Department of Biochemistry, Institute of Science, Banaras Hindu University, Varanasi, Uttar Pradesh, India

**Vandana Gautam** Department of Botanical and Environmental Sciences, Guru Nanak Dev University, Amritsar, Punjab, India

**Shikha Gupta** Amity Institute of Organic Agriculture, Amity University Uttar Pradesh, Noida, Uttar Pradesh, India

**Yasir Hamid** Ministry of Education (MOE) Key Lab of Environmental Remediation and Ecosystem Health, College of Environmental and Resources Science, Zhejiang University, Hangzhou, P.R. China

**Yousef Alhaj Hamoud** The National Key Laboratory of Water Disaster Prevention, College of Hydrology and Water Resources, Hohai University, Nanjing, P.R. China

**Ofelia Adriana Hernández-Rodríguez** Universidad Autónoma de Chihuahua, Facultad de Ciencias Agrotecnológicas, Chihuahua, Mexico

**Akbar Hossain** Division of Soil Science, Bangladesh Wheat and Maize Research Institute, Dinajpur, Bangladesh

**Sajad Hussain** National Research Center of Intercropping, The Islamic University of Bahawalpur, Pakistan

**Mohd Ibrahim** Department of Botanical and Environmental Sciences, Guru Nanak Dev University, Amritsar, Punjab, India

**Md. Riazul Islam** Department of Pathology, Regional Spices Research Center, BARI, Magura, Bangladesh

**Victoria J** Department of Biotechnology, Motilal Nehru National Institute of Technology Allahabad, Prayagraj, Uttar Pradesh, India

**Ksenija Jakovljević** University of Belgrade, Faculty of Biology, Institute of Botany and Botanical Garden, Belgrade, Serbia

**Izabela Jośko** Institute of Plant Genetics, Breeding and Biotechnology, Faculty of Agrobioengineering, University of Life Sciences, Lublin, Poland

**Nidhi Kandhol** Crop Nanobiology and Molecular Stress Physiology Laboratory, Amity Institute of Organic Agriculture, Amity University Uttar Pradesh, Noida, Uttar Pradesh, India

**Nitika Kapoor** Department of Botanical and Environmental Sciences, Guru Nanak Dev University, Amritsar, Punjab, India

**Gurvarinder Kaur** Department of Botany, Punjabi University, Patiala, Punjab, India

**Rupinder Kaur** Department of Biotechnology, DAV College, Amritsar, Punjab, India

**Ali Raza Khan** School of Emergency Management, Institute of Environmental Health and Ecological Security, School of Environment and Safety Engineering, Jiangsu Province Engineering Research Center of Green Technology and Contingency Management for Emerging Pollutants, Jiangsu University, Zhenjiang, P.R. China

**Kanika Khanna** Department of Botanical and Environmental Sciences, Guru Nanak Dev University, Amritsar, Punjab, India

**Shubhra Khare** Department of Applied Sciences and Humanities, Invertis University, Bareilly, Uttar Pradesh, India

**Vashista Kotra** Amity Institute of Organic Agriculture, Amity University Uttar Pradesh, Noida, Uttar Pradesh, India

**Jaspreet Kour** Department of Botanical and Environmental Sciences, Guru Nanak Dev University, Amritsar, Punjab, India

**Ján Kováč** Department of Plant Physiology, Faculty of Natural Sciences, Comenius University in Bratislava, Bratislava, Slovakia; Department of Phytology, Faculty of Forestry, Technical University in Zvolen, Zvolen, Slovakia

**Pardeep Kumar** Department of Botanical and Environmental Sciences, Guru Nanak Dev University, Amritsar, Punjab, India

**Zuzana Lukačová** Department of Plant Physiology, Faculty of Natural Sciences, Comenius University in Bratislava, Bratislava, Slovakia

**Isha Madaan** Department of Botany, Punjabi University, Patiala, Punjab, India; Science Department, Government College of Education, Jalandhar, Punjab, India

**Shivani Mahra** Department of Biotechnology, Motilal Nehru National Institute of Technology Allahabad, Prayagraj, Uttar Pradesh, India

**Avani Maurya** Crop Nanobiology and Molecular Stress Physiology Laboratory, Amity Institute of Organic Agriculture, Amity University Uttar Pradesh, Noida, Uttar Pradesh, India

**Bilal Ahmad Mir** Department of Botany, School of Life Science, University of Kashmir, Satellite campus, Kargil, Jammu and Kashmir, India

**Tomica Mišljenović** University of Belgrade, Faculty of Biology, Institute of Botany and Botanical Garden, Belgrade, Serbia

**Niharika** Plant Physiology Laboratory, Department of Botany, University of Allahabad, Prayagraj, Uttar Pradesh, India

**Puja Ohri** Department of Zoology, Guru Nanak Dev University, Amritsar, Punjab, India

**Dámaris Leopoldina Ojeda-Barrios** Universidad Autónoma de Chihuahua, Facultad de Ciencias Agrotecnológicas, Chihuahua, Mexico

**Yuridia Ortiz-Rivera** Universidad Autónoma de Ciudad Juárez Chihuahua, Mexico

**Akhilesh Kumar Pandey** Department of Biotechnology, Faculty of Biosciences and Biotechnology, Invertis University, Bareilly, Uttar Pradesh, India

**Sangeeta Pandey** Amity Institute of Organic Agriculture, Amity University Uttar Pradesh, Noida, Uttar Pradesh, India

**Jose Peralta-Videa** Department of Chemistry and Biochemistry, The University of Texas at El Paso, El Paso, TX, United States

**Dorina Podar** Department of Molecular Biology and Biotechnology, Faculty of Biology and Geology, Babeş-Bolyai University, Cluj-Napoca, Romania

**Ved Prakash** Department of Biotechnology, Motilal Nehru National Institute of Technology Allahabad, Prayagraj, Uttar Pradesh, India

**Md. Atikur Rahman** Department of Soil Science, Spices Research Center, Bangladesh Agricultural Research Institute (BARI), Bogura, Bangladesh

**Shivendra Sahi** Department of Biology, Saint Joseph's University, Philadelphia, PA, United States

**Abdul Salam** Institute of Crop Science, Department of Agronomy, Ministry of Agriculture and Rural Affairs Key Lab of Spectroscopy Sensing, Zhejiang University, Hangzhou, P.R. China; Key Laboratory of Natural Pesticide and Chemical Biology, Ministry of Education, South China Agricultural University, Guangzhou, P.R. China

**Hiba Shaghaleh** Key Lab of Integrated Regulation and Resource Development on Shallow Lakes, Ministry of Education, Hohai University, Nanjing, P.R. China

**Anket Sharma** University of Maryland, College Park, MD, United States

**Ashutosh Sharma** Faculty of Agricultural Sciences, DAV University, Jalandhar, Punjab, India

**Indu Sharma** University Institute of Sciences, Sant Baba Bhag Singh University, Jalandhar, Punjab, India

**Neerja Sharma** Department of Botanical and Environmental Sciences, Guru Nanak Dev University, Amritsar, Punjab, India

**Pooja Sharma** Department of Botanical and Environmental Sciences, Guru Nanak Dev University, Amritsar, Punjab, India; Department of Microbiology, DAV University, Jalandhar, Punjab, India

**Samarth Sharma** Department of Biotechnology, Motilal Nehru National Institute of Technology Allahabad, Prayagraj, Uttar Pradesh, India

**Shivesh Sharma** Department of Biotechnology, Motilal Nehru National Institute of Technology Allahabad, Prayagraj, Uttar Pradesh, India

**Vishakha Sharma** Crop Nanobiology and Molecular Stress Physiology Laboratory, Amity Institute of Organic Agriculture, Amity University Uttar Pradesh, Noida, Uttar Pradesh, India

**Mohamed Salah Sheteiwy** Department of Applied Biology, Faculty of Science, University of Sharjah, Sharjah, United Arab Emirates; Department of Agronomy, Faculty of Agriculture, Mansoura University, Mansoura, Egypt

**Amrit Pal Singh** Department of Pharmaceutical Sciences, Guru Nanak Dev University, Amritsar, Punjab, India

**Arun Dev Singh** Department of Botanical and Environmental Sciences, Guru Nanak Dev University, Amritsar, Punjab, India

**Monika Singh** G.L. Bajaj Institute of Technology and Management, Greater Noida, Uttar Pradesh, India

**N.B. Singh** Plant Physiology Laboratory, Department of Botany, University of Allahabad, Prayagraj, Uttar Pradesh, India

**Vijay Pratap Singh** Department of Botany, C.M.P. Degree College, University of Allahabad, Prayagraj, Uttar Pradesh, India

**Geetika Sirhindi** Department of Botany, Punjabi University, Patiala, Punjab, India

**Milan Skalicky** Department of Botany and Plant Physiology, Faculty of Agrobiology, Food, and Natural Resources, Czech University of Life Sciences Prague, Prague, Czech Republic

**Aakriti Srivastava** Crop Nanobiology and Molecular Stress Physiology Laboratory, Amity Institute of Organic Agriculture, Amity University Uttar Pradesh, Noida, Uttar Pradesh, India; Amity Institute of Food Technology, Amity University Uttar Pradesh, Sector-125, Noida, Uttar Pradesh, India

**Shubhangi Suri** Department of Biotechnology, Motilal Nehru National Institute of Technology Allahabad, Prayagraj, Uttar Pradesh, India

**Monika Thakur** Amity Institute of Food Technology, Amity University Uttar Pradesh, Sector-125, Noida, Uttar Pradesh, India

**Kavita Tiwari** Department of Biotechnology, Motilal Nehru National Institute of Technology Allahabad, Prayagraj, Uttar Pradesh, India

**Durgesh Kumar Tripathi** Crop Nanobiology and Molecular Stress Physiology Laboratory, Amity Institute of Organic Agriculture, Amity University Uttar Pradesh, Noida, Uttar Pradesh, India

**Sneha Tripathi** Department of Biotechnology, Motilal Nehru National Institute of Technology Allahabad, Prayagraj, Uttar Pradesh, India

**Zaid Ulhassan** Institute of Crop Science, Department of Agronomy, Ministry of Agriculture and Rural Affairs Key Lab of Spectroscopy Sensing, Zhejiang University, Hangzhou, P.R. China

**Marek Vaculík** Department of Plant Physiology, Faculty of Natural Sciences, Comenius University in Bratislava, Bratislava, Slovakia; Department of Experimental Plant Biology, Institute of Botany, Plant Science and Biodiversity Centre, Slovak Academy of Sciences, Bratislava, Slovakia

**Vijay Bahadur Yadav** Department of Botany, Rashtriya Snatkottar Mahavidyalaya, Jaunpur, Uttar Pradesh, India

**Weijun Zhou** Institute of Crop Science, Department of Agronomy, Ministry of Agriculture and Rural Affairs Key Lab of Spectroscopy Sensing, Zhejiang University, Hangzhou, P.R. China

Chapter 1

# Zinc hyperaccumulation in plants: mechanisms and principles

**Marek Vaculík[1,2], Tomica Mišljenović[3], Zuzana Lukačová[1], Ksenija Jakovljević[3], Dorina Podar[4] and Ján Kováč[1,5]**

[1]*Department of Plant Physiology, Faculty of Natural Sciences, Comenius University in Bratislava, Bratislava, Slovakia,* [2]*Department of Experimental Plant Biology, Institute of Botany, Plant Science and Biodiversity Centre, Slovak Academy of Sciences, Bratislava, Slovakia,* [3]*University of Belgrade, Faculty of Biology, Institute of Botany and Botanical Garden, Belgrade, Serbia,* [4]*Department of Molecular Biology and Biotechnology, Faculty of Biology and Geology, Babeş-Bolyai University, Cluj-Napoca, Romania,* [5]*Department of Phytology, Faculty of Forestry, Technical University in Zvolen, Zvolen, Slovakia*

## 1.1 Introduction

Zinc (Zn) is considered an essential element for all living organisms. It is taken in various extent by plants from the soil solution. Its importance has been shown in various physiological processes, for example, it serves as a cofactor of several enzymes, and has an important role in metabolism of carbohydrates; it regulates the activity, conformation stability, and folding of proteins, and also plays an important role in plant detoxification and membrane stability. Plants usually take up this element by the roots and transfer it to the other plant part in a certain amount. However, when present in excess, it might cause a toxicity symptom. There is relatively broad knowledge about the harmful effect of Zn excess, which can negatively affect the growth and production of agricultural plants. However, there is a group of plants that can take up and actively translocate the extreme doses of Zn to their aerial parts without visible symptoms of Zn toxicity. These extremely Zn-tolerant plants are called hyperaccumulators. Up to now, 31 plant species belonging to 11 families and distributed over the world have been identified to perform this living strategy. Plant species listed as Zn hyperaccumulators display several adaptations resulting from the activated hyperaccumulation mechanisms. Moreover, these plant properties differ on the biochemical,

**Zinc in Plants. DOI: https://doi.org/10.1016/B978-0-323-91314-0.00017-X**

physiological, and molecular levels between the plant populations or species. The present chapter summarizes the knowledge about the availability, essentiality, and toxicity of Zn for common nontolerant plant species, as well as revealed the mysterious phenomenon of plant Zn hyperaccumulation. A detailed list of Zn hyperaccumulating species, and mechanisms of Zn uptake, cellular and tissue localization as well as various physiological and metabolic adaptations of Zn hyperaccumulating species are here described, to better understand the extreme and fascinating life of these unique plants.

## 1.2 Behavior of Zn in soils

Zn is a natural component of soils all over the world with typical background concentrations of 10–100 mg/kg (Alloway, 1995, 2013). It is the 24th most abundant element of the earth's crust and in nature occurs as a part of parental rocks or in Zn-containing ores. Most of Zn is getting to the soils by natural processes, like weathering of parental rocks or volcanic activity. The most important minerals containing Zn in their crystalline structure belong to Zn carbonates, like smithsonite, Zn silicates, like hemimorphite, or Zn sulfides, such as sphalerite and wurzite. The last ones also belong to the most economically important Zn ores (Mertens and Smolders, 2013). Although the concentration of Zn varies between various types of soils, the usually enriched concentration of Zn can be found in mining, ore-processing, and industrial sites. Although the emissions of atmospheric Zn have been dramatically reduced in the last decades, anthropogenic activities still represent one of the most important sources of Zn input to the soils around the world (Mertens and Smolders, 2013). Besides, important sources of soil Zn input are fossil fuel combustion, traffic, and agriculture (Alloway, 2013; Balafrej et al., 2020).

There are soils characterized by unexceptional high content of Zn. One of these soils represents calamine soils that have been developed in a former mining and smelting area near La Calamine in Belgium. The active mining of Zn and Zn-Pb containing ores from the early 19th century and flowed ore processing by smelting led to local anomaly of Zn in these soils (Coppola et al., 2008; Van Damme et al., 2010). On the other hand, it has been reported that around 30% of agricultural soils are Zn deficient, mostly due to lower Zn bioavailability, which is influenced by various physical–chemical and biological factors, such as soil pH, soil organic matter, interaction with other elements—mostly phosphorus—activity of soil microbial communities, or exudation of plant roots (Mertens and Smolders, 2013; Kwon et al., 2017; Balafrej et al., 2020).

A small proportion of Zn total content is in the soil present in the solution (Alloway, 2013). Either in the soil solution or in the rhizosphere, it can be present in different forms, mainly as free ions ($Zn^{2+}$ or $ZnOH^{+}$), in a crystalline form in iron-magnesium minerals, such as sulfides (ZnS), or it couples with iron, aluminum hydrated oxides and humic compounds (Chahal et al., 2005; Moreira et al., 2018; Balafrej et al., 2020). In the colloidal fraction, Zn is predominantly in the form of neutral sulfate $ZnSO_4$ or phosphate $ZnHPO_4$ (Sadeghzadeh and Rengel, 2011).

## 1.3 Zinc essentiality and its importance for plants

Zinc is an essential element for all living organisms such as bacteria, fungi, plants, animals, and humans (King et al., 2000; Stefanidou et al., 2006; Andresen et al., 2018). The requirement of zinc for plant life was questioned until 1926 when Sommer and Lipman (Sommer and Lipman, 1926) showed that Zn was needed for the growth and development of sunflowers and barley. This finding resulted in Zn being generally recognized as essential for higher plants (Broadley et al., 2007). Zinc essentiality for experimental animals was shown in 1934 (Todd et al., 1934), and for humans in 1961 (Prasad, 2013). In plants, even if zinc is needed only in trace amounts, it is crucial for plant development as it plays a significant part in a wide range of processes (Sharma et al., 2013).

There is a variability of Zn requirement for each plant. If 10–20 ppm of Zn in DW is considered as a Zn deficit and 100–500 ppm of Zn as a range of first toxicity visible symptoms, normal concentration of Zn in many crop tissues is in a range of 30–50 ppm of DW (Kabata-Pendias and Pendias, 2001). Zinc is in general absorbed by the roots and transferred to the aboveground plant part under tight control thanks to a network of barriers, transporters, chelators, and cell wall compartments (Caldelas and Weiss, 2017; Olsen and Palmgren, 2014; Sinclair and Krämer, 2012). The absorption of Zn through roots is generally in the form of $Zn^{2+}$ ions or hydrated Zn except at high pH of soil solution, where it is absorbed in the form of ZnOH. However, different factors alter its bioavailability: from the plant side, the root exudates, and from the soil, it is pH, microbial communities, and organic matter (Balafrej et al., 2020) promoting or limiting its toxicity. After entering the root rhizodermis, ions are radially transported through apoplasm or symplasm into the central cylinder, where Zn is pumped into the stellar apoplasm and to the xylem (Caldelas and Weiss, 2017).

The symplasmic route in the ion radial root transport is contributed by plasmodesmata in the neighboring cell walls forming a continuum without barriers; such a pathway does not require any energy (Gupta et al., 2016). In the apoplasmic space which is formed by the cell walls and intercellular spaces, Zn is partially blocked in the flow by binding to hydroxyl groups of cellulose and hemicellulose, and by carboxyl or hydroxyl groups of pectin and also by binding to the cell wall proteins (Weiss et al., 2005; Sarret et al., 2009). This extracellular space for the ions movement is used only in the areas where no apoplasmic barriers are developed, especially in the apical and lateral root initiation regions (White et al., 2002). The apoplasmic route is most often blocked in the endodermis, or, even more distally in the root exodermis by Casparian strip, a mixture of lignin and amorphous suberin deposited in the primary cell wall. This, and the next stages of apoplasmic barrier development, like suberin lamellae or tertiary cell wall, makes it impermeable (Gupta et al., 2016). In some species of Brassicaceae, the anatomists described another unusual root feature contributing to the restriction of the movement of ions—the cell layer of so-called periendodermis with irregularly lignin-thickened inner tangential walls. Zelko et al. (2008) described such root cortex modification in a hyperaccumulating *Noccaea* (former *Thlaspi) caerulescens*. Cells formed a compact cylinder with no similar structure in the nonhyperaccumulator *T. arvense*. More recently, this specific root structure of *N. caerulescens* has been more deeply characterized, and it is suggested that it might play a role in controlling ion transport toward the root tissues (Kováč et al., 2020). van de Mortel et al. (2006) described two layers of endodermis in *N. caerulescens* roots and high lignin biosynthesis genes under Zn supply which corresponds to the lignin deposition in the endodermis. However, the phenomenon of the double-layered endodermis or such cortical modifications has not been confirmed or satisfyingly explained by other studies. Anyway, plants use lignin deposits to control metal influx or to prevent metal efflux from the central cylinder (van de Mortel et al., 2006). On the other side, also inorganic compounds like silicon (Si) contribute to cell wall incrustation and, resulting in the bounded Zn-complexes (Lux et al., 2020).

In general, Zn is considered an essential elements for all living organisms, including plants. Zn plays an important role in plants, and an appropriate concentration of Zn is required for optimal plant growth and development. Its importance has been shown for various physiological processes: it is a cofactor of several enzymes, has an important role in the metabolism of carbohydrates, stabilizes the cell membranes, and plays an important role in plant detoxification. Plants suffering from Zn deficiency show significant symptoms such as growth reduction or leaf chlorosis (Mengel and Kirkby, 2001; Broadley et al., 2007; Marschner and Marschner, 2012). Zinc is a cofactor of more than 300 proteins, and is involved in Zinc finger proteins: RNA and DNA polymerases (Lopez Millan et al., 2005). It is the only element present in all six groups of enzymes (oxidoreductases, transferases, hydrolases, lyases, isomerases, and ligases). Zinc is also a part of structural and catalytic units and regulates the activity, conformation stability, and folding of proteins. Several studies also described the role of Zn in processes connected with the stability of cell membranes, alleviation of oxidative stress, and secondary cell signalization (Cakmak, 2000; Yamasaki et al., 2007; Disante et al., 2010). Zinc is also important for the chloroplast and cytoplasm of plant cells. In these cell compartments, there are Zn-dependent enzymes that are involved in processes related to photosynthesis, repair of photosystem II during photoinhibition, and maintaining the suitable $CO_2$ concentration in leaf mesophyll cells (Bailey et al., 2002; Lin et al., 2005; Hansch and Mendel, 2009). Peck and McDonald (2010) confirmed the role of Zn in regulation of Rubisco activity together with the alleviation of temperature stress. Most plant enzymes that contain Zn are involved in carbohydrate metabolism and regulate the synthesis of auxin, while Zn is necessary for the biosynthesis of tryptophan, a precursor of auxins (Coleman, 1992; Marschner and Marschner, 2012).

Zinc finger proteins contain cysteine and histidine-rich motifs. These amino acids coordinate Zn ions that are involved in peptide structures for specific biological functions (Hall, 2005). Through the bounded Zn, the motifs develop protuberances that form tandem connections with molecules. Zn fingers are bound to DNA or other proteins in a variable way, according to their sequence and structure. Zinc fingers play an important role in gene transcription, translation, cytoskeleton organization, cellular adhesion, protein folding, chromatin remodeling, and cellular signalization (Laity et al., 2001). It has been widely confirmed that many Zn fingers are important for the regulation of plant growth and development (Ciftci-Yilmaz and Mittler, 2008), and are involved in plant reactions to biotic and abiotic stresses (Cui et al., 2002; Sakamoto et al., 2004; Giri et al., 2011). Additionally, plant enzymes activated by Zn are involved in the maintenance of the integrity of cellular membranes (Broadley et al., 2012), formation of chlorophyll (Samreen et al., 2017), and Zn presence in plants plays fundamental roles in resistance against the drought stress and helps the plant to withstand cold temperatures (Hassan et al., 2020).

## 1.4 Problems with Zn excess in nontolerant plants

Zinc toxicity is a less common problem than Zn deficit, and occurs mostly on artificially contaminated sites (mining areas, cities, industrial sites) (Broadley et al., 2007). In some cases, excess or deficit of other elements may lead to Zn excess in plants. Maize (*Zea mays* L.) growing in low iron (Fe) conditions accumulates 15 times more Zn compared to normal Zn and normal Fe conditions in the substrate (Kanai et al., 2009). Due to the low cation transporters specificity for Zn uptake, the excess of Zn may be induced also by the deficit of copper (Cu) or magnesium (Mg) (Sinclair and Krämer, 2012). Contrary, the excess of Zn induced the deficit of Fe (Becher et al., 2004). This Zn-induced Fe deficiency caused overexpression of membrane localized Fe reductases—FRO2 and Fe transporter—IRT1. Moreover, excess of Zn and Fe deficit are visible as a chlorosis on leaves. If plants are growing in excess of Zn, but well-supplied with Fe, the chlorosis is minimized, and this supports the role of Zn excess on Fe deficiency (Fukao et al., 2011). The excess of Zn decreases the availability and amount of phosphorus (P) in the root, and changes the translocation of other elements. For example, increased manganese (Mn) and limited magnesium (Mg) translocation from roots to the leaves has been observed under Zn excess in tea plants (Venkatesan et al., 2006). In sugar beet (*Beta vulgaris* L.) excess of Zn decreased nitrogen (N), Mg, potassium (K), and manganese (Mn) concentrations in the whole plant, and increased phosphorus (P) and calcium (Ca) content in shoots. Also, Fe deficit-like symptoms including chlorosis were observed (Sagardoy et al., 2009). Any toxicity of Zn is necessarily connected with other symptoms caused by excess or deficit of other elements. Nevertheless, some characteristic symptoms still occur.

The increase in tissue Zn concentration leads to a reduction of plant growth and causes anatomical alterations in plant roots and shoots (Yadav et al., 2021). Stoláriková-Vaculíková et al. (2015) have reported modified development of root apoplasmic barriers in *Populus deltoides* as a direct response to increased Zn concentration in the growth medium. Furthermore, excessive Zn concentration decreased the root area and diameter of vascular bundles and influenced the formation of Casparian bands in endodermal layer as well as increased the lignification of xylem tissue. Furthermore, Stoláriková et al. (2012) also reported that Zn induced the formation of apoplasmic barriers in *Populus* x *euramericana* clone I-214. Likewise, Vaculik et al. (2012) also found that in roots of *Salix caprea* excessive Zn induced the formation of Casparian bands and suberin lamellae in exodermis and endodermis. Päivöke (1983) studied the impact of Zn toxicity on root anatomy of *Pisum sativum* and stated that Zn reduced the number of root hairs, and induced the lignification of exodermis and endodermis; moreover, Zn also increased the number of calcium oxalate crystals in the pericycle of the root. Zn also alters the intercellular spaces and breakdown of vascular bundles in the Zn-containing plant stem (Sridhar et al., 2005). Parallel to root and stem, leaf lamina also showed alterations upon exposure to Zn, such as palisade cell distorted with large intercellular spaces. Moreover, Zn exposure also modified the spongy parenchyma, damaged the chloroplast, and decreased the number of thylakoids (Stoláriková-Vaculíková et al., 2015). Excessive presence of Zn reduced the number and area of starch grains in the chloroplast of *Populus alba* clone Villafranca (Todeschini et al., 2011), and induced the breakage of epidermal cell, palisade cell, and spongy cell in *Brassica juncea*. Furthermore, excess of Zn also induced the loss of cell, and decreased intercellular space and starch content (Sridhar et al., 2005). Zinc was predominantly found in the rhizodermis of hybrid poplar (*Populus* × *euramericana* clone I-214) exposed to elevated Zn, and the concentration of Zn decreased gradually from the surface to the inner part of the root (Stoláriková et al., 2012). The same pattern of Zn localization has also been documented in other secondarily thickened roots, such as gray mangroves (McFarlane and Burchett, 2000), spruce (Brunner et al., 2008), or species from the Salicaceae family (Di Baccio et al., 2009; Vaculík et al., 2012).

Excess of Zn in a bean (*Phaseolus vulgaris* L.) inhibits the carboxylation activity of Rubisco, probably because of the substitution of Mg by Zn in the reaction center, which influences the ability of this enzyme to bind $CO_2$ (Van Assche and Clijsters, 1986). The substitution of Mn in water oxidation complex (WOC) of photosystem II (PSII) by Zn is common for cyanobacterium *Spirulina platensis* (Ranjani et al., 2014). A similar trend is observed in maize, the negative impact of Zn on photosystem II is through light-harvesting complex and water oxidation complex. This inhibition is preferentially on photosystem II and less on photosystems (Adam and Murthy, 2014).

Even if Zn is not active in redox reactions (Marschner and Marschner, 2012), its presence in excess amounts may lead to a generation of reactive oxygen species (ROS). One of the main factors causing ROS production is the higher activity of lipoxygenases which subsequently oxidize fatty acids in membranes and form toxic malondialdehyde (Barrameda-Medina et al., 2014). Also, displacement of other metals from active sites of enzymes may cause an imbalance in ROS production and scavenging (Sagardoy et al., 2009). It is also important to note the impairment of photosystems and water-splitting complexes as they generate many ROS. Subsequent changes at the ultrastructure level include damage to the cell membranes, as well as collapse of thylakoids in chloroplasts and cristae in mitochondria (Mukhopadhyay and Mondal, 2014). The visual symptoms such as browning, wilting, or leave burning (Venkatesan et al., 2006) are only a part of overall toxicity together with excessive ROS production, destabilization of elements homeostasis, decreased stomatal conductivity, decreased transpiration rate, and overall photosynthesis (Subba et al., 2014).

Variability in demand for Zn is clearly visible in different reactions of two crop plants, *Lactuca sativa* and *Brassica oleracea*, to Zn excess. Both plants accumulate Zn in roots with the same toxicity symptoms visible as a similar reduction in biomass production. In green parts, the concentration of Zn in *B. oleracea* exceeded the concentration in *L. sativa*, however without any marked symptoms. Overall better tolerance of *B. oleracea* to Zn was connected with an increase in superoxide dismutase, glutathione reductase, and ascorbate peroxidase; glutathione and ascorbate were kept in high amount and in a reduced state, active in detoxification. Decreased ROS together with lower lipoxygenase activity promote better growth in toxic environment. Contrary, *L. sativa* with higher ROS production due to lower activity of antioxidant enzymes and enhanced lipoxygenase activity showed reduced growth (Barrameda-Medina et al., 2014).

In young leaves of common bean (*Phaseolus vulgaris*), ascorbate in the root increases shortly after Zn exposure, and thanks to ethylene, the signal from the root is transferred to shoot before Zn enters the leaves (Cuypers et al., 2001). A similar occurs in sugar beet, enhanced Zn in medium modifies the enzymes of the Krebs cycle in root cells. It causes an increase in citric and malic acid concentration in xylem and leaves where they supply limited products of photosynthates (Sagardoy et al., 2011) or maybe precursors for chelators. This is the case of increasing proline concentration which may improve the osmotic environment same as a chelator eliminating Zn in excess (Subba et al., 2014). In the roots of Arabidopsis, Zn bound by chelators is transported to the vacuoles through vacuolar transporters ZIF1 which are overexpressed in Zn excess (Haydon and Cobbett, 2007). Also, transporter AtHMA3 decreases the amount of Zn in cytoplasm and prevents the transport to the shoots (Morel et al., 2008). Overexpression of vacuolar transporters is connected with a decrease in the expression of transporters in epidermal cells responsible for Zn uptake like AtZIP4 (van de Mortel et al., 2008) or AtHMA4. The former might be responsible also for the transport of Zn out of root back to the rhizosphere (Mills et al., 2003).

## 1.5 Hyperaccumulation of Zn

### 1.5.1 Definition of Zn hyperaccumulation

The search for plants growing on naturally metal-enriched sites is probably as old as ore mining itself. Nevertheless, for scientific research, these species were introduced much recently. Brooks (1994) and Reeves (2006) reviewing this topic, put the period of the first discoveries of metallophytes, plants that were able to grow on metalliferous or metallicolous sites, to the mid and late 19th century in Germany. The works of Riss, Sachs, Baumann, and Forchhammer described *Viola calaminaria* and *Thlaspi calaminare* as plant species growing on calamine (Zn bearing) soils. These plants were able to accumulate concentrations of Zn in leaves or stems far above the limit typical for "trace element" or toxicity levels (Brooks, 1994; Reeves, 2006). During the following decades, new species with such characteristics were found all over the world. Mainly because of the growing number of Ni-accumulating species and variable terminology used, the term "hyperaccumulator" was introduced to characterize those plant species with the ability to accumulate metals in high amounts (Jaffré et al., 1976; Brooks et al., 1977). In contrast to numerous species that act as indicators or excluders in metalliferous habitats, hyperaccumulators actively take up potentially toxic elements and transport them to their aboveground parts, especially to the leaves, where they accumulate in concentrations 100–1000 times higher than surrounding species without toxicity symptoms. Generally, most known plant species take up and accumulate elements in the roots, and translocate only part of them to their aerial parts. However, the accumulation of metals in the leaves is dependent on metabolic activity, and exceeding the "normal" tissue metal concentration requires some abnormal physiological adjustment.

To be considered a hyperaccumulator, in addition to active transport to the aboveground parts, the species has to take up a specific element in concentrations exceeding the nominal threshold, for which Zn is 3000 mg/kg (van der Ent et al., 2012, 2013). However, the concentration of 10,000 mg/kg proposed by Baker and Brooks (1989) is still commonly used as a delimiting criterion for hyperaccumulation (Baker and Brooks, 1989), although it has been shown that this value is too restrictive and that a concentration of 3000 mg/kg in plants is quite sufficient to indicate specific physiological mechanisms characteristic of hyperaccumulators (Reeves and Baker, 2000; Krämer, 2010; van der Ent et al., 2013). To date, the phenomenon of hyperaccumulation has been noted in more than 700 taxa, most of which (about 500 taxa) accumulate Ni (Purwadi et al., 2021). Hyperaccumulation of Ni has been confirmed in more than 70% of hyperaccumulating species (Nkrumah et al., 2018), while Zn is much less frequently accumulated at very high concentrations (van de Mortel et al., 2008).

## 1.5.2 Plants that hyperaccumulate zinc

So far, hyperaccumulation of Zn has been confirmed in 31 taxa (Table 1.1). Most of these belong to the Brassicaceae family (Fig. 1.1), which is known for the largest number of hyperaccumulators discovered, with just over 100 taxa, most of which hyperaccumulate Ni (Cappa et al., 2015). Of the 16 taxa in the family Brassicaceae that hyperaccumulate Zn, 12 belong to *Noccaea* and two taxa belong to the genera *Arabis* and *Arabidopsis* (Fig. 1.2). In addition to *Viola*, the genus *Thlaspi* (part of which was later placed in the genus *Noccaea*) was the first to be included in research on Zn accumulation in plants (Reeves and Baker, 2000). In terms of metal accumulation, *Noccaea caerulescens* has been most intensively studied as a hyperaccumulator of Ni, Cd, and Zn (Callahan et al., 2016; Tang et al., 2019; Kozhevnikova et al., 2020). Hyperaccumulation of Zn is considered a constitutive feature of *N. caerulescens*, considering that excessive Zn concentrations (8890 mg/kg) were found even in plants growing in soil containing only 139 mg/kg Zn (Reeves et al., 2001; Nkrumah et al., 2018). On the other hand, foliar Zn concentrations reach 53,450 mg/kg in metalliferous sites, rich in Zn and Pb (Reeves et al., 2001). Anomalous Zn concentrations have also been found in other *Noccaea* species, but predominantly in those growing on metalliferous soil (Brooks, 1994). A significant potential for Zn hyperaccumulation was also found in *N. kovatsii* (Heuff.) F. K. Mey growing on schists in Mt. Kopaonik in Serbia, considering shoot concentrations of 4920 mg/kg, with a strong tendency for Zn uptake in the shoot (Mišljenović et al., 2020). A comparable high concentration (4850 mg/kg) was previously reported by Reeves and Brooks (1983). Despite exceeding the hyperaccumulation threshold and showing a strong tendency to accumulate, due to the small number of populations with high Zn concentrations, *N. kovatsii* can only be considered a doubtful hyperaccumulator until more detailed analysis confirms or refutes this potential. Similar to *N. caerulescens*, *Arabidopsis halleri* is a good model species to study hyperaccumulation, with Zn hyperaccumulation being a constitutive trait, despite some intraspecific variability (Huguet et al., 2012).

**TABLE 1.1** List of Zn hyperaccumulating species with the highest recorded Zn concentrations in mg/kg.

| Taxon | Family | Zn (shoot) | Zn (soil) | References |
|---|---|---|---|---|
| *Rostellularia procumbens* (L.) Nees | Acanthaceae | 11,071 | 51,793 t, 1495 e | Phaenark et al. (2009) |
| *Gynura pseudochina* (L.) DC. | Asteraceae | 6171 | 16,740 t 1682 e | Phaenark et al. (2009) |
| *Picris divaricata* Vaniot | Asteraceae | 18,339 | – | Tang et al. (2009) |
| *Arabidopsis halleri* (L.) O'Kane & Al-Shehbaz | Brassicaceae | 53,900 | 342 t 210 e | Stein et al. (2017) |
| *A. halleri* subsp. *gemmifera* (Matsum.) O'Kane & Al-Shehbaz | Brassicaceae | 20,300 | 2880 t | Kubota and Takenaka (2003) |
| *Arabis alpina* L. | Brassicaceae | 7693 | 4178 t | Li et al. (2019) |
| *Arabis paniculata* Franch | Brassicaceae | 20,700 | 179,000 t 2170 e | Tang et al. (2009) |
| *Noccaea bulbosa* (Spruner) Al-Shehbaz | Brassicaceae | 10,500 | – | Brooks (1994) |
| *Noccaea caerulescens* (J.Presl & C.Presl) F.K. Mey. | Brassicaceae | 53,450 | 64,360 t | Reeves et al. (2001) |
| *N. caerulescens* subsp. *brachypetala* (Jord.) Tzvelev | Brassicaceae | 15,300 | – | Reeves (2006) |
| *N. caerulescens* subsp. *calaminaris* (Lej.) Holub | Brassicaceae | 39,600 | – | Brooks (1994) |
| *N. caerulescens* subsp. *tatrense* (Zapal.) Holub | Brassicaceae | 27,000 | – | Brooks (1994) |
| *Noccaea cepaeifolia* (Wulfen) Rchb. | Brassicaceae | 18,500 | – | Reeves (2006) |
| *Noccaea eburneosa* F.K. Mey | Brassicaceae | 10,500 | – | Reeves and Brooks (1983) |
| *Noccaea goesingensis* (Halácsy) F.K.Mey. | Brassicaceae | 14,500 | 1180 t | Reeves and Brooks (1983) |
| | Brassicaceae | 11,000 | – | Brooks (1994) |

*(Continued)*

**TABLE 1.1 (Continued)**

| Taxon | Family | Zn (shoot) | Zn (soil) | References |
|---|---|---|---|---|
| *Noccaea limosellifolia* (Reut. ex Burnat) F.K.Mey. | | | | |
| *Noccaea ochroleuca* (Boiss. & Heldr.) F.K. Mey. | Brassicaceae | 6310 | – | Reeves and Brooks (1983) |
| *Noccaea praecox* (Wulfen) F.K.Mey. | Brassicaceae | 21,000 | – | Brooks (1994) |
| *Noccaea stenoptera* (Boiss. & Reut.) F.K. Mey. | Brassicaceae | 16,000 | – | Brooks (1994) |
| *Polycarpaea synandra* F. Muell. | Caryophyllaceae | 6960 | – | Cole et al. (1968) |
| *Sabulina verna* (L.) Rchb. | Caryophyllaceae | 11,400 | – | Reeves (2006) |
| *Sedum alfredii* Hance | Crassulaceae | 5000 | 2571 t144 e | Yang et al. (2002) |
| *Sedum plumbizincicola* X.H.Guo & S.B.Zhou ex L.H.Wu | Crassulaceae | 18,400 | – | Cao et al. (2014) |
| *Dichapetalum gelonioides* subsp. *pilosum* Leenh. | Dichapetalaceae | 26,360 | – | Baker et al. (1992) |
| *D. gelonioides* subsp. *sumatranum* Leenh. | Dichapetalaceae | 15,660 | – | Baker et al. (1992) |
| *Anthyllis vulneraria* L. | Fabaceae | 17,428 | | Ievinsh et al. (2020) |
| *Crotalaria montana* Roxb. ex Roth | Fabaceae | 4883 | 19,549 t 1822 e | Phaenark et al. (2009) |
| *Corydalis davidii* Franch | Papaveraceae | 14,000 | 26700 t 700 e | Lin et al. (2012) |
| *Potentilla griffithii* Hook.f. | Rosaceae | 22,991 | – | Qiu et al. (2006) |
| *Viola baoshanensis* W. S.Shu, W.Liu & C.Y.Lan | Violaceae | 3428 | 8849 t 257 e | Wu et al. (2010) |
| *Arabidopsis arenosa* | Brassicaceae | 13,700 | 18,700 t | Gieron et al. (2021) |

t-total Zn; e-extractable Zn.

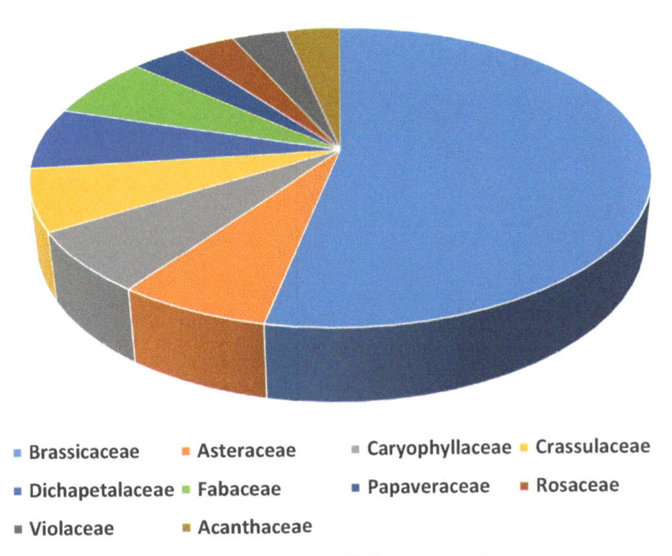


**FIGURE 1.1** The proportion of families among Zn hyperaccumulators.

Thus, foliar Zn concentrations of up to 53,900 mg/kg were found in *A. halleri* at a nonmetalliferous site, which is almost 20 times higher than the nominal hyperaccumulation threshold. The strong shoot-to-root ratio confirms the hyperaccumulation potential of this species (Stein et al., 2017). High Zn concentrations were also observed in *A. halleri* subsp. *gemmifera*, *Arabis alpina*, and *A. paniculata*, which are commonly associated with calamine soil (Table 1.1). Particularly strong hyperaccumulation was observed in the shoot of *A. paniculata* under field conditions (71,900 and 20,700 mg/kg, as maximum and average concentrations, respectively; Tang et al., 2009). Recently, *A. arenosa* has been described as a novel Zn hyperaccumulator (Szopinski et al., 2020; Gieron et al., 2021), although several previous studies did not indicate the hyperaccumulation potential of this species (e.g., Szarek-Łukaszewska and Niklinska, 2002, Peer et al., 2006).

Hyperaccumulation of Zn on nonmetalliferous substrate was also found in *Dichapetalum gelonioides* subsp. *sumatranum* and *D. gelonioides* subsp. *pilosum* with leaf concentrations of 15,660 and 26,360 mg/kg, respectively (Baker et al., 1992). However, unlike *N. caerulescens* and *A. halleri*, these two taxa grow exclusively in nonmetalliferous soil that has a low Zn content. For example, 10,730 mg/kg Zn was found in the dry leaves of *D. gelonioides* subsp. *sumatranum*, despite a concentration of only 20 mg/kg Zn in the soil (Nkrumah et al., 2018).

Within the Violaceae family, a considerable number of hyperaccumulating species have been identified, particularly among *Hybanthus*, *Rinorea*, and *Viola* species. Although they predominantly hyperaccumulate Ni (Paul et al., 2020; van der Ent et al., 2020), hyperaccumulation of As and Tl has

**FIGURE 1.2** Flowers (A) and siliques (B) of common model of Zn hyperaccumulators, *Noccaea caerulescens*, pictures from population of Banská Štiavnica, Slovakia; similar plant species, *N. kovatsii* from Mt. Kopaonik, Serbia showing hyperaccumulation propereties (C); old-known metalophyte and recently described as Zn hyperaccumulating species *Arabidopsis arenosa* from Banská Štiavnica, Slovakia (D).

also been observed in *Viola* species (Tomović et al., 2021). *Viola baoshanensis* is the only species that can be considered a hyperaccumulator of Zn, with 3428 mg/kg Zn in shoots under field conditions with a shoot-to-root ratio greater than 1 in plants under natural conditions and in most experimentally grown plants. A hyperaccumulation concentration of Zn (10,000 mg/kg) was also found in plant tissue of *Viola lutea* subsp. *calaminaria* (Ging.) Nauenb (known as *V. calaminaria*) (Reeves and Baker, 2000). However, similar to *N. kovatsii*, this species can only be considered a "doubtful hyperaccumulator" (Macnair, 2003) due to the small number of samples with high Zn concentrations.

Caryophyllaceae is one of the families with a high proportion of hyperaccumulators (Prasad et al., 2003). However, hyperaccumulation of Zn was observed only in *Sabulina verna* (syn. *Minuartia verna*) and *Polycarpaea synandra* (Reeves, 2006; Paul et al., 2018). The association of *S. verna* with metalliferous habitats was recognized as early as the 16th century (Thalius, 1588), and later studies confirmed its capacity for uptake of Cd, Cu, Pb, and Zn (Baker and Proctor, 1990; Wenzel and Jockwer, 1999). *Polycarpea synandra*, the Zn hyperaccumulator from northern Australia, is associated with even more extreme habitats and grows on a substrate with anomalous concentrations of Zn and Pb (50,000 and 5000 mg/kg, respectively) (Paul et al., 2018).

The Fabaceae and Crassulaceae families are represented by two species among Zn hyperaccumulators. Zinc concentrations in the aboveground tissues of *Anthyllis vulneraria* exceeded the nominal hyperaccumulation threshold (17,428 mg/kg), indicating that this widespread species is a potentially very good candidate for phytoextraction (Ievinsh et al., 2020). Significantly lower Zn concentrations were found in *Crotalaria montana*, but still above the nominal hyperaccumulation threshold (Phaenark et al., 2009). Two *Sedum* species (*Sedum alfredii* and *S. plumbizincicola*) have also been shown to be Zn hyperaccumulators (Yang et al., 2002; Cao et al., 2014). Particularly high concentrations were found in the shoots of *S. plumbizincicola* (18,400 mg/kg) (Cao et al., 2014). *Sedum alfredii* could be advantageous for application in phytoextraction, considering that the shoots not only contain a significant amount of accumulated Zn but also have a larger biomass compared to some of the Zn hyperaccumulators (Yang et al., 2002). Significant potential for phytoextraction was also found in *Corydalis davidii* (family Papaveraceae), due to its high concentration of accumulated Zn (up to 14,000 mg/kg), shoot-to-root ratio greater than 1, and relatively large aboveground biomass, especially compared to known hyperaccumulating plant species such as those in the genus *Noccaea* (Lin et al., 2012).

Two Asian species, *Picris divaricata* and *Gynura pseudochina*, represented the Asteraceae family among the hyperaccumulators of Zn. The highest Zn concentration determined in the shoot of *P. divaricata* was 18,339 mg/kg, with a mean value of 5911 mg/kg (Tang et al., 2009). A slightly lower value was found in the shoots of *G. pseudochina* (Phaenark et al., 2009). Due to rapid growth and Zn concentrations above the nominal hyperaccumulation threshold, this species shows good potential for use in phytoremediation. *Rostellularia procumbens* (syn. *Justicia procumbens* L., family Acanthaceae) shows similar characteristics. However, the use of these two species in phytoremediation is feasible exclusively in the rainy season when they exhibit intense growth (Phaenark et al., 2009).

*Potentilla griffithii* Hook. f. is the only representative of the Rosaceae family among Zn hyperaccumulators. Concentrations above the hyperaccumulation threshold were determined both in the natural population growing in Zn-rich mining soils and under experimental conditions. The highest foliar concentration determined was 22,991 mg/kg, with an average value of 17,062 mg/kg. A slightly higher concentration was found in the shoot under the experimental conditions (26,700 mg/kg; Qiu et al., 2006). In the study by Hu et al. (2009), petioles were analyzed separately and the results showed that they accumulated 19,600 mg/kg Zn, more than the leaves and roots. Although in *P. griffithii* the shoot-to-root ratio is predominantly greater than one, which is an additional criterion to distinguish hyperaccumulator species (van der Ent et al., 2013), in some populations of this species, Zn concentrations were lower in shoots compared with those in soil, most likely due to an anomalous amount of Zn in the soil (up to 193,000 mg/kg) and the existence of an internal mechanism preventing the uptake of excessive amounts of this element (Qiu et al., 2006).

According to Brooks (1994), *Haumaniastrum katangense* (S.Moore) P.A. Duvign. and Plancke is also considered a hyperaccumulator of Zn, with a foliar concentration of 19,800 mg/kg Zn. However, since this concentration was only found in a single specimen of this species, the hyperaccumulation of Zn in this species is doubtful and most likely due to the local presence of Zn-rich ore, so these data should be omitted (Paton and Brooks, 1996).

Although hyperaccumulation is usually associated with single elements, there are species capable of accumulating multiple elements in amounts above the nominal threshold (Mišljenović et al., 2020). Geochemically related elements are usually taken up simultaneously, as is the case of Zn and Cd in the species *Gynura pseudochina*, *Rostellularia procumbens*, *Picris divaricata*, *Arabidopsis halleri* subsp. *gemmifera*, *Sedum alfredii*, *S. plumbizincicola*, *Viola baoshanensis* (Kubota and Takenaka, 2003; Yang et al., 2004; Phaenark et al., 2009; Tang et al., 2009; Wu et al., 2010; Deinlein et al., 2012; Cao et al., 2014; Dinh et al., 2015; Xv et al., 2020). At the same time, Cd hyperaccumulators are highly tolerant to elevated Zn concentrations since they are predominantly bound to calamine soils that are rich in Cd, Pb, and Zn (Bert et al., 2002; van der Ent et al., 2013). Moreover, the fact that Cd hyperaccumulators simultaneously take up Zn, a strong correlation between root and shoot factors, and between concentrations of these two elements in different species of the same genus, such as *Noccaea praecox* and *N. caerulescens*, largely indicate the presence of certain common genetic traits, while in *S. alfredii*, this trait is related to the metal-rich substrate (Verbruggen et al., 2009). In addition to Zn and Cd, hyperaccumulation of

Pb has also been found in *Arabis alpina* and *A. paniculata* (Tang et al., 2009; Li et al., 2019), while *Polycarpaea synandra* accumulates Zn and Pb above the nominal threshold (Paul et al., 2018). Hyperaccumulation of Zn and Ni is extremely rare and uptake of only one of these two elements is much more common in nature (Nkrumah et al., 2018). However, in *Noccaea* species, hyperaccumulation of Ni and Zn, as well as other geochemically related elements (Cd and Pb), is not so uncommon. Hyperaccumulation of Zn and Ni has been noted in *Noccaea bulbosa*, *N. goesingense*, and *N. ochroleuca* (Reeves and Brooks, 1983), while *N. praecox* and *N. caerulescens* also accumulate Cd above the nominal threshold in addition to these two elements (Tolra et al., 2006; Kozhevnikova et al., 2020). In contrast to the hyperaccumulation of Zn, which is considered a constitutive trait in *N. caerulescens*, the accumulation of Cd and Ni proved to be population-specific under both field and experimental conditions (Kozhevnikova et al., 2020).

### 1.5.3 Geographical distribution of Zn hyperaccumulators

Three main groups of Zn hyperaccumulators can be identified based on their geographical distribution across the world. The first group is represented by species distributed in the eastern, southern, and south-eastern parts of Asia, Australia, and parts of Africa. These include *Arabidopsis halleri* subsp. *gemmifera*, *Arabis paniculata*, *Corydalis davidii*, *Crotalaria montana*, *Dichapetalum gelonioides*, *Gynura pseudochina*, *Picris divaricata*, *Polycarpaea synandra*, *Potentilla griffithii*, *Rostellularia procumbens*, *Sedum alfredii*, *S. plumbizincicola*, and *Viola baoshanensis*. The second group consists of species whose distribution is related to Europe, and these are mainly species of the genus *Noccaea*. Most of them inhabit parts of southern, somewhat less frequently of central Europe, while only *N. caerulescens* reach northern Europe, that is, the Baltic countries. In contrast to these species, whose distribution is restricted either to parts of Asia, Australia, and Africa or to European countries, the ranges of a third group covering *Arabidopsis halleri*, *Arabis alpina*, and *Sabulina verna*, although predominantly associated with Europe, also include parts of Africa and especially wider or narrower parts of Asia (Fig. 1.3; POWO, 2021).

Endemism is a phenomenon very commonly associated with hyperaccumulators, as these species are often exclusively associated with a particular type of metalliferous habitat. Due to the hostile conditions, only a small number of species survive in such habitats, and there is less competition with other species, which, along with the potentially higher demand for heavy metals, facilitates survival (Baker et al., 2000). Endemic metallophytes can be paleoendemic and neoendemic. Paleoendemic metallophytes are species that were formerly more widespread but whose range has narrowed over time and whose distribution is now exclusively in metalliferous habitats

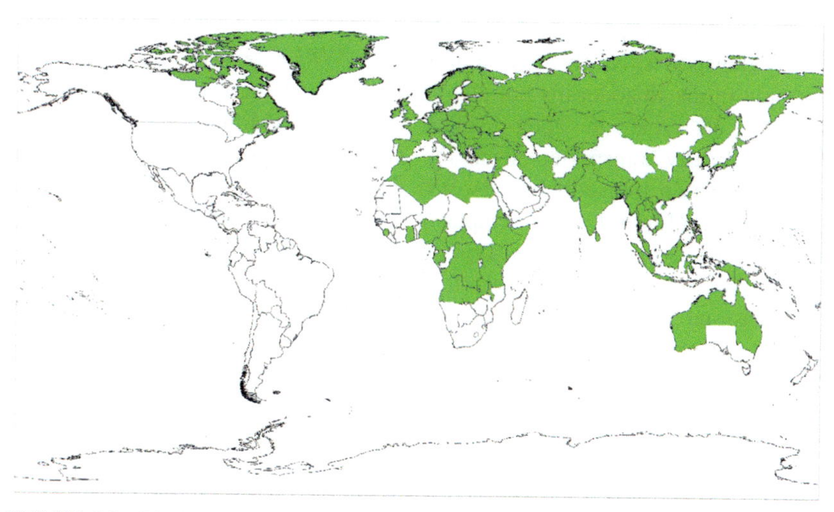

**FIGURE 1.3** Distribution of Zn hyperaccumulators around the world. Map lines delineate study areas and do not necessarily depict accepted national boundaries.

where competitive pressure is lower. These species generally have no close relatives, which is the case for phylogenetically isolated hyperaccumulator species in distant New Caledonia (Reeves et al., 2021). On the other hand, neoendemic metallophytes evolved from intolerant ancestors that reached metal-rich areas, and such species usually have relatives in near or distant environments (Baker et al., 2000; Reeves et al., 2021). This is the case for phylogenetically diverse genera such as *Noccaea* and *Alyssum*.

### 1.5.4 Molecular mechanism of Zn uptake in hyperaccumulators

Zinc is generally absorbed by the roots and transferred to the aboveground plant parts under tight control of a network of barriers, transporters, chelators, and cell wall compartments (Sinclair and Krämer, 2012; Olsen and Palmgren, 2014; Caldelas and Weiss, 2017). The phenomenon of hyperaccumulation in the view of the metal tolerance has been studied for several decades. However, in recent years, several candidate genes and their protein products have been identified and characterized as contributing to excessive uptake of Zn in hyperaccumulating plants. The hot candidates are members of transition metal membrane transport proteins, the cation diffusion facilitators (CDF), and ZRT-IRT-like protein family members (ZIP; zinc-regulated transporters, iron-regulated transporters). In the nonhyperaccumulating plant species, like rice, wheat, barley, and others, high- and low-affinity transport phenomena have been described depending on the Zn content in the cultivation medium or in the soil (Hacisalihoglu et al., 2001; Meng et al., 2014). Milner et al. (2012) confirmed the presence of such a high-affinity transport system in the hyperaccumulating *N. caerulescens* represented by the ZIP family transporter, NcZNT1. Behind this, *NcZNT1* expression was confirmed not only in the rhizodermis but also in the root and shoot vasculature, which suggests its role in long-distance transport (Milner et al., 2012). Nonspecifically transporters from the ZIP family, like IRT1, facilitate not only Fe and Zn transport but also the uptake of toxic cations such as cadmium or nickel into the root symplasm (Baxter et al., 2008).

The carriers facilitating the saturable uptake of various elements, including Zn, are located mostly on the rhizodermal and endodermal plasma membranes (Caldelas and Weiss, 2017; Kaur and Garg, 2021). A major contributor to Zn transport across the endodermal plasma membrane in *N. caerulescens* is NcZNT (van de Mortel et al., 2006). NcZNT2 and 5 are the orthologs of *Arabidopsis thaliana* iron-regulated transporter 3 (IRT3) and ZIP5 that are responsible for Zn transport in the roots (van de Mortel et al., 2006; Fasani, 2012). Several authors confirmed subcellular plasma membrane localization of IRT1 and IRT3 involved in Zn transport in the roots of *A. halleri* or *N. caerulescens* (Talke et al., 2006; Shanmugam et al., 2013; Lin et al., 2016; Moreira et al., 2018). This confirms the importance of these transporters in the uptake of Zn ions to the intracellular space. Other transporters from the ZIP family (ZIP19, ZIP23) were induced in the presence of Zn in *A. halleri* and *N. caerulescens* (Lin et al., 2016), and, contrary to this, in *A. thaliana*, an up-regulation of *ZIP19* and *ZIP23* expression levels was observed in plants grown without Zn in the cultivation medium (Humayan Kabir et al., 2014). These results suggest that the up-regulation of the mentioned genes in the hyperaccumulating plants can be correlated with the low Zn rhizosphere concentration due to higher activity of the heavy metal ATPase 4 (HMA4) translocating Zn from the xylem to the shoot (Lin et al., 2016; Merlot et al., 2018). In *T. caerulescens*, endogenous overexpression of the gene for Zn transporter *TcZNT1*, closely related to *AtZIP4* was also found to be associated with increased Zn uptake into the roots in comparison with the nonaccumulating species *T. arvense* (Pence et al., 2000). In the roots of nonaccumulating plants, a significant role in reducing Zn ions flux to the stele plays the retention of metals in root cell vacuoles, and many transporters are known at the tonoplast contributing to the ion detoxification. This mechanism is, however, suppressed in the roots of hyperaccumulators; for example, Lasat et al. (1998) confirmed in *N. caerulescens* 2.5 times less Zn in the vacuoles of the root cells in comparison to *N. arvense*. Yang et al. (2018) contributed to the elucidation of the Zn root uptake and the shoot distribution in the hyperaccumulating ecotype of *S. alfredii*. The results suggested that another member of ZIP family, like SaZIP4, is an important Zn uptake transporter.

A typical feature for hyperaccumulating plants is the accumulation of Zn in the shoot. The ions are intricately allocated within the plant body (Kisko et al., 2015; Kaur and Garg, 2021), and, again, both apoplasmic and symplasmic pathways are involved in Zn fluxes and deposition to the aerial plant parts (Noulas et al., 2018). To be translocated to the shoot, Zn needs to leave the root symplasmic space and enters the conductive tissue formed by the cell walls of xylem dead cells. Key components contributing to Zn hyperaccumulation and hypertolerance are plasma membrane transporters of the Heavy Metal ATPases (HMA) family, expressed in the root vasculature and exporting Zn from the adjacent cells to the xylem. HMA2 transports Zn-ligand complexes and HMA4 recognizes $Zn^{2+}$ ions (Eren and Argüello, 2004; Verret et al., 2004; Hanikenne et al., 2008; Frérot et al., 2018). A strong induction of their genes which protein products are localized at the root pericycle plasma membrane, enables Zn loading into the xylem cells (Hussain et al., 2004). To prevent the Zn spreading by surrounding cell walls, Zn is chelated in the surrounding xylem parenchyma cells (Sinclair and Krämer, 2012; Moreira et al., 2018). According to Thomine et al. (2003), a predominant role in the root-to-shoot Zn mobility is in *A. halleri* nicotianamine synthase 2 (NAS2) and the natural resistance-associated macrophage protein family (Nramp3). Zn loading into the xylem followed by its accumulation in the aerial part in hyperaccumulating plants is provided by forming complexes with several molecules comprising organic molecules such as citrate, malate, and amino acid histidine (Cornu et al., 2015). Moreover, exogenous histidine supply increases the xylem Zn loading process (Kozhevnikova et al., 2014). Exclusion of Zn in the root cell vacuoles restricts its shoot translocation while it is being unloaded into the aboveground plant parts, and its sequestration in the vacuoles of the shoot cells may be responsible for Zn detoxifying processes in the hyperaccumulating plants (Verbruggen et al., 2009), while mostly metallothioneins (MTs) and phytate assist in the chelating processes. This was confirmed, for example, in *N. caerulescens* (Monsant et al., 2011). ZIP family of transporters is also involved in the Zn translocation processes; likely ZIP4 and IRT3 with strong evidence of involvement in Zn accumulation in the shoots was described (Lin et al., 2005). In *S. alfredii*, Yang et al. (2018) showed *SaZIP4* as a key gene for Zn transport; ZIP4 is in the plasma membrane contributing to Zn uptake not only in the shoots but also in the roots. In the shoot of *A. halleri*, another member of the ZIP family, ZIP6, located at the plasma membrane, was confirmed (Becher et al., 2004; Gupta et al., 2016).

Yellow stripe-like (YSL) protein family represents another group of transporters responsible not only for Zn but also for Mn, Cu, Cd, Fe, and Ni transfer (Curie et al., 2009). Other essential regulators of Zn excess are MATE (multidrug and toxin efflux) and FRD3 (ferric reductase defective 3) transporters (Pineau et al., 2012). Gene expression for FRD3, an essential transporter for Fe and Zn, was exclusively confirmed in the pericycle; it is an essential protein for Zn and Fe translocation into the shoot (Hassan and Aarts, 2011; Kutrowska and Szelag, 2014). Plant cadmium resistance 2 (PCR2) was confirmed to have an efflux activity in *A. thaliana* contributing also to Zn translocation (Song et al., 2010), however, its role in Zn hyperaccumulation is not fully confirmed yet.

A higher amount of free Zn ions can disrupt the cell metabolism and also restrict its transport to satisfy the demands of sink plant organs, tissues, cells, or single organelles (Caldelas and Weiss, 2017). Especially the vital organelles such as chloroplasts need to be protected by an excess of free $Zn^{2+}$ through the Zn vacuole sequestration (Sitko et al., 2017). ATPase superfamily plays an important role in protecting these cell organelles against Zn toxicity; P1B-type ATPase was identified on the chloroplast envelope in *A. thaliana* (Sitko et al., 2017), however, the implication of HMA1 in transport processes of hyperaccumulating plants has not been confirmed (Balafrej et al., 2020). The main organelle in the leaves responsible for Zn storage is vacuole, where several members of MTP/ABCC (metal tolerance proteins/ ATP binding cassete C) or CDF were confirmed (Dräger et al., 2004). Also, Ricachenevsky et al. (2013) concluded that the CDF family, known as the MTPs (MTP1 and MTP3), is the main protein facilitating the efflux of $Zn^{2+}$ into the vacuole where it stays stable at the acidic pH of the vacuolar sap. Assunção et al. (2001) isolated cDNA for TcZTP, the most closely related to *A. thaliana* ZAT (a member of the CDF family) that has been proposed to mediate a cellular Zn detoxification (van der Zaal et al., 1999). Constitutive up-regulation of the expression of *TcZTP1* was confirmed in *T. caerulescens* when compared to *T. arvense* (Assunção et al., 2001). A distinct and tissue-specific regulation of different members of the *ZIP* and *NAS* metal homeostasis gene families in the Zn hyperaccumulator *A. halleri* was identified. This is probably a reflection of cellular tasks common to roots and shoots of the hyperaccumulating species (Becher et al., 2004; Weber et al., 2004). Weber et al. (2004) confirmed consistently higher NA levels in *A. halleri* roots and thus demonstrated that formation of NA confers $Zn^{2+}$ tolerance and NA plays an important role in the hyperaccumulation phenomenon of Zn by *A. halleri*. The creation of the NA-Zn complexes keeps Zn ions mobile for the translocation to the shoot and confers the tolerance (Weber et al., 2004).

The predominant role of ZIP9 in roots, and ZIP6 in shoots, and the necessity for chelator biosynthesis achieving buffering of free cytoplasmic Zn have been confirmed for *A. halleri*. The major role for these processes was confirmed also for *NAS2* in the roots, but for *NAS3* in the shoot, while different transcript profiles confirmed different roles of the roots and the shoots in metal hyperaccumulation (Becher et al., 2004). An up-regulation of *MTP8* and *MTP11* after Zn exposure was observed in the shoots of *N. caerulescens* and *A. halleri* (Hanikenne and Nouet, 2011). A key role in Zn sequestration in the shoots of *A. halleri* and *N. goesingense* play MTP1 and TgMTP1 proteins, respectively (Dräger et al., 2004; Gustin et al., 2009). Previously described xylem transporters, HMA4 and HMA3 (P1B-typ of ATPase superfamily), are studied recently regarding their implication in Zn/Cd sequestration in the shoots (Mishra et al., 2017). As mentioned earlier, different hyperaccumulating plants store Zn in different shoot tissues which is reflected in the differences of the sites with up-regulation of the specific transporters. In *A. halleri*, the highest expression levels of *HMA3* and *HMA4* were documented in leaf mesophyll, and, in *N. caerulescens*, *HMA3* and *HMA4* were with the strongest up-regulation confirmed in the bundle sheath of the vein; HMA3 was confirmed in the tonoplast contributing to the Zn vacuolar sequestration (Mishra et al., 2017). Zn detoxification and metal sequestration also contribute to *ZAT/CDF1* and *HMA3*, dominantly expressed in *A. halleri* shoot (Becher et al., 2004). The HMA transporters evidently cooperate here with Nramp family in Zn sequestration in the leaf vacuoles. These proteins require the transmembrane proton gradient produced by $H^+$ ATPase to facilitate the transportation of many divalent cations toward the membrane (Weber et al., 2004; Krämer et al., 2007; Ishida et al., 2018), however, still they roles and involvement in processes of Zn hyperaccumulation need to be studied in deep.

### 1.5.5 Tissue and cellular accumulation of Zn in hyperaccumulators

Hyperaccumulators can store metals in extremely high concentrations without experiencing toxicity symptoms. This remarkable trait is possible due to the outstanding potential of hyperaccumulating plants to detoxify metals accumulated in their aboveground tissues (Krämer, 2010) and maintain homeostasis. The predominant strategy of most Zn hyperaccumulators to store Zn at high concentrations without affecting vital processes such as photosynthesis is to store Zn in vacuoles, especially in the vacuoles of large epidermal cells. Water-storage cells of the epidermis are the dominant site of Zn deposition. Also, stomatal-guard cells are rich in accumulated Zn but subsidiary cells surrounding stomata contain only a small amount of Zn. Important Zn-storage sites are also vascular bundles, but mesophyll is used only sparsely (Kozhevnikova et al., 2017). Cells of the epidermis may vary in size according to Zn stored in their vacuoles (Küpper et al., 1999). The mechanisms of accumulation predetermine some epidermal cells to be storage sites for Zn, or random selection of some epidermal cells causes an increase in size due to increased volume of vacuoles containing Zn. Nevertheless, the variability in size of epidermal cells is higher than mesophyll cells which are not the storage cells for Zn (Leitenmaier and Küpper, 2011).

Although there is a general tendency of plants to lower the Zn content in leaf mesophyll cells, these cells in hyperaccumulators are also much more tolerant to elevated metal content and highly resistant compared to nonaccumulating species (Küpper et al., 1999). In *S. plumbizincicola*, the main storage site of Zn is the epidermis of young leaves, and some Zn is accumulated also in the mesophyll cells with a lower degree. In mature leaves, only the epidermis is used for Zn accumulation (Cao et al., 2014). The higher metabolic activity of mesophyll cells of young leaves is probably enough to confer Zn toxicity. Accumulation of Zn into epidermal cells seems to be a specific feature of some hyperaccumulators, at least in the case of *S. alfredii*. Two ecotypes with contrasting accumulation (HE—hyperaccumulating, NHE—nonaccumulating) show preferential accumulation of Zn into epidermal cells, although the Zn accumulation in the epidermis of HE is much greater. Over time of exposure, the importance of the epidermis as a storage site of Zn in HE increases, and mesophyll cells are used for this purpose only partially. In NHE, mesophyll is used as a storage site to a great extend (Tian et al., 2009). Contrary, mesophyll cells in *A. halleri* leaves are the most important storage site of Zn. Epidermal cells in this species are used only partially, except trichomes. They contained the largest amount of Zn accumulated by the leaves. This is because small epidermal cells are probably less suitable than large mesophyll cells in *A. halleri* (Küpper et al., 2000; Zhao et al., 2000). It was also shown that a part of the accumulated Zn in *A. halleri* was stored in the trichomes, more specifically in a narrow ring at the trichome base (Küpper et al., 2000). Moreover, Fukuda et al. (2020) also showed that Zn can be stored in the midrib of *A. halleri* leaves. In *A. halleri* as well, small vesicles containing Zn may be observed as a part of the vacuole. These vesicles may transport Zn from cytoplasm and after degradation release it into the vacuole. Before compartmentation in vacuole, Zn-silicate transiently accumulated precipitates were observed; they were continuously degraded to Zn and $SiO_2$, and sequestered to the vacuoles (Neumann and zur Nieden, 2001).

In *Potentilla griffithii*, Zn was detected in root cell walls and in cytoplasm and vacuoles of epidermal and bundle sheet cells in the leaves (Hu et al., 2009). Later Qiu et al. (2011) realized that from 94% of a whole Zn accumulated in the protoplast of this plant species, 96% of it was located in the vacuoles. Vacuoles in roots are only transiently used for Zn compartmentation also in *S. alfredii*. Around 2.7 times less Zn is sequestered into root vacuoles in HE than in NHE ecotypes. For both ecotypes, vacuolar sequestration is an important strategy of tolerance, but NHE used vacuoles in roots, and HE in stems (Yang et al., 2005). Transient is also an accumulation of Zn in *N. caerulescens* root vacuoles. Contrary to nonaccumulating *T. arvense*, only a small fraction of Zn is stored in roots in vacuoles, and the majority of Zn is released back to cytoplasm to be transported to shoot (Lasat et al., 1998). In *S. plumbizincicola* young leaves, Zn is equally distributed in the cytoplasm, cell wall, and cell organelles. In mature leaves, Zn is predominantly in the soluble fraction of cytoplasm and the cell wall accumulates the lowest concentration of Zn (Cao et al., 2014).

### 1.5.6 Adaptation mechanisms of Zn hyperaccumulators

Nonhyperaccumulating plants are known to produce various spectra of chelating agents to decrease excessive concentrations of free ions in the cytoplasm when exposed to excess metals or metalloids. This is also the case of Zn. Various metal-binding peptides, such as phytochelatins (PC), metallothioneins (MT), carboxyl acids like citric or malic acid, or amino acids like histidine, or nicotinamine (NA) have been reported to bind, sequester and detoxify excess of Zn in many plant species (Hall, 2002; Broadley et al., 2007; Milner and Kochian, 2008; Sinclair and Krämer, 2012).

It has been shown that PC-based sequestration is not essential for the hypertolerance of Zn as well as Cu, Ni, and Co (Schat et al., 2002). Similarly, there is no significant role of PC in Zn toxicity in *S. alfredii* (Sun et al., 2005) or *N. caerulescens* (Wójcik et al., 2006). These plants produce 2–3 times less PC compounds compared to nonhyperaccumulating *T. arvense*. It is hypothesized that the synthesis of PC and effective metal binding is energetically high in tissues of plants with exceptionally high metal concentrations, and so is inefficient when compared to alternative detoxication mechanisms (Ebbs et al., 2002). Although PC production is not enhanced in *S. alfredii*, larger Zn dosages were almost linearly associated with an increase in glutathione (GSH). This thiol compound is a precursor for PC and may be able to chelate Zn or also detoxify generated reactive oxygen species (ROS). In the study of Sun et al. (2005), a significant increase in GSH concentration was observed in the leaves of *S. alfredii* from the metalicolous population upon exposure to Zn, while this was not the case in the plants from the control site. Similarly, in the study by Jin et al. (2008), an increase in GSH concentration was observed in the leaves of the Zn-hyperaccumulating ecotype of *S. alfredii* upon exposure to high Zn concentrations, while no change in GSH concentration was observed in the nonhyperaccumulating ecotype, which was less tolerant to Zn. The higher Zn tolerance of the metalicolous *S. alfredii* population could be attributed to the increase in GSH concentrations, which could be involved in the chelation of the metal or the antioxidant response of the plants. Genus *Thlaspi* has a constitutively higher GSH level (Freeman et al., 2004). This is caused by constitutively elevated salicylic acid, which normally increases the level of GSH in resistance to biotic stress. In *T. goesingense* is salicylic acid elevated without pathogens, and expression of resistance genes is also stopped. Elevated GSH as a result of elevated salicylic acid may diminish oxidative stress. Salicylic acid, moreover, inhibits the synthesis of PC (Freeman et al., 2005).

Metalothioneins, cysteine-rich binding peptides, have an elevated expression in *N. caerulescens*. However, as yeast complementation showed, their affinity for Zn is reduced. The mentioned reduction is caused by the loss of three cysteine residues from the conserved Cys-rich domain. This loss still ensures the proper capacity to bind Cu. Such modified MT in *N. caerulescens* are expressed in shoots, contrary to common expression in nonaccumulating species in roots (Roosens et al., 2005). This suggests possible diversification of chelators. By modifying binding site and lowering the affinity for Zn but preserving the affinity for Cu, MT are detached from direct influence on Zn-hyperaccumulation and possible interaction with other chelators, and they are more effectively focused on other metals. This specification may keep metal homeostasis (presumably Cu) in proper status as hyperaccumulation not only elevates Zn but also other elements are imbalanced or taken up in excess. In the study by Hassinen et al. (2009) on the crosses of *N. caerulescens* with different Zn tolerance, it was also shown that MT is not primarily involved in Zn accumulation.

Zinc is better bound to oxygen and nitrogen-containing ligands than to thiol (sulfur-containing) groups (Salt et al., 2002). The distance between Zn and O or N in organic acids or histidine is shorter than between Zn and S in thiols (Lefèvre et al., 2016). Organic acids might have an especially important role in the detoxification and storage of Zn in *N. caerulescens* shoots (Salt et al., 1999), and high constitutive concentrations of malate were recorded in the shoots of *N. caerulescens* (Wójcik et al., 2006; Shen et al., 1997). In the study of Wójcik et al. (2006), high concentrations of citrate were also determined in the shoots of *N. caeulescens*. An important role of organic acids in Zn detoxification was also confirmed by Sarret et al. (2002) showing that a large proportion of Zn in the shoots of *A. halleri* was associated with malate. In *S. alfredii*, malate was suggested also as metal-ligand in leaves. In roots, complexation to citrate occurred (Lu et al., 2014), which is also present in xylem conduits during Zn transport but contributes only partially as a larger amount of Zn is in aqueous form (Lu et al., 2013).

The increased synthesis of nicotianamine (NA) in Zn-hyperaccumulating plants is essential for Zn hyperaccumulation, as shown by Uraguchi et al. (2019) on *A. halleri*. Namely, elevated NA production in roots of *A. halleri* is a Zn hyperaccumulation factor regardless of the edaphic environment, and it contributes to Zn hyperaccumulation in soils with different Zn availability (Uraguchi et al., 2019). Deinlein et al. (2012) also suggested that NA forms complexes with Zn in root cells and facilitates the symplasmic passage of Zn toward the xylem. Foroughi et al. (2014) highlighted that it is unlikely that NA or other small molecules are involved in the storage of Zn but are more likely involved in the transport and redistribution of micronutrients. Complexation of Zn with histidine is also of particular importance in Zn hyperaccumulation. Kozhevnikova et al. (2014) concluded that Zn loading of xylem appears to be a species-wide trait in *N. caerulescens*, and that high endogenous L-histidine concentrations are likely responsible for the hyperaccumulation of Ni and Zn by this plant species (Kozhevnikova et al., 2021).

Observed binding of Zn on oxygen-containing ligands may however include water and carboxyl groups (Küpper et al., 2004), which predict Zn as a free cation in solution or bind to some polymers like cellulose or pectins in cell walls. Aqueous forms of Zn were measured also by Salt et al. who predicted Zn as a form of hydrated cation (Salt et al., 1999). Küpper et al. (2000) also suggested the presence of Zn oxides in solid form in extraordinary excess in epidermal trichomes of *A. halleri*. Nevertheless, 60% of Zn in shoot of *A. halleri* was in a water-soluble form (Zhao et al., 2000). In *N. caerulescens*, 80% of shoot-located Zn was found in a soluble form (Zhao et al., 1998). In *P. griffithii*, 71%–85% of Zn is in vacuoles of epidermal cells as a soluble form. Free atoms are bound to organic acid and rather soluble in vacuoles than complexed on some precipitates (Qiu et al., 2011). In *S. alfredii* during transport in xylem, almost 60% of Zn was found to be in an aqueous form (Lu et al., 2013).

Although the sequestration of Zn in the metabolically less active compartments of the plant is a very efficient mechanism of protection of vital processes, free metal ions in the cytoplasm at different concentrations can trigger oxidative stress via the generation of reactive oxygen species (ROS), including hydrogen peroxide ($H_2O_2$) and superoxide radicals, leading to lipid peroxidation (Lin and Aarts, 2012; Barrameda-Medina et al., 2014). Zinc itself is not a redox-active element but is still able to be responsible for elevated ROS production. Moreover, Zn is an essential element for functioning of one of most abundant antioxidant enzymes, superoxide dismutase, Cu-Zn-SOD, which means that excess of Zn induces ROS production but at the same time enhances the activity of the antioxidant system (Marschner and Marschner, 2012). Antioxidant mechanisms in cells of Zn hyperaccumulators can vary and include an increase in the concentration of antioxidant enzymes, increased activity of antioxidant enzymes, or increased accumulation of nonenzymatic antioxidants, such as GSH (Barrameda-Medina et al., 2014). A significant increase in ascorbic acid content was observed in the leaves of hyperaccumulating and nonhyperaccumulating ecotypes of *S. alfredii*, while the activities of antioxidant enzymes superoxide dismutase (SOD), catalase (CAT), guaiacol peroxidase (GPX), ascorbate peroxidase (APX), dehydroascorbate reductase (DHAR) and glutathione reductase (GR) were increased only in the leaves of hyperaccumulating ecotype at high Zn concentrations, which may be responsible for its better growth (Jin et al., 2008). It was also shown that higher peroxidase activities contribute to the metal tolerance in *A. halleri* by alleviating the ROS damage (Chiang et al., 2006), whereas high concentrations of SOD, APX, and CAT were also recorded in *N. caerulescens* upon exposure to high Zn concentrations (Wójcik et al., 2006).

## 1.6 Conclusions and remarks

Zinc belongs to the elements that are actively taken up by plants and utilized in various spectra of plant-based internal processes. The importance of Zn is highlighted also due to the fact that it is listed as an essential plant micronutrient. Although the Zn deficiency in agricultural soils worldwide is a more common problem, there are also sites with elevated soil Zn content. The tolerance of plants to Zn significantly varies between the species. Most plants actively acquire and take up Zn, however, they cannot tolerate it when excessively present. On the other hand, we have shown that a relatively small group of plant species do the opposite. Zinc hyperaccumulators, as these plants are called, have developed in different regions of the world and have some common attributes, although the exact way of their physiological adaptation can differ between the plant populations or species. In general, a common characteristic of these plants is the fact that they can actively take up and store Zn in their aerial parts in concentrations exceeding 3000 mg/kg Zn in the dry aboveground biomass.

Active uptake and translocation of Zn to the aerial parts is no doubt an important feature and makes these plant species utilizable for the phytoremediation of Zn-contaminated substrates. Phytoremediation is a complex of methods and approaches that use properties of plants for cleaning or remediation of polluted environment. This plant-based concept is getting more popular due to the increasing number of polluted sites in the world, especially as a result of permanently growing human activities, like mining and ore processing, fossil fuel combustion, industrial production, intense traffic, and nonoptimal practices in agriculture. Although many Zn hyperaccumulators have been identified and grow on artificially polluted former sites with a long history of mining, the idea is not to use these species for remediation of such heavily contaminated localities. Oppositely, we may better consider the protection of such sites to be able to follow the adaptation of these plants to unfavorable environmental conditions also in the future and to study the processes related to the adaptation of these plants to changing environmental conditions (e.g., climate change effect). The potential of Zn hyperaccumulators to remediate Zn-contaminated sites could be better used in less contaminated soil substrates, where the initial Zn concentration makes it feasible to phytoextract Zn in horizon of next decades, not centuries. In the last years, several projects focused on the use of plants in the phytomining of metals have been developed. This topic is intensively studied especially between Ni hyperaccumulators, which also represent the highest amount of known species within this specific group of plants. However, in the case of Zn, the phytomining is probably not feasible, and this technique could be better considered for those elements with a rare occurrence in the Earth crust and with considerably higher prices at the market. There is also one thing that should not be omitted. All recently known Zn hyperaccumulators are herbaceous species that produce only a limited amount of aboveground biomass. Another fact is that many of these species show intraspecific variability in Zn uptake. Therefore some populations or ecotypes cannot be classified as true hyperaccumulators, although the concentration of Zn in their tissues can be still high, and, therefore, they can still be considered suitable for cleaning of Zn polluted sites.

Another important issue that can be partially solved by the use of Zn hyperaccumulators is the problem of Zn deficiency. This is a serious problem in many countries, resulting from low initial Zn content in the soils or its low bioavailability. Although Zn hyperaccumulators cannot be considered edible plants, they can become a part of fodder mixture, and in this way increase the bioavailable portion of Zn from animal products. Moreover, a detailed understanding of mechanisms of Zn uptake, translocation, and deposition in hyperaccumulators could help improve the properties of common agriculturally grown species using the whole spectra of biotechnology tools.

## Acknowledgments

The work is a result of scientific cooperation supported by COST Action Nr. COST CA19116. The work was also supported by Slovak Research and Development Agency APVV under the contracts Nr. APVV-17–0164 APVV-23-0318, and APVV-SK-CN-21-0034, and by grant Nr. VEGA 1/0472/22.

## References

Adam, S., Murthy, S.D.S., 2014. Characterization of alterations in photosynthetic electron transport activities in maize thylakoid membranes under zinc stress. European Journal of Experimental Biology 4, 25–29.

Alloway, B.J., 1995. Soil processes and the behavior of metals. In: Alloway, B.J. (Ed.), Heavy Metals in Soils. Blackie, London, pp. 38–57.

Alloway, B.J., 2013. Sources of heavy metals and metalloids in soils. In: Alloway, B.J. (Ed.), Heavy Metals in Soils. Springer, pp. 11–50.

Andresen, E., Peiter, E., Küpper, H., 2018. Trace metal metabolism in plants. Journal of Experimental Botany 69, 909–954.

Van Assche, F., Clijsters, H., 1986. Inhibition of photosynthesis in Phaseolus vulgaris by treatment with toxic concentration of zinc: effect on ribulose-1,5-bisphosphate carboxylase/oxygenase. Journal of Plant Physiology 125, 355–360.

Assunção, A.G.L., Martins, P.D.C., De Folter, S., Vooijs, R., Schat, H., Aarts, M.G.M., 2001. Elevated expression of metal transporter genes in three accessions of the metal hyperaccumulator Thlaspi caerulescens. Plant, Cell & Environment 24, 217–226.

Di Baccio, D., Tognetti, R., Minnocci, A., Sebastiani, L., 2009. Responses of the Populus × euramericana clone I–214 to excess zinc: carbon assimilation, structural modifications, metal distribution and cellular localization. Environmental and Experimental Botany 67, 153–163.

Bailey, S., Thompson, E., Nixon, P.J., Horton, P., Mullineaux, C.W., Robinson, C., et al., 2002. A critical role for the Var2 FtsH homologue of *Arabidopsis thaliana* in the photosystem II repair cycle in vivo. Journal of Biological Chemistry 277, 2006–2011.

Baker, A.J., Brooks, R., 1989. Terrestrial higher plants which hyperaccumulate metallic elements. A review of their distribution, ecology and phytochemistry. Biorecovery 1 (2), 81–126.

Baker, A.J., McGrath, S.P., Reeves, R.D., Smith, J.A.C., 2000. Metal hyperaccumulator plants: a review of the ecology and physiology of a biological resource for phytoremediation of metal-polluted soils. In: Terry, N., Bañuelos, G. (Eds.), Phytoremediation of Contaminated Soil and Water. CRC Press, pp. 85–107.

Baker, A.J.M., Proctor, J., 1990. The influence of cadmium, copper, lead, and zinc on the distribution and evolution of metallophytes in the British Isles. Plant Systematics and Evolution 173 (1), 91–108.

Baker, A., Proctor, J., van Balgooy, M., Reeves, R., 1992. Hyperaccumulation of nickel by the flora of the ultramafics of Palawan, Republic of the Philippines. In: Baker, A.J.M., Proctor, J., Reeves, R.D. (Eds.), The Vegetation of Ultramafic (Serpentine) Soils. Intercept Ltd., Andover, Hants, UK, pp. 291–304.

Balafrej, H., Bogusz, D., Triqui, Z.-E.A., Guedira, A., Bendaou, N., Smouni, A., et al., 2020. Zinc hyperaccumulation in. Plants: A Review. Plants 9, 562.

Barrameda-Medina, Y., Montesinos-Pereira, D., Romero, L., Blasco, B., Ruiz, J.M., 2014. Role of GSH homeostasis under Zn toxicity in plants with different Zn tolerance. Plant Science 227, 110–121.

Baxter, I.R., Vitek, O., Lahner, B., Muthukumar, B., Borghi, M., Morrissey, J., et al., 2008. The leaf ionome as a multivariable system to detect a plant's physiological status. PNAS 105, 12081–12086. Available from: https://doi.org/10.1073/pnas.0804175105.

Becher, M., Talke, I.N., Krall, L., Krämer, U., 2004. Cross-species microarray transcript profiling reveals high constitutive expression of metal homeostasis genes in shoots of the zinc hyperaccumulator Arabidopsis halleri. The Plant Journal 37, 251–268.

Bert, V., Bonnin, I., Saumitou-Laprade, P., de Laguerie, P., Petit, D., 2002. Do *Arabidopsis halleri* from nonmetallicolous populations accumulate zinc and cadmium more effectively than those from metallicolous populations? New Phytologist 155, 47–57.

Broadley, M., Brown, P., Cakmak, I., Rengel, Z., Zhao, F., 2012. Function of nutrients: micronutrients. In: Marschner, P. (Ed.), Marschner's Mineral Nutrition of Higher Plants, Third edn Academic Press, London, pp. 191–248. ISBN 978-0-12–384905-2.

Broadley, M.R., White, P.J., Hammond, J.P., Zelko, I., Lux, A., 2007. Zinc in plants. New Phytologist 173, 677–702.

Brooks, R.R., 1994. Plants that hyperaccumulate heavy metals. In: Farago, M.E. (Ed.), Plants and the Chemical Elements: Biochemistry, Uptake, Tolerance and Toxicity. VCH Verlagsgesellschaft mbH, pp. 87–105.

Brooks, R.R., Lee, J., Reeves, R.D., Jaffré, T., 1977. Detection of nickeliferous rocks by analysis of herbarium specimens of indicator plants. Journal of Geochemical Exploration 7, 49–57.

Brunner, I., Luster, J., Günthardt-Goerg, M.S., Frey, B., 2008. Heavy metal accumulation and phytostabilisation potential of tree fine roots in a contaminated soil. Environmental Pollution (Barking, Essex: 1987) 152, 559–568.

Cakmak, I., 2000. Possible roles of zinc in protecting plant cells from damage by reactive oxygen species. New Phytologist 146, 185–205.

Caldelas, C., Weiss, D.J., 2017. Zinc Homeostasis and isotopic fractionation in plants: a review. Plant and Soil 411, 17–46.

Callahan, D.L., Hare, D.J., Bishop, D.P., Doble, P.A., Roessner, U., 2016. Elemental imaging of leaves from the metal hyperaccumulating plant *Noccaea caerulescens* shows different spatial distribution of Ni, Zn and Cd. RSC Advances 6 (3), 2337–2344.

Cao, D., Zhang, H., Wang, Y., Zheng, L., 2014. Accumulation and distribution characteristics of zinc and cadmium in the hyperaccumulator plant *Sedum plumbizincicola*. Bulletin of Environmental Contamination and Toxicology 93 (2), 171–176.

Cappa, J.J., Yetter, C., Fakra, S., Cappa, P.J., deTar, R., Landes, C., et al., 2015. Evolution of selenium hyperaccumulation in *Stanleya* (Brassicaceae) as inferred from phylogeny, physiology and X-ray microprobe analysis. New Phytologist 205 (2), 583–595.

Chahal, D.S., Sharma, B.D., Singh, P.K., 2005. Distribution of forms of zinc and their association with soil properties and uptake in different soil orders in semi-arid soils of Punjab, India. Communications in Soil Science and Plant Analysis 36, 2857–2874.

Chiang, H.C., Lo, J.C., Yeh, K.C., 2006. Genes associated with heavy metal tolerance and accumulation in Zn/Cd hyperaccumulator *Arabidopsis halleri*: a genomic survey with cDNA microarray. Environmental Science & Technology 40 (21), 6792–6798.

Ciftci-Yilmaz, S., Mittler, R., 2008. The zinc finger network of plants. Cellular and Molecular Life Sciences 65, 1150–1160.

Cole, M.M., Provan, D.M., Tooms, J.S., 1968. Geobotany, biogeochemistry and geochemistry in the Bulman-Waimuna springs area, Northern territory, Australia. Transactions of the Institute of Mining and Metallurgy 77, B81–B103.

Coleman, J.E., 1992. Zinc proteins: enzymes storage proteins transcription factors, and replication proteins. Annual Review of Biochemistry 61, 897–946.

Coppola, V., Boni, M., Gilg, H.A., Balassone, G., Dejonghe, L., 2008. The "calamine" nonsulfide Zn-Pb deposits of Belgium: petrographical, mineralogical and geochemical characterization. Ore Geology Reviews 33, 187–210.

Cornu, J.-Y., Deinlein, U., Höreth, S., Braun, M., Schmidt, H., Weber, M., et al., 2015. Contrasting effects of nicotianamine synthase knockdown on zinc and nickel tolerance and accumulation in the zinc/cadmium hyperaccumulator Arabidopsis halleri. New Phytologist 206, 738–750.

Cui, J., Jander, G., Racki, L.R., Kim, P.D., Pierce, N.E., Ausubel, F.M., 2002. Signals involved in Arabidopsis resistance to Trichoplusiani caterpillars induced by virulent and avirulent strains of the phytopathogen Pseudomonas syringae. Plant Physiology 129, 551–564.

Curie, C., Cassin, G., Couch, D., Divol, F., Higuchi, K., Le Jean, M., et al., 2009. Metal movement within the plant: contribution of nicotianamine and yellow stripe 1-like transporters. Annals of Botany 103, 1–11.

Cuypers, A., Vangronsveld, J., Clijsters, H., 2001. The redox status of plant cells (AsA and GSH) is sensitive to zinc imposed oxidative stress in roots and primary leaves of Phaseolus vulgaris. Plant Physiology and Biochemistry 39, 657–664.

Van Damme, A., Degryse, F., Smolders, E., Sarret, G., Dewit, J., Swennen, R., et al., 2010. Zinc speciation in mining and smelter contaminated overbank sediments by EXAFS spectroscopy. Geochimica et Cosmochimica Acta 74, 3707–3720.

Deinlein, U., Weber, M., Schmidt, H., Rensch, S., Trampczynska, A., Hansen, T.H., et al., 2012. Elevated nicotianamine levels in *Arabidopsis halleri* roots play a key role in zinc hyperaccumulation. The Plant Cell 24 (2), 708–723.

Dinh, N.T., Vu, D.T., Mulligan, D., Nguyen, A.V., 2015. Accumulation and distribution of zinc in the leaves and roots of the hyperaccumulator *Noccaea caerulescens*. Environmental and Experimental Botany 110, 85–95.

Disante, K.B., Fuentes, D., Cortina, J., 2010. Response to drought of Zn-stressed Quercus suber L. seedlings. Environmental and Experimental Botany 70, 96–103.

Dräger, D.B., Desbrosses-Fonrouge, A.-G., Krach, C., Chardonnens, A.N., Meyer, R.C., Saumitou-Laprade, P., et al., 2004. Two genes encoding Arabidopsis halleri MTP1 metal transport proteins co-segregate with zinc tolerance and account for high MTP1 transcript levels. The Plant Journal 39, 425–439.

Ebbs, S., Ahner, B., Kochian, L., Lau, I., 2002. Phytochelatin synthesis is not responsible for Cd tolerance in the Zn/Cd hyperaccumulator Thlaspi caerulescens (J. & C. Presl). Planta 214, 635–640.

van der Ent, A., Baker, A.J.M., Reeves, R.D., Pollard, A.J., Schat, H., 2013. Hyperaccumulators of metal and metalloid trace elements: facts and fiction. Plant and Soil 362 (1), 319–334.

van der Ent, A., de Jonge, M.D., Mak, R., Mesjasz-Przybyłowicz, J., Przybyłowicz, W.J., Barnabas, A.D., et al., 2020. X-ray fluorescence elemental mapping of roots, stems and leaves of the nickel hyperaccumulators *Rinorea* cf. *bengalensis* and *Rinorea* cf. *javanica* (Violaceae) from Sabah (Malaysia), Borneo. Plant and Soil 448 (1), 15–36.

Eren, E., Argüello, J.M., 2004. Arabidopsis HMA2, a Divalent Heavy Metal-Transporting PIB-Type ATPase, Is Involved in Cytoplasmic Zn2 + Homeostasis. Plant Physiology 136, 3712–3723.

Fasani, E., 2012. Plants that Hyperaccumulate Heavy Metals. In: Furini, A. (Ed.), Plants and Heavy Metals, SpringerBriefs in Molecular Science. Springer, Netherlands, Dordrecht, pp. 55–74.

Foroughi, S., Baker, A.J., Roessner, U., Johnson, A.A., Bacic, A., Callahan, D.L., 2014. Hyperaccumulation of zinc by *Noccaea caerulescens* results in a cascade of stress responses and changes in the elemental profile. Metallomics 6 (9), 1671–1682.

Freeman, J.L., Garcia, D., Kim, D., Hopf, A., Salt, D.E., 2005. Constitutively elevated salicylic acid signals glutathione-mediated nickel tolerance in thlaspi nickel hyperaccumulators. Plant Physiology 137, 1082–1091.

Freeman, J.L., Persans, M.W., Nieman, K., et al., 2004. Increased glutathione biosynthesis plays a role in nickel tolerance in Thlaspi nickel hyperaccumulators. The Plant Cell Online 16, 2176–2191.

Frérot, H., Hautekeete, N., Decombeix, I., Bouchet, M.-H., CREACH, A., Saumitou-Laprade, P., et al., 2018. Habitat heterogeneity in the pseudometallophyte Arabidopsis halleri and its structuring effect on natural variation of zinc and cadmium hyperaccumulation. Plant and Soil 423.

Fukao, Y., Ferjani, A., Tomioka, R., et al., 2011. iTRAQ analysis reveals mechanisms of growth defects due to excess zinc in Arabidopsis. Plant Physiology 155, 1893–1907.

Fukuda, N., Kitajima, N., Terada, Y., et al., 2020. Visible cellular distribution of cadmium and zinc in the hyperaccumulator *Arabidopsis halleri* ssp. *gemmifera* determined by 2-D X-ray fluorescence imaging using high-energy synchrotron radiation. Metallomics 12, 193–203.

Gieron, Z., Sitko, K., Zieleznik-Rusinowska, P., Szopinski, M., Rojek-Jelonek, M., Rostanski, A., et al., 2021. Ecophysiology of Arabidopsis arenosa, a new hyperaccumulator of Cd and Zn. Journal of Hazardous Materials 412, 125052.

Giri, J., Vij, S., Dansana, P.K., Tyagi, A.K., 2011. Rice A20/ AN1 zinc-finger containing stress-associated proteins (SAP1/11) and a receptor-like cytoplasmic kinase (OsRLCK253) interact via A20 zinc-finger and confer abiotic stress tolerance in transgenic Arabidopsis plants. New Phytologist 191, 721–732.

Gupta, N., Ram, H., Kumar, B., 2016. Mechanism of Zinc absorption in plants: uptake, transport, translocation and accumulation. Reviews in Environmental Science and Biotechnology 15, 89–109.

Gustin, J.L., Loureiro, M.E., Kim, D., Na, G., Tikhonova, M., Salt, D.E., 2009. MTP1-dependent Zn sequestration into shoot vacuoles suggests dual roles in Zn tolerance and accumulation in Zn-hyperaccumulating plants. The Plant Journal 57, 1116–1127.

Hacisalihoglu, G., Hart, J.J., Kochian, L.V., 2001. High- and low-affinity zinc transport systems and their possible role in zinc efficiency in bread wheat1. Plant Physiology 125, 456–463.

Hall, J.L., 2002. Cellular mechanisms for heavy metal detoxification and tolerance. Journal of Experimental Botany 53, 1–11.

Hall, T.M., 2005. Multiple modes of RNA recognition by zinc finger proteins. Current Opinion in Structural Biology 15, 367–373.

Hanikenne, M., Nouet, C., 2011. Metal hyperaccumulation and hypertolerance: a model for plant evolutionary genomics. Current Opinion in Plant Biology 14, 252–259.

Hanikenne, M., Talke, I.N., Haydon, M.J., Lanz, C., Nolte, A., Motte, P., et al., 2008. Evolution of metal hyperaccumulation required cis-regulatory changes and triplication of HMA4. Nature 453, 391–395.

Hansch, R., Mendel, R.R., 2009. Physiological functions of mineral micronutrients (Cu, Zn, Mn, Fe, Ni, Mo, B, Cl). Current Opinion in Plant Biology 12, 259–266.

Hassan, M.U., Aamer, M., Chattha, M.U., Haiying, T., Shahzad, B., Barbanti, L., et al., 2020. The critical role of zinc in plants facing the drought stress. Agriculture 10, 396.

Hassan, Z., Aarts, M.G.M., 2011. Opportunities and feasibilities for biotechnological improvement of Zn, Cd or Ni tolerance and accumulation in plants. Environmental and Experimental Botany 72, 53–63.

Hassinen, V.H., Tuomainen, M., Peräniemi, S., Schat, H., Kärenlampi, S.O., Tervahauta, A., 2009. Metallothioneins 2 and 3 contribute to the metal-adapted phenotype but are not directly linked to Zn accumulation in the metal hyperaccumulator, *Thlaspi caerulescens*. Journal of Experimental Botany 60 (1), 187–196.

Haydon, M.J., Cobbett, C.S., 2007. A novel major facilitator superfamily protein at the tonoplast influences zinc tolerance and accumulation in Arabidopsis. Plant Physiology 143, 1705–1719.

Huguet, S., Bert, V., Laboudigue, A., Barthès, V., Isaure, M.P., Llorens, I., et al., 2012. Cd speciation and localization in the hyperaccumulator *Arabidopsis halleri*. Environmental and Experimental Botany 82, 54–65.

Humayan Kabir, A., Swaraz, A.M., Stangoulis, J., 2014. Zinc-deficiency resistance and biofortification in plants. Journal of Plant Nutrition and Soil Science 177, 311–319.

Hussain, D., Haydon, M.J., Wang, Y., Wong, E., Sherson, S.M., Young, J., et al., 2004. P-type ATPase heavy metal transporters with roles in essential zinc homeostasis in arabidopsis. The Plant Cell 16, 1327–1339.

Hu, P.J., Qiu, R.L., Senthilkumar, P., Jiang, D., Chen, Z.W., Tang, Y.T., et al., 2009. Tolerance, accumulation and distribution of zinc and cadmium in hyperaccumulator *Potentilla griffithii*. Environmental and Experimental Botany 66 (2), 317–325.

Ievinsh, G., Andersone-Ozola, U., Landorfa-Svalbe, Z., Karlsons, A., Osvalde, A., 2020. Wild plants from coastal habitats as a potential resource for soil remediation. In: Giri, B., Varma, A. (Eds.), Soil Health. Soil Biology, vol 59. Springer, Cham, pp. 121–144.

Ishida, J.K., Caldas, D.G.G., Oliveira, L.R., Frederici, G.C., Leite, L.M.P., Mui, T.S., 2018. Genome-wide characterization of the NRAMP gene family in *Phaseolus vulgaris* provides insights into functional implications during common bean development. Genetics and Molecular Biology 41, 820–833.

Jaffré, T., Brooks, R.R., Lee, J., Reeves, R.D., 1976. Sebertia acuminata: a hyperaccumulator of nickel from New Caledonia. Science 193, 579–580.

Jin, X., Yang, X., Mahmood, Q., et al., 2008. Response of antioxidant enzymes, ascorbate and glutathione metabolism towards cadmium in hyperaccumulator and nonhyperaccumulator ecotypes of Sedum alfredii H. Environmental Toxicology 23, 517–529.

Kabata-Pendias, A., Pendias, H., 2001. Trace Elements in Plants and Soils, 3rd ed. CRC Press, p. 331.

Kanai, M., Hirai, M., Yoshiba, M., Tadano, T., Higuchi, K., 2009. Iron deficiency causes zinc excess in Zea mays. Soil Science & Plant Nutrition 55, 271–276.

Kaur, H., Garg, N., 2021. Zinc toxicity in plants: a review. Planta 253, 129.

King, J.C., Shames, D.M., Woodhouse, L.R., 2000. Zinc homeostasis in humans. Journal of Nutrition 130, 1360S–1366S.

Kisko, M., Bouain, N., Rouached, A., Choudhary, S.P., Rouached, H., 2015. Molecular mechanisms of phosphate and zinc signalling crosstalk in plants: phosphate and zinc loading into root xylem in Arabidopsis. Environmental and Experimental Botany, Plant Signalling Mechanisms in Response to the Environment 114, 57–64.

Kováč, J., Lux, A., Soukup, M., Weidinger, M., Gruber, D., Lichtscheidl, I., et al., 2020. A new insight on structural and some functional aspects of peri-endodermal thickenings, a specific layer in *Noccaea caerulescens* roots. Annals of Botany 126, 423–434.

Kozhevnikova, A.D., Seregin, I.V., Aarts, M.G., Schat, H., 2020. Intra-specific variation in zinc, cadmium and nickel hypertolerance and hyperaccumulation capacities in *Noccaea caerulescens*. Plant and Soil 452 (1), 479–498.

Kozhevnikova, A.D., Seregin, I.V., Erlikh, N.T., Shevyreva, T.A., Andreev, I.M., Verweij, R., et al., 2014. Histidine-mediated xylem loading of zinc is a species-wide character in Noccaea caerulescens. New Phytologist 203, 508–519.

Kozhevnikova, A.D., Seregin, I.V., Gosti, F., Schat, H., 2017. Zinc accumulation and distribution over tissues in Noccaea caerulescens in nature and in hydroponics: a comparison. Plant and Soil 411, 5–16.

Kozhevnikova, A.D., Seregin, I.V., Zhukovskaya, N.V., Kartashov, A.V., Schat, H., 2021. Histidine-mediated nickel and zinc translocation in intact plants of the hyperaccumulator *Noccaea caerulescens*. Russian Journal of Plant Physiology 68 (1), S37–S50.

Krämer, U., 2010. Metal hyperaccumulation in plants. Annual Review of Plant Biology 61, 517–534.

Krämer, U., Talke, I.N., Hanikenne, M., 2007. Transition metal transport. FEBS Letters 581, 2263–2272.

Kubota, H., Takenaka, C., 2003. Field Note: *Arabis gemmifera* is a hyperaccumulator of Cd and Zn. International Journal of Phytoremediation 5 (3), 197–201.

Kutrowska, A., Szelag, M., 2014. Low-molecular weight organic acids and peptides involved in the long-distance transport of trace metals. Acta Physiologiae Plantarum 36, 1957–1968.

Kwon, M.J., Boyanov, M.I., Yang, J.S., Lee, S., Hwang, Y.H., Lee, J.Y., et al., 2017. Transformation of zinc-concentrate in surface and subsurface environments: implication for assesing zinc mobility/toxicity and choosing an optimal remediation strategy. Environmentall Pollution 226, 346–355.

Küpper, H., Lombi, E., Zhao, F.J., McGrath, S.P., 2000. Cellular compartmentation of cadmium and zinc in relation to other elements in the hyperaccumulator Arabidopsis halleri. Planta 212, 75–84.

Küpper, H., Mijovilovich, A., Meyer-Klaucke, W., Kroneck, P.M.H., 2004. Tissue- and Age-Dependent Differences in the Complexation of Cadmium and Zinc in the Cadmium/Zinc Hyperaccumulator *Thlaspi caerulescens* (Ganges Ecotype) Revealed by X-Ray Absorption Spectroscopy. Plant Physiology 134, 748–757.

Küpper, H., Zhao, F.J., McGrath, S.P., 1999. Cellular compartmentation of zinc in leaves of the hyperaccumulator Thlaspi caerulescens. Plant Physiology 119, 305–312.

Laity, J.H., Lee, B.M., Wright, P.E., 2001. Zinc finger proteins: new insights into structural and functional diversity. Currrent Opinion in Structural Biology 11, 39–46.

Lasat, M.M., Baker, A.J.M., Kochian, L.V., 1998. Altered Zn compartmentation in the root symplasm and stimulated Zn absorption into the leaf as mechanisms involved in Zn hyperaccumulation in Thlaspi caerulescens. Plant Physiology 118, 875–883.

Lefèvre, I., Vogel-Mikuš, K., Arčon, I., Lutts, S., 2016. How do roots of the metal-resistant perennial bush Zygophyllum fabago cope with cadmium and zinc toxicities? Plant and Soil 404, 193–207.

Leitenmaier, B., Küpper, H., 2011. Cadmium uptake and sequestration kinetics in individual leaf cell protoplasts of the Cd/Zn hyperaccumulator Thlaspi caerulescens. Plant, Cell & Environment 34, 208–219.

Lin, Y.F., Aarts, M.G., 2012. The molecular mechanism of zinc and cadmium stress response in plants. Cellular and Molecular Life Sciences 69 (19), 3187–3206.

Lin, C.W., Chang, H.B., Huang, H.J., 2005. Zinc induces mitogenactivated protein kinase activation mediated by reactive oxygen species in rice roots. Plant Physiology and Biochemistry 43, 963–968.

Lin, Y.-F., Hassan, Z., Talukdar, S., Schat, H., Aarts, M.G.M., 2016. Expression of the ZNT1 zinc transporter from the metal hyperaccumulator Noccaea caerulescens confers enhanced zinc and cadmium tolerance and accumulation to Arabidopsis thaliana. PLoS ONE 11, e0149750.

Lin, W., Xiao, T., Wu, Y., Ao, Z., Ning, Z., 2012. Hyperaccumulation of zinc by *Corydalis davidii* in Zn-polluted soils. Chemosphere 86 (8), 837–842.

Li, Z., Colinet, G., Zu, Y., Wang, J., An, L., Li, Q., et al., 2019. Species diversity of *Arabis alpina* L. communities in two Pb/Zn mining areas with different smelting history in Yunnan Province, China. Chemosphere 233, 603–614.

Lopez Millan, A.F., Ellis, D.R., Grusak, M.A., 2005. Effect of zinc and manganese supply on the activities of superoxide dismutase and carbonic anhydrase in Medicago truncatula wild type and raz mutant plants. Plant Science 168, 1015–1022.

Lux, A., Lukačová, Z., Vaculík, M., Švubová, R., Kohanová, J., Soukup, M., et al., 2020. Silicification of Root Tissues. Plants 9, 111.

Lu, L., Liao, X., Labavitch, J., et al., 2014. Speciation and localization of Zn in the hyperaccumulator Sedum alfredii by extended X-ray absorption fine structure and micro-X-ray fluorescence. Plant Physiology and Biochemistry 84, 224–232.

Lu, L., Tian, S., Zhang, J., et al., 2013. Efficient xylem transport and phloem remobilization of Z n in the hyperaccumulator plant species *S edum alfredii*. New Phytologist 198, 721–731.

Macnair, M.R., 2003. The hyperaccumulation of metals by plants. Advances in Botanical Research 40, 63–105.

Marschner, H., Marschner, P., 2012. Marschner's Mineral Nutrition of Higher Plants., 3rd ed. Academic Press, London, Waltham, MA.

McFarlane, G.R., Burchett, M.D., 2000. Cellular distribution of copper, lead and zinc in the grey mangrove, Avicena marina (Forsk.) Vierh. Aquatic Bot 68, 45–59.

Mengel, K., Kirkby, E.A., 2001. Principles of Plant Nutrition, 5th ed. Kluwer Academic Publishers, Dordrecht, p. 849.

Meng, F., Liu, D., Yang, X., Shohag, M.J.I., Yang, J., Li, T., et al., 2014. Zinc uptake kinetics in the low and high-affinity systems of two contrasting rice genotypes. Journal of Plant Nutrition and Soil Science 177, 412–420.

Merlot, S., Sanchez Garcia de la Torre, V., Hanikenne, M., 2018. Physiology and Molecular Biology of Trace Element Hyperaccumulation. In: Van der Ent, A., Echevarria, G., Baker, A.J.M., Morel, J.L. (Eds.), Agromining: Farming for Metals: Extracting Unconventional Resources Using Plants, Mineral Resource Reviews. Springer International Publishing, Cham, pp. 93–116.

Mertens, J., Smolders, E., 2013. Zinc. In: Alloway, B.J. (Ed.), Heavy Metals in Soils. Springer, pp. 11–50.

Mills, R.F., Krijger, G.C., Baccarini, P.J., Hall, J.L., Williams, L.E., 2003. Functional expression of AtHMA4, a P1B-type ATPase of the Zn/Co/Cd/Pb subclass. The Plant Journal 35, 164–176.

Milner, M.J., Craft, E., Yamaji, N., Koyama, E., Ma, J.F., Kochian, L.V., 2012. Characterization of the high affinity Zn transporter from Noccaea caerulescens, NcZNT1, and dissection of its promoter for its role in Zn uptake and hyperaccumulation. New Phytologist 195, 113–123.

Milner, M.J., Kochian, L.V., 2008. Investigating heavy-metal hyperaccumulation using Thlaspi caerulescens as a model system. Annals of Botany 102, 3–13.

Mishra, S., Mishra, A., Küpper, H., 2017. Protein biochemistry and expression regulation of Cadmium/Zinc Pumping ATPases in the hyperaccumulator plants Arabidopsis halleri and Noccaea caerulescens. Frontiers in Plant Science 8, 835.

Mišljenović, T., Jovanović, S., Mihailović, N., Gajić, B., Tomović, G., Baker, A.J.M., et al., 2020. Natural variation of nickel, zinc and cadmium (hyper) accumulation in facultative serpentinophytes *Noccaea kovatsii* and *N. praecox*. Plant and Soil 447 (1), 475–495.

Monsant, A.C., Kappen, P., Wang, Y., Pigram, P.J., Baker, A.J.M., Tang, C., 2011. In vivo speciation of zinc in Noccaea caerulescens in response to nitrogen form and zinc exposure. Plant and Soil 348, 167.

Moreira, A., Moraes, L.A.C., dos Reis, A.R., 2018. The molecular genetics of zinc uptake and utilization efficiency in crop plants. Plant Micronutrient Use Efficiency. Elsevier, pp. 87–108.

Morel, M., Crouzet, J., Gravot, A., et al., 2008. AtHMA3, a P1B-ATPase allowing Cd/Zn/Co/Pb vacuolar storage in Arabidopsis. Plant Physiology 149, 894–904.

van de Mortel, J.E., Almar Villanueva, L., Schat, H., Kwekkeboom, J., Coughlan, S., Moerland, P.D., et al., 2006. Large expression differences in genes for iron and zinc homeostasis, stress response, and lignin biosynthesis distinguish roots of *Arabidopsis thaliana* and the related metal hyperaccumulator *Thlaspi caerulescens*. Plant Physiology 142, 1127–1147.

van de Mortel, J.E., Schat, H., Moerland, P.D., Van Themaat, E.V., van der Ent, S.J., Blankestijn, H., et al., 2008. Expression differences for genes involved in lignin, glutathione and sulphate metabolism in response to cadmium in *Arabidopsis thaliana* and the related Zn/Cd-hyperaccumulator *Thlaspi caerulescens*. Plant, Cell & Environment 31 (3), 301–324.

Mukhopadhyay, M., Mondal, T.K., 2014. The physio-chemical responses of camellia plants to abiotic stresses. Journal of Plant Science & Research 1, 1–12.

Neumann, D., zur Nieden, U., 2001. Silicon and heavy metal tolerance of higher plants. Phytochemistry 56, 685–692.

Nkrumah, P.N., Echevarria, G., Erskine, P.D., van der Ent, A., 2018. Contrasting nickel and zinc hyperaccumulation in subspecies of *Dichapetalum gelonioides* from Southeast Asia. Scientific Reports 8 (1), 1–15.

Noulas, C., Tziouvalekas, M., Karyotis, T., 2018. Zinc in soils, water and food crops. Journal of Trace Elements in Medicine and Biology 49, 252–260.

Olsen, L., Palmgren, M., 2014. Many rivers to cross: the journey of zinc from soil to seed. Frontiers in Plant Science 5, 30.

Paton, A., Brooks, R.R., 1996. A re-evaluation of *Haumaniastrum* species as geobotanical indicators of copper and cobalt. Journal of Geochemical Exploration 56 (1), 37–45.

Paul, A.L.D., Erskine, P.D., van der Ent, A., 2018. Metallophytes on Zn-Pb mineralised soils and mining wastes in Broken Hill, NSW, Australia. Australian Journal of Botany 66 (2), 124–133.

Paul, A.L., Gei, V., Isnard, S., Fogliani, B., Echevarria, G., Erskine, P.D., et al., 2020. Nickel hyperaccumulation in New Caledonian *Hybanthus* (Violaceae) and occurrence of nickel-rich phloem in *Hybanthus austrocaledonicus*. Annals of Botany 126 (5), 905–914.

Peck, A.W., McDonald, G.K., 2010. Adequate zinc nutrition alleviates the adverse effects of heat stress in bread wheat. Plant and Soil 337, 355–374.

Peer, W.A., Mahmoudian, M., Freeman, J.L., Lahner, B., Richards, E.L., Reeves, R.D., et al., 2006. Assessment of plants from the Brassicaceae family as genetic models for the study of nickel and zinc hyperaccumulation. New Phytologist 172, 248–260.

Pence, N.S., Larsen, P.B., Ebbs, S.D., Letham, D.L.D., Lasat, M.M., Garvin, D.F., et al., 2000. The molecular physiology of heavy metal transport in the Zn/Cd hyperaccumulator Thlaspi caerulescens. Proceedings of the National Academy of Sciences 97, 4956–4960.

Phaenark, C., Pokethitiyook, P., Kruatrachue, M., Ngernsansaruay, C., 2009. Cd and Zn accumulation in plants from the Padaeng zinc mine area. International Journal of Phytoremediation 11 (5), 479–495.

Pineau, C., Loubet, S., Lefoulon, C., Chalies, C., Fizames, C., Lacombe, B., et al., 2012. Natural variation at the FRD3 MATE transporter locus reveals cross-talk between Fe homeostasis and Zn tolerance in Arabidopsis thaliana. PLoS Genetics 8, e1003120.

POWO. 2021. Plants of the World Online. Facilitated by the Royal Botanic Gardens, Kew. Published on the Internet; http://www.plantsoftheworldonline.org/ Retrieved 04 November 2021.

Prasad, A.S., 2013. Discovery of human zinc deficiency: its impact on human health and disease. Advances in Nutrition 4, 176–190.

Prasad, M.N.V., de Oliveira, Freitas, H.M., 2003. Metal hyperaccumulation in plants—biodiversity prospecting for phytoremediation technology. Electronic Journal of Biotechnology 6 (3), 110–146.

Purwadi, I., Gei, V., Echevarria, G., Erskine, P.D., Mesjasz-Przybyłowicz, J., Przybyłowicz, W. J., et al., 2021. Tools for the discovery of hyperaccumulator plant species in the field and in the herbarium. In: van der Ent, A., Baker, A.J.M., Echevarria, G., Simonnot, M.O., Morel, J. L. (Eds.), Agromining: Farming for Metals. Springer, Cham, pp. 183–195. Mineral Resource Reviews.

Päivöke, A., 1983. The short-term effects of zinc on the growth, anatomy and acid phosphatase activity of pea seedlings. Annales Botanici Fennici 20, 197–203.

Qiu, R., Fang, X., Tang, Y., Du, S., Zeng, X., Brewer, E., 2006. Zinc hyperaccumulation and uptake by *Potentilla griffithii* Hook. International Journal of Phytoremediation 8 (4), 299–310.

Qiu, R.-L., Thangavel, P., Hu, P.-J., et al., 2011. Interaction of cadmium and zinc on accumulation and sub-cellular distribution in leaves of hyperaccumulator Potentilla griffithii. Journal of Hazardous Materials 186, 1425–1430.

Ranjani, R., Adam, S., Murthy, S., 2014. Zinc induced alterations in the photosystem II mediated photochemistry of cyanobacterium Spirulina platensis. Research Journal of Pharmaceutical, Biological and Chemical Sciences 5, 1039–1044.

Reeves, R.D., Baker, A.J.M., 2000. Metal-accumulating plants. In: Raskin, I., Ensley, B.D. (Eds.), Phytoremediation of Toxic Metals: Using Plants to Clean Up the Environment. Wiley, New York, pp. 193–229.

Reeves, R.D., Brooks, R.R., 1983. European species of *Thlaspi* L. (Cruciferae) as indicators of nickel and zinc. Journal of Geochemical Exploration 18 (3), 275–283.

Reeves, R.D., 2006. Hyperaccumulation of trace elements by plants. In: Morel, J.L., Echevarria, G., Goncharova, N. (Eds.), Phytoremediation of Metal-Contaminated Soils. Springer, Dordrecht, pp. 25–52.

Reeves, R.D., Schwartz, C., Morel, J.L., Edmondson, J., 2001. Distribution and metal-accumulating behavior of *Thlaspi caerulescens* and associated Metallophytes in France. International Journal of Phytoremediation 3 (2), 145–172.

Reeves, R.D., van der Ent, A., Echevarria, G., Isnard, S., Baker, A.J.M., 2021. Global distribution and ecology of hyperaccumulator plants. In: van der Ent, A., Baker, A.J.M., Echevarria, G., Simonnot, M.O., Morel, J.L. (Eds.), Agromining: Farming for Metals. Springer, Cham, pp. 133–154.

Ricachenevsky, F.K., Menguer, P.K., Sperotto, R.A., Williams, L.E., Fett, J.P., 2013. Roles of plant metal tolerance proteins (MTP) in metal storage and potential use in biofortification strategies. Frontiers in Plant Science 4, 144.

Roosens, N.H., Leplae, R., Bernard, C., Verbruggen, N., 2005. Variations in plant metallothioneins: the heavy metal hyperaccumulator Thlaspi caerulescens as a study case. Planta 222, 716–729.

Sadeghzadeh, B., Rengel, Z., 2011. Zinc in Soils and Crop Nutrition. The Molecular and Physiological Basis of Nutrient Use Efficiency in Crops. John Wiley & Sons, Ltd., pp. 335–375.

Sagardoy, R., Morales, F., López-Millán, A.-F., Abadía, A., Abadía, J., 2009. Effects of zinc toxicity on sugar beet (Beta vulgaris L.) plants grown in hydroponics. Plant Biology 11, 339–350.

Sagardoy, R., Morales, F., Rellán-Álvarez, R., et al., 2011. Carboxylate metabolism in sugar beet plants grown with excess Zn. Journal of Plant Physiology 168, 730–733.

Sakamoto, H., Maruyama, K., Sakuma, Y., Meshi, T., Iwabuchi, M., Shinozaki, K., et al., 2004. Arabidopsis Cys2/His2-type zinc-finger proteins function as transcription repressors under drought, cold, and high-salinity stress conditions. Plant Physiology 136, 2734–2746.

Salt, D.E., Prince, R.C., Baker, A.J., Raskin, I., Pickering, I.J., 1999. Zinc ligands in the metal hyperaccumulator *Thlaspi caerulescens* as determined using X-ray absorption spectroscopy. Environmental Science & Technology 33 (5), 713–717.

Salt, D.E., Prince, R.C., Pickering, I.J., 2002. Chemical speciation of accumulated metals in plants: evidence from X-ray absorption spectroscopy. Microchemical Journal 71, 255–259.

Samreen, T., Humaira, Shah, H.U., Ullah, S., Javid, M., 2017. Zinc effect on growth rate, chlorophyll, protein and mineral contents of hydroponically grown mungbeans plant (*Vigna radiata*). Arabian Journal of Chemistry 10, 1802–1807.

Sarret, G., Saumitou-Laprade, P., Bert, V., Proux, O., Hazemann, J.L., Traverse, A., et al., 2002. Forms of zinc accumulated in the hyperaccumulator. Arabidopsis halleri Plant Physiology 130 (4), 1815–1826.

Sarret, G., Willems, G., Isaure, M.-P., Marcus, M.A., Fakra, S.C., Frérot, H., et al., 2009. Zinc distribution and speciation in Arabidopsis halleri × Arabidopsis lyrata progenies presenting various zinc accumulation capacities. New Phytologist 184, 581–595.

Schat, H., Llugany, M., Vooijs, R., Hartley-Whitaker, J., Bleeker, P.M., 2002. The role of phytochelatins in constitutive and adaptive heavy metal tolerances in hyperaccumulator and non-hyperaccumulator metallophytes. Journal of Experimental Botany 53 (379), 2381–2392.

Shanmugam, V., Lo, J.-C., Yeh, K.-C., 2013. Control of Zn uptake in Arabidopsis halleri: a balance between Zn and Fe. Frontiers in Plant Science 4, 281. Available from: https://doi.org/10.3389/fpls.2013.00281.

Sharma, A., Patni, B., Shankhdhar, D., Shankhdhar, S.C., 2013. Zinc - an indispensable micronutrient. Physiology and Molecular Biology of Plants 19, 11–20.

Shen, Z.G., Zhao, F.J., McGrath, S.P., 1997. Uptake and transport of zinc in the hyperaccumulator *Thlaspi caerulescens* and the non-hyperaccumulator *Thlaspi ochroleucum*. Plant, Cell & Environment 20 (7), 898–906.

Sinclair, S.A., Krämer, U., 2012. The zinc homeostasis network of land plants. Biochimica et Biophysica Acta (BBA) - Molecular Cell Research. Cell Biology of Metals 1823, 1553–1567.

Sitko, K., Rusinowski, S., Kalaji, H.M., Szopiński, M., Małkowski, E., 2017. Photosynthetic efficiency as bioindicator of environmental pressure in A. halleri. Plant Physiology 175, 290–302.

Sommer, A.L., Lipman, C.B., 1926. Evidence on the indispensable nature of zinc and boron for higher green plants. Plant Physiology 1, 231–249.

Song, W.-Y., Choi, K.S., Kim, D.Y., Geisler, M., Park, J., Vincenzetti, V., et al., 2010. *Arabidopsis* PCR2 Is a zinc exporter involved in both zinc extrusion and long-distance zinc transport. The Plant Cell 22, 2237–2252.

Sridhar, B.B.M., Diehl, S.V., Han, F.X., Monts, D.L., Su, Y., 2005. Anatomical changes due to uptake and accumulation of Zn and Cd in Indian mustard (Brassica juncea). Environmental and Experimental Botany 54, 131–141.

Stefanidou, M., Maravelias, C., Dona, A., Spiliopoulou, C., 2006. Zinc: a multipurpose trace element. Archives in Toxicology 80, 1–9.

Stein, R.J., Höreth, S., de Melo, J.R.F., Syllwasschy, L., Lee, G., Garbin, M.L., et al., 2017. Relationships between soil and leaf mineral composition are element-specific, environment-dependent and geographically structured in the emerging model *Arabidopsis halleri*. New Phytologist 213 (3), 1274–1286.

Stoláriková-Vaculíková, M., Romeo, S., Minnoci, A., Luxová, M., Vaculík, M., Lux, A., et al., 2015. Anatomical, biochemical and morphological responses of poplar *Populus deltoides* clone Lux to Zn excess. Environmental and Experimental Botany 109, 235–243.

Stoláriková, M., Vaculík, M., Lux, A., Di Baccio, D., Minnocci, A., Andreucci, A., et al., 2012. Anatomical differences of poplar *(Populus × euramericana* clone I-214) roots exposed to zinc excess. Biologia 67/3, 483–489.

Subba, P., Mukhopadhyay, M., Mahato, S.K., et al., 2014. Zinc stress induces physiological, ultra-structural and biochemical changes in mandarin orange (*Citrus reticulata* Blanco) seedlings. Physiology and Molecular Biology of Plants 20, 461–473.

Sun, Q., Ye, Z.H., Wang, X.R., Wong, M.H., 2005. Increase of glutathione in mine population of *Sedum alfredii*: a Zn hyperaccumulator and Pb accumulator. Phytochemistry 66 (21), 2549–2556.

Szarek-Łukaszewska, G., Niklinska, M., 2002. Concentration of alkaline and heavy metals in Biscutella laevigata L. and Plantago lanceolate L. growing on calamine spoils (S. Poland). Acta Biologica Cracoviensia s. Botanica 44, 29–38.

Szopinski, M., Sitko, K., Rusinowski, S., Zieleznik-Rusinowska, P., Corso, M., Rostanski, A., et al., 2020. Different strategies of Cd tolerance and accumulation in Arabidopsis halleri and Arabidopsis arenosa. Plant, Cell and Environment 43, 3002–3019.

Tang, Y., Qiu, R., Zeng, X., Fang, X., Yu, F., Zhou, X., et al., 2009. Zn and Cd hyperaccumulating characteristics of *Picris divaricata* Vant. International Journal of Environment and Pollution 38 (1–2), 26–38.

Tang, Y.T., Sterckeman, T., Echevarria, G., Morel, J.L., Qiu, R.L., 2019. Effects of the interactions between nickel and other trace metals on their accumulation in the hyperaccumulator *Noccaea caerulescens*. Environmental and Experimental Botany 158, 73–79.

Thalius J. 1588. *Flora of the Harz Mountains, or an Enumeration of Indigenous Plant Species in the Mountains and Their Surroundings*. Frankfurt a.M.

Thomine, S., Lelièvre, F., Debarbieux, E., Schroeder, J.I., Barbier-Brygoo, H., 2003. AtNRAMP3, a multispecific vacuolar metal transporter involved in plant responses to iron deficiency. The Plant Journal 34, 685–695.

Tian, S., Lu, L., Yang, X., et al., 2009. Stem and leaf sequestration of zinc at the cellular level in the hyperaccumulator *Sedum alfredii*. New Phytologist 182, 116–126.

Todd, W.R., Elvehjem, C.A., Hart, E.B., 1934. Zinc in the nutrition of the rat. American Journal of Physiology 107, 146.

Todeschini, V., Lingua, G., D'Agostino, G., et al., 2011. Effects of high zinc concentration on poplar leaves: A morphological and biochemical study. Environmental and Experimental Botany 71, 50–56.

Tolra, R., Pongrac, P., Poschenrieder, C., Vogel-Mikuš, K., Regvar, M., Barcelo, J., 2006. Distinctive effects of cadmium on glucosinolate profiles in Cd hyperaccumulator *Thlaspi praecox* and non-hyperaccumulator *Thlaspi arvense*. Plant and Soil 288 (1), 333–341.

Tomović, G., Đurović, S., Buzurović, U., Niketić, M., Milanović, Đ., Mihailović, N., et al., 2021. Accumulation of Potentially Toxic Elements in *Viola* L.(Sect. *Melanium* Ging.) from the Ultramafic and Non-ultramafic Soils of the Balkan Peninsula. Water, Air, & Soil Pollution 232 (2), 1–18.

Uraguchi, S., Weber, M., Clemens, S., 2019. Elevated root nicotianamine concentrations are critical for Zn hyperaccumulation across diverse edaphic environments. Plant, Cell & Environment 42 (6), 2003–2014.

Vaculík, M., Konlechner, C., Langer, I., Adlassnig, W., Puschenreiter, M., Lux, A., et al., 2012. Root anatomy and element distribution vary between two *Salix caprea* isolates with different Cd accumulation capacities. Environmental Pollution 163, 117–126.

Venkatesan, S., Hemalatha, K.V., Jayaganesh, S., 2006. Zinc toxicity and its influence on nutrient uptake in tea. American Journal of Plant Physiology 1, 185–192.

Verbruggen, N., Hermans, C., Schat, H., 2009. Molecular mechanisms of metal hyperaccumulation in plants. New Phytologist 181 (4), 759–776.

Verret, F., Gravot, A., Auroy, P., Leonhardt, N., David, P., Nussaume, L., et al., 2004. Overexpression of AtHMA4 enhances root-to-shoot translocation of zinc and cadmium and plant metal tolerance. FEBS Letters 576, 306–312.

Weber, M., Harada, E., Vess, C., Roepenack-Lahaye, E.v, Clemens, S., 2004. Comparative microarray analysis of Arabidopsis thaliana and Arabidopsis halleri roots identifies nicotianamine synthase, a ZIP transporter and other genes as potential metal hyperaccumulation factors. The Plant Journal 37, 269–281.

Weiss, D.J., Mason, T.F.D., Zhao, F.J., Kirk, G.J.D., Coles, B.J., Horstwood, M.S.A., 2005. Isotopic discrimination of zinc in higher plants. The New Phytologist 165, 703–710.

Wenzel, W.W., Jockwer, F., 1999. Accumulation of heavy metals in plants grown on mineralised soils of the Austrian Alps. Environmental Pollution 104 (1), 145–155.

White, P.J., Whiting, S.N., Baker, A.J.M., Broadley, M.R., 2002. Does zinc move apoplastically to the xylem in roots of Thlaspi caerulescens? The New Phytologist 153, 201–207.

Wu, C., Liao, B., Wang, S.L., Zhang, J., Li, J.T., 2010. Pb and Zn accumulation in a Cd-hyperaccumulator (*Viola baoshanensis*). International Journal of Phytoremediation 12 (6), 574–585.

Wójcik, M., Skórzyńska-Polit, E., Tukiendorf, A., 2006. Organic acids accumulation and antioxidant enzyme activities in *Thlaspi caerulescens* under Zn and Cd stress. Plant Growth Regulation 48 (2), 145–155.

Xv, L., Ge, J., Tian, S., Wang, H., Yu, H., Zhao, J., et al., 2020. A Cd/Zn Co-hyperaccumulator and Pb accumulator, *Sedum alfredii*, is of high Cu tolerance. Environmental Pollution 263, 114401.

Yadav, V., Arif, N., Kováč, J., Singh, V.P., Tripathi, D.K., Chauhan, D.K., et al., 2021. Structural modifications of plant organs and tissues by metals and metalloids in the environment: a review. Plant Physiology and Biochemistry 159, 100–112.

Yamasaki, S., Sakata-Sogawa, K., Hasegava, A., Suzuki, T., Kabu, K., Sato, E., et al., 2007. Zinc is a novel intracellular second messenger. The Journal of Cell Biology 177, 637–645.

Yang, X., Li, T., Yang, J., et al., 2005. Zinc compartmentation in root, transport into xylem, and absorption into leaf cells in the hyperaccumulating species of Sedum alfredii Hance. Planta 224, 185–195.

Yang, X.E., Long, X., Ni, W., Fu, C., 2002. *Sedum alfredii* H: a new Zn hyperaccumulating plant first found in China. Chinese Science Bulletin 47 (19), 1634–1637.

Yang, X.E., Long, X.X., Ye, H.B., He, Z.L., Calvert, D.V., Stoffella, P.J., 2004. Cadmium tolerance and hyperaccumulation in a new Zn-hyperaccumulating plant species (*Sedum alfredii* Hance). Plant and Soil 259 (1), 181–189.

Yang, Q., Ma, X., Luo, S., Gao, J., Yang, X., Feng, Y., 2018. SaZIP4, an uptake transporter of Zn/Cd hyperaccumulator Sedum alfredii Hance. Environmental and Experimental Botany 155, 107–117.

van der Zaal, B.J., Neuteboom, L.W., Pinas, J.E., Chardonnens, A.N., Schat, H., Verkleij, J.A. C., et al., 1999. Overexpression of a novel Arabidopsis gene related to putative zinc-transporter genes from animals can lead to enhanced zinc resistance and accumulation. Plant Physiology 119, 1047–1056.

Zelko, I., Lux, A., Czibula, K., 2008. Difference in the root structure of hyperaccumulator Thlaspi caerulescens and non-hyperaccumulator Thlaspi arvense. IJEP 33, 123.

Zhao, F.J., Lombi, E., Breedon, T., M, S.P., 2000. Zinc hyperaccumulation and cellular distribution in Arabidopsis halleri. Plant, Cell and Environment 23, 507–514.

Zhao, F.J., Shen, Z.G., McGrath, S.P., 1998. Solubility of zinc and interactions between zinc and phosphorus in the hyperaccumulator Thlaspi caerulescens. Plant, Cell & Environment 21, 108–114.

Chapter 2

# Mechanisms of zinc in the soil-plant interphase: role and regulation

Nimisha Amist[1], Shubhra Khare[2], Niharika[3], Zeba Azim[3] and N.B. Singh[3]

[1]*Department of Botany, Ewing Christian College, Prayagraj, Uttar Pradesh, India,* [2]*Department of Applied Sciences and Humanities, Invertis University, Bareilly, Uttar Pradesh, India,* [3]*Plant Physiology Laboratory, Department of Botany, University of Allahabad, Prayagraj, Uttar Pradesh, India*

## 2.1 Introduction

Zinc (Zn) in nature is reported from all the soil types, generally in the range of 10–100 mg/kg. Zn is a transition metal with an atomic number of 30, and five stable isotopes viz $^{70}Zn$ (0.62%), $^{68}Zn$ (18.75%), $^{67}Zn$ (4.90%), $^{66}Zn$ (27.90%) and $^{64}Zn$ (48.63%) have been reported. Both heavy and light isotopes of Zn have been reported in plants' roots and shoots (Weiss et al., 2005). The Zn radioisotopes have been categorized based on their half-lives, and there are 30 short-lived reported within the atomic mass range 54–83, with $^{65}Zn$ the longest-lived, normally used as a Zn tracer in plants. However, in solution, Zn exists in the +2 oxidation state and exhibits redox stability in physiological conditions due to a completely filled *d-shell* of electrons (Barak and Helmke, 1993; Auld, 2001). Zn interacts with other molecules to generate a variety of soluble salts, including alkali metal acetates, formates, nitrates, sulphates, halides, thiocyanates, perchlorates, zincates and fluosilicates. Zn also reacts with ammonia to synthesize Zn-ammonia salts along with compounds with moderate solublility like hydroxide, carbonate and Zn-ammonium phosphate. Varied ranges of soluble and insoluble organic complexes are also formed by Zn interaction with organic compounds (Lindsay, 1979; Barak and Helmke, 1993).

The Zn has a very long historical association with humans. It was an integral part of human life. In 200 BC, Zn alloy (brass) was used by the Romans in their daily life activities. Metallic Zn production was initiated in Asia

**Zinc in Plants. DOI: https://doi.org/10.1016/B978-0-323-91314-0.00009-0**
 

around the 20th century. It seems that during the end of the Middle Ages, traders from the East brought metallic zinc to Europe, where it was industrially synthesised in Bristol, England, in the middle of the eighteenth century. The production of metallic Zn has increased tremendously. The higher concentrations Zn than the expected might be due to previous atmospheric deposition. The elevated level of Zn in the soil parent rock along with sewage sludge applications, Zn mining areas and release of Zn from smelters contribute towards the higher concentrations of Zn in local areas.The past production of Zn and smelter activities have polluted huge areas worldwide through the atmospheric deposition of airborne Zn.

Zn is an essential micronutrient for all biota. High Zn levels are responsible for the lethal impact on plants, soil Rhizophora, and microorganisms, eventually imposing ecotoxicological effects on humans. However, no case of prolonged poisoning of Zn to people or wildlife through the food chain has been acknowledged. It seems that the phytotoxicity barrier efficiently restricts Zn transfer. As a result, more attention is directed toward the effect of Zn on soil biota and soil functioning when evaluating the risk. A lower concentration of Zn causes deficiency in plants and, consequently to humans. One third of the world's agricultural land and people are Zn-deficient, and this is mostly due to the element's decreased content and bioavailability. Due to poor soil conditions, inadequate availability, and immobility, plants experience zinc deficiencies. Consequently, despite the high concentration of zinc present in soil, a shortfall is noted. A zinc shortage in the soil affects the food chain in the area, which makes zinc scarce for humans. Growth hindrance, along with late sexual and skeletal maturation and behavioral effects, are some of the Zn deficiency symptoms developed in humans. This generates a need for the development of more advanced Zn-efficient cultivars along with biodegradable Zn fertilizers to address the Zn deficiency problems, along with the provision of Zn additives. This chapter aims to understand the significance of Zn, its bioavailability, and its uptake from soil to plants.

## 2.2 Biochemical and physiological importance of zinc

One essential element for plants is Zn (Marschner 1993). The majority of species, including humans, depend on it for their existence (Hambidge 2000). More than 300 proteins require Zn as a cofactor; the majority of these proteins are members of the Zn-finger protein family, as well as RNA and DNA polymerases (Coleman, 1998; Lopez Millan et al., 2005). Zn is the sole metal linked to all six major types of enzymes: hydrolases, lyases, isomerases, ligases, oxidoreductases, and transferases. Thus, Zn controls all the processes involving activities, conformational stability, and folding of proteins as it is a component of structural or catalytic units. Zn has also been reported to be involved in the maintenance of membrane integrity and stability (Cakmak, 2000; Disante et al., 2010) and the mitigation of oxidative stress (Cakmak, 2000). Zn also works as a secondary messenger in intercellular signal transduction systems (Yamasaki et al., 2007). Zn participates in several metabolic processes including regulation of hormone synthesis and activity (e.g., tryptophan synthesis, a precursor of IAA) and governs signal transduction, specifically mitogen-activated protein kinases (Lin et al., 2005; Hansch and Mendel, 2009). Zn also participates in the PS II complex repair mechanism after photoinhibition (Bailey et al., 2002; Hansch and Mendel, 2009). Zn regulates RUBISCO activity along with mitigating the antagonistic effects of heat stress in wheat (Peck and McDonald, 2010). Unfortunately, due its low availability and immobility, deficiency of this metal ion has become a major agricultural problem these days. Therefore, Zn deficiency is detected even though a high aggregate is available in the soil. The primary factor influencing Zn uptake is the root-shoot interaction, and deviations in the structure of conducting tissue as well as soil parameters like pH, clay or calcium carbonate concentration, etc., have a major impact on this root-shoot barrier. As a result, a Zn deficit causes significant output losses and, in extreme situations, plant mortality. Zn absorption is a complex process. It is chiefly controlled by Zn transporters and metal chelators in plant systems. The Zn efficiency toward plants is also dependent upon the stages of plant growth along with edaphic and abiotic factors. A finding based on the molecular analysis of Zn hyperaccumulators revealed that Zn homeostasis is being maintained by the active participation of Zn transporters and the detoxification mechanism with vacuolar sequestration (Claus et al., 2013).

## 2.3 Bioavailability of zinc in soil

Heavy metal accumulation in the rhizosphere needs attention because it may have effects on the quality of foodstuffs, soil health, and ecosystems (Khan et al., 2013; Rutkowska et al., 2015). The bioavailability and mobility of heavy metals have become a threat to mankind and its surrounding environment (Kabata-Pendias, 2004). Plants can uptake only chelated metal species found in soluble and transferable forms present in the soil (Kabata-Pendias, 1993). Zn plays a very significant role in plants due to its high transportability and bioavailability in the rhizosphere. However, due to high accumulation in the soil, it may have phytotoxic outcomes and decrease the quality and production of crop plants (Baran et al., 2018). Moreover, morphological developments and metabolic processes of soil microbes are reduced with the increased quantity of Zn (Olaniran et al., 2013). Zn is found in the rhizosphere in two forms, i.e., $Zn^{2+}$ and in organometallic complexes. Many physicochemical soil elements, including organic matter content, pH, oxidation-reducing potential, granulometric composition, iron and manganese hydroxides, absorptive capacity and moisture influences the binding of Zn and its bioavailability in soil (Kabata-Pendias 2004; Kim et al. 2015). Baran et al. (2018) reported that some other factors, such as organic C, sand, silt, and clay constituents, are also related to the quantity of available forms of Zn present in the soil. The bioavailability of Zn in soils is also controlled by the organic form of C, in reference to pyrrolidone carboxylic acid. In many areas, Zn deficiency is reported, especially in calcareous soils, and Zn deficiency inhibits $\pm 30\%$ of crop yields globally (Alloway, 2009). Zn and phosphorus deficiency co-occur because both elements are affected by the increased pH (Duffner et al., 2012). The release of protons causes acidification, which is required for the mobilization of Zn (Straczek and Hinsinger, 2004). Previous studies reported some contrasting results where other compounds in the solution such as citrate forms complex with $Zn^{2+}$ that alters its activity by reducing functionality of $Zn^{2+}$ in the soil solution. As a consequence, $Zn^{2+}$ withdraws from soil surfaces to maintain $Zn^{2+}$ content in solution (Yin et al., 2012).

Rose et al. (2011) reported more exudation of malate than citrate ions in response from $Zn^{2+}$ deficient rice genotypes. The study reported that in response to $Zn^{2+}$-deficient conditions, the resistance potency of rice plants results in the release of malate ions (Widodo et al., 2010). In a calcareous soil, the exudation of carboxylates from the roots of spinach (*Spinacia oleracea* L.) and tomato (*Solanum lycopersicum* L.) significantly mobilized Cu and Zn form (Degryse et al., 2008). Moreover, other experiments performed by Gao et al. (2009) proved that in a rice field, malate concentrations had a minimum impact on the concentration of Zn in the soil solution.

Zn diffuses slowly in the soil; hence, it is less available in the soil. Some forms of Zn, such as $ZnSO_4$ and Zn-EDTA, are found to be more soluble, whereas chelated forms are less concentrated and are expensive (Mortvedt and Gilkes, 1993). In soils, Zn fertilizers can be given to crops either before sowing or by foliar spray technique (Mortvedt and Gilkes, 1993). For proper fertilization of Zn in the soil, Zn can be winded under the seed, and with the right approach, the appropriate amount and time of fertilizer treatment play a very significant role in the uptake of $Zn^{2+}$ (Martens and Westermann, 1991). Consumption of cereal-based food in developing countries is the chief cause of Zn deficiency. More amounts of phytate compounds are found in the starchy endosperm of cereal grains, which reduces mineral contents as well as their bioavailability (Brinch-Pedersen et al., 2007). Biofortification is a very useful cost-effective approach used to solve these nutrition defects in cereal grains related to Zn (Bouis et al., 2011). Basic knowledge about Zn absorption and its sequestration to retain metal homeostasis is a must for the execution of this biofortification mechanism.

## 2.4 Uptake of zinc from soil

Plants uptake most of the rhizospheric mineral nutrients *via* their roots in their ionic forms, where Zn is taken up mainly as a divalent cation ($Zn^{2+}$ ion) through mass flow and diffusion mechanisms. The weathering of parent rocks by chemical and physical methods provides reservoirs for these nutrients in the rhizosphere. The total amount of Zn present in the soil is not completely utilized by plants; only some of its fractions are available to plants, and the bioavailability of metal ions depends on various edaphic and non-edaphic properties such as the physicochemical properties of the soil, the action of plant roots, and microflora activities found in the soil (Degryse et al., 2006). The uptake of trace metals by plants is generally based on the activity of free metal ions and not on their total dissolved concentration. Heavy metals such as Cu and Zn are absorbed preferentially via active uptake or a combination of both active and passive uptakes through symplastic and apoplastic movements (Degryse et al., 2006). In previous studies, it was reported that the actions of free metals as well as the concentration of metal-ligand complexes were responsible for the uptake of Zn in barley and lettuce (McLaughlin et al., 1997; Bell et al., 1991). Some root mechanisms are also responsible for making these metals in their exchangeable form through the exchange of ions and the exudation of organic acids. Edaphic factors such as the pH of soil play a crucial role in the uptake of Zn, where an increased pH decreases Zn availability by enhancing its absorption at cation exchange sites of soil constituents. Therefore, it was reported that at pH ranges of 5.5–7.0, the soluble form of Zn in the soil decreases by 30–45 folds per unit increase in the pH of the soil, and this leads to Zn deficiency in plants (Marschner, 1993). Another edaphic factor, i.e., soil moisture, plays a significant role in Zn deficiency and affects its uptake by plant roots via diffusion (Gupta et al., 2016). In soils with less organic matter, a low pH increases the solubility of Zn and the ratio of $Zn^{2+}$ to organic-Zn ligand complexes. However, in some studies, it was found that Zn complexes formed with organic ligands are also taken up by the roots of crop plants. Two mechanisms based on ligands synthesized by roots are found to be crucial for the uptake of $Zn^{2+}$. In the first one, the solubility of Zn phosphates, Zn hydroxides, and other complexes increases by efflux of reductants, protons, and organic acids; this strategy results in the release of $Zn^{2+}$ available for absorption via epidermal cells of the root. The second mechanism is found only in the roots of cereal, where it involves the efflux of phytosiderophores, which acquire a high binding affinity for metals and form stable complexes with Zn; this complex results in their chelation and leads to an influx of Zn into the epidermal cells of the root (Gupta et al., 2016).

Passive uptake of $Zn^{2+}$ is based on available water molecules and differences generated by the Zn gradient in the root cell plasma membrane (RCPM). Hyperpolarization of RCPM takes place by the action of the RCPM $H^+$-ATPase system, which pumps $H^+$ ions extracellularly with the hydrolysis of ATP, and this hyperpolarization is essential for the uptake of $Zn^{2+}$ (Gupta et al., 2016). However, this hyperpolarized state also results in a low pH in the soil, which favors the uptake of cations. Cell membranes are generally restricted for the passage of cations; therefore, some carrier proteins are required for further transportation of cations. On the other hand, some nonselective channels were also reported to take part in the uptake of cations passively (Gupta et al., 2016).

## 2.5 Transport of zinc in plant

The intricate process of metal ion transportation in plants involves a series of sequential processes, including ion intake from soil to root, movement from root to shoot, and ultimately ion transference from shoot to different plant sink organs. Specific transporter proteins that regulate both intercellular and intracellular transport execute these sequential events. Plants can develop high-affinity metal transporters in response to a small abundance of metal ions/trace elements in the soil. Zn is a partially mobile essential micronutrient present in plants for several metabolic processes in the cells.

The plant tissues, such as the epidermis, cortex, endodermis, and pericycle, are involved in the transportation of $Zn^{2+}$ from root to shoot. The thickness of these tissues increases when moving from the stellar center to the periphery. Water and minerals are radially transported towards the xylem, resulting in an increase in concentration, which helps in the large accretion of metal cations in the pericycle, which finally loads the xylem vessel (Claus et al., 2013).

$Zn^{2+}$ is transported to the xylem by symplastic as well as apoplastic pathways. In symplastic pathways, no membrane barriers are present, and the cytoplasms of neighboring cells in root tissues are connected by plasmodesmata. On the other hand, Zn ions are also transported in the root by apoplastic pathways *via* cell walls and intercellular spaces (Steudle, 1994). This transportation is interrupted in the endodermis region due to the presence of suberin in Casparian strips. These pathways are also responsible for the mobilization of Zn in the shoot of the plant, where the Zn transport *via* the apoplastic route is less selective and comprises the entry of Zn from the cell wall into the cytosol, whereas the symplastic route is selective and controls the extent of nutrient transport (Clemens et al., 2002). Xylem loading is another barrier to $Zn^{2+}$ transport that occurs at the time of nutrient transfer. Xylem loading is the chief determining phase in the export of metal cations in roots.

The activation of $H^+$-ATPase in the xylem parenchyma causes the membrane to become hyperpolarized, which restricts the flow of cations from the cytosol. Accordingly, $Zn^{2+}$ is actively loaded from xylem parenchyma to apoplastic xylem (Sondergaard et al., 2004). The P1B-type ATPase family and the HMA family of transporters are implicated in the active efflux of symplastic $Zn^{2+}$ through pericycle cells and xylem parenchyma. Zn is delivered by both short- and long-distance channels to a variety of plant organs and developing sinks after it enters the phloem. According to reports, the phloem moves Zn more than the xylem because of the higher concentration of chelating solutes (peptides, organic acids, etc.) in the sap of the phloem. In phloem tissues, Zn is typically transmitted as an ionic form or as a compound of zinc-malate, zinc-nicotianamine, and zinc-histidine. Though the xylem appears to be a secondary method of transport for Zn, it is essential for nutrient transfer to various organs. In new storage tissues such as developing grains and tubers, Zn transport occurs mainly through the phloem.

In plants, the mobility of zinc varies from species to species (Peaslee et al., 1981). The families of transporter protein that participate in Zn transport across membranes are ZIP (Zinc-, Iron- Permease family/ZRT-, IRT- like proteins) that assist Zn inflow in the cytosol and, probably, facilitate the uptake of Zn available in the soil (Olsen and Palmgren, 2014; White and Broadley, 2011), vacuolar scavenging (Morel et al., 2009), control of Zn efflux from plastids (Kim et al., 2009), HMA (Heavy Metal ATPases) /P-type ATPase family mediate xylem loading (Hanikenne et al., 2008; Hussain et al., 2004), Cation diffusion facilitators (CDFs)/ metal tolerance proteins (MTPs) involved in Zn transport to the vacuoles (Kobae et al., 2004). Vacuolar Iron Transporter (VIT), Natural resistance-associated macrophage protein (NRAMP), Plant Cadmium Resistance (PCR), Yellow Stripe like (YSL) etc., are also in charge of Zn transport.

Zn is taken in the form of a Zn-phytosiderophore complex or as $Zn^{2+}$ ions by plants (Broadley et al., 2007). Although the ZIP proteins (ZIP1, ZIP3, and ZIP4) are utilized in Zn uptake and transportation by plant roots and sometimes by $Ca^{2+}$ channels, which are present in the plasma membrane, also mediate the transportation of $Zn^{2+}$ ions (Palmgren et al., 2008). In *Arabidopsis* (Milner et al., 2013), barley (Tiong et al., 2014), and rice (Ishimaru et al., 2007), many genes of this family are expressed in the epidermal and cortical cells along with stelar cells. This indicates that Zn absorption and movement from the roots to the shoots are important processes that these transporters contribute significantly (Bashir et al., 2012; Tiong et al., 2015). The amino and carboxyl-terminal motifs of the ZIP proteins are found on the plasma membrane's outer surface. Structurally, the proteins are made up of eight possible transmembrane domains. The ZIP proteins range in length from 309 to 476 amino acids. This portion of the ZIP protein has a "variable region," which is a polypeptide sequence that varies in length between transmembrane domains III and IV. A metal-binding domain abundant in histidine residues can be found in the variable region.

ZIP proteins are present at almost all phylogenetic levels, including bacteria, fungi, plants (monocots and dicots), and mammals (Eide, 2006; Pence et al., 2000). Under the conditions of Zn deficiency as well as Zn fertilization, the expression level (high or low) of various Zn transporters such as ZIP1-ZIP6, ZIP9, ZIP10, ZIP12, and IRT3 has been discovered in *A. thaliana, T. caerulescens, A. halleri*, etc (Van De Mortel et al., 2006; Guerinot, 2000; Talke et al., 2006). In plants, the $Zn^{++}$ influx to the leaves is transported by the ZIP family (Ishimaru et al., 2005). The mineral elements were transported to the xylem through roots (the source) and transferred to the above-ground parts of the plant. ZIP1, ZIP2, ZIP9, and IRT3 are the primary Zn influx transporters in root cells (Grotz et al., 1998), whereas ZIP4 is implicated in Zn transport across chloroplast membranes (Guerinot, 2000). The binding affinities and transport specificities of these zinc-transporting proteins vary. ZIP proteins play important role in the transportation of Zn into the cytoplasm of root cells (Olsen and Palmgren, 2014; White and Broadley, 2011). The complex process by which ZIP proteins take up zinc in cells has been deciphered using the expression of Arabidopsis thaliana ZIP1, ZIP2, ZIP3, and ZIP4 cDNAs in the zrt1zrt2 double mutant yeast strain, which lacks Zn uptake transporters (ZRT1 and ZRT2). All the proteins were able to mediate Zn uptake into the cells across the plasma membrane (Grotz et al., 1998).

IRT proteins also belong to the ZIP family and are present in the plasma membrane. IRT1 and IRT3 have been studied as putative Zn transporters in *Arabidopsis*. Besides the Zn transporter, it acts as a mediator for Fe transporters and their transcriptional activation under Fe-deficiency conditions. It has been reported that at the time of Zn deficiency, the expression of *IRT3* is elevated in the roots of Zn hyper-accumulating in plant species like *A. halleri* and *Noccaea caerulescens* (Talke et al., 2006). The AtIRT3 gene over expression in *A. thaliana* lead to accretion of Zn and Fe in the shoots and roots, respectively (Lin et al., 2009).

The members of the HMA family are known as P1B-type ATPases and are related to the transportation of a variety of mono- or divalent cations such as $Na^+$, $K^+$, $H^+$, $Ca^{2+}$, $Zn^{2+}$, $Cd^{2+}$ and $Mg^{2+}$across membranes. The presence of highly conserved amino acid sequences in three transmembrane regions is a characteristic feature of the members of the HMA family, which enter into active metal export (Arguello et al., 2007). In *A. thaliana*, eight HMAs have been documented, out of which only four (HMA1, HMA2, HMA3, and HMA4) are involved in $Zn^{2+}$ and $Cd^{2+}$ transport (Cobbett et al., 2003). In yeast and *Arabidopsis, the* HMA3 transporter protein was confined in the vacuolar membrane and responsible for heavy metal sequestration into the vacuole under excessive concentrations of these ions in the cytoplasm (Gravot et al., 2004; Morel et al., 2009). The difference between HMA1 and the other HMAs (HMA2, HMA3, and HMA4) lies in the amino-acid sequences that are involved in $Zn^{2+}$ transport. The amino acid sequence defines substrate specificity, exhibiting broader specificity. In excess, $Zn^{2+}$ is transported from the chloroplasts to the cytoplasm by the HMA1, preventing organelle damage (Kim et al., 2009). The transporter proteins HMA2 and HMA4, which are restricted to the plasma membrane of the cells that make up the root pericycle, help $Zn^{2+}$ to exit these cells and are also involved in the loading of xylem in roots (Hussain et al., 2004; Sinclair et al., 2007).

Metal tolerance proteins (MTPs) are proteins belonging to the CDF protein family. These homodimeric membrane proteins are widely distributed and aid in the cytosolic outflow of divalent cations in conjunction with the entry of $H^+$ or $Na^+$. In contrast to the transport carried out by ZIP proteins, which transport $Zn^{2+}$ and other metal ions from the extracellular space or organelle lumen into the cytoplasm, it has been linked to the process of zinc accumulation in the vacuole of plants cells and plays a significant role in the uptake of $Zn^{2+}$ and other metal ions into the lumen of intracellular organelles in or out of the cell (Eide, 2006; Ricachenevsky et al., 2013). MTP3 is regarded as a crucial protein during xylem loading, and in nontoxic Zn overload conditions, their expression is elevated in root epidermal and cortical tissues. Transgenic mice with RNAi-silencing of the MTP3 gene have higher amounts of zinc in the shoot and lower biomass, indicating that MTP3 limits the flow of zinc from the root to the shoot (Arrivault et al., 2006).

In addition, Zn may be transported from the phloem to the leaf or sink tissue by YSL proteins in the form of a complex with micro proteins. The mobility of Zn in phloem is mostly low, but sometimes it varies. However, Zn is stored as an organic acid complex inside the vacuole and may move as $Zn^{2+}$ or as complex organic acid, nicotinamide, or histidine (Leitenmaier and Küpper, 2013).

There have been reports of several PCR family members in various plant species. Twelve members of the multi-gene PCR family are found in *Arabidopsis thaliana*, while 21 members of the PCR family have been found in *Oryza sativa*. According to (Song et al., 2010), AtPCR2 exhibits Cd resistance and aids in the transfer of Zn and Fe. AtPCR2 is involved in Zn transport and xylem loading in the root elongation zone. As a result, this protein provides metal tolerance and mediates Zn extrusion in the root hair zone (older portion).

Besides the transportation of heavy metals across tonoplasts, VIT family members also mediate the transport of other micronutrients like Zn and Mn. According to Zhang et al. (2012) the two OsVIT1 and OsVIT2, members of the VITs family mediate Fe and Zn sinking in flag leaves by vacuolar storage, which in turn controls the metal translocation from flag leaf to developing grains in rice plants (Fig. 2.1).

## 2.6 Factors controlling zinc uptake

There are several factors like parental material, soil chemical properties (pH, organic matter, clay minerals, sesquioxides, and carbonates), and nutrient interactions that control the supplementation of Zn from soil to plants (Alloway, 2009; Mousavi, 2011). Besides these factors, Zn uptake is also controlled at genetic levels through various processes such as transcription, posttranscription, and translation.

In calcareous soils, the fixation of Zn occurs through two pertinent processes involving the co-precipitation on calcite and chemisorptions on Fe oxides (Rengel, 2015; Ryan et al., 2013). The availability of Zn and organic acid is usually limited, with a limited capacity to replenish the nutrients absorbed from soil solutions (Alloway, 2009; Rashid and Ryan, 2004; Ryan

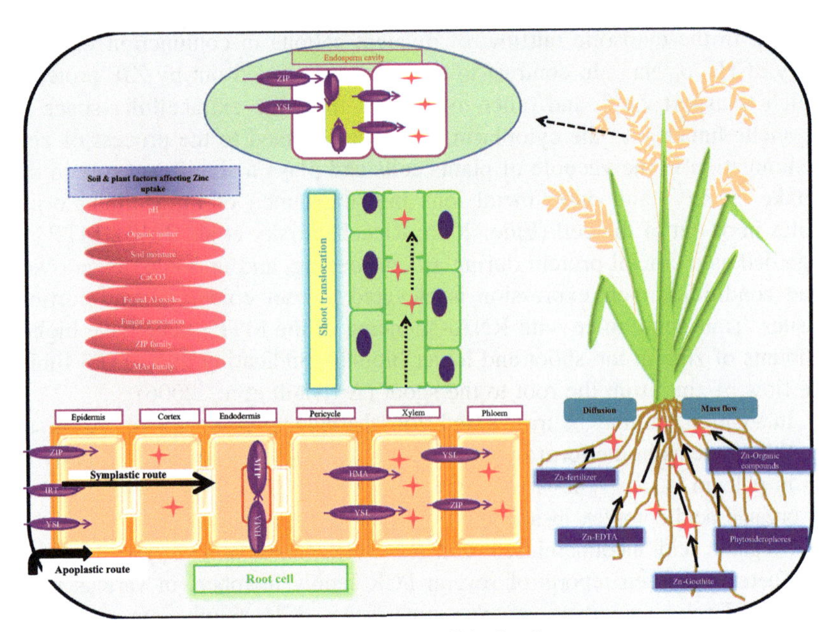


**FIGURE 2.1** Uptake and transport of zinc from soil in the plants.

et al., 2013). The uptake of Zn in plants declines with the rise of clay content and Olsen P in calcareous soil, whereas in noncalcareous soil, Zn uptake depends upon crystalline iron oxide content (Aurora Moreno-Lora and Antonio Delgado, 2020). Fe oxides adsorb Zn in an irreversible process at pH > 6 via surface complexation and non-specific adsorption, which is pH-dependent because rising pH causes an increase in negative surface charges (Buekers et al., 2007, 2008; Ryan et al., 2013; Stahl and James, 2010). Despite this, biological elements that affect Zn uptake by plants include microbial activity and exudation of organic anion in the rhizosphere (Aurora Moreno-Lora and Antonio Delgado, 2020).

In yeast, Zn absorption is regulated at the transcriptional and post-translational levels. Zinc-responsive transcriptional activator protein (ZAP1), which binds upstream to a region known as the zinc responsive element (ZRE) and controls the expression of several zinc-regulated genes, including ZRT1 and ZRT2, is the transcriptional product of the ZAP1 gene. High levels of extracellular zinc have been demonstrated to trigger post-translational regulation of ZRT1 protein by ubiquitin-mediated endocytosis, which is subsequently followed by its destruction in the vacuole.

## 2.7 Role of zinc in plants

Zn is a vital micronutrient for all living organisms, playing a significant role in growth, development, and defensive processes in living systems. Zn plays an important role in supporting the nutrient requirements of biological systems by influencing diverse enzymatic activities. They have been shown to mediate catalytic reactions, signal transduction, and protein synthesis (Palmer and Guerinot, 2009). Zn is found to activate enzymes like hydrogenase, superoxide dismutase, alcohol dehydrogenase, and carbonic anhydrase (Srivastava and Gupta, 1996). Zn as a cofactor is required in mitochondrial biogenesis and the proteolytic cleavage of proteins is required for mitochondrial functioning (Taylor et al., 2001). A reduction in Zn levels could lead to a lowered synthesis of mitochondrial enzymes and proteins, ultimately initiating impairment in mitochondrial processing. Zn can regulate various enzymes involved in glycolysis, the Krebs cycle, and mitochondrial ATP production (Srivastava and Gupta, 1996).

Apart from this, they have been found to maintain the structural integrity of biological membranes and aid in the mobilization of ions across the cellular system. Zn interacts with phospholipids and sulfhydryl groups of membrane proteins for the proper functionality of the membrane (Welch and Norvell, 1993). In fact, Zn is predominantly found to protect cells from toxic free radicals causing peroxidation. Sulphydryl groups in the cellular membranes are sensitive to the endogenous levels of Zn and showed a reduction under low Zn supply (Cakmak, 2000). In the plasma membrane of root cells, Zn can influence the ion-gated channels and membrane transporter proteins for nutrient mobilization.

The content of Zn in plants can alter the chloroplast structure and final output of photosynthetic efficiency in plants (Kasim, 2007). The decreased activity of carbonic anhydrase due to low levels of Zn could be attributed to reduced photosynthesis output (Cakmak and Engels, 1999). Zn can influence carbohydrate and starch metabolism due to impaired sucrose mobilization in plant cells (Marschner et al., 1996). Moreover, Zn can mediate proteolytic activity in the chloroplast, including repairing photosystem II due to injury caused by photolytic damage to the D1 protein (Hansch and Mendel, 2009).

A Zn-dependent change in the mineral nutrient content of plants is observed. Reduced levels of Zn are correlated with the gradual increase in phosphorous uptake by plant roots (Parker, 1997). On the other hand, a decrease in Zn levels was endorsed by the reduction in phosphorus content, ultimately disturbing the nitrogen metabolism of plants (Norvell and Welch, 1993). Additionally, Zn can influence the sodium uptake in plants, and it could also regulate salinity-induced damage in the membrane system (Norvell and Welch, 1993). Furthermore, Norvell and Welch (1993) also reported that Zn sufficiency is closely related to chlorine ions; a reduction in Zn content could greatly affect chloride ion channels in root cells.

Protein synthesis was greatly influenced by the cellular Zn content due to structural modifications in the ribosomes and RNA content of plants. With regard to this, Brown et al. (1993) observed that the rice meristem showed a declining level of RNA and the number of free ribosomes under low Zn content. Transcription factors associated with Zn play a regulatory role in the formation of reproductive parts and fruit development in plants (Kobayashi et al., 1998). Reduced Zn content can severely inhibit the physiology and developmental processes of the reproductive structures of flowers. Insufficient supply of Zn to maize plants inhibited the development of male reproductive structures, thereby reducing pollen production and ultimately influencing the fertility of male flowers (Sharma et al., 1990). Zn is essentially required for the synthesis of growth hormone like auxin and declination in Zn content could lead to disturbed auxin synthesis as a consequence of inhibited tryptophan (precursor of auxin) content in cells (Brennan, 2005). All these factors suggest that Zn plays a critical role in controlling diverse cellular activities for the proper growth and metabolism of plants.

## 2.8 Conclusion

The uptake of Zn is a complex process and is conducted by a group of transporter proteins and chelating ligands (phytosiderophores, organic acids) whose diverse expression are accountable for dissimilar interspecific and intraspecific variances among plants. Zn is found in the rhizosphere in two forms, i.e., $Zn^{2+}$ and in organometallic complexes. The binding of Zn and its bioavailability in soil depends upon various physicochemical soil components such as granulometric composition, organic matter content, pH, oxidation-reducing potential, iron and manganese hydroxides, absorptive capacity and moisture (Kabata-Pendias, 2004; Kim et al., 2015). Plants uptake most of the rhizospheric mineral nutrients via their roots in their ionic forms, where Zn is taken up mainly as the divalent cation ($Zn^{2+}$ ion) through mass flow and diffusion mechanisms. Conversely, the machinery of Zn absorption is strongly controlled inside the plant body at each hierarchy level. The major absorption of Zn at the organ level is through shoots (leaves, stem, and flower/fruit) rather than roots. However, in leaves, Zn deposits are observed in mesophyll tissues. The transport and movement of Zn with the tissues are restricted to vascular tissues and cortical parenchyma, irrespective of organs. In the cell, a high concentration of Zn is sequestrated in the vacuole in comparison to the cytosol. Abiotic factors like heat and drought influence the levels of mineral nutrients and heavy metals (P, N, Cd, Fe) in the soil rhizosphere and the microbial population in specific soil types, and they also control the Zn absorption by plants due to altered soil Eh–pH equilibrium. The main reason behind Zn deficiency in plants is low availability due to its immobile nature and, to some extent, adverse soil conditions.

Using new methods from biotechnology and conventional breeding, recent studies have aimed to reveal the genetic differences present in agricultural plants and to use them to better crop nutrition. To address the zinc deficiency that affects humans and plants, more research is necessary. An encouraging avenue to further understand the mechanism of zinc absorption by plants and transport in plant systems is the expression of transporters protein and binding ligands in conjunction with secondary genes and messengers (such as phosphatases, transcription factors, kinase cascades, etc.) in modulating Zn uptake, accumulation, and translocation. Understanding and preserving Zn equilibrium between soil and plants will be made easier with the aid of future research.

## References

Alloway, B.J., 2009. Soil factors associated with zinc deficiency in crops and humans. Environmental Geochemistry and Health 31 (5), 537–548.

Arguello, J.M., Eren, E., Gonzalez-Guerrero, M., 2007. The structure and function of heavy metal transport P1B-ATPases. Biometals: An International Journal on the Role of Metal Ions in Biology, Biochemistry, and Medicine 20, 233–24.

Arrivault, S., Senger, T., Kramer, U., 2006. The *Arabidopsis* metal tolerance protein AtMTP3 maintains metal homeostasis by mediating Zn exclusion from the shoot under Fe deficiency and Zn oversupply. The Plant Journal: For Cell and Molecular Biology 46, 861–879.

Auld, D.S., 2001. Zinc coordination sphere in biochemical zinc sites. Biometals: An International Journal on the Role of Metal Ions in Biology, Biochemistry, and Medicine 14, 271–313.

Aurora Moreno-Lora, Antonio Delgado, 2020. Factors determining Zn availability and uptake by plants in soils developed under Mediterranean climate. Geoderma 376, 114509.

Bailey, S., Thompson, E., Nixon, P.J., Horton, P., Mullineaux, C.W., Robinson, C., et al., 2002. A critical role for the Var2 FtsH homologue of *Arabidopsis thaliana* in the photosystem II repair cycle in vivo. The Journal of Biological Chemistry 277, 2006–2011.

Barak, P., Helmke, P.A., 1993. The chemistry of zinc. In: Robson, A.D. (Ed.), Zinc in Soil and Plants. Kluwer Academic Publishers, Dordrecht, the Netherlands, pp. 1–13.

Baran, A., Wieczorek, J., Mazurek, R., Urbański, K., Klimkowicz-Pawlas, A., 2018. Potential ecological risk assessment and predicting zinc accumulation in soils. Environmental Geochemistry and Health 40 (1), 435–450.

Bashir, K., Ishimaru, Y., Nishizawa, N.K., 2012. Molecular mechanisms of zinc uptake and translocation in rice. Plant and Soil 361, 189–201.

Bell, P.F., Chaney, R.L., Angle, J.S., 1991. Free metal activity and total metal concentrations as indices of micronutrient availability to barley [*Hordeum-vulgare* (L.) 'Klages']. Plant and Soil 130, 51–62.

Bouis, H.E., Hotz, C., Mc Clafferty, B., Meenakshi, J.V., Pfeiffer, W.H., 2011. Biofortification: a new tool to reduce micronutrient malnutrition, Food and Nutrition Bulletin, 32. pp. S31–S40.

Brennan, R.F., 2005. Zinc Application and Its Availability to Plants. Ph.D. dissertation, School of Environmental Science, Division of Science and Engineering, Murdoch University.

Brinch-Pedersen, H., Borg, S., Tauris, B., Holm, P.B., 2007. Molecular genetic approaches to increasing mineral availability and vitamin content of cereals. Journal of Cereal Science 46 (3), 308–326.

Broadley, M.R., White, P.J., Hammond, J.P., Zelko, I., Lux, A., 2007. Zn in plants. The New Phytologist 173, 677–702.

Brown, P.H., Cakmak, I., Zhang, Q., 1993. Form and function of zinc in plants. In: Robson, A. D. (Ed.), Zinc in Soils and Plants. Kluwer Academic Publishers, Dordrecht, pp. 90–106.

Buekers, J., Van Laer, L., Amery, F., Van Buggenhout, S., Maes, A., Smolders, E., 2007. Role of soil constituents in fixation of soluble Zn, Cu, Ni and Cd added to soils. European Journal of Soil Science 58, 1514–1524.

Buekers, J., Degryse, F., Maes, A., Smolders, E., 2008. Modelling the effects of ageing on Cd, Zn, Ni and Cu solubility in soils using an assemblage model. European Journal of Soil Science 59, 1160–1170.

Cakmak, I., 2000. Possible roles of zinc in protecting plant cells from damage by reactive oxygen species. The New Phytologist 146, 185–205.

Cakmak, I., Engels, C., 1999. Role of mineral nutrients in photosynthesis and yield formation. In: Rengel, Z. (Ed.), Mineral Nutrition of Crops. Haworth Press, New York, USA, pp. 141–168.

Claus, J., Bohmann, A., Chavarria Krauser, A., 2013. Zinc uptake and radial transport in roots of *Arabidopsis thaliana:* a modelling approach to understand accumulation. Annals of Botany 112, 369–380.

Clemens, S., Palmgren, M.G., Kramer, U., 2002. A long way ahead: understanding and engineering plant metal accumulation. Trends in Plant Science 7, 309–315.

Cobbett, C.S., Hussain, D., Haydon, M.J., 2003. Structural and functional relationships between type 1B heavy metal-transporting P-type ATPases in *Arabidopsis*. The New Phytologist 159, 315–321.

Coleman, J.E., 1998. Zinc enzymes. Current Opinion in Chemical Biology 2, 222–234.

Degryse, F., Smolders, E., Parker, D.R., 2006. Metal complexes increase uptake of Zn and Cu by plants: implications for uptake and deficiency studies in chelator-buffered solutions. Plant and Soil 289 (1), 171–185.

Degryse, F., Verma, V.K., Smolders, E., 2008. Mobilization of Cu and Zn by root exudates of dicotyledonous plants in resin buffered solutions and in soil. Plant and Soil 306, 69–84.

Disante, K.B., Fuentes, D., Cortina, J., 2010. Response to drought of Zn-stressed *Quercus suber* L. seedlings. Environmental and Experimental Botany 70, 96–103.

Duffner, A., Hoffland, E., Temminghoff, E.J., 2012. Bioavailability of zinc and phosphorus in calcareous soils as affected by citrate exudation. Plant and Soil 361 (1), 165–175.

Eide, D.J., 2006. Zinc transporters and the cellular trafficking of zinc. Biochimica et Biophysica Acta 1763, 711–722.

Gao, X., Zhang, F., Hoffland, E., 2009. Malate exudation by six aerobic rice genotypes varying in zinc uptake efficiency. Journal of Environmental Quality 38, 2315–2321.

Gravot, A., Lieutaud, A., Verret, F., Auroy, P., Vavasseur, A., Richaud, P., 2004. AtHMA3, a plant P1B-ATPase, functions as a Cd/Pb transporter in yeast. FEBS Letters 561, 22–28.

Grotz, N., Fox, T., Connolly, E., Park, W., Guerinot, M.L., Eide, D., 1998. Identification of a family of zinc transporter genes from *Arabidopsis* that respond to zinc deficiency. Proceedings of the National Academy of Sciences of the United States of America 95, 7220–7224.

Guerinot, M.L., 2000. The ZIP family of metal transporters. Biochimica et Biophysica Acta (BBA)-Biomembranes 1465 (1–2), 190–198.

Gupta, N., Ram, H., Kumar, B., 2016. Mechanism of zinc absorption in plants: uptake, transport, translocation and accumulation. Reviews in Environmental Science and Bio/Technology 15 (1), 89–109.

Hambidge, M., 2000. Human zinc deficiency. The Journal of Nutrition 130, 1344–1349.

Hanikenne, M., Talke, I.N., Haydon, M.J., Lanz, C., Nolte, A., Motte, P., et al., 2008. Evolution of metal hyperaccumulation required cis-regulatory changes and triplication of HMA4. Nature 453, 391–395.

Hansch, R., Mendel, R.R., 2009. Physiological functions of mineral micronutrients (Cu, Zn, Mn, Fe, Ni, Mo, B, Cl). Current Opinion in Plant Biology 12, 259–266.

Hussain, D., Haydon, M.J., Wang, Y., Wong, E., Sherson, S.M., Young, J., et al., 2004. P-type ATPase heavy metal transporters with roles in essential zinc homeostasis in *Arabidopsis*. The Plant Cell 16, 1327–1339.

Ishimaru, Y., Suzuki, M., Kobayashi, T., Takahashi, M., Nakanishi, H., Mori, S., et al., 2005. OsZIP4, a novel Zn-regulated Zn transporter in rice. Journal of Experimental Botany 56, 3207–3214.

Ishimaru, Y., Masuda, H., Suzuki, M., Bashir, K., Takahashi, M., Nakanishi, H., et al., 2007. Overexpression of the OsZIP4 zinc transporter confers disarrangement of zinc distribution in rice plants. Journal of Experimental Botany 58, 2909–2915.

Kabata-Pendias, A., 1993. Behavioural properties of trace metals in soils. Applied Geochemistry: Journal of the International Association of Geochemistry and Cosmochemistry 2, 3–9.

Kabata-Pendias, A., 2004. Soil–plant transfer of trace elements—an environmental issue. Geoderma 122, 143–149.

Kasim, W.A., 2007. Physiological consequences of structural and ultra-structural changes induced by Zn stress in *Phaseolus vulgaris*. I. Growth and photosynthetic apparatus. International Journal of Botany 3 (1), 15–22.

Khan, M.U., Muhammad, S., Malik, R.N., 2013. Potential risk assessment of metal consumption in food crops irrigated with wastewater. CLEAN – Soil, Air, Water 42 (10), 1415–1422.

Kim, Y.Y., Choi, H., Segami, S., Cho, H.T., Martinoia, E., Maeshima, M., et al., 2009. AtHMA1 contributes to the detoxification of excess Zn(II) in *Arabidopsis*. The Plant Journal: for Cell and Molecular Biology 58, 737–753.

Kim, R.Y., Yoon, J.K., Kim, T.S., Yang, J.E., Owens, G., Kim, K.R., 2015. Bioavailability of heavy metals in soils:Definitions and practical implementation: a critical review. Environmental Geochemistry and Health 37 (6), 1041–1061.

Kobae, Y., Uemura, T., Sato, M.H., Ohnishi, M., Mimura, T., Nakagawa, T., et al., 2004. Zinc transporter of *Arabidopsis thaliana* AtMTP1 is localized to vacuolar membranes and implicated in zinc homeostasis. Plant & Cell Physiology 45, 1749–1758.

Kobayashi, A., Sakamoto, A., Kubo, K., Rybka, Z., Kanno, Y., Takatsuji, H., 1998. Seven zinc-finger transcription factors are expressed sequentially during the development of anthers in petunia. The Plant Journal: For Cell and Molecular Biology 13, 571–576.

Leitenmaier, B., Küpper, H., 2013. Compartmentation and complexation of metals in hyperaccumulator plants. Frontiers in Plant Science 4, 374–380.

Lin, C.W., Chang, H.B., Huang, H.J., 2005. Zinc induces mitogen activated protein kinase activation mediated by reactive oxygen species in rice roots. Plant Physiology and Biochemistry: PPB/Societe Francaise de Physiologie Vegetale 43, 963–968.

Lin, Y.F., Liang, H, M., Yang, S.Y., Boch, A., Clemens, S., et al., 2009. *Arabidopsis* IRT3 is a zinc-regulated and plasma membrane localized zinc/iron transporter. The New Phytologist 182, 392–404.

Lindsay, W.L., 1979. Chemical Equilibria in Soils. John Wiley & Sons, Inc, New York, NY, USA.

Lopez Millan, A.F., Ellis, D.R., Grusak, M.A., 2005. Effect of zinc and manganese supply on the activities of superoxide dismutase and carbonic anhydrase in *Medicago truncatula* wild type and raz mutant plants. Plant Science (Shannon, Ireland) 168, 1015–1022.

Marschner, H., 1993. Zinc uptake from soils. In: Robson, A.D. (Ed.), Zinc in Soils and Plants. Kluwer, Dordrecht, pp. 59–77.

Marschner, H., Kirkby, E.A., Cakmak, I., 1996. Effect of mineral nutritional status on shoot-root partitioning of photo assimilates and cycling of mineral nutrients. Journal of Experimental Botany 47, 1255–1263.

Martens, D.C., Westermann, D.T., 1991. Fertilizer applications for correcting micronutrient deficiencies. In: Mortvedt, J.J. (Ed.), Micronutrients in Agriculture. Soil Sci Soc Am, Madison (WI), pp. 549–592.

McLaughlin, M.J., Smolders, E., Merckx, R., Maes, A., 1997. Plant uptake of Cd and Zn in chelator-buffered nutrient solution depends on ligand type. In: Ando, T., et al., (Eds.), Plant Nutrition—For Sustainable Food Production and Environment. Kluwer Academic Publishers, Dordrecht, The Netherlands, pp. 113–118.

Milner, M.J., Seamon, J., Craft, E., Kochian, L.V., 2013. Transport properties of members of the ZIP family in plants and their role in Zn and Mn homeostasis. Journal of Experimental Botany 64, 369–381.

Morel, M., Crouzet, J., Gravot, A., Auroy, P., Leonhardt, N., Vavasseur, A., et al., 2009. AtHMA3, a P1BATPase allowing Cd/Zn/Co/Pb vacuolar storage in *Arabidopsis*. Plant Physiology 149, 894–904.

Mortvedt, J.J., Gilkes, R.J., 1993. Zinc fertilizers. In: Robson, A.D. (Ed.), Zinc in Soils and Plants. Kluwer, Dordrecht, pp. 33–44.

Mousavi, S.R., 2011. Zinc in crop production and interaction with phosphorus. Australian Journal of Basic and Applied Sciences 5, 1503–1509.

Norvell, A.W., Welch, R.M., 1993. Growth and nutrient uptake by barley (*Hordeum vulgare* L. cv. Herta): studies using an *N*-(2- hydroxyethyl) ethylenedinitrilotriacetic acid-buffered nutrient solution technique. I. Zinc ion requirements. Plant Physiology 101, 619–625.

Olaniran, A.O., Balgobind, A., Pillay, B., 2013. Bioavailability of heavy metals in soil: Impact on microbial biodegradation of organic compounds and possible improvement strategies. International Journal of Molecular Sciences 14, 10197–10228.

Olsen, L.I., Palmgren, M.G., 2014. Many rivers to cross: the journey of zinc from soil to seed. Frontiers in Plant Science . 5, 30.

Palmer, C.M., Guerinot, M.L., 2009. Facing the challenges of Cu, Fe and Zn homeostasis in plants. Nature Chemical Biology 5, 333–340.

Palmgren, M.G., Clemens, S., Williams, L.E., Krämer, U., Borg, S., Schjørring, J.K., et al., 2008. Zinc biofortification of cereals: problems and solutions. Trends in Plant Science 13, 464–473.

Parker, D.R., 1997. Responses of six crop species to solution $Zn^{2+}$ activities buffered with HEDTA. Soil Science Society of America Journal 61, 167–176.

Peaslee, D.E., Isarangkura, R., Legget, J.E., 1981. Accumulation and translocation of zinc by two corn cultivars. Agronomy Journal 73, 729–732.

Peck, A.W., McDonald, G.K., 2010. Adequate zinc nutrition alleviates the adverse effects of heat stress in bread wheat. Plant and soil 337, 355–374.

Pence, N.S., Larsen, P.B., Ebbs, S.D., Letham, D.L.D., Lasat, M.M., Garvin, D.F., et al., 2000. The molecular physiology of heavy metal transport in the Zn/Cd hyperaccumulator *Thlaspi caerulescens*. Proceedings of the National Academy of Sciences of the United States of America 97, 4956–4960.

Rashid, A., Ryan, J., 2004. Micronutrient constraints to crop production in soils with mediterranean-type characteristics. Journal of Plant Nutrition 27, 959–975.

Rengel, Z., 2015. Availability of Mn, Zn and Fe in the rhizosphere. Journal of Soil Science and Plant Nutrition 15, 3970–4409.

Ricachenevsky, F.K., Menguer, P.K., Sperotto, R.A., Williams, L.E., Fett, J.P., 2013. Roles of plant metal tolerance proteins (MTP) in metal storage and potential use in biofortification strategies. Frontiers in Plant Science 4, 144.

Rose, M.T., Rose, T.J., Pariasca-Tanaka, J., Widodo, B., Wissuwa, M., 2011. Revisiting the role of organic acids in the bicarbonate tolerance of zinc-efficient rice genotypes. Functional Plant Biology: FPB 38, 493–504.

Rutkowska, B., Szulc, W., Bomze, K., Gozdowski, D., Spychaj- Fabisiak, E., 2015. Soil factors affecting solubility and mobility of zinc in contaminated soils. International Journal of Environmental Science and Technology 12, 1687–1694.

Ryan, J., Rashid, A., Torrent, J., Yau, S.K., Ibrikci, H., Sommer, R., et al., 2013. Micronutrient constraints to crop production in the middle east-west Asia region: Significance, research, and management. Advances in Agronomy 122, 1–84.

Sharma, P.N., Chatterjee, C., Agarwala, S.C., Sharma, C.P., 1990. Zinc deficiency and pollen fertility in maize (*Zea mays*). Plant and Soil 124, 221–225.

Sinclair, S.A., Sherson, S.M., Jarvis, R., Camakaris, J., Cobbett, C.S., 2007. The use of the zinc-fluorophore, Zinpyr-1, in the study of zinc homeostasis in *Arabidopsis* roots. The New Phytologist 174, 39–4.

Sondergaard, T.E., Schulz, A., Palmgren, M.G., 2004. Energization of transport processes in plants. Roles of the plasma membrane $H^+$-ATPase. Plant Physiology 136, 2475–2482.

Song, W.Y., Choi, K.S., Kim, D.Y., Geisler, M., Park, J., Vincenzetti, V., et al., 2010. *Arabidopsis* PCR2 is a zinc exporter involved in both zinc extrusion and long-distance zinc transport. The Plant Cell 22, 2237–2252.

Srivastava, P.C., Gupta, U.C., 1996. Trace Elements in Crop Production. Science Publishers, Lebanon, NH, p. 356.

Stahl, R.S., James, B.R., 2010. Zinc sorption by B horizon soils as a function of pH. Soil Science Society of America Journal. Soil Science Society of America 55, 1592.

Steudle, E., 1994. Water transport across roots. Plant and Soil 167, 79–90.

Straczek, A., Hinsinger, P., 2004. Zinc mobilisation from a contaminated soil by three genotypes of tobacco as affected by soil and rhizosphere pH. Plant and Soil 260 (1), 19–32.

Talke, I.N., Hanikenne, M., Kräme, r U., 2006. Zn-dependent global transcriptional control, transcriptional de-regulation and higher gene copy number for genes in metal homeostasis of the hyper accumulator *Arabidopsis halleri*. Plant Physiology 142, 148–167.

Taylor, A.B., Smith, B.S., Kitada, S., et al., 2001. Crystal structures of mitochondrial processing peptidase reveal the mode for specific cleavage of import signal sequences. Str 9 (7), 615–625.

Tiong, J., McDonald, G.K., Genc, Y., Pedas, P., Hayes, J.E., Toubia, J., et al., 2014. HvZIP7 mediates zinc accumulation in barley (*Hordeum vulgare*) at moderately high zinc supply. The New Phytologist 201, 131–143.

Tiong, J., McDonald, G., Genc, Y., Shirley, N., Langridge, P., Huang, C.Y., 2015. Increased expression of six ZIP family genes by zinc (Zn) deficiency is associated with enhanced uptake and root-to-shoot translocation of Zn in barley (*Hordeum vulgare*). New Phytologist 207 (4), 1097–1109.

Van De Mortel, J.E., Almar Villanueva, L., Schat, H., Kwekkeboom, J., Coughlan, S., Moerland, P.D., Ver Loren van Themaat, E., Koornneef, M., Aarts, M.G., 2006. Large expression differences in genes for iron and zinc homeostasis, stress response, and lignin biosynthesis distinguish roots of Arabidopsis thaliana and the related metal hyperaccumulator *Thlaspi caerulescens*. Plant physiology 142 (3), 1127–1147.

Weiss, D.J., Mason, T.F.D., Zhao, F.J., Kirk, G.J.D., Coles, B.J., Horstwood, M.S.A., 2005. Isotopic discrimination of zinc in higher plants. The New Phytologist 165, 703–710.

Welch, R.M., Norvell, W.A., 1993. Growth and nutrient uptake by barley (*Hordeum vulgare* L. cv Herta): studies using an N-(2-Hydroxyethyl) ethylenedinitrilotriacetic acid-buffered nutrient solution technique (II. Role of zinc in the uptake and root leakage of mineral nutrients). Plant Physiology 101 (2), 627–631.

White, P.J., Broadley, M.R., 2011. Physiological limits to zinc biofortification of edible crops. Frontiers in Plant Science 2, 80.

Widodo, B., Broadley, M.R., Rose, T., Frei, M., Pariasca-Tanaka, J., Yoshihashi, T., et al., 2010. Response to zinc deficiency of two rice lines with contrasting tolerance is determined by root growth maintenance and organic acid exudation rates, and not by zinc-transporter activity. The New Phytologist 186, 400–414.

Yamasaki, S., Sakata-Sogawa, K., Hasegawa, A., Suzuki, T., Kabu, K., Sato, E., et al., 2007. Zinc is a novel intracellular second messenger. The Journal of Cell Biology 177, 637–645.

Yin, H., Tan, W., Zheng, L., Cui, H., Qiu, G., Liu, F., et al., 2012. Characterization of Ni-rich hexagonal birnessite and its geochemical effects on aqueous $Pb^{2+}/Zn^{2+}$ and As (III). Geochimica et Cosmochimica Acta 93, 47–62.

Zhang, Y., Xu, Y.H., Yi, H.Y., Gong, J.M., 2012. Vacuolar membrane transporters OsVIT1 and OsVIT2 modulate iron translocation between flag leaves and seeds in rice. The Plant Journal: For Cell and Molecular Biology 72, 400–410.

Chapter 3

# Zinc deficiency and toxicity in soil and plants: causes and remediation

**Akhilesh Kumar Pandey[1] and Arti Gautam[2]**
*[1]Department of Biotechnology, Faculty of Biosciences and Biotechnology, Invertis University, Bareilly, Uttar Pradesh, India, [2]Department of Biochemistry, Institute of Science, Banaras Hindu University, Varanasi, Uttar Pradesh, India*

## 3.1 Introduction

Zinc (Zn) is an essential trace element required for the proper growth, development, and reproduction of plants (Sharma et al., 2013). Minerals such as Zn sulfide (ZnS), zincite (ZnO), sphalerite (ZnFeS), and smithsonite ($ZnCO_3$) contribute to Zn content in the soil (Kaya et al., 2020). Zn levels in uncontaminated soil range from 50 to 55 mg/kg, and when present in excess, it is reported to contaminate the soil and water (Mertens and Smolders 2013). The Food and Agricultural Organization reports that Zn content exceeding 3 mg/L in water makes it opalescent, which develops a greasy film on boiling and becomes astringent in taste (Mertens and Smolders 2013). Most of the plant typically contains 30–100 mg of Zn/kg dry matter above which is considered to induce toxicity (Mertens and Smolders 2013). Zn occurs in soil in three forms, which include water-soluble Zn, adsorbed and exchangeable Zn that are associated with clay, organic matter, and Al and Fe hydroxides, and lastly, Zn in the form of insoluble complexes and minerals (Gupta et al., 2016). Water-soluble Zn is present in about $4 \times 10^{-10}$ and $4 \times 10^{-6}$ M, which is a very low quantity in the soil, while insoluble Zn, which is present in the form of complexes and not uptaken by the plants, comprises more than 90% of the total Zn in the soil (Broadley et al., 2007). Exchangeable Zn uptake by the plants ranges from 0.1 to 2 μg Zn/g (Broadley et al., 2007). Zn distribution in the soil is mainly determined by the precipitation of Zn, complexation, and adsorption. pH is a dominant factor that governs the availability of Zn to the plant. Types of soils, moisture content, interaction with nitrogen, clay content, and presence of

**Zinc in Plants. DOI: https://doi.org/10.1016/B978-0-323-91314-0.00013-2**
 

microbiota affect Zn uptake by the root of the plant (Gao et al., 2019; Gautam and Pandey, 2022).

People in developed and developing countries mostly rely on cereals for their food consumption, which possess poor Zn content (Zaman et al., 2018). Being an important micronutrient, it is required in small amounts by plants and animals, including humans (Noulas et al., 2018). It is required by all six classes of enzymes, which are oxidoreductase, hydrolase, isomerase, transferase, ligase, and lyases, and approximately 3000 proteins contain the Zn prosthetic group in higher animals (Akhtar et al., 2019). In plants, Zn plays a crucial role in metabolic processes such as the activation of enzymes like RNA polymerases, alcohol dehydrogenase, superoxide dismutase, and carbonic anhydrase (Pandey et al., 2020; Sharma et al., 2013). Zn is an integral part of transcription factors belonging to the Zn finger family that control cell proliferation and differentiation (Gautam et al., 2020; Sharma et al., 2013). Zn also contributes to the development and functioning of chloroplasts in plants; therefore, Zn deficiency in crops reduces yield and nutrition (Sharma et al., 2013). Zn deficiency-induced abiotic stress leads to decreased plant growth, small leaves and stunted internodes, leaf chlorosis, delayed maturity, and tissue death in prolonged deficiency cases (Sharma et al., 2013). Therefore, an adequate amount of Zn is crucial for the proper growth and development of the plant. Moreover, the use of Zn-containing synthetic fertilizers is often insufficient to maintain an adequate amount of Zn in the soil (Hafeez et al., 2013). Zn toxicity prevails very less in comparison to Zn deficiency; however, in Zn mining and smelting areas, the soil gets enriched by Zn anthropogenically, and Zn toxicity occurs in plants growing in such areas (Gong et al., 2020). Zn toxicity leads to stunted growth and chloroplast degradation due to iron deficiency-induced reduced chlorophyll synthesis and interference with the uptake of other mineral nutrients such as Mg and Mn (Chen et al., 2018). To avoid toxicity conditions, plants have evolved a complex homeostatic network that coordinates the uptake, transportation, and storage of Zn in the plants (Gautam et al., 2020; Sinclair and Krämer, 2012). This homeostatic network comprises different Zn transporters that exhibit substrate specificity and increase the uptake and distribution of Zn during deficiency conditions in plants (Sinclair and Krämer, 2012). In other words, it coordinates the external supply and internal requirements of Zn in normal as well as stressful conditions.

As Zn is an essential nutrient mineral for human nutrition, it is important to understand the distribution of Zn in the soil and the factors deciding its bioavailability and efficient uptake by plants. Further, the Zn deficiency and phytotoxicity-induced effects on plant growth and development are also discussed, along with the bioremediation techniques applicable to them. This chapter also summarizes the advancement and current knowledge regarding the bioremediation and tolerance mechanisms of Zn deficiency and phytotoxicity and its future research directions.

## 3.2 Basic soil chemistry and bioavailability of zinc

About 50% of the soil in the world is naturally Zn-deficient, and countries such as Bangladesh, Pakistan, Sudan, the Philippines, Brazil, and Sub-Saharan Africa have the majority of Zn-deficient soils (Hacisalihoglu 2020). India's 50% cultivable lands are Zn deficient, while Turkey, Western Australia, and China have approximately 50%, 45%, and 33% Zn deficient lands, respectively (Hacisalihoglu 2020). Additionally, about 17% of people worldwide are facing the risk of Zn deficiency (Wessells and Brown 2012). Zn deficiency often does not occur due to the presence of lower Zn content in the soil but due to its lower bioavailability in the plant. It is said to be a primary deficiency when the total Zn content present in the soil is low, while deficiencies due to lower Zn availability with higher Zn content in the soil are called secondary deficiencies (Alloway 2009). Lower Zn content is usually found in sandy soils, strongly leached soils, highly cropped soils, and calcareous soils (Alloway 2009). Calcareous soil has high levels of calcium carbonate that affect the bioavailability of nutrients, including Zn (Fig. 3.1) (He et al., 2021). These soils occupy more than 30% of the earth's surface and are largely present in arid and semiarid regions (He et al., 2021). It largely comprises limestones or shell beds because it is generally formed by

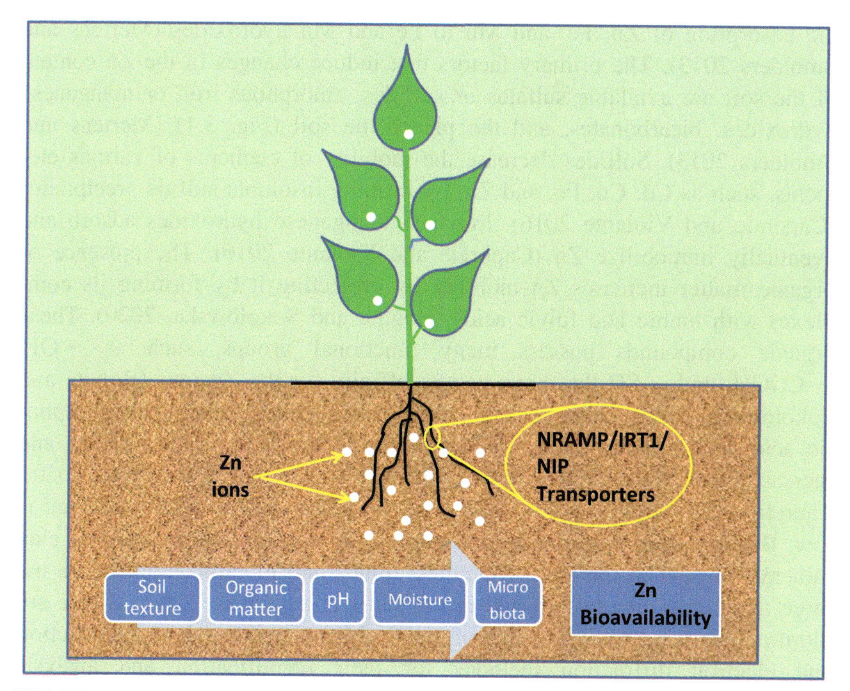


**FIGURE 3.1** Physical and chemical factors of the soils that affect the bioavailability of zinc in the soil and its absorption via various transporters inside the roots.

the weathering of calcareous rocks (He et al., 2021). The limestones are naturally present in the form of chalk and marl, which contain almost 3%–25% calcium carbonate by weight. These soils affect the soil water relations and availability of plant nutrients mainly due to their low water holding capacity and high infiltration rate (volume of water that flows per unit area of soil surface). Surface cracking of soil, the presence of less organic matter and clay content, nutrient loss due to deep percolation or leaching, the alkalinity of the soil, and the lower availability of different micronutrients, including Zn, are some other factors that characterize calcareous soils (He et al., 2021). Soil pH is an important factor that affects the Zn bioavailability of plants (Noulas et al., 2018). A decrease in soil pH leads to an increase in Zn solubility, and hence its mobility increases. Activities of Zn ions are directly proportional to the square of proton activity (Noulas et al., 2018). This can be explained by some reports that Zn induces phytotoxicity in agricultural soils when soil pH falls below 5 and contributes to the reduction of plant growth and yield. An increase in pH from 5.5 to 7 can lead to a reduction of Zn availability to 45% (Mertens and Smolders 2013). Calcareous soils are alkaline soil, as the pH of these soils is generally above 7 and may be up to 8.5. The presence of sodium carbonate in these soils further increases their pH above 9 (Mertens and Smolders 2013). The pH of soil plays a pivotal role in the formation of metal complexes with organic carbon and promotes the adsorption of Zn, Fe, and Mn to Fe and Mn hydroxides (Mertens and Smolders 2013). The primary factors that induce changes in the Zn content of the soil are available sulfates or sulfides, amorphous iron or manganese hydroxides, bicarbonates, and the pH of the soil (Fig. 3.1) (Mertens and Smolders 2013). Sulfides decrease the mobility of elements of various elements, such as Cd, Cu, Fe, and Zn, by forming insoluble sulfide precipitates (Caporale and Violante 2016). Iron and manganese hydroxides adsorb and eventually immobilize Zn (Caporale and Violante 2016). The presence of organic matter increases Zn mobility by protecting it by forming its complexes with humic and fulvic acids (Boguta and Sokołowska, 2020). These organic compounds possess many functional groups, such as $-OH$, $-COOH$, and $-SH$ that have greater affinity for the Zn ions (Boguta and Sokołowska, 2020). Other organic compounds such as amino acids, phosphoric acid, and hydroxy acids frequently form complexes with Zn ions and increase their solubility and mobility in soil (Boguta and Sokołowska, 2020). Therefore, Zn deficiency generally arises in places where organic content is low. Besides iron, manganese, and organic matter, Zn reacts with the clay minerals present in the soil. Kaolinites, illites, and montmorillonites are the three main clay minerals founds in the soil (Refaey et al., 2017). These are ultrafine particles having a size of $\leq 2\,\mu M$ and require X-ray diffraction and electron diffraction methods for their identification and analysis (Krupskaya et al., 2017). Clays are phyllosilicates and are characterized by the presence of sheets of aluminum and silica and have net negative charges that hold various positive charge minerals and nutrients such as $K^+$, $Ca^{2+}$, $Mg^{2+}$, and $Zn^{2+}$ (Krupskaya et al., 2017). Plants exchange these nutrient ions with the hydrogen ions ($H^+$) released from their roots and utilize them for their nutritional requirements (Krupskaya et al., 2017). Thus, the greater the cation exchange capability, the greater the ion adsorption, and the more it is beneficial for the plants (Balyan and Pandey, 2024; Krupskaya et al., 2017).

Besides these, Zn and nitrogen interactions have gained widespread attention in relation to Zn uptake by plants in recent years. Nitrogen is reported to increase the availability of Zn in plants. Nitrogen can be present in the form of ammonium, which undergoes biological oxidation and releases hydrogen ions in the surrounding soils (Ercoli et al., 2017). Zn also plays a role in nitrogen metabolism in plants, and its deficiency causes decreased protein metabolism. Zn interaction with nitrogen leads to its improved uptake and mobilization, which consequently enhances the growth and yield of different crops (Ercoli et al., 2017). Moisture is another factor that governs the availability of many nutrients because it controls the root architecture and hence the diffusion of nutrients, including Zn (Cakmak and Kutman 2018).

## 3.3 Zinc deficiency-induced changes in plants

Zn is among the 17 essential minerals that play a key role in the growth of plants, auxin metabolism, gene expression, and synthesis of proteins, including enzymes, the development of pollens, signal transduction, and carrying out different physiological processes (Gautam et al., 2020; Hacisalihoglu, 2020). Plants rely on soils for their Zn uptake, and their deficiency in the soil gives rise to abiotic stress in plants (Fig. 3.2) (Moreira et al., 2018). Zn deficiency impairs stem elongation and decreases the synthesis of protein and starch in tomato plants (Akther et al., 2020). The sugar content remained unaltered during Zn deficiency (Candan et al., 2018). Severe deficiency includes symptoms such as necrosis, interveinal chlorosis, the development of bronze tints, and shortening of the internodes (Akhtar et al., 2019). Besides tomatoes, Zn deficiencies are reported in a wide range of crops, including cotton, soybeans, and dry beans (Ahmad et al., 2020; Pawlowski et al., 2019). Zn deficiency resulted in stunted growth and shortening of leaves, followed by whitening in corn plants (Khatun et al., 2018). Brown spots are reported in rice, which then coalesce to turn the whole leaf brown (Bandyopadhyay et al., 2017). Zn-induced reduced activity of RNA polymerase leads to declined RNA synthesis and subsequently inhibition of protein synthesis (Bandyopadhyay et al., 2017). This inhibition of protein synthesis leads to the accumulation of amino acids and amides in leaves (Bandyopadhyay et al., 2017). Adequate Zn is also necessary for the formation of pollen tubes, which are also important sites of protein synthesis (Sharafi 2019). Zn deficiency markedly enhanced phosphorus accumulation in wheat (Chattha et al., 2017). It was reported that mild deficiency reduced

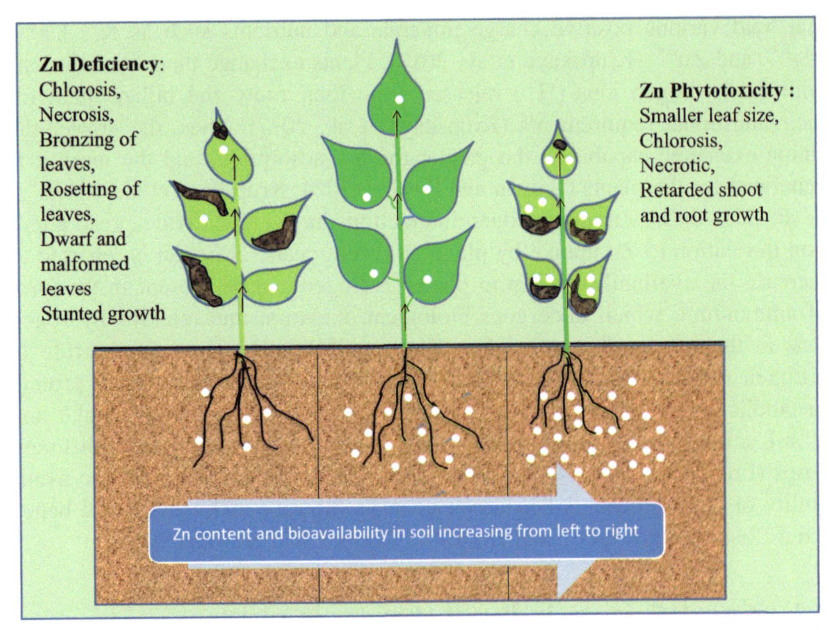


**FIGURE 3.2** The impact of different concentration of Zn in soil on plant growth. Zinc deficiency and toxicity symptoms exhibited by crops.

the dry matter of the shoot but increased the dry matter of the root, and severe Zn deficiency enhanced the phosphorus content in leaves. Necrotic symptoms were seen due to phosphorus toxicity as well as Zn deficiency (Chattha et al., 2017). Zn deficiency is most widespread in rice crops. Leaves are seen to develop brown blotches and streaks on the surface of leaves, while other leaves show late maturity or are small and die in prolonged conditions (Bandyopadhyay et al., 2017). This leads to a reduced yield of the crops. In maize plants, Zn deficiency significantly reduced pollen viability, and by supplying an adequate amount of Zn to the Zn-deficient plants, pollen fertility was recovered (Sharma et al., 1990). Alfalfa plants also showed Zn deficiency symptoms, including interveinal chlorosis, necrosis on leaf blades and margins, stunted leaves, reduction in growth, and hence yield (He et al., 2021). In chickpeas, the Zn-deficient plant showed reduced enzyme activity, disturbed ribosomal stabilization, and reduced protein synthesis (Ullah et al., 2020). Additionally, Zn deficiency induces flower abortion and infertility (Ullah et al., 2020). Groundnut shows irregular mottling and yellow coloration of veins of leaves in mild deficiency, which turns into the yellow coloration of entire leaves during severe conditions (Radhika and Meena 2021). Barley crops showed inhibition of shoot, necrosis, and reddish brown lesions on the edges and tips of leaves (Batova et al., 2019) (Table 3.1).

**TABLE 3.1** Details of concentration-wise deficiency and toxicity symptoms of the zinc in crops.

| Sr. No. | Plant species | Critical concentration | Deficiency symptoms | Toxicity symptoms | References |
|---|---|---|---|---|---|
| 1. | Maize | 1.2–1.7 mg/kg | Chlorotic streaks and purple hues on sheaths and margins of leaf, lowered activity of superoxide dismutase | Severe chlorosis and stunting, inhibition of iron transport in plants | Mattiello et al. (2015); Hong and Jin (2007) |
| 2. | Alfalfa | 2 mg/kg | Necrosis of leaf blades and margins, chlorosis, and reduced growth | Increased root diameter, no significant alteration in nodule development | He et al. (2021); Petković et al. (2019) |
| 3. | Chickpea | 0.48–2.5 mg/kg | Delayed crop maturity, reduced root and shoot growth. malformed leaves, shortening of internodes, and petioles,reduced protein synthesis, flower abortion, ovule infertility, and low seed production | Chlorosis, necrosis, stunted root and shoot growth, reduced biomass | Ullah et al. (2020); Hidoto et al. (2017) |
| 4. | Groundnut | 0.75 ppm | Interveinal chlorosis, decreased internode length, and restricted development of new leaves | Chlorosis, stunning, purple coloration of petioles and stem, and necrosis | Aboyeji et al. (2019); Noman et al. (2020) |
| 5. | Sorghum | 25–30 kg/ha | Internode elongation, crowding of leaves at the top | Chlorosis, reduced biomass | Ohki (1984); Hopkins et al. (1998) |
| 6. | Wheat | 0.38–2 mg/kg | Yellow patches in the mid vein of the leaves, the low release of phytosiderophores, stunted plant, and delayed maturity | Chlorosis, stunting growth | Chattha et al. (2017); Khobra and Singh (2018) |

(*Continued*)

**TABLE 3.1** (Continued)

| Sr. No. | Plant species | Critical concentration | Deficiency symptoms | Toxicity symptoms | References |
|---|---|---|---|---|---|
| 7. | Rice | 0.75 mg/kg | Decreased shoot growth | Chlorosis, stunting of plants | Rattan and Shukla (1984); Lee et al. (2017) |
| 8. | Barley | 0.08 mg/kg | Decreased shoot growth, development of chlorotic area on leaves | Chlorosis and stunted growth of plants | Versieren et al. (2017) |
| 9. | Tomato | 5–100 kg/ hectare | Decreased shoot growth, chlorosis of leaves | Chlorosis, necrosis, and stunted growth | Kaya and Higgs (2002); Vijayarengan and Mahalakshmi (2013) |
| 10. | Soybeans | 0.95–40 mg/kg | Decreased shoot growth | Reduced biomass, chlorosis, and necrosis of leaf blades | Pawlowski et al. (2019); Pascual et al. (2016) |

## 3.4 Physiological and biochemical aspects of zinc phytotoxicity

Though Zn is an essential micronutrient in excessive quantity, it induces toxicity in plants. Different anthropogenic activities such as smelting, mining, incinerators, galvanized products, and excessive use of fertilizers and pesticides lead to excessive dispersal of Zn in soil, inducing phytotoxicity to the plants (Gong et al., 2020). Phytotoxicity is induced at about 300–1000 mg Zn/kg dry weight of plants (Chaney 1993). General symptoms include reduced development of the plant in relation to root and shoot growth, curling and rolling of leaves, chlorosis, and death of leaf tips (Fig. 3.2) (Gong et al., 2020). In an acidic condition, excess Zn inhibits iron uptake and translocation and interferes with chlorophyll biosynthesis, which results in chlorosis (Adele et al., 2018). Crops differ significantly in relation to toxicity in acidic and alkaline soils (Chaney 2010). The graminaceous species are less susceptible to Zn toxicity than dicots in acidic soils, while in alkaline soils, dicots are less susceptible to Zn toxicity than graminaceous crops (Chaney 1993). In graminaceous crops, alkaline pH induces the natural secretion of phytosiderophores, enhancing the dissolved Zn in the soil, and the diffusion of Zn to the roots, which leads to increased susceptibility to Zn in comparison to dicots (Chaney 1993). Zn toxicity reduces the root length, due to which the uptake of various nutrients, including phosphorus, is decreased (Gautam and Pandey, 2021). Reduction of phosphorus uptake leads to decreased production of NADPH (Bouain et al., 2014). The NADPH acts as an electron quencher, which, with decreased production, is not available for electron quenching. Therefore, increased Zn uptake leads to oxidative stress by producing free radicles (Marreiro et al., 2017). These free radicles provoked an excessive generation of reactive oxygen species (Pandey et al., 2019). Zn toxicity reduces the activity of photosystem II and the ribulose bisphosphate carboxylase enzyme in addition to reducing ATP synthesis, chloroplast activity, and hence photosynthesis (Bouain et al., 2014). A higher amount of Zn reduces iron and phosphorus uptake. Crops such as spinach and beet, which are usually rich in Zn due to their inherent Zn uptake capacity, are more sensitive to Zn toxicity (Pongrac et al., 2019). Effects of Zn toxicity were studied in sugar beet, in which decreased growth of root and shoot was observed along with damaged root systems, decreased chlorophyll content, and lowered efficiency of photosystem II (Papazoglou and Fernando, 2017). The level of ascorbate and glutathione was also increased due to enhanced reactive oxygen species in sugar beet grown in Zn-excess soils (Papazoglou and Fernando, 2017). Soybean plants showed lowered dry matter, chlorosis, and a change in periodic movement as the main symptoms of Zn toxicity (Ganesh and Sundaramoorthy 2018) (Table 3.1).

## 3.5 Zinc homeostasis network

Plants maintain an adequate amount of Zn by modulating the level of supply and consumption throughout the plants (Zlobin 2021). Thus, acquisition, redistribution, sequestration of Zn, and maintenance of internal Zn demand in fluctuating environmental conditions are known as homeostasis (Zlobin 2021). Homeostasis begins with the uptake of Zn from the soil via roots, from where it reaches the cell wall (Recena et al., 2021). Zn is precipitated inside the cell wall in the form of phosphates, carbonates, or hydroxides (Sinclair and Krämer, 2012). Inside the cytoplasm, the Zn is prevented precipitating and being bonded to other biomolecules through chelation by low-molecular-weight ligands (Sinclair and Krämer, 2012). Cytosol stores Zn by forming complexes with metallothionein proteins (Pandey et al., 2019; Sinclair and Krämer, 2012). Zn is also stored inside the vacuoles, which makes them a major storage place for excess Zn (Sinclair and Krämer, 2012). During times of deficiency, the Zn immobilized in vacuoles gets remobilized to fulfill the demands of the plant system (Sinclair and Krämer, 2012). Besides cell vacuoles, extra Zn is also stored in the epidermis and trichomes. The internal concentration of Zn is about 100 μM Zn in eukaryotic cells, while it is in femtomolar ranges in the case of prokaryotes. Strict control of Zn concentrations is a requirement of every living organism (Zlobin 2021). This need has led to the evolution of a complex homeostatic network of Zn transporters, metallochaperon proteins, and low-molecular-weight ligands that ensure correct acquisition of Zn in every cellular compartment (Hara et al., 2017). Some of the Zn transporters transport $Zn^{2+}$ ions, while others transport the Zn-ligand complex. Transporters involved in $Zn^{2+}$ uptake are ZRT-IRT (Zn-regulated transporters, iron-regulated transporters like proteins), NRAMP (natural resistance-associated macrophage protein family), HMA1, HMA2, HMA3, HMA4 proteins (heavy metal ATPase), MTP (metal tolerance proteins), cation exchangers (CAX), ZAT (Zn transporters of *Arabidopsis thalliana*), and MHX ($Mg^{2+}/H^{+}$ exchanger) (Fig. 3.3) (Sinclair and Krämer, 2012). ZRT-IRT-like proteins, or ZIP family transporters, are present in most organism, such as archaea, bacteria, fungi, plants, and animals, and maintain Zn homeostasis (Lehtovirta-Morley et al., 2017). It transports Zn cations into the cytoplasm. Increased expression of ZIP in Arabidopsis led to increased accumulation of Zn and Fe in shoots and roots, respectively (Lin et al., 2009). The NRAMP transporters are metal transporters and are known to transport Fe, Zn, and Mn (Ahmadi 2017). Out of six *NRAMP* genes, *NRAMP4* is known to transport Zn cations. The *NRAMP* gene present in groundnuts (*AhNRAMP1*) was expressed in tobacco and rice, and transgenics were produced that were more tolerant toward Zn deficiency (Qin et al., 2017). HMA transporters play a key role in the detoxification of Zn in the plant. *OsHMA2* transcripts were reported in the roots of rice plants and are involved in the transportation of Zn (Qin et al., 2017). The

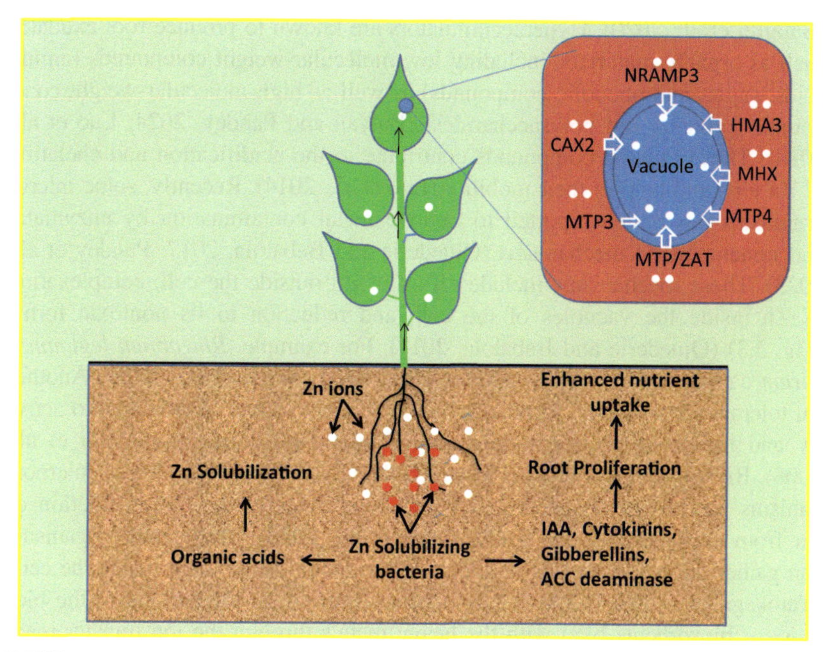


**FIGURE 3.3** Localization of various zinc transporters and role of soil microbiota in maintaining zinc homeostasis inside the plants.

concentration of Zn was found to be decreased in *OsHMA2*-suppressed rice plants in comparison with the wild varieties (Qin et al., 2017). OsHMA2 plays a key role in the loading of Zn into the xylem tissues and its translocation to the shoots (Qin et al., 2017). MTP plays a key role in the sequestration of excess Zn ions from the cytoplasm to the vacuoles and helps avoid its interaction with the metabolic and physiological processes of the plant (Vatansever et al., 2017).

## 3.6 Bioremediation and tolerance of zinc deficiency and phytotoxicity

Some plant species are reported to have developed many strategies to overcome the toxicity of Zn. Plants that can accumulate up to 6300 ppm of Zn are called hyperaccumulators (Balafrej et al., 2020). Plants such as *Arabidopsis halleri* and *Noccaeac aerulescens* can accumulate 13,620–43,710 ppm of Zn in their leaves (Balafrej et al., 2020). Plants such as *Dichapetalum* subsp. *Sumatranum* and *D.pilosum* are strong accumulators that can accumulate about 15,660–26,360 ppm of Zn (Balafrej et al., 2020). These plants do not show any signs of toxicity, like chlorosis or growth inhibition (Peng et al., 2020). The Zn tolerance capability in *Noccaeac aerulescens* was marked by stable PSII, which was impaired in Zn-toxicity-exposed sensitive plants

(Balafrej et al., 2020). Hyperaccumulators are known to produce root exudates such as organic materials, including low-molecular-weight compounds (amino acids, sugar, and phenolic compounds) as well as high-molecular-weight compounds (proteins and polysaccharides) (Balyan and Pandey, 2024; Luo et al., 2014). These organic compounds contribute to the acidification and chelation of Zn ions and help in their mobility (Luo et al., 2014). Recently, some microorganisms have been reported to remove metal contamination by enzymatic and nonenzymatic mechanisms (Ojuederie and Babalola, 2017; Pandey et al., 2023). These mechanisms include efflux of Zn outside the cell, complexation of Zn inside the vacuoles of the cell, and reduction to its nontoxic forms (Fig. 3.3) (Ojuederie and Babalola, 2017). For example, *Rhizobium leguminosarum* can tolerate Zn up to a level of 92.9 μM (Wani et al., 2008). Another Zn tolerance strain, RL9, produces indole acetic acid and showed sidero activity and increased plant growth during Zn-deficient conditions (Wani et al., 2008). Biosorption is a process in which the alive or dead biomass of microorganisms absorbs heavy metals and also contributes to the bioremediation of Zn from contaminated lands (Tarekegn et al., 2020). These microorganisms can gather the heavy metals on their surfaces or infiltrate them inside the cells (Tarekegn et al., 2020). Active compounds present on the surface of the biomass or biosorbents bind with the heavy metals through the ion transfer reaction between the active groups and metal cations (Balafrej et al., 2020). At pH 5–7, metal ions, including Zn ions, are more strongly linked on the biomass surface in comparison with the other pHs (Balafrej et al., 2020). Lowering the pH leads to the release of the metals from the biosorbants. Bioaccumulation comes into play when the rate of accumulation of metals inside the biosorbants is greater than their release into the environment (Balafrej et al., 2020). Bioaccumulators gather a high concentration of metal ions inside the cells and exhibit biotransformational capabilities that could change toxic ions to their less toxic or nontoxic forms (Pandey et al., 2023; Tarekegn et al., 2020). Some of the Zn hyperaccumulators are *Acer pseudoplatanus* (family Sapindaceae), *Arenaria patula* (family Caryophyllaceae), *Arabidopsis arenosa* (family Brassicaceae), *Arabidopsis halleri* (family Brassicaceae), *Biscutella laevigata* (family Brassicaceae), *Cochlearia pyrenaica* (family Brassicaceae), *Dianthus* (family Caryophyllaceae), *Dichapetalum gelonioides* (family Dichapetalaceae), *Galium mollugo* (family Rubiaceae), and *Gamhena canescens* (family Amaranthaceae) (Balafrej et al., 2020; Manara et al., 2020; Shi and Cai 2010)

Biotechnological applications are implied to develop biofortified plants that can combat Zn deficiency (Hefferon, 2019; Pandey et al., 2020). Rice, oil seeds, cassava, and potatoes are some of the crops whose transgenic lines are developed that possess enhanced levels of Zn ions in their endosperm (Malik and Maqbool 2020). A large number of populations, including many poor people, use rice, potatoes, and oils as a part of their daily consumption (Hefferon, 2019; Pandey et al., 2020, 2021). Rice transgenic lines showed high expression of ferritin (iron storage protein) and iron transporter OsYSL2 in the endosperm, which led to increased accumulation of Fe (4.4 folds) as well as Zn (1.6 folds) (Masuda et al., 2013). Similarly, Aung et al. (2013) developed a transgenic line of rice that overexpressed the nicotianamine synthase gene isolated from *Hordeum vulgare* (*HvNAS1*) and exhibited an accumulation of 3.4 folds higher iron and 1.3 folds higher Zn concentrations in comparison to conventional varieties.

## 3.7 Conclusion

All these recent advancements have enhanced our current knowledge of soil chemistry and different parameters affecting Zn bioavailability in plants. The plethora of research has contributed toward the comprehensive understanding of Zn-associated nutritional deficiency as well as the toxicity induced in the plants that degrade the quality and yield of the plants, including several economically important crops. Several years of evolution have triggered the ability of plants to adjust the huge alterations in Zn availability and storage inside the vacuoles, but the mechanism is little explored. Indeed, some further studies using molecular biology and transcriptomics tools are needed to elucidate the mechanism and associated signaling pathways that regulate transporters and hence Zn homeostasis in plants. Biotechnological tools, molecular breeding approaches, and agronomic practices can lead to biofortification, where researchers can increase the Zn content in edible parts of the plants, and it will also be a cost-effective way to eradicate malnutrition from the rural populations of the areas where such Zn-associated nutritional problems are prevalent. Nanotechnology is a new-age approach that has immense potential to develop nanofertilizers and sensors that contribute toward plant health management, nutrient utilization, improvement of soils, crop yield, and hence food security in the future.

## References

Aboyeji, C., Dunsin, O., Adekiya, A.O., Chinedum, C., Suleiman, K.O., Okunlola, F.O., et al., 2019. Zinc sulphate and boron-based foliar fertilizer effect on growth, yield, minerals, and heavy metal composition of groundnut (*Arachishypogaea* L) grown on an Alfisol. International Journal of Agronomy 2019.

Adele, N.C., Ngwenya, B.T., Heal, K.V., Mosselmans, J.F.W., 2018. Soil bacteria override speciation effects on zinc phytotoxicity in zinc-contaminated soils. Environmental Science & Technology 52 (6), 3412–3421.

Ahmadi, H., 2017. Functional analysis of the metal hyperaccumulation and hypertolerance candidate genes NAS4, ZIP6, CAX1 and NRAMP3 in Arabidopsis halleri (Doctoral dissertation).

Ahmad, P., Alyemeni, M.N., Al-Huqail, A.A., Alqahtani, M.A., Wijaya, L., Ashraf, M., Kaya, C., Bajguz, A., 2020. Zinc oxide nanoparticles application alleviates arsenic (As) toxicity in soybean plants by restricting the uptake of as and modulating key biochemical attributes, antioxidant enzymes, ascorbate-glutathione cycle and glyoxalase system. Plants 9 (7), 825.

Akhtar, M., Yousaf, S., Sarwar, N., Hussain, S., 2019. Zinc biofortification of cereals—role of phosphorus and other impediments in alkaline calcareous soils. Environmental Geochemistry and Health 41 (5), 2365–2379.

Akther, M.S., Das, U., Tahura, S., Prity, S.A., Islam, M., Kabir, A.H., 2020. Regulation of Zn uptake and redox status confers Zn deficiency tolerance in tomato. Scientia Horticulturae 273, 109624.

Alloway, B.J., 2009. Soil factors associated with zinc deficiency in crops and humans. Environmental Geochemistry and Health 31 (5), 537–548.

Aung, M.S., Masuda, H., Kobayashi, T., Nakanishi, H., Yamakawa, T., Nishizawa, N.K., 2013. Iron biofortification of Myanmar rice. Frontiers in Plant Science 4, 158.

Balafrej, H., Bogusz, D., Triqui, Z.E.A., Guedira, A., Bendaou, N., Smouni, A., et al., 2020. Zinc hyperaccumulation in plants: a review. Plants 9 (5), 562.

Balyan, G., Pandey, A.K., 2024. Root exudates, the warrior of plant life: revolution below the ground. South African Journal of Botany 164, 280–287.

Bandyopadhyay, T., Mehra, P., Hairat, S., Giri, J., 2017. Morpho-physiological and transcriptome profiling reveal novel zinc deficiency-responsive genes in rice. Functional & Integrative Genomics 17 (5), 565–581.

Batova, Y., Kaznina, N., Kholoptseva, E., Titov, A., 2019. Effect of zinc deficiency on the photosynthetic apparatus of barley plants at different phases of development. Photosynthesis and Hydrogen Energy Research for Sustainability 162.

Boguta, P., Sokołowska, Z., 2020. Zinc binding to fulvic acids: assessing the impact of pH, metal concentrations and chemical properties of fulvic acids on the mechanism and stability of formed soluble complexes. Molecules (Basel, Switzerland) 25 (6), 1297.

Bouain, N., Kisko, M., Rouached, A., Dauzat, M., Lacombe, B., Belgaroui, N., et al., 2014. Phosphate/zinc interaction analysis in two lettuce varieties reveals contrasting effects on biomass, photosynthesis, and dynamics of Pi transport. BioMed research international 2014.

Broadley, M.R., White, P.J., Hammond, J.P., Zelko, I., Lux, A., 2007. Zinc in plants. New Phytologist 173 (4), 677–702.

Cakmak, I., Kutman, U.Á., 2018. Agronomic biofortification of cereals with zinc: a review. European Journal of Soil Science 69 (1), 172–180.

Candan, N., Cakmak, I., Ozturk, L., 2018. Zinc-biofortified seeds improved seedling growth under zinc deficiency and drought stress in durum wheat. Journal of Plant Nutrition and Soil Science 181 (3), 388–395.

Caporale, A.G., Violante, A., 2016. Chemical processes affecting the mobility of heavy metals and metalloids in soil environments. Current Pollution Reports 2, 15–27. Available from: https://doi.org/10.1007/s40726-015-0024-y.

Chaney, R.Á., 1993. Zinc phytotoxicity. Zinc in Soils and Plants. Springer, Dordrecht, pp. 135–150.

Chaney, R.L., 2010. Cadmium and zinc. Trace elements in soils, pp. 409–440.

Chattha, M.U., Hassan, M.U., Khan, I., Chattha, M.B., Mahmood, A., Nawaz, M., et al., 2017. Biofortification of wheat cultivars to combat zinc deficiency. Frontiers in Plant Science 8, 281.

Chen, J., Dou, R., Yang, Z., You, T., Gao, X., Wang, L., 2018. Phytotoxicity and bioaccumulation of zinc oxide nanoparticles in rice (*Oryza sativa* L.). Plant Physiology and Biochemistry 130, 604–612.

Ercoli, L., Schüßler, A., Arduini, I., Pellegrino, E., 2017. Strong increase of durum wheat iron and zinc content by field-inoculation with arbuscular mycorrhizal fungi at different soil nitrogen availabilities. Plant and Soil 419 (1), 153–167.

Ganesh, S.K., Sundaramoorthy, P., 2018. Copper and zinc induced changes in soybean (*Glycine max* (L.) Merr.). InnovAgri 1 (1), 9–12.

Gao, X., Rodrigues, S.M., Spielman-Sun, E., Lopes, S., Rodrigues, S., Zhang, Y., et al., 2019. Effect of soil organic matter, soil pH, and moisture content on solubility and dissolution rate of CuO NPs in soil. Environmental Science & Technology 53 (9), 4959–4967.

Gautam, A., Pandey, A.K., 2021. Aquaporins responses under challenging environmental conditions and abiotic stress tolerance in plants. The Botanical Review 1–29.

Gautam, A., Pandey, A.K., 2022. Microbial management of crop abiotic stress: current trends and prospects. In: *Mitigation of plant abiotic stress by microorganisms*, Academic Press, pp. 53–75.

Gautam, A., Pandey, A.K., Dubey, R.S., 2020. Unravelling molecular mechanisms for enhancing arsenic tolerance in plants: a review. Plant Gene 23, 100240.

Gautam, A., Pandey, P., Pandey, A.K., 2020. Proteomics in relation to abiotic stress tolerance in plants. In: *Plant life under changing environment*, Academic Press, pp. 513–541.

Gong, B., He, E., Qiu, H., Van Gestel, C.A., Romero-Freire, A., Zhao, L., et al., 2020. Interactions of arsenic, copper, and zinc in soil-plant system: partition, uptake and phytotoxicity. Science of the Total Environment 745, 140926.

Gupta, N., Ram, H., Kumar, B., 2016. Mechanism of zinc absorption in plants: uptake, transport, translocation and accumulation. Reviews in Environmental Science and Bio/Technology 15 (1), 89–109.

Hacisalihoglu, G., 2020. Zinc (Zn): the last nutrient in the alphabet and shedding light on Zn efficiency for the future of crop production under suboptimal Zn. Plants 9 (11), 1471.

Hafeez, B., Khanif, Y.M., Saleem, M., 2013. Role of zinc in plant nutrition—a review. Journal of Experimental Agriculture International 374–391.

Hara, T., Takeda, T.A., Takagishi, T., Fukue, K., Kambe, T., Fukada, T., 2017. Physiological roles of zinc transporters: molecular and genetic importance in zinc homeostasis. The Journal of Physiological Sciences 67 (2), 283–301.

He, H., Wu, M., Su, R., Zhang, Z., Chang, C., Peng, Q., et al., 2021. Strong phosphorus (P)-zinc (Zn) interactions in a calcareous soil-alfalfa system suggest that rational P fertilization should be considered for Zn biofortification on Zn-deficient soils and phytoremediation of Zn-contaminated soils. Plant and Soil 1–15.

Hefferon, K., 2019. Biotechnological approaches for generating zinc-enriched crops to combat malnutrition. Nutrients 11 (2), 253.

Hidoto, L., Worku, W., Mohammed, H., Bunyamin, T., 2017. Effects of zinc application strategy on zinc content and productivity of chickpea grown under zinc deficient soils. Journal of soil science and plant nutrition 17 (1), 112–126.

Hong, W.A.N.G., Jin, J.Y., 2007. Effects of zinc deficiency and drought on plant growth and metabolism of reactive oxygen species in maize (*Zea mays* L). Agricultural Sciences in China 6 (8), 988–995.

Hopkins, B.G., Whitney, D.A., Lamond, R.E., Jolley, V.D., 1998. Phytosiderophore release by sorghum, wheat, and corn under zinc deficiency. Journal of Plant Nutrition 21 (12), 2623–2637.

Kaya, C., Higgs, D., 2002. Response of tomato (*Lycopersicon esculentum* L.) cultivars to foliar application of zinc when grown in sand culture at low zinc. Scientia Horticulturae 93 (1), 53–64.

Kaya, M., Hussaini, S., Kursunoglu, S., 2020. Critical review on secondary zinc resources and their recycling technologies. Hydrometallurgy 195, 105362.

Khatun, M.A., Hossain, M.M., Bari, M.A., Abdullahil, K.M., Parvez, M.S., Alam, M.F., et al., 2018. Zinc deficiency tolerance in maize is associated with the up-regulation of Zn transporter genes and antioxidant activities. Plant Biology 20 (4), 765–770.

Khobra, R., Singh, B., 2018. Phytosiderophore release in relation to multiple micronutrient metal deficiency in wheat. Journal of Plant Nutrition 41 (6), 679–688.

Krupskaya, V.V., Zakusin, S.V., Tyupina, E.A., Dorzhieva, O.V., Zhukhlistov, A.P., Belousov, P.E., et al., 2017. Experimental study of montmorillonite structure and transformation of its properties under treatment with inorganic acid solutions. Minerals 7 (4), 49.

Lee, J.S., Wissuwa, M., Zamora, O.B., Ismail, A.M., 2017. Biochemical indicators of root damage in rice (*Oryza sativa*) genotypes under zinc deficiency stress. Journal of Plant Research 130 (6), 1071–1077.

Lehtovirta-Morley, L.E., Alsarraf, M., Wilson, D., 2017. Pan-domain analysis of ZIP zinc transporters. International Journal of Molecular Sciences 18 (12), 2631.

Lin, Y.F., Liang, H.M., Yang, S.Y., Boch, A., Clemens, S., Chen, C.C., et al., 2009. Arabidopsis IRT3 is a zinc-regulated and plasma membrane localized zinc/iron transporter. New Phytologist 182 (2), 392–404.

Luo, Q., Sun, L., Hu, X., Zhou, R., 2014. The variation of root exudates from the hyperaccumulator *Sedum alfredii* under cadmium stress: metabonomics analysis. PLoS One 9 (12), e115581.

Malik, K.A., Maqbool, A., 2020. Transgenic crops for biofortification. Frontiers in Sustainable Food Systems 4, 182.

Manara, A., Fasani, E., Furini, A., DalCorso, G., 2020. Evolution of the metal hyperaccumulation and hypertolerance traits. Plant, Cell & Environment 43 (12), 2969–2986.

Marreiro, D.D.N., Cruz, K.J.C., Morais, J.B.S., Beserra, J.B., Severo, J.S., De Oliveira, A.R.S., 2017. Zinc and oxidative stress: current mechanisms. Antioxidants 6 (2), 24.

Masuda, H., Aung, M.S., Nishizawa, N.K., 2013. Iron biofortification of rice using different transgenic approaches. Rice 6, 1–12.

Mattiello, E.M., Ruiz, H.A., Neves, J.C., Ventrella, M.C., Araújo, W.L., 2015. Zinc deficiency affects physiological and anatomical characteristics in maize leaves. Journal of plant physiology 183, 138–143.

Mertens, J., Smolders, E., 2013. Zinc. In: Alloway, B. (Ed.), Heavy Metals in Soils. Environmental Pollution, 22. Springer, Dordrecht. Available from: https://doi.org/10.1007/978-94-007-4470-7_17.

Moreira, A., Moraes, L.A., dos Reis, A.R., 2018. The molecular genetics of zinc uptake and utilization efficiency in crop plants. Plant Micronutrient Use Efficiency 87–108.

Noman, H.M., Rana, D.S., Choudhary, A.K., Dass, A., Rajanna, G.A., Pande, P., 2020. Improving productivity, quality and biofortification in groundnut (*Arachis hypogaea* L.) through sulfur and zinc nutrition in alluvial soils of the semi-arid region of India. Journal of Plant Nutrition 1–24.

Noulas, C., Tziouvalekas, M., Karyotis, T., 2018. Zinc in soils, water and food crops. Journal of Trace Elements in Medicine and Biology 49, 252–260.

Ohki, K., 1984. Zinc nutrition related to critical deficiency and toxicity levels for sorghum 1. Agronomy Journal 76 (2), 253–256.

Ojuederie, O.B., Babalola, O.O., 2017. Microbial and plant-assisted bioremediation of heavy metal polluted environments: a review. International Journal of Environmental Research and Public Health 14 (12), 1504.

Pandey, A.K., Gautam, A., Dubey, R.S., 2019. Transport and detoxification of metalloids in plants in relation to plant-metalloid tolerance. Plant Gene 17, 100171.

Pandey, A.K., Gautam, A., Dubey, R.S., 2021. Effect of chromium on protein oxidation, protease activity, photosynthetic parameters and alleviation of toxicity in growing rice seedlings. Indian Journal of Agricultural Biochemistry 34 (1), 39–44.

Pandey, A.K., Gautam, A., Pandey, P., Dubey, R.S., 2019. Alleviation of chromium toxicity in rice seedling using *Phyllanthus emblica* aqueous extract in relation to metal uptake and modulation of antioxidative defense. South African Journal of Botany 121, 306–316.

Pandey, A.K., Gautam, A., Singh, A.K., 2023. Insight to chromium homeostasis for combating chromium contamination of soil: phytoaccumulators-based approach. Environmental Pollution 121163.

Pandey, A.K., Gedda, M.R., Verma, A.K., 2020. Effect of arsenic stress on expression pattern of a rice specific miR156j at various developmental stages and their allied co-expression target networks. Frontiers in Plant Science 11, 752.

Pandey, P., Srivastava, S., Pandey, A.K., Dubey, R.S., 2020. Abiotic-stress tolerance in plants-system biology approach. In: *Plant life under changing environment*, Academic Press, pp. 577–609.

Papazoglou, E.G., Fernando, A.L., 2017. Preliminary studies on the growth, tolerance and phytoremediation ability of sugarbeet (*Beta vulgaris* L.) grown on heavy metal contaminated soil. Industrial Crops and Products 107, 463–471.

Pascual, M.B., Echevarria, V., Gonzalo, M.J., Hernández-Apaolaza, L., 2016. Silicon addition to soybean (*Glycine max* L.) plants alleviate zinc deficiency. Plant Physiology and Biochemistry 108, 132–138.

Pawlowski, M.L., Helfenstein, J., Frossard, E., Hartman, G.L., 2019. Boron and zinc deficiencies and toxicities and their interactions with other nutrients in soybean roots, leaves, and seeds. Journal of Plant Nutrition 42 (6), 634–649.

Peng, J.S., Guan, Y.H., Lin, X.J., Xu, X.J., Xiao, L., Wang, H.H., et al., 2020. Comparative understanding of metal hyperaccumulation in plants: a mini-review. Environmental Geochemistry and Health 1–9.

Petković, K., Manojlović, M., Lombnæs, P., Čabilovski, R., Lončarić, Z., 2019. Foliar application of selenium, zinc and copper in alfalfa (*Medicago sativa* L.) biofortification. Turkish Journal of Field Crops 24 (1), 81–90.

Pongrac, P., McNicol, J.W., Lilly, A., Thompson, J.A., Wright, G., Hillier, S., et al., 2019. Mineral element composition of cabbage as affected by soil type and phosphorus and zinc fertilisation. Plant and Soil 434 (1), 151–165.

Qin, L., Han, P., Chen, L., Walk, T.C., Li, Y., Hu, X., et al., 2017. Genome-wide identification and expression analysis of NRAMP family genes in soybean (*Glycine max* L.). Frontiers in Plant Science 8, 1436.

Radhika, K., & Meena, S., 2021. Effect of zinc on growth, yield, nutrient uptake and quality of groundnut: A review.

Rattan, R.K., Shukla, L.M., 1984. Critical limits of deficiency and toxicity of zinc in paddy in a TypicUstipsamment. Communications in Soil Science and Plant Analysis 15 (9), 1041–1050.

Recena, R., García-López, A.M., Delgado, A., 2021. Zinc uptake by plants as affected by fertilization with Zn sulfate, phosphorus availability, and soil properties. Agronomy 11 (2), 390.

Refaey, Y., Jansen, B., Parsons, J.R., de Voogt, P., Bagnis, S., Markus, A., et al., 2017. Effects of clay minerals, hydroxides, and timing of dissolved organic matter addition on the competitive sorption of copper, nickel, and zinc: a column experiment. Journal of Environmental Management 187, 273–285.

Sharafi, Y., 2019. Effects of zinc on pollen gamete penetration to pistils in some apple crosses assessed by fluorescence microscopy. Caryologia: International Journal of Cytology, Cytosystematics and Cytogenetics 72 (3), 63–73.

Sharma, P.N., Chatterjee, C., Agarwala, S.C., Sharma, C.P., 1990. Zinc deficiency and pollen fertility in maize (*Zea mays*). Plant Nutrition—Physiology and Applications. Springer, Dordrecht, pp. 261–265.

Sharma, A., Patni, B., Shankhdhar, D., Shankhdhar, S.C., 2013. Zinc—an indispensable micronutrient. Physiology and Molecular Biology of Plants 19 (1), 11–20.

Shi, G., Cai, Q., 2010. Zinc tolerance and accumulation in eight oil crops. Journal of Plant Nutrition 33 (7), 982–997.

Sinclair, S.A., Krämer, U., 2012. The zinc homeostasis network of land plants. Biochimicaet Biophysica Acta (BBA)-Molecular Cell Research 1823 (9), 1553–1567.

Tarekegn, M.M., Salilih, F.Z., Ishetu, A.I., 2020. Microbes used as a tool for bioremediation of heavy metal from the environment. Cogent Food & Agriculture 6 (1), 1783174.

Ullah, A., Farooq, M., Rehman, A., Hussain, M., Siddique, K.H., 2020. Zinc nutrition in chickpea (*Cicer arietinum*): a review. Crop and Pasture Science 71 (3), 199–218.

Vatansever, R., Filiz, E., Eroglu, S., 2017. Genome-wide exploration of metal tolerance protein (MTP) genes in common wheat (*Triticum aestivum*): insights into metal homeostasis and biofortification. Biometals: An International Journal on the Role of Metal Ions in Biology, Biochemistry, and Medicine 30 (2), 217–235.

Versieren, L., Evers, S., AbdElgawad, H., Asard, H., Smolders, E., 2017. Mixture toxicity of copper, cadmium, and zinc to barley seedlings is not explained by antioxidant and oxidative stress biomarkers. Environmental Toxicology and Chemistry 36 (1), 220–230.

Vijayarengan, P., Mahalakshmi, G., 2013. Zinc toxicity in tomato plants. World Applied Sciences Journal 24 (5), 649–653.

Wani, P.A., Khan, M.S., Zaidi, A., 2008. Impact of zinc-tolerant plant growth-promoting rhizobacteria on lentil grown in zinc-amended soil. Agronomy for Sustainable Development 28 (3), 449–455.

Wessells, K.R., Brown, K.H., 2012. Estimating the global prevalence of zinc deficiency: results based on zinc availability in national food supplies and the prevalence of stunting. PLoS One 7 (11), e50568.

Zaman, Q.U., Aslam, Z., Yaseen, M., Ihsan, M.Z., Khaliq, A., Fahad, S., et al., 2018. Zinc biofortification in rice: leveraging agriculture to moderate hidden hunger in developing countries. Archives of Agronomy and Soil Science 64 (2), 147–161.

Zlobin, I.E., 2021. Current understanding of plant zinc homeostasis regulation mechanisms. Plant Physiology and Biochemistry.

Chapter 4

# Zinc toxicity in plants: a brief overview on recent developments

**Zaid Ulhassan[1], Mohamed Salah Sheteiwy[2,3], Ali Raza Khan[4], Yasir Hamid[5], Wardah Azhar[1], Sajad Hussain[6], Abdul Salam[1,7] and Weijun Zhou[1]**

*[1]Institute of Crop Science, Department of Agronomy, Ministry of Agriculture and Rural Affairs Key Lab of Spectroscopy Sensing, Zhejiang University, Hangzhou, P.R. China, [2]Department of Applied Biology, Faculty of Science, University of Sharjah, Sharjah, United Arab Emirates, [3]Department of Agronomy, Faculty of Agriculture, Mansoura University, Mansoura, Egypt, [4]School of Emergency Management, Institute of Environmental Health and Ecological Security, School of Environment and Safety Engineering, Jiangsu Province Engineering Research Center of Green Technology and Contingency Management for Emerging Pollutants, Jiangsu University, Zhenjiang, P.R. China, [5]Ministry of Education (MOE) Key Lab of Environmental Remediation and Ecosystem Health, College of Environmental and Resources Science, Zhejiang University, Hangzhou, P.R. China, [6]National Research Center of Intercropping, The Islamic University of Bahawalpur, Pakistan, [7]Key Laboratory of Natural Pesticide and Chemical Biology, Ministry of Education, South China Agricultural University, Guangzhou, P.R. China*

## 4.1 Zinc uptake by plants

Roots are the entry points of Zn (ionic form) in plants via soil medium, mainly by complex formation with organic acids governed chelators (Palmgren et al., 2008). After entering plant roots, Zn ions are translocated into plant aerial parts via the xylem. Heavy metals (HMs) linked protein transporters such as the ZIP family are involved in the uptake or transfer of Zn from soil to plant roots (Milner et al., 2013; Tiong et al., 2014). Plasma membrane and vacuole tonoplasts are the key sites of these transporters (Milner et al., 2013). The specialized protein transporters of HMs such as *HMA2* and *HMA4* (HMs ATPs) mediated the transport of Zn from rhizodermal and cortex into the xylem (Hussain et al., 2004). The transportation of Zn-positive ions can be carried out by utilizing extracellular apoplastic pathways via a weak Casparian strip (White et al., 2002). The transport of Zn

Zinc in Plants. DOI: https://doi.org/10.1016/B978-0-323-91314-0.00008-9
 

ions from shoots to seeds (mainly) ensured in phloem and symplastic Zn is interconnected with nicotianamine through chelation (Deinlein et al., 2012). Oligopeptide transporters are considered to be involved in the protein transport of phloem cells (Curie et al., 2009). Zn is absorbed by plants via leaves, but still, uptake mechanisms are not fully understood (Fernández and Brown, 2013). Various surface properties such as a waxy outside layer, cuticle composition or structure, and stomatal or trichome density affect the nutrient absorption in leaves (Eichert and Goldbach, 2008). In addition, physiological attributes such as phenology, stress level (Fernández and Brown, 2013), and environmental factors (temperature, light intensity, humidity, etc.) (Fernández and Eichert, 2009) also influence the plant's health. Generally, the absorption of Zn ions occurs through stomates and/or cuticles. Usually, the diffusion of lipophilic compounds (nonpolar) occurs through cuticles, while the penetration of hydrophilic compounds is still unknown (Fernández and Brown, 2013). Previously, these hydrophilic compounds played key roles in the transport of nutrients between apoplast and cellular spaces. Elemental nutrients such as Zn, Fe, and Ca exist as + ve charged cations and apoplast is full of −ve charged due to the presence of a free carboxyl group containing galacturonic acids (Shomer et al., 2003) that mediates the binding and gathering of + ve charged cations in apoplasts. In this way, their transportation to other cellular organs is restricted (Fernández and Brown, 2013).

## 4.2 Zinc transport in plants

Besides the formation of nicotianamine complexes in the symplasts of mesophyll cells, Zn can occur in diverse forms based on the cell type. In earlier investigations on alpine penny-cress (*Noccaea caerulescens*), it was observed that Zn in epidermis cells was retained in complex formation along with citrate or malate (Schneider et al., 2013). The entry and exit information of Zn ions via mitochondria and chloroplast membranes are still not complete (Nouet et al., 2011). The studies on the transport of Zin ions in vacuoles are well documented because they work as a storage substance for metabolites and a storehouse for harmful agents such as metal ions. Elemental nutrients keep the physiological functions of plants by maintaining the molecular mechanisms at optimal levels within organelles. The storage of metal ions in the vacuole is carried out by two mechanisms. One is the uninterrupted pumping via tonoplast and the other is the vacuole-directed transportation in vesicles (Sharma et al., 2016). In the first mechanism, metal homeostasis depends upon the tonoplast, proton pumps, and particular protein transporters which are meticulous by proton motive forces (pmf) (Sharma et al., 2016). Metal transporter proteins (MTPs) related to the cation diffusion facilitator (CDF) protein family play a crucial role in sustaining metal homeostasis, detoxification, and accumulation, for instance, the involvement of Zn transportor *AtMTP1* in *Arabidopsis thaliana* (Mäser et al., 2001). Different types of metal ions such as *OsMTP1* are being utilized by plant for the transport of Zn in rice (*Oryza sativa*) (Menguer et al., 2013). Zinc-induced facilitators (ZIF) are also involved in the transport of organic molecules and enhanced Zn accumulation as noticed in the shoots of *Arabidopsis thaliana* (Haydon and Cobbett, 2007). The functions of HMs-associated protein transporters located on the tonoplast of vacuoles are dependent on the differentiation in membrane electrical potential and pmf (Seidel et al., 2013). Only a few studies have been documented concerning the transport of Zn species in the vesicles (vacuole). Zn abundant vesicles have been noticed in the root cell of *Arabidopsis thaliana* (Kawachi et al., 2009). The same types of vesicles were found in the root hairs of the empress tree (*Paulownia tomentosa*) (Azzarello et al., 2012).

## 4.3 Functions of zinc in plant

Zn ions displayed dual (positive and negative) effects on plant cellular organs. These ions can stimulate enzymes such as carbonic anhydrase, superoxide dismutase, and RNA polymerase which are involved in carbohydrate metabolism, nucleic acid, lipids, and proteo-synthesis (Palmer and Guerinot, 2009). Zn ions are also part of "Zn fingers," a transcription factor that regulates cellular propagation and discrepancies. Zn ions are also involved in the development and a few functions (such as PSII repairment) of chloroplasts (Hänsch and Mendel, 2009). The basic principle behind the excessive Zn ions-induced toxic effects in plants is that these ions compete for the availability of binding sites proposed for other biological dynamic sites. Excessive Zn levels cause leaf chlorosis, mainly in young leaves leads to growth inhibition and biomass reduction (Broadley et al., 2007) and lower doses promote plant growth and development by minimizing the environmental stress (UlHassan et al., 2017; Salam et al., 2021). On the other hand, Zn deficiency disrupts enzyme activities which causes inhibition in photosynthesis. Moreover, Zn increases the $CO_2$ levels in chloroplast and thus boosts the carboxylation capacity of Rubisco enzymes (Salama et al., 2006). Externally, the shortage of Zn ions also impaired the plant functions during the development process and caused abnormalities. In the case of intensive Zn deficiency, the following damaging effects may occur such as dwarf growth, chlorosis, and spikelet sterility in *Poaceae* family (Sharma et al., 2013).

## 4.4 Zinc-induced phytotoxic effects

### 4.4.1 Seed germination, evident toxic symptoms, and growth traits

The reduction in germination rate and biomass are the key physiological alterations under Zn stress, which eventually declined the yield and productivity. Surplus Zn ions negatively affect the amylases activity and transport of sugar to the embryo axis, resulting in the reduction of seed germination in *Cucumus sativus* (cucumber) (Aydin et al., 2012), *Medicago sativa* (alfalfa) (Yahaghi et al., 2019), *Helianthus* (sunflower) (Huang et al., 2021), *Triticum* (wheat) (Vasyuk et al., 2019). These findings suggested that the reduction in seed germination was dose-dependent.

Chlorosis followed by necrosis occurred which is associated with the reduction in ATPase levels or activity as observed in *Arabidopsis thaliana* (Fukao et al., 2011). Excessive Zn levels greatly decreased the leaf number and leaf area which eventually declined the cell division or elongation (Glinska et al., 2016). The elevated levels of Zn ions lowered the energy metabolism by damaging mitochondrial organization, decreasing nicotinamide, and increasing $NAD^+$ breakage (Zhang et al., 2017) which leads to growth reduction. Zn caused leaf/shoot reddening which can be linked to the synthesis of anthocyanins, indicating phenolics synthesis as a stress marker of Zn as observed in lettuce, coriander, and *Arabidopsis thaliana* (Sofo et al., 2018). In addition, Zn toxicity has been found in root systems such as a reduction in root diameter, surface area, and volume in various plants (Bernardy et al., 2016). Even the morphology of root hairs was severely affected by Zn stress probably by decreasing the protein synthesis (Fukao et al., 2011). Also, root growth reduction disrupts the balance of mineral nutrients and rate of water absorption which also disturbs the shoot growth. Zn ions inhibited the root elongation but improved the later root growth in *Arabidopsis thaliana* and *Brassica napus* (Zhang et al., 2018; Feigl et al., 2019). Higher concentrations of Zn caused callose deposition in the root meristem of *B. napus* and *B. juncea* which mediates growth inhibition (Feigl et al., 2015). Increased pectin contents were also noticed under Zn stress which binds Zn in cell wall, immobilize callose, and ensure to restrict the entry of Zn into cytoplasm, ultimately causing growth inhibition (Feigl et al., 2019). Lignin biosynthesis genes were also induced under Zn stress which can cause the deposition of lignin on root epidermis and hampers root growth (Li et al., 2012). The above-mentioned reports suggested that Zn toxicity is dependent on the type of plant species and applied Zn concentrations.

### 4.4.2 Photosynthesis and cellular respiration

Zn-induced toxicity effects reduce the net photosynthesis, but it is not clear whether it was due to reduced photochemistry, enhanced respiration rate or inhibition in stomatal conductance, mesophyll conductance, or the combination of the above processes or actions (Vassilev et al., 2011). The application of elevated Zn level (600 μM) retarded the $CO_2$ assimilation rate in the second leaf stage of wheat seedlings and thus affect the photosynthetic carbon fixation. The chlorophyll fluorescence analysis revealed that Zn greatly disturbed the photosynthesis electron transport process and perturbed the complex formation or activity in reaction centers that suppressed the transformation of energy to PSII (reduced PSII efficiency) as observed in *Triticum durum* (Paunov et al., 2018) and *A. arenosa* (Szopinski et al., 2019). The excessive Zn concentrations decreased the stomatal conductance and increased the stomatal closure leading to a reduction in the photosynthetic rate and biomass of *B. rapa* (Fatemi et al., 2020). In *limoniastrum monopetalum*, Zn decreased the $CO_2$ assimilation by displacing Mg ions from Rubisco, thus leading to inhibition in carboxylation activity (Cambrollé et al., 2013). Photosynthetic pigments (Chl a and Chl b) were notably reduced under Zn stress and Chl a was more susceptible to Zn toxicity than Chl b (Garg and Singh, 2018). There is a possibility that Zn may inhibited the enzyme activities that were involved in the biosynthesis of chlorophylls. Another possibility is that Zn may increase the activities of enzymes involved in the degradation of chlorophyll synthesis as noticed in *P. monspeliensis* (Ouni et al., 2016). A recent study proposed that Zn synthesized ethylene responsive factor proteins may be involved in restricting the genes involved in chlorophyll degradation (Caronare et al., 2019).

Respiration and photosynthesis are interrelated and altered oxidase activity (AOX) that diminishes photosynthesis (Hussain et al., 2021b). Zn stress increased the activities of citrate, fumarate, and isocitrate dehydrogenase in sugar beet (Sagardoy et al., 2011), indicating a break-down of respiratory agents which are required by plants (as energy source) to handle Zn toxicity. There are conflicting reports concerning the toxic effects of Zn on cellular respiration. In vivo investigations proposed that mitochondrial respiration was increased with the increase in Zn in sunflower plants (Ismail and Azooz, 2005). The increase in respiration is normally associated with cytochrome pathways that leads to the increase in ATP synthesis, while AOX is responsible for sustaining respiration (Vanlerberghe, 2013). In this way, Zn disturbs the mitochondrial activity resulting in abnormal respiration. In another study, it was proposed that Zn stress enhanced cellular respiration as well as the synthesis of organic acids such as oxalic acid and fumaric acid via tricarboxylic acid and largely perquisites for vacuoles sequestration (Igamberdiev and Eprintsev, 2016). Usually, Zn ions compete with iron oxide for the active sites on plant cysteine oxidase and restrict their activities which causes stabilization of ERF transcription factor (Carbonare et al., 2019) and subsequent stimulation in hypoxia-responsive genes may be related to plant adaptation against Zn stress. Overall findings suggested that Zn stress target or trigger numerous events related to photosynthesis and respiration processes. Zn-induced inhibition in photosynthesis and respiratory modifications are not with consistent with the type of plant specie and its toxic effects on agricultural plants have been less documented. Zn facilitated toxic effects have been rarely reported (in detail) at cellular and/or molecular levels.

### 4.4.3 Water levels, nutrient uptake, and nitrogen metabolism

Zn stress disturbs the water relations or relative water content (RWC) by downregulating the water-linked transporters, reducing transpiration rates, and severe dehydration (Garg and Singh, 2018) that is possibly due to the extra accumulation of Zn in aerial plant parts. Most probably, excess Zn diminishes the roots penetrating capacity or vascular system governed by the obstruction in the xylem, ultimately restricting the escalation process to sap (Koleva et al., 2010). Mir et al. (2015) reported that the reduction in leaf water potential in response to Zn stress is due to the reduction in water penetration. Decreased water levels suggest the deprivation of turgidity or turgor pressure which consequently minimizes cellular development and thus reduces growth and biomass. Zn alters the water balance by restricting its transportation within plant parts or restricting tonoplast activity related to water transport (Kholodova et al., 2011). Few studies describe the molecular basis of Zn-facilitated decline in water transport in roots. It was observed that Zn downregulated the expression of aquaporin genes existing in plasma membrane such as *McPIP1;1*, *McPIP2;1* and *McPIP2;3* or tonoplast such as *McTIP1;2* and *McTIP2;2* in Zn-stressed *Mesembryanthemum crystallinum* (Lucini and Bernardo, 2015). In barley roots, the exposure of high Zn restricts the water transport properties which were correlated with the reduced shoot hydraulic conductivity and root aquaporins as illustrated in the downregulating aquaporins (*HvPIP1;2; HvPIP2;4* and *HvPIP2;5*) related genes (Gitto and Fricke, 2018). Recent investigations explained that the reduction in root hydraulic conductivity and participation of PIP1 isoforms are the key markers of long-term Zn stress responses in *B. rapa* (Fatemi et al., 2020). These findings suggested that Zn ions impair the morpho-physiological traits by disturbing the water absorption and transport systems.

Regarding nutrient status, Zn competes with mineral nutrients for binding at particular sites. For instance, Zn inhibits root growth and damages root elongation (cracking) with the lowest capacity to penetrate mineral nutrients, thus leading to cell death (Šimon et al., 2021). The modifications in the functions of ion channels, membrane transporters, and depolarization of membrane penetrations (Bazihizina et al., 2014). Excess Zn not only limits the nutrients transported with plant parts but also alters their allocation in plant organs (Garg and Singh, 2018). The alterations or differences in the accumulation of nutrients in plant parts caused the increase or decrease of their contents depending upon the particular plant species (Di Baccio et al., 2011; Souza et al., 2020). For example, Zn stress enhanced the Fe levels by improving the root-to-shoot translocation and maintaining carbon assimilation process (Yang et al., 2011). Contrary to this, Zn reduced the Fe levels in the shoots of wheat plants possibly due to the competition between Zn and Fe for their transportation and associated chelation process within plant tissues (Glińska et al., 2016). Another study revealed that nicotianamine (NA) form chelates with Fe in symplast and then transports through the phloem to aerial parts indicating the involvement of NA in Fe homeostasis. The competition between Zn and Fe for NA-mediated chelation may result in the reduction of Fe contents in the shoots of soybeans (Ibiang et al., 2017). On the contrary, Zn stress impairs the Fe-linked morpho-physiological and molecular attributes in *Arabidopsis thaliana* and Zn did not inhibit Fe uptake via iron-regulated transporter (IRT1). Alternatively, Zn upregulated the Fe-regulated transcription of genes such as ferric reduction oxidase2 (*FRO2*) and ferric chelate reductase (*FCR*), indicating that Zn impairs the Fe-homeostasis (Lešková et al., 2017). Recent investigations evidenced that IRT1 proteins were involved in Fe-deficiency under Zn stress in wheat (Erenoğlu, 2019), although it should be explored in other plants. The intervention of Zn with Mn and N causes chlorosis and reduction in net photosynthetic rate, indicating Zn-induced inhibition in chlorophyll synthetic pathways. In addition, Zn stress reduced the K contents which may be due to the disturbances in water balance, whereas higher transport of K to aerial parts may help plants to improve photosynthesis and synthesis of ATP as noticed in *Zygophyllum fabago* (Lefèvre et al., 2014). Zn-mediated Fe or Mg deficiency can limit the production of photosynthetic pigments as Fe is a perquisite for heme-based synthesis and Mg takes part in the formation of chlorophylls (Tripathy and Pattanayak, 2012). These outcomes confirmed the cross-talks between Zn and other mineral nutrients, but the exact mechanisms behind the interferences are still lacking.

Nitrogen metabolism was greatly disturbed by excess Zn as noticed in *Brassica* species (Blasco et al., 2019). The exposure of Zn disorganizes the chloroplasts and/or restricts the nitrate supply at the active sites of nitrate reductase (NR) synthesis which restricted its activity. The decreased levels of NR minimize the production of free amino acids such as glutamate and asparagine which consequently restricts photosynthesis (Sidhu, 2016). The activity of NR relies on photosynthesis synthesis due to its dependence on the photosynthesis of carbon (Solanki and Dhankhar, 2011). In this way, Zn stress governed the inhibition in carbon dioxide fixation and chlorophyll synthesis may lead to impairment in nitrogen metabolism and, ultimately reducing NR activity. The applications of higher Zn levels may reduce the NR activity by its direct effects on protein synthesis as reported in Zn-stressed *B. rapa* (Ramakrishna and Rao, 2015). In the roots of *Arabidopsis thaliana* (wild-type), Zn stress induced the expression of nitrate transporter (*NRT1.1*) in both elongation zone and meristem that can cause an increase in nitrate uptake and accumulation of Zn in plant parts (Pan et al., 2020). The possible roles of NRT in nitrate-facilitated Zn accumulation in different plants are yet to be explored. Zn stress also impairs the nodulation as well as nitrogen fixation capacity in legume plants (Souza et al., 2020) which suggested that Zn displayed harmful effects on roots functioning and rhizobial organisms that restricted the nodule formation. Zn stress also affect the symbiotic plasmids (containing genes for nodulation or nitrogen fixation), thereby disturbing the ability of bacteria to form nodule in legumes (Wang et al., 2018). The complete understanding of Zn and rhizobial interactions should be further investigated.

### 4.4.4 Redox balance

Due to the nonredox nature, Zn cannot induce reactive oxygen species (ROS) through hydroxyl radical, and superoxide radical via Haber-Weiss reactions. The extra generation of ROS in plants due to Zn toxicity was possibly due to the changes in metabolic pathways or mechanisms involved in electron transport processes. Earlier studies investigated that excess Zn can increase the activity of lipoxygenase (LOX) and lipids peroxidation as MDA, which impairs membrane integrity and water absorption (Goodarzi et al., 2020). The elevated Zn concentrations stimulate leaf senescence by enhancing the protein degradation that can facilitate the Zn-binding with the sulfhydryl cluster of proteins, leading to abnormalities in protein structure (Singh et al., 2018). It is proposed that excessive generation of ROS impairs the proteins functioning either by the byproducts of lipid peroxidation or oxidation of amino acids or production of carbonyl group in proteins (Madian and Regnier, 2010). These mechanisms could be involved in the Zn-induced ROS and its induced oxidative damage on proteins. Another study revealed that Zn stress induced the synthesis of protein tyrosine nitration in *B. napus* roots which was possibly due to the imbalance in the homeostasis of ROS and RNS (Feigl et al., 2019). Recent proteomic experiments on Zn-exposed rice roots revealed that Zn enhanced proteins synthesis which was responsible for proteolysis, redox balance, defense mechanisms, and ion homeostasis (Šimon et al., 2021). In addition, Zn caused chromosome abnormalities which involves uneven scattering, formation of dicentric chromosomes, inability of scatter chromosome to remerge, incapability of chromatin to arrive at poles, and formation of micronucleus (Oladele et al., 2013). The profiles of Random Amplification of Polymorphic DNA (RAPD) indicated the damages in DNA due to loss in particular bands, chromosome rearrangements, alteration in oligonucleotide primer sites, and recombination of homologous at the exposure of Zn in *Zea mays* (Erturk et al., 2015). These observations exposed that Zn toxicity severely damages the proteins, cellular membranes, metabolic functioning, and ultimately cell death.

### 4.4.5 Antioxidants defense responses

To handle the excessive generation of ROS, maintain the ROS balance, and prevent the oxidative damages, plants active their intrinsic antioxidants defense systems including enzymatic or nonenzymatic components such as superoxide dismutase (SOD), peroxidase (POD), catalase (CAT), ascorbate peroxidase (APX), guaicol peroxidase (GPOX), glutathione reductase (GR), dehydroascorbate reductase (DHAR), monodehydroascorbate reductase (MDHAR), osmoprotectants, carotenoids, ascorbate (AsA) and reduced glutathione (GSH) (Ali et al., 2018; Ulhassan et al., 2018, 2019a,b,c; Mwamba et al., 2020; Nazir et al., 2020; Hussain et al., 2021a; Tripathi et al., 2020; Sheteiwy et al., 2021; Yang et al., 2021). It was observed that Zn toxicity downregulated these antioxidant enzymes activities, indicating overproduction of ROS (Song et al., 2011). In most of the previous reports, excessive Zn levels increased in the activities of antioxidants such as SOD, POD, and CAT in *Glycine max* (dos Santos et al., 2020), *Capsicum annuum* (Kaya et al., 2018), *Carthamus tinctorius* (Namdjoyan et al., 2017), *Zea mays* (Salam et al., 2021), and *Plantago major* (Branch et al., 2017), which suggested that these plants upregulated their tolerance responses against Zn stress by sustaining the redox balance and curbing the additional oxidative damages. Among nonenzymatic antioxidants, glutathione (GSH) and ascorbate (AsA) form complexes with Zn, mediate the removal of cytosolic Zn and thus reduce the Zn toxicity. Probably, the binding of Zn with thiolic compounds weakens the extra production of ROS and thus plays crucial roles in decreasing the oxidative damages with cellular organs. Metallothioneins (MTs) also have implications in Zn homeostasis and defense strategies in order to promote plant tolerance (Pramanick et al., 2017). MTs actively participate in binding with metal ions using carboxyl (C) terminals and cysteine-loaded amino (N) terminal sites. N-sites established equilibrium with elements such as Zn and C-sites displayed great affinity to toxic metal ins and facilitated their sequestrations (Barbosa et al., 2016). These studies suggested that plants maintain the metabolic equilibrium of the above-reported compounds via synthesis or complexation to cope with the Zn stress. Further research should be carried on hyperaccumulators for breeding programs to develop Zn tolerant genotypes. Besides enzymatic and nonenzymatic antioxidants, Zn stress assists in the accumulation of different metabolites such as polyamines, free amino acids, proline, sugars, proteins, and glycine betaine having important roles in osmotic adjustments and plant tolerance against Zn stress (Kaya et al., 2018; AbdElgawad et al., 2020; Souza et al., 2020). The accumulation of these osmoprotectants/metabolites enhances the plant tolerance against Zn toxicity and can play important roles in plant adaptation to severe environmental conditions (Fig. 4.1).

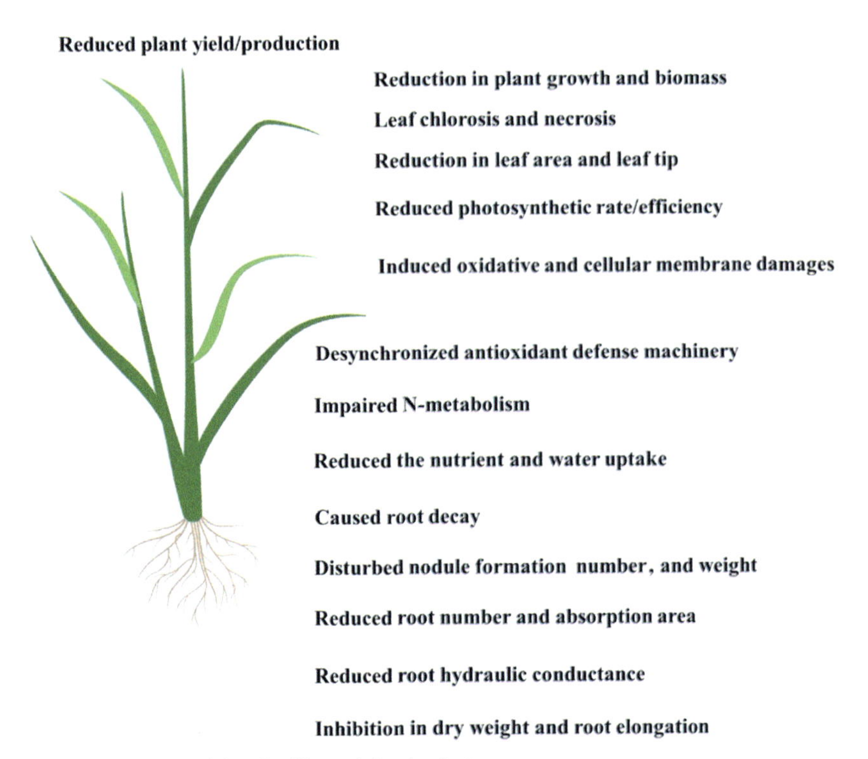


**FIGURE 4.1** Potential toxic effects of zinc in plants.

## 4.5 Conclusions

Although Zn is an essential micronutrient for normal plant growth and development, higher levels of Zn can be phytotoxic affecting the morphological, physiological, biochemical, cellular, and molecular attributes. The information related to Zn-bioavailability is important to understand its binding capacity and transport capability in soil-plant systems and then its accumulation into human food chains. Zn toxicity is directly related to Zn-homeostasis in plants via different transporter gene families. Various transporters are involved in the homeostasis of Zn within plant parts. The mechanistic interface between transporter proteins and Zn-transferring within cellular organs may further enhance our understanding concerning the Zn-transport in plants. The complete knowledge of molecular mechanisms (via genome editing approaches) of Zn uptake, transport, and accumulation can help plant scientists to develop Zn-tolerant hyperaccumulators for the possible remediation of Zn-contaminated soils and crop cultivation in these contaminated soils. In a nutshell, the comprehension of Zn as a signaling or stress marker can allure plant scientists to develop stress mitigation strategies for the sustainable development of the agriculture sector.

## Acknowledgments

This work was supported by the Science and Technology Department of Zhejiang Province (2021C02064-2), and Collaborative Innovation Center for Modern Crop Production co-sponsored by Province and Ministry (CIC-MCP). We thank Rui Sun from Agricultural Experiment Station of Zhejiang University for his assistance.

## References

AbdElgawad, H., Zinta, G., Hamed, B.A., Selim, S., Beemster, G., Hozzein, W.N., et al., 2020. Maize roots and shoots show distinct profiles of oxidative stress and antioxidant defense under heavy metal toxicity. Environmental Pollution 258, 113705.

Ali, S., Gill, R.A., Ulhassan, Z., Najeeb, U., Kanwar, M.K., Abid, M., et al., 2018. Insights on the responses of Brassica napus cultivars against the cobalt-stress as revealed by carbon assimilation, anatomical changes and secondary metabolites. Environmental and Experimental Botany 156, 183–196.

Aydin, S.S., Gökçe, E., Büyük, İ., Aras, S., 2012. Characterization of stress induced by copper and zinc on cucumber (Cucumis sativus L.) seedlings by means of molecular and population parameters. Mutation Research/Genetic Toxicology and Environmental Mutagenesis 746 (1), 49–55.

Azzarello, E., Pandolfi, C., Giordano, C., Rossi, M., Mugnai, S., Mancuso, S., 2012. Ultramorphological and physiological modifications induced by high zinc levels in Paulownia tomentosa. Environmental and Experimental Botany 81, 11–17.

Barbosa, B.C.F., Silva, S.C., de Oliveira, R.R., Chalfun Jr, A., 2016. Zinc supply impacts on the relative expression of a metallothioneinlike gene in Cofea arabica plants. Plant and Soil 411, 179–191.

Bazihizina, N., Taiti, C., Marti, L., Rodrigo-Moreno, A., Spinelli, F., Giordano, C., et al., 2014. Zn2 + -induced changes at the root level account for the increased tolerance of acclimated tobacco plants. Journal of Experimental Botany 65 (17), 4931–4942.

Bernardy, K., Farias, J.G., Dorneles, A.O.S., Pereira, A.S., Schorr, M.R.W., Thewes, F.R., et al., 2016. Changes in root morphology and dry matter production in Pfaffia glomerata (Spreng.) Pedersen accessions in response to excessive zinc. Revista Brasileira de Plantas Medicinais 18, 613–620.

Blasco, B., Navarro-León, E., Ruiz, J.M., 2019. Study of Zn accumulation and tolerance of HMA4 TILLING mutants of Brassica rapa grown under Zn deficiency and Zn toxicity. Plant Science 287, 110201.

Branch, D., Damghan, I., Branch, Q., Qaemshahr, I., 2017. Protective role of exogenous nitric oxide against zinc toxicity in Plantago major L. Applied Ecology and Environmental Research 15, 511–524.

Broadley, M.R., White, P.J., Hammond, J.P., Zelko, I., Lux, A., 2007. Zinc in plants. New Phytologist 173 (4), 677–702.

Cambrollé, J., Mancilla-Leytón, J.M., Muñoz-Vallés, S., Figueroa-Luque, E., Luque, T., Figueroa, M.E., 2013. Evaluation of zinc tolerance and accumulation potential of the coastal shrub Limoniastrum monopetalum (L.) Boiss. Environmental and Experimental Botany 85, 50–57.

Carbonare, L.D., White, M.D., Shukla, V., Francini, A., Perata, P., Flashman, E., et al., 2019. Zinc excess induces a hypoxia-like response by inhibiting cysteine oxidases in poplar roots. Plant Physiology 180 (3), 1614–1628.

Curie, C., Cassin, G., Couch, D., Divol, F., Higuchi, K., Le Jean, M., et al., 2009. Metal movement within the plant: contribution of nicotianamine and yellow stripe 1-like transporters. Annals of Botany 103 (1), 1–11.

Deinlein, U., Weber, M., Schmidt, H., Rensch, S., Trampczynska, A., Hansen, T.H., et al., 2012. Elevated nicotianamine levels in Arabidopsis halleri roots play a key role in zinc hyperaccumulation. The Plant Cell 24 (2), 708–723.

Di Baccio, D., Galla, G., Bracci, T., Andreucci, A., Barcaccia, G., Tognetti, R., et al., 2011. Transcriptome analyses of Populus × euramericana clone I-214 leaves exposed to excess zinc. Tree Physiology 31 (12), 1293–1308.

dos Santos, L.R., da Silva, B.R.S., Pedron, T., Batista, B.L., da Silva Lobato, A.K., 2020. 24-Epibrassinolide improves root anatomy and antioxidant enzymes in soybean plants subjected to zinc stress. Journal of Soil Science and Plant Nutrition 20 (1), 105–124.

Eichert, T., Goldbach, H.E., 2008. Equivalent pore radii of hydrophilic foliar uptake routes in stomatous and astomatous leaf surfaces–further evidence for a stomatal pathway. Physiologia Plantarum 132 (4), 491–502.

Erenoğlu, E.B., 2019. Iron deficiency-induced zinc uptake by bread wheat. Journal of Plant Nutrition and Soil Science 182 (3), 496–501.

Erturk, F.A., Nardemir, G., Hilal, A.Y., Arslan, E., Agar, G., 2015. Determination of genotoxic effects of boron and zinc on *Zea mays* using protein and random amplification of polymorphic DNA analyses. Toxicology and Industrial Health 31 (11), 1015–1023.

Fatemi, H., Zaghdoud, C., Nortes, P.A., Carvajal, M., Martínez-Ballesta, M.D.C., 2020. Differential aquaporin response to distinct effects of two Zn concentrations after foliar application in pak choi (Brassica rapa L.) plants. Agronomy 10 (3), 450.

Feigl, G., Lehotai, N., Molnár, Á., Ördög, A., Rodríguez-Ruiz, M., Palma, J.M., et al., 2015. Zinc induces distinct changes in the metabolism of reactive oxygen and nitrogen species (ROS and RNS) in the roots of two Brassica species with different sensitivity to zinc stress. Annals of Botany 116 (4), 613–625.

Feigl, G., Molnár, Á., Szőllősi, R., Ördög, A., Törőcsik, K., Oláh, D., et al., 2019. Zinc-induced root architectural changes of rhizotron-grown B. napus correlate with a differential nitro-oxidative response. Nitric Oxide: Biology and Chemistry / Official Journal of the Nitric Oxide Society 90, 55–65.

Fernández, V., Brown, P.H., 2013. From plant surface to plant metabolism: the uncertain fate of foliar-applied nutrients. Frontiers in Plant Science 4, 289.

Fernández, V., Eichert, T., 2009. Uptake of hydrophilic solutes through plant leaves: current state of knowledge and perspectives of foliar fertilization. Critical Reviews in Plant Sciences 28 (1–2), 36–68.

Fukao, Y., Ferjani, A., Tomioka, R., Nagasaki, N., Kurata, R., Nishimori, Y., et al., 2011. iTRAQ analysis reveals mechanisms of growth defects due to excess zinc in Arabidopsis. Plant Physiology 155 (4), 1893–1907.

Garg, N., Singh, S., 2018. Arbuscular mycorrhiza Rhizophagus irregularis and silicon modulate growth, proline biosynthesis and yield in Cajanus cajan L. Millsp. (pigeonpea) genotypes under cadmium and zinc stress. Journal of Plant Growth Regulation 37 (1), 46–63.

Gitto, A., Fricke, W., 2018. Zinc treatment of hydroponically grown barley plants causes a reduction in root and cell hydraulic conductivity and isoform-dependent decrease in aquaporin gene expression. Physiologia Plantarum 164 (2), 176–190.

Glińska, S., Gapińska, M., Michlewska, S., Skiba, E., Kubicki, J., 2016. Analysis of Triticum aestivum seedling response to the excess of zinc. Protoplasma 253 (2), 367–377.

Goodarzi, A., Namdjoyan, S., Soorki, A.A., 2020. Effects of exogenous melatonin and glutathione on zinc toxicity in safflower (Carthamus tinctorius L.) seedlings. Ecotoxicology and Environmental Safety 201, 110853.

Haydon, M.J., Cobbett, C.S., 2007. A novel major facilitator superfamily protein at the tonoplast influences zinc tolerance and accumulation in Arabidopsis. Plant Physiology 143 (4), 1705–1719.

Huang, Y.T., Cai, S.Y., Ruan, X.L., Chen, S.Y., Mei, G.F., Ruan, G.H., et al., 2021. Salicylic acid enhances sunflower seed germination under Zn2 + stress via involvement in $Zn^{2+}$ metabolic balance and phytohormone interactions. Scientia Horticulturae 275, 109702.

Hussain, D., Haydon, M.J., Wang, Y., Wong, E., Sherson, S.M., Young, J., et al., 2004. P-type ATPase heavy metal transporters with roles in essential zinc homeostasis in Arabidopsis. The Plant Cell 16 (5), 1327–1339.

Hussain, S., Mumtaz, M., Manzoor, S., Shuxian, L., Ahmed, I., Skalicky, M., et al., 2021a. Foliar application of silicon improves growth of soybean by enhancing carbon metabolism under shading conditions. Plant Physiology and Biochemistry: PPB / Societe Francaise de Physiologie Vegetale 159, 43–52.

Hussain, S., Ulhassan, Z., Brestic, M., Zivcak, M., Zhou, W., Allakhverdiev, S.I., et al., 2021b. Photosynthesis research under climate change. Photosynthesis Research 1–15.

Hänsch, R., Mendel, R.R., 2009. Physiological functions of mineral micronutrients (cu, Zn, Mn, Fe, Ni, Mo, B, cl). Current Opinion in Plant Biology 12 (3), 259–266.

Ibiang, Y.B., Mitsumoto, H., Sakamoto, K., 2017. Bradyrhizobia and arbuscular mycorrhizal fungi modulate manganese, iron, phosphorus, and polyphenols in soybean (*Glycine max* (L.) Merr.) under excess zinc. Environmental and Experimental Botany 137, 1–13.

Igamberdiev, A.U., Eprintsev, A.T., 2016. Organic acids: the pools of fixed carbon involved in redox regulation and energy balance in higher plants. Frontiers in Plant Science 7, 1042.

Ismail, A.M., Azooz, M.M., 2005. Effect of zinc supply on growth and some metabolic characteristics of safflower and sunflower plants. Indian Journal of Plant Physiology 10 (3), 260.

Kawachi, M., Kobae, Y., Mori, H., Tomioka, R., Lee, Y., Maeshima, M., 2009. A mutant strain Arabidopsis thaliana that lacks vacuolar membrane zinc transporter MTP1 revealed the latent tolerance to excessive zinc. Plant and Cell Physiology 50 (6), 1156–1170.

Kaya, C., Ashraf, M., Akram, N.A., 2018. Hydrogen sulfide regulates the levels of key metabolites and antioxidant defense system to counteract oxidative stress in pepper (Capsicum annuum L.) plants exposed to high zinc regime. Environmental Science and Pollution Research 25 (13), 12612–12618.

Kholodova, V., Volkov, K., Abdeyeva, A., Kuznetsov, V., 2011. Water status in Mesembryanthemum crystallinum under heavy metal stress. Environmental and Experimental Botany 71 (3), 382–389.

Koleva, L., Semerdjieva, I., Nikolova, A., Vassilev, A., 2010. Comparative morphological and histological study on zinc-and cadmium-treated durum wheat plants with similar growth inhibition. *General and Applied.* Plant Physiology 36 (1/2), 8–11.

Lefèvre, I., Vogel-Mikuš, K.A.T.A.R.I.N.A., Jeromel, L., Vavpetič, P., Planchon, S., Arčon, I., et al., 2014. Differential cadmium and zinc distribution in relation to their physiological impact in the leaves of the accumulating Z ygophyllum fabago L. Plant, Cell & Environment 37 (6), 1299–1320.

Lešková, A., Giehl, R.F., Hartmann, A., Fargašová, A., von Wirén, N., 2017. Heavy metals induce iron deficiency responses at different hierarchic and regulatory levels. Plant Physiology 174 (3), 1648–1668.

Li, X., Yang, Y., Zhang, J., Jia, L., Li, Q., Zhang, T., et al., 2012. Zinc induced phytotoxicity mechanism involved in root growth of Triticum aestivum L. Ecotoxicology and Environmental Safety 86, 198–203.

Lucini, L., Bernardo, L., 2015. Comparison of proteome response to saline and zinc stress in lettuce. Frontiers in Plant Science 6, 240.

Madian, A.G., Regnier, F.E., 2010. Proteomic identification of carbonylated proteins and their oxidation sites. Journal of Proteome Research 9 (8), 3766–3780.

Mäser, P., Thomine, S., Schroeder, J.I., Ward, J.M., Hirschi, K., Sze, H., et al., 2001. Phylogenetic relationships within cation transporter families of Arabidopsis. Plant Physiology 126 (4), 1646–1667.

Menguer, P.K., Farthing, E., Peaston, K.A., Ricachenevsky, F.K., Fett, J.P., Williams, L.E., 2013. Functional analysis of the rice vacuolar zinc transporter OsMTP1. Journal of Experimental Botany 64 (10), 2871–2883.

Milner, M.J., Seamon, J., Craft, E., Kochian, L.V., 2013. Transport properties of members of the ZIP family in plants and their role in Zn and Mn homeostasis. Journal of Experimental Botany 64 (1), 369–381.

Mir, B.A., Khan, T.A., Fariduddin, Q., 2015. 24-epibrassinolide and spermidine modulate photosynthesis and antioxidant systems in Vigna radiata under salt and zinc stress. International Journal of Advanced Research 3 (5), 592–608.

Mwamba, T.M., Islam, F., Ali, B., Lwalaba, J.L.W., Gill, R.A., Zhang, F., et al., 2020. Comparative metabolomic responses of low-and high-cadmium accumulating genotypes reveal the cadmium adaptive mechanism in Brassica napus. Chemosphere 250, 126308.

Namdjoyan, S., Kermanian, H., Soorki, A.A., Tabatabaei, S.M., Elyasi, N., 2017. Interactive effects of salicylic acid and nitric oxide in alleviating zinc toxicity of Safflower (Carthamus tinctorius L.). Ecotoxicology (London, England) 26 (6), 752–761.

Nazir, M.M., Ulhassan, Z., Zeeshan, M., Ali, S., Gill, M.B., 2020. Toxic Metals/Metalloids Accumulation, Tolerance, and Homeostasis in Brassica Oilseed Species. The Plant Family Brassicaceae. Springer, Singapore, pp. 379–408.

Nouet, C., Motte, P., Hanikenne, M., 2011. Chloroplastic and mitochondrial metal homeostasis. Trends in Plant Science 16 (7), 395–404.

Oladele, E.O., Odeigah, P.G.C., Taiwo, I.A., 2013. The genotoxic effect of lead and zinc on bambara groundnut (*Vigna subterranean*). African Journal of Environmental Science and Technology 7 (1), 9–13.

Ouni, Y., Mateos-Naranjo, E., Abdelly, C., Lakhdar, A., 2016. Interactive effect of salinity and zinc stress on growth and photosynthetic responses of the perennial grass, Polypogon monspeliensis. Ecological Engineering 95, 171–179.

Palmer, C.M., Guerinot, M.L., 2009. Facing the challenges of Cu, Fe and Zn homeostasis in plants. Nature Chemical Biology 5 (5), 333–340.

Palmgren, M.G., Clemens, S., Williams, L.E., Krämer, U., Borg, S., Schjørring, J.K., et al., 2008. Zinc biofortification of cereals: problems and solutions. Trends in Plant Science 13 (9), 464–473.

Pan, W., You, Y., Weng, Y.N., Shentu, J.L., Lu, Q., Xu, Q.R., et al., 2020. Zn stress facilitates nitrate transporter 1.1-mediated nitrate uptake aggravating Zn accumulation in Arabidopsis plants. Ecotoxicology and Environmental Safety 190, 110104.

Paunov, M., Koleva, L., Vassilev, A., Vangronsveld, J., Goltsev, V., 2018. Effects of different metals on photosynthesis: cadmium and zinc affect chlorophyll fluorescence in durum wheat. International Journal of Molecular Sciences 19 (3), 787.

Pramanick, P., Chakraborty, A., Raychaudhuri, S.S., 2017. Phenotypic and biochemical alterations in relation to MT2 gene expression in Plantago ovata Forsk under zinc stress. Biometals: an International Journal on the Role of Metal Ions in Biology, Biochemistry, and Medicine 30, 171–184.

Ramakrishna, B., Rao, S.S.R., 2015. Foliar application of brassinosteroids alleviates adverse effects of zinc toxicity in radish (*Raphanus sativus* L.) plants. Protoplasma 252 (2), 665–677.

Sagardoy, R., Morales, F., Rellán-Álvarez, R., Abadía, A., Abadía, J., López-Millán, A.F., 2011. Carboxylate metabolism in sugar beet plants grown with excess Zn. Journal of Plant Physiology 168 (7), 730–733.

Salama, Z.A., El-Fouly, M.M., Lazova, G., Popova, L.P., 2006. Carboxylating enzymes and carbonic anhydrase functions were suppressed by zinc deficiency in maize and chickpea plants. Acta Physiologiae Plantarum 28 (5), 445–451.

Salam, A., Khan, A.R., Liu, L., Yang, S., Azhar, W., Ulhassan, Z., et al., 2021. Seed priming with zinc oxide nanoparticles downplayed ultrastructural damage and improved photosynthetic apparatus in maize under cobalt stress. Journal of Hazardous Materials 127021.

Schneider, T., Persson, D.P., Husted, S., Schellenberg, M., Gehrig, P., Lee, Y., et al., 2013. A proteomics approach to investigate the process of Z n hyperaccumulation in N occaea caerulescens (J & C. P resl) FK M eyer. The Plant Journal 73 (1), 131–142.

Seidel, T., Siek, M., Marg, B., Dietz, K.J., 2013. Energization of vacuolar transport in plant cells and its significance under stress. International Review of Cell and Molecular Biology 304, 57–131.

Sharma, A., Patni, B., Shankhdhar, D., Shankhdhar, S.C., 2013. Zinc—an indispensable micronutrient. Physiology and Molecular Biology of Plants 19 (1), 11–20.

Sharma, S.S., Dietz, K.J., Mimura, T., 2016. Vacuolar compartmentalization as indispensable component of heavy metal detoxification in plants. Plant, Cell & Environment 39 (5), 1112–1126.

Sheteiwy, M.S., Ali, D.F.I., Xiong, Y.C., Brestic, M., Skalicky, M., Hamoud, Y.A., et al., 2021. Physiological and biochemical responses of soybean plants inoculated with Arbuscular mycorrhizal fungi and Bradyrhizobium under drought stress. BMC Plant Biology 21 (1), 1–21.

Shomer, I., Novacky, A.J., Pike, S.M., Yermiyahu, U., Kinraide, T.B., 2003. Electrical potentials of plant cell walls in response to the ionic environment. Plant Physiology 133 (1), 411–422.

Sidhu, G.P.S., 2016. Physiological, biochemical and molecular mechanisms of zinc uptake, toxicity and tolerance in plants. Journal of Global Biosciences 5 (9), 4603–4633.

Šimon, M., Shen, Z.J., Ghoto, K., Chen, J., Liu, X., Gao, G.F., et al., 2021. Proteomic investigation of Zn-challenged rice roots reveals adverse effects and root physiological adaptation. Plant and Soil 460 (1), 69–88.

Singh, R.P., Mishra, S., Jha, P., Raghuvanshi, S., Jha, P.N., 2018. Effect of inoculation of zinc-resistant bacterium Enterobacter ludwigii CDP-14 on growth, biochemical parameters and zinc uptake in wheat (*Triticum aestivum* L.) plant. Ecological Engineering 116, 163–173.

Sofo, A., Moreira, I., Gattullo, C.E., Martins, L.L., Mourato, M., 2018. Antioxidant responses of edible and model plant species subjected to subtoxic zinc concentrations. Journal of Trace Elements in Medicine and Biology 49, 261–268.

Solanki, R., Dhankhar, R., 2011. Zinc and copper induced changes in physiological characteristics of Vigna mungo (L.). Journal of Environmental Biology 32 (6), 747.

Song, A., Li, P., Li, Z., Fan, F., Nikolic, M., Lian, Y., 2011. The alleviation of zinc toxicity by silicon is related to zinc transport and antioxidative reactions in rice. Plant and Soil 344, 319–333.

Souza, S.C., Souza, L.A., Schiavinato, M.A., de Oliveira Silva, F.M., de Andrade, S.A., 2020. Zinc toxicity in seedlings of three trees from the Fabaceae associated with arbuscular mycorrhizal fungi. Ecotoxicology and Environmental Safety 195, 110450.

Szopinski, M., Sitko, K., Gieron, Z., Rusinowski, S., Corso, M., Hermans, C., et al., 2019. Toxic efects of Cd and Zn on the photosynthetic apparatus of the Arabidopsis halleri and Arabidopsis arenosa pseudo-metallophytes. Frontiers in Plant Science 10, 748.

Tiong, J., McDonald, G.K., Genc, Y., Pedas, P., Hayes, J.E., Toubia, J., et al., 2014. H v ZIP 7 mediates zinc accumulation in barley (H ordeum vulgare) at moderately high zinc supply. New Phytologist 201 (1), 131–143.

Tripathi, D.K., Varma, R.K., Singh, S., Sachan, M., Guerriero, G., Kushwaha, B.K., et al., 2020. Silicon tackles butachlor toxicity in rice seedlings by regulating anatomical characteristics, ascorbate-glutathione cycle, proline metabolism and levels of nutrients. Scientific Report. 10, 1–15.

Tripathy, B.C., Pattanayak, G.K., 2012. Chlorophyll biosynthesis in higher plants. Photosynthesis. Springer, Dordrecht, pp. 63–94.

UlHassan, Z., Ali, S., Rizwan, M., Hussain, A., Akbar, Z., Rasool, N., Abbas, F., 2017. Role of Zinc in alleviating heavy metal stress. In: Naeem, A.G.M., et al., (Eds.), Essential Plant Nutrients. Springer International Publishing. Available from: https://doi.org/10.1007/978-3-319-58841-4_14.

Ulhassan, Z., Gill, R.A., Mwamba, T.M., Abid, M., Li, L., Zhang, N., et al., 2018. Comparative orchestrating response of four oilseed rape (Brassica napus) cultivars against the selenium stress as revealed by physio-chemical, ultrastructural and molecular profiling. Ecotoxicology and Environmental Safety 161, 634–647.

Ulhassan, Z., Gill, R.A., Ali, S., Mwamba, T.M., Ali, B., Wang, J., et al., 2019a. Dual behavior of selenium: Insights into physio-biochemical, anatomical and molecular analyses of four *Brassica napus* cultivars. Chemosphere 225, 329–341.

Ulhassan, Z., Huang, Q., Gill, R.A., Ali, S., Mwamba, T.M., Ali, B., et al., 2019b. Protective mechanisms of melatonin against selenium toxicity in *Brassica napus*: insights into physiological traits, thiol biosynthesis, and antioxidant machinery. BMC Plant Biology 19, 507. Available from: https://doi.org/10.1186/s12870-019-2110-6.

Ulhassan, Z., Gill, R.A., Huang, H., Ali, S., Mwamba, T.M., Ali, B., et al., 2019c. Selenium mitigates the chromium toxicity in *Brassica napus* L. by ameliorating nutrients uptake, amino acids metabolism and antioxidant defense system. Plant Physiology and Biochemistry: PPB / Societe Francaise de Physiologie Vegetale 145, 142–152.

Vanlerberghe, G.C., 2013. Alternative oxidase: a mitochondrial respiratory pathway to maintain metabolic and signaling homeostasis during abiotic and biotic stress in plants. International Journal of Molecular Sciences 14 (4), 6805–6847.

Vasyuk, V.A., Voytenko, L.V., Shcherbatiuk, M.M., Kosakivska, I.V., 2019. Effect of exogenous abscisic acid on seed germination and growth of winter wheat seedlings under zinc stress. Journal of Stress Physiology & Biochemistry 15 (2).

Wang, X., Zhao, L., Zhang, L., Wu, Y., Chou, M., Wei, G., 2018. Comparative symbiotic plasmid analysis indicates that symbiosis gene ancestor type affects plasmid genetic evolution. Letters in Applied Microbiology 67 (1), 22–31.

White, P.J., Whiting, S.N., Baker, A.J., Broadley, M.R., 2002. Does zinc move apoplastically to the xylem in roots of Thlaspi caerulescens? New Phytologist 201–207.

Yahaghi, Z., Shirvani, M., Nourbakhsh, F., Pueyo, J.J., 2019. Uptake and effects of lead and zinc on alfalfa (Medicago sativa L.) seed germination and seedling growth: Role of plant growth promoting bacteria. South African Journal of Botany 124, 573–582.
Yang, Y., Sun, C., Yao, Y., Zhang, Y., Achal, V., 2011. Growth and physiological responses of grape (Vitis vinifera "Combier") to excess zinc. Acta Physiologiae Plantarum 33 (4), 1483–1491.
Yang, S., Ulhassan, Z., Shah, A.M., Khan, A.R., Azhar, W., Hamid, Y., et al., 2021. Salicylic acid underpins silicon in ameliorating chromium toxicity in rice by modulating antioxidant defense, ion homeostasis and cellular ultrastructure. Plant Physiology and Biochemistry 166, 1001–1013.
Zhang, Y., Wang, Y., Ding, Z., Wang, H., Song, L., Jia, S., et al., 2017. Zinc stress affects ionome and metabolome in tea plants. Plant Physiology and Biochemistry 111, 318–328.
Zhang, P., Sun, L., Qin, J., Wan, J., Wang, R., Li, S., et al., 2018. cGMP is involved in Zn tolerance through the modulation of auxin redistribution in root tips. Environmental and Experimental Botany 147, 22–30.

Chapter 5

# Role of zinc for abiotic stress tolerance in plants

Md. Atikur Rahman[1], Md. Riazul Islam[2], Md. Abdul Azim[3], Milan Skalicky[4] and Akbar Hossain[5]

[1]*Department of Soil Science, Spices Research Center, Bangladesh Agricultural Research Institute (BARI), Bogura, Bangladesh,* [2]*Department of Pathology, Regional Spices Research Center, BARI, Magura, Bangladesh,* [3]*Biotechnology Division, Bangladesh Sugarcrop Research Institute, Pabna, Bangladesh,* [4]*Department of Botany and Plant Physiology, Faculty of Agrobiology, Food, and Natural Resources, Czech University of Life Sciences Prague, Prague, Czech Republic,* [5]*Division of Soil Science, Bangladesh Wheat and Maize Research Institute, Dinajpur, Bangladesh*

## 5.1 Introduction

Food, oil, fiber, and other crops that yield byproducts are becoming more and more necessary as the world's population grows (Condon et al., 2004; Morison et al., 2008). Security of food supply and nutrition are essential for the growing population's health (Das et al., 2020; Bhatt et al., 2020; Praharaj et al., 2021). However, environmental stressors, in contrast to this rising demand, reduce productivity and nutritional quality. Abiotic stressors can reduce crop yields by more than half in some cases (Wang et al., 2003). In comparison to earlier decades, models of climate prediction show that many abiotic conditions, including salt, drought, extreme heat, flooding, Ultra violate radiation, ozone layer and metal toxicity have grown significantly (Bates et al., 2008; Mittler and Blumwald, 2010). During their life cycles, plants undergo a range of physiological and metabolic alterations when they are stressed, from germination to harvest. Germination inhibition, reduced growth, curtailed photosynthesis, lesser biomass absorption, imbalance of water, poor uptake of nutrients and ultimately reduced quality and yield are all examples of these alterations (Hasanuzzaman et al., 2012a,b, 2013a,b,c,d, 2014a,b, 2015, 2016).

The concentration of Zn element in grain is highly dependent on the remobilization from vegetative tissues (Figs. 5.1 and 5.2) (Zaman et al., 2018; Guo et al., 2009) under unfavorable field conditions (Kutman et al., 2010;

**Zinc in Plants. DOI: https://doi.org/10.1016/B978-0-323-91314-0.00007-7**
 

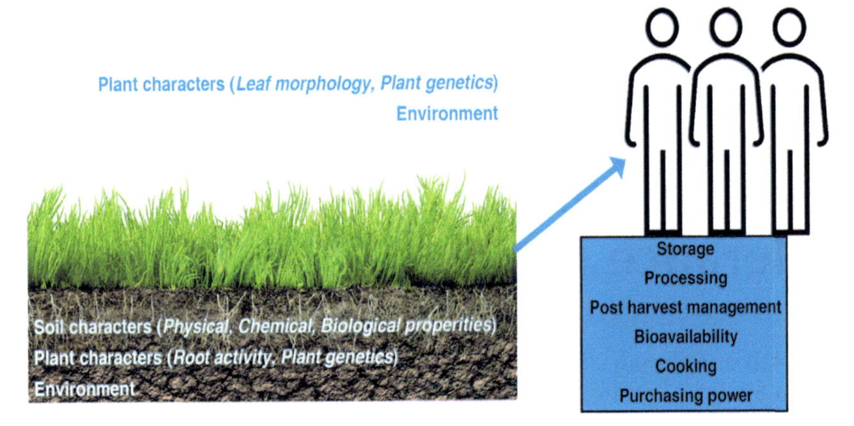


**FIGURE 5.1** Movement of Zn from soil to plants to the human body.

**FIGURE 5.2** Uptake, translocation and remobilization of Zn in plants (Praharaj et al., 2021).

Pearson and Rengel, 1994). However, unless there is enough zinc in the soil that can be absorbed, plants' ability to derive enough important micronutrients, including zinc, from the soil will not be sufficient (Cakmak, 2008). Therefore, when provided with adequate plant-available zinc, a genotype that is suited under favorable climatic conditions is capable of efficiently absorbing and

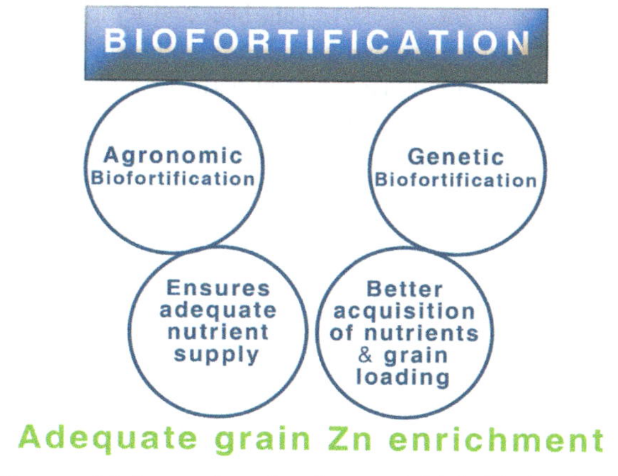


**FIGURE 5.3** The synergistic impact of genetic and agronomic biofortification (Praharaj et al., 2021).

translocating zinc towards the grain, and this should yield the greatest possible outcome (Fig. 5.3) (Malesh et al., 2016).

Understanding plants responses in abiotic stresses is critical for creating strategies for enhancing the tolerance of plants in stress conditions which is a prerequisite for attaining food security globally (Condon et al., 2004; Morison et al., 2008). Antioxidants, osmoprotectants, hormones and many more critical metabolites are changed in reply to abiotic stresses and these metabolites play important roles in abiotic stress defense. Traditional crop development procedures are time-consuming, costly and occasionally ineffective and might result in decrease crop production due to evolution or genetic degradation.

Exogenous phyto-protectants such as micronutrients or trace elements are effective methods for increasing the tolerance of plants from abiotic stress resulting in enhanced productivity and quality (Hasanuzzaman et al., 2011a,b, 2013a,b,c; Nahar et al., 2015a). The trace elements nutrients are those which are needed in minute amounts for a variety of physiological processes. These ingredients help to promote growth but sometimes they are species and conditions specific. Copper (Cu), manganese (Mn), iron (Fe), zinc (Zn), chromium (Cr), molybdenum (Mo), cobalt (Co), selenium (Se), nickel (Ni), silicon (Si) and other micronutrients or trace elements may be present in plants. Plants benefit from micronutrients not merely in terms of processes related to physiology and growth, but for stress resistance, indeed. Among them, exogenous Zn, for example, enhanced tolerance of cadmium (Cd) and boosted dry-matter buildup by lowering membrane damage and Cd uptake caused by Cd (Wu and Zhang, 2002) as a heavy metal. Zn supplementation decreased sodium absorption, increased antioxidants and decreased salt-induced peroxidation of lipids and leakage of electrolytes (Aktaş et al., 2007; Tavallali et al., 2010). Zn improved seed output and weight of 1000 kernels in drought-stricken areas (Monjezi et al., 2013). Plants are recurrently subjected to diverse environmental challenges including drought, excessive salt, heat and temperature fluctuations. Abiotic stress limits growth, resulting in lower yield and severe agricultural losses around the world. Salinity and drought harm 10% or more of agricultural land resulting in a reduction (50%) in mean yields of essential crops around the world (Bray, 2000). Susceptibility or tolerance to these strains is likewise treated as a complex process, for example, stress can distress various phases of the development of a plant and many stresses can influence at the same time (Chinnusamy et al., 2004). The functions related to the physiology of trace elements in providing tolerance against abiotic stress, on the other hand, are currently being investigated. Consequently, substantial experiments based on the mechanisms of adaptation and abiotic stress tolerance were conducted. This chapter primarily discusses Zn's functions as a trace element in plants' responses to abiotic stress tolerance and focus on recent scientific discoveries.

## 5.2 Essence of Zn for completing the life cycle of plants

Plants can't grow without zinc (Zn) where the first effect of Zn shortage is lower amounts of cell RNA which causes protein synthesis to stop resulting in an accumulation of free amino acids (Malakouti, 2007). The process of producing photosynthetic material (the source) and the process of storing the photosynthesized material in storage tissues (the sink) are two important elements of plant physiology. Zn has a favorable effect on both the source and the sink. Plants use Zn in a variety of ways. It is vital for the production and breakdown of carbohydrates, proteins, lipids and nucleic acids and as a structural component of numerous enzymes related to energy and metabolism. It also plays a significant influence in the genetic arena (Oliver, 1997). Carbonic anhydrase ($CO_2$ transport in photosynthesis), ribulose biphosphate carboxylase (starch formation), RNA polymerase (protein synthesis), superoxide dismutase (converting superoxide radicals to hydrogen peroxide and water) and several dehydrogenases such as alcohol dehydrogenase and glutamic dehydrogenase are all enzymes that contain zinc (Alloway, 2004). According to Marschner (1995), Zn's metabolic roles are based on its great proclivity for forming tetrahedral multiplexes with *O*-, *N*- and especially ligands of S and it accordingly serves as both functional and mechanical components of enzyme processes. Even though more than 70 metalloenzymes containing Zn have been discovered, they only account for a minor part of total Zn in plants. Based on the categories of plant and the level of the deficit, a Zn deficit may reduce net photosynthesis by 50%–70%. This decreased photosynthetic efficiency could be attributed at least in part to a reduction in the action of the enzyme carbonic anhydrase, particularly in the $C_4$ plants. Zn deficiency has an unfavorable effect on enzymes involved in the production of sucrose such as aldolase (Malakouti, 2007). The reduced activity of sucrose synthetase causes a decrease in sucrose levels in sugar beet and maize. In a Zn-deficient plant, the amount of protein is considerably reduced but the composition is essentially unaltered. The concentration of free amino acids in Zn-deficient leaves was 6.5 times higher than in controls but after 3 days, it declined and the protein content increased (Malakouti, 2007). The reduction in RNA and the distortion and decrease of ribosomes are thought to be the mechanisms through which Zn shortage affects protein production. Zn deficiency was discovered to drastically diminish the amount of RNA and the number of unbound ribosomes in the meristem of rice seedlings (Malakouti, 2007). Zn is obligatory owing to the function of the RNA polymerase enzyme where it protects RNA related to ribosome against ribonuclease attack. In higher plants, elevated levels of ribonuclease activity are a common symptom of Zn shortage (Malakouti, 2007). As a result, a significant fall in the level of RNA is the first causative effect of Zn shortage. The decrease in RNA can, however, occur before the increase in ribonuclease activity. The role of Zn in protein synthesis suggests that meristematic tissue where cell division and nucleic acid synthesis occur, requires relatively high Zn concentrations. Zn is also known to be essential for membrane maintenance via interactions with phospholipids and membrane protein sulphydryl groups (Malakouti, 2007). Some believe that the loss of membrane integrity is the first metabolic alteration produced by Zn deficiency (Alloway, 2004). Many researches have reported on the effect of Zn in increasing the protein content of cereals (Malakouti, 2000). The grain's average protein content was found to be 13.05%. The rate of RNA polymerase enzyme drops dramatically with Zn shortage, as does the rate of amino acid transport. The rate of RNA breakdown as well as the activity of the RNAase increases culminating in a significant decrease in the protein synthesis rate. Zn is also required for the formation of ribosomes (Marschner, 1995). The magnitude and importance of Zn shortage in soils, plants, foods and humans were raised by Zn deficit in cereals (Cakmak, 2008; Malakouti et al., 2006). Soil will be treated as short in Zn when DTPA-extractable concentration is curtailed to 1.00 mg/kg of soil. Zn deficiency can be caused by soil calcareousness, high pH, low organic matter, dryness, excessive bicarbonate in irrigation water, misuse of P fertilizer and a lack of Zn fertilizer application (Malakouti, 2007). Zn deficiency in soils is more widespread than other micronutrient deficits, according to research (Malakouti, 1998).

It is difficult to understand the physiological underpinnings of micronutrient accumulation in grains in order to effectively implement biofortification. The concentration of micronutrient in the edible part of the crop depends on the process of nutrient absorption, translocation and redistribution within the plant. Accumulation of micronutrients in a sufficient amounts are regulated by internal mechanisms of plant (Welch and Graham, 2004).

## 5.3 Consequences of abiotic stresses in plants

Plants, unlike mammals, are sessile organisms that must develop in the same habitat while battling abiotic stress from numerous environmental conditions.

Due to climate change, plants are frequently exposed to abiotic stresses such as drought, salt, extreme heat, metal toxicity, air pollution, flooding, chilling etc. Due to the rapid change of climate, these occurrences are becoming increasingly prevalent. These stressors damage plants individually or combinedly and subsequently, it is frequently erratic and more havoc than anyone anticipated. Those impacts vary widely instead of the type, period and intensity of stressor, plant species and genotypes (Pandey, 2015). As a result, generalizing the common effect in plants by abiotic stress is extremely difficult. Abiotic stressors in crops, on the other hand, cause loss in seed germination, inhibition of growth, photosynthesis and energy assimilation disruption, metabolic disparity, homeostasis in nutrition and uptake of nutrients, crop yield lessening and degradation of crop quality. One of the most familiar results of abiotic stress on plants is oxidative stress, which is covered in detail later in this chapter.

One of the most destructive abiotic pressures lowering land area and agricultural production is salinity/salt stress (Yamaguchi and Blumwald, 2005; Shahbaz and Ashraf, 2013). Soil salinity hinders the germination, seedling vigor and yield of crops, (Munns and Tester, 2008). Genetic constitution, development phase, stress concentration and stress extent are some of the most essential elements that influence the plant's reaction to salt stress. Ionic specificity, osmotic stress, nutritional and problems related to hormone-changed metabolic and physiological processes and eventually oxidative damage happen when crops are exposed to excessive salt (Munns, 2002; Huang et al., 2007). Salinity stress in plants causes disruption of cellular membranes, reduction in photosynthesis, creation of various toxic metabolites, nutritional inaccessibility and ultimately death of plants (Mahajan and Tuteja, 2005). Salt stress can affect plants in three main ways: it can cause an ionic imbalance within the cell, lower the amount of certain ions, especially potassium ion ($K^+$) and calcium ion ($Ca^{2+}$), and eventually cause sodium ($Na^+$) and chloride ($Cl^-$) ion toxicity. It can also develop osmotic stress through a reduction in soil water potential. Protein synthesis is hampered by low $K^+$ concentrations because elevated $Na^+$ inhibits uptake of potassium ion, which is essential for protein synthesis since it serves as a holding material under salt stress (Blaha et al., 2000). Many enzymatic activities are disrupted by an amplified ratio of $Na^+$ - $K^+$ and decreased $K^+$ obtainability. Salt stress is caused by two primary strains known as osmotic and ionic stressors. Growth inhibition is a general response of crop to salt stress that is induced by changes in solvable salt concentration and the soil solutions' osmotic potential (Tavakkoli et al., 2011). Because of the acute osmotic stress that occurs in the rhizosphere, salt hinders water uptake and reduces cell enlargement and lateral growth (Munns and Tester, 2008). In comparison to the rhizosphere, shoot development is significantly hampered by salt-induced osmotic stress. This could be because leaf area development is slower and less than root area development (Munns and Tester, 2008).

Chlorotic toxicity is caused by a reduction in chlorophyll (chl) production caused by the Cl ion. Excess Cl accumulation also results in a significant loss in growth and water consumption efficiency. One of the most important stages of a plant's life cycle is germination since it establishes the ultimate plant density. Salinity has been shown to reduce seedling germination and establishment which is one of the key issues in salt-affected areas. Salinity directly inhibits seed germination by postponing germination, reducing germination percentages and restricting plant growth (Rahman et al., 2000). While seedlings subjected to modest levels of salt stress became dormant, those exposed to extended salinity were unable to germinate (Khan et al., 2006). It was likewise discovered that salt stress inhibited seed germination due to the growth medium's reduced osmotic potential. Salt stress also has an impact on the metabolism of nucleic acid (Gomes-Filho et al., 2008), protein metabolism (Dantas et al., 2007) and food reserve usage of seeds (Othman et al., 2006). Kaveh et al. (2011) steered research using *Solanum lycopersicum* who discovered salinity drastically inhibited germination. An increase in NaCl concentrations resulted in the steady hindrance and decreased germination of *Hordeum secalinum*. It was also discovered that treating seeds with 400 and 500 mM NaCl reduced germination rates by 40% and 38%, respectively (Lombardi and Lupi, 2006). It was found by Bordi (2010) that varied NaCl treatments impaired *Brassica napus* germination percentages significantly. Osmotic stress, ionic imbalance, and decreased seed water intake are a few potential reasons for the germination suppression caused by salt stress that was previously described. Four different rice varieties were evaluated in different salt levels; germination was significantly lower in sensitive varieties than in tolerant varieties (Hasanuzzaman et al., 2009). Plants under salt stress typically experience a slowing of growth. First, the salt causes a reduction in the plant's ability to absorb water, which in turn causes the plant's growth rate to drop off quickly. There are two phases to the growth reduction brought on by salt. The initial stage of growth inhibition is produced by osmotic stressor whereas the second phase is caused by excessive salt accumulation and associated toxicity to crops (Munns, 2002). Henceforth, the second phenomenon reduces photosynthate levels in plants and causes leaf mortality, which has an impact on production. That loss of growth instigated by salt stressor results in delayed flowering and poorer crop production (Munns and Tester, 2008). Furthermore, salinity alters the plant-water relationship, subsequent loss in cell turgor. In addition, a larger negative water potential brought on by elevated salts in the root zone makes it harder to take in water (Munns, 2002). One of the best strategies for osmotically challenged plants to control water loss is stomatal closure (James et al., 2008). The leaf's ability to absorb $CO_2$ is reduced when its stomata close, and the carboxylation reaction is constrained. Consequently, during photosynthesis, less carbohydrates are produced, which has an adverse effect on crop output and plant development.

Reduced photosynthesis and lower levels of chl a, chl b, and carotenoid (Car) pigments were seen in numerous crops under salt stress, including rice (Amirjani, 2011), mustard (Ahmed et al., 2012), and many others (Hasanuzzaman et al., 2012a). Under salt stress plants are not capable to absorb water and nutrients from the soil that is known as physiological dryness, which prevents stomatal closure, photosynthesis, cell division, and leaf growth. Because of the increasing concentration of harmful ions, salt stress promotes early senescence, yellowing and eventual mortality of tissues of older leaves which has a negative impact on photosynthesis (Munns, 2002; Munns and Tester, 2008). Therefore, a notable impact of salt stress is reduced photosynthesis (Leisner et al., 2010; Raziuddin et al., 2011). The photosynthetic process is also hampered by increased toxicity in chloroplasts by $Na^+$ and $K^+$ ions. Salinity has been found to cause an optimistic growth impediment that is linked with a significant reduction in photosynthesis (Fisarakis et al., 2001). Salt stressors can also have an impact on the photophosphorylation and metabolism of carbon (Sudhir and Murthy, 2004). Because of inadequate physiological development and growth, high salt stress significantly reduces crop production. Under high salinity, there was a decrease in seed weight, seed per pod and pod per plant in Vigna radiata (Nahar and Hasanuzzaman, 2009).

Due to the rising scarcity of water, drought stressors have been considered the key barrier to the productivity of crops around the globe. Drought stress significantly reduces plant development, albeit the amount of growth decrease varies depending on the concentration of osmotic stress, stage of the plant and variety (Budak et al., 2013). Because tolerant soybean genotypes can hold more water than susceptible genotypes, the decrease in shoot and root growth is larger in susceptible genotypes of soybean (Hossain et al., 2014). A study on six bedding plants under drought stress found that species differences are significant in plant dry weight, leaf area, and height. Under scarcity of water, the biomass of Petunia × hybrida, *Cineraria maritima* L and *Plumbago auriculata* decreased significantly but decreased area of leaf in Plumbago and plant height decreased in Plumbago and Vinca (Niu et al., 2006). Drought stress affected the growing progression (stem length, leaf number, stem diameter, etc.) of two *Jatropha curcas*, in accordance with Sapeta et al. (2013). Similarly, both genotypes showed drought tolerance after replenishment. Another study found that when two kinds of barley were subjected to stress like drought (50% field capacity), their appearance rate of leaf and area of leaf decreased significantly (Thameur et al., 2012). Under moisture shortage conditions, reduced water intake and supply of energy, decreased activity of enzymes, turgor loss and reduced expansion and division of cells are all treated as explanations for stunted growth (Taiz and Zeiger, 2006; Hasanuzzaman et al., 2014c). The physiological underpinning of crop output namely photosynthesis is considered since it is the most sensitive physiological activity to dryness (Luo et al., 2016). Factors influencing photosynthesis include stomatal opening and closure, water status in leaf tissues, rate of $CO_2$ assimilation, electron transport, $CO_2$ assimilation processes and ribulose bisphosphate (RuBP) production. Photosynthesis is affected by changes in any of these components during a dehydration state (Ahmad et al., 2014; Hasanuzzaman et al., 2014b). Due to stomatal closure, reduced $CO_2$ absorption and increased generation of ROS, photosynthesis in potatoes is reduced when soil moisture is low (Li et al., 2017; Obidiegwu et al., 2015). Under water shortage conditions, compared to the susceptible Trihybrid 321 cultivar, the drought-tolerant Giza 2 cultivar of maize exhibited increased photosynthetic activity, according to a comparative study of two maize cultivars (Moussa and Abdel-Aziz, 2008). According to Ashraf and Harris (2013), frequent reactions that inhibit photosynthesis in water shortage situations include a depletion in photosynthetic pigments, alteration in gas exchange characteristics, damage to photosystems, and reduced activity of enzymes related to photosynthesis. Drought stress reduced chlorophyll concentration and photosynthesis in wheat, according to Marcińska et al. (2013).

Excess water or floods which create submersion/waterlogging among the abiotic stresses is also damaging and even fatal to plants (Nishiuchi et al., 2012). When a plant is fully submerged or stays submerged and is compelled to transition from an aerobic to anaerobic condition, this is known as submergence (Jackson and Ram, 2003; Nishiuchi et al., 2012). Plants under submerged anaerobic conditions are additionally subjected to secondary stressors like poor light intensity, impeded exchange of gas and nutritional deficiency (Ram et al., 1999). Water logging, moreover, is a situation of soil in which surplus water prevents diffusion of gas in the rhizosphere of crops (Setter and Waters, 2003). Plants were subjected to complex changes in various environmental factors as a result of waterlogging or excess water induced through high precipitation, inadequate irrigation and drainage methods and they suffered from oxygen, light and nutrient deficiencies. Because stagnant water restricts gas transport and soil microorganisms quickly consume oxygen within hours, oxygen deprivation and anoxia are the results. One of the initial responses of waterlogging caused by hypoxia or anoxia is the stomatal closure to prevent loss of water which limits breathing (García-Sánchez et al., 2007). As a outcome of the reduced respiration, the photosynthetic machinery is down regulated, lowering the chlorophyll content (Damanik et al., 2010). According to Arbona et al. (2008), waterlogging and submergence affect the amounts of photosynthetic pigment by reducing gas exchange, stomatal conductance, and leaf water potential. Moreover, the plant accumulates volatile ethylene as a result of oxygen deprivation, which when submerged has negative effects as well as signaling effects (Steffens et al., 2011). Hypoxia induces oxidative stress because it produces too many reactive oxygen species (ROS), albeit the exact mechanism is unknown (Kumutha et al., 2009; Sairam et al., 2011). Most crops are damaged by too much water, except for rice and a few other aquatic plants (Bailey-Serres and Colmer, 2014). Prompter pseudostem elongation and yellowing of older leaves were evident damage indicators in rice seedlings under long-term waterlogging conditions (Jackson and Ram, 2003). After 7 days of submergence, Banerjee et al. (2015) noticed biochemical changes in rice seedlings, including an increase in protein oxidation, phenol content, and flavonoid content and a decrease in lipid peroxidation. Because of a massive oxygen burst and higher light intensity than when immersed, rice seedlings are also negatively impacted by oxidative stress in post-submergence situations. According to Ella et al. (2003) 7 days of submergence in the recovery stage, rice seedlings showed reduced chlorophyll content and increased lipid peroxidation. A brief (24-hour) waterlogging event resulted in a temporary N deficit that reduced wheat development and yield (Robertson et al., 2009). At the three-leaf stage, summer maize development and yield parameters declined in field conditions following 72 hours of waterlogging (Ren et al., 2014). Due to the activation of DPI-sensitive NADPH oxidase, Kumutha et al. (2009) observed enhanced lipid peroxidation and growth inhibition in pigeon pea seedlings under 4 days of waterlogging. Smaller, chlorotic leaves that withered earlier were another effect of it. Kumar et al. (2013) examined the morphological and physiological reactions of tolerant and sensitive varieties of V. radiata after 3, 6, and 9 days of waterlogging. More growth inhibition and yield loss were seen in sensitive cultivars (Pusa Baisakhi & MH−1 K−24) than in tolerant cultivars (T44 & MH−96−1) due to poor leaf respiration rate maintenance and decreased adventitious root development. Waterlogging and post-waterlogging circumstances reduced the amount of photosynthesis and produced oxidative damage in citrus seedlings (Hossain et al., 2009a,b).

In the last few decades, rising global temperatures have caused substantial loss of crops in numerous regions around the globe (Long and Ort, 2010). By 2100, global temperatures are anticipated to rise by 2.5°C−5.4°C (Ciscar, 2012). The intensity, length and temperature rate rise are all grave factors in instigating plant mortality (Wahid et al., 2007). High temperatures can cause decreased biomass and germination as well as rising chlorosis, tillering, necrosis, premature mortality, fruit senescence and early senescence of floral buds (HT, Wahid et al., 2007). Heat stress results in physiological disorders and structural collapse in seeds which lowers germination and vigor as well as the emergence and establishment of seedlings (Akman, 2009). High temperatures inhibited the germination of rice seeds by limiting a collection of proteins associated in the metabolism of methionine, amino acid biosynthesis, energy metabolism, reserve breakdown, and protein folding (Liu et al., 2014a,b). Protein synthesis disruption causes maize germination to decrease at 37°C, but above 45°C, coleoptile growth completely ceases (Rilkey, 1981; Akman, 2009). Elevating the soil temperature over 45°C in sorghum causes epicotyl emergence and germination failure (Peacock et al., 1990). Seedling emergence in tomatoes is halted once the temperature reaches 30°C (Camejo et al., 2005). Germination was 90% in rice in a temperature span of 15°C and 37°C. At 8°C and 45°C, however, no germination occurred (Hartmann and Kester, 1975). At temperatures of 20°C, 25°C, 30°C, 35°C and 40°C, in *Phacelia tanacetifolia* seed germination was decreased by 2.7%, 5.8%, 84.0%, 89.0% and 91.5%, respectively, compared to that of 15°C (optimum for germination) (Tiryaki and Keles, 2012). Germination in Phacelia seeds is decreased under extreme temperatures because of a lack of plasma membrane rearrangement, maintenance of cytoplasmic ion channel and restriction of the reactivation of metabolic pathways for the mobilization of ions from a bound to a free state (Pirovano et al., 1997). High temperatures disrupt the source-sink balance, impede assimilate translocation, damage the soil-plant-water interaction, intake of nutrients, photosynthetic activities, transpiration, respiration and inhibit growth and development (Prasad et al., 2011). One of the most common signs of HT damage in plants is a reduction in water content and an accumulation of osmoprotectant molecules. Trifoliate orange seedlings water content was lowered by the rise in temperature (Fu et al., 2014). In transgenic plants grown under HT conditions, a rise in Pro level was observed (Cvikrová et al., 2013). In wheat, high temperatures (day/night, 31/18°C and 34/22°C) were observed to reduce harvest index and total biomass (Prasad et al., 2011). Thermolability is a property of enzymes. Most enzymes lose their catalytic capabilities as a result of HT which causes them to denature. Under HT, protein and enzyme biosynthesis is impeded. Severe HT induces cell death by causing total protein denaturation, function of membrane and enzymes (Allakhverdiev et al., 2008). Damages caused by HT include structural thylakoids' disarray and enlargement of lamellae of stroma (Zhang et al., 2005). Wheat exposed to heat stress (34/22°C, and 31/18°C night/day) lost 19% of its chlorophyll concentration (Prasad et al., 2011). The photosynthesis rate and transpiration decrease as the atmospheric temperature rises from 22°C to 32°C (Zhang et al., 2010). Under HT, the solubility of oxygen increases greater than that of carbon dioxide. At HT, oxygen concentration rises resulting in RuBP oxygenation which is referred to as photorespiration. Gross photosynthesis is slowed by HT stress while respiration and photorespiration increase resulting in a decrease in net photosynthesis (Allakhverdiev et al., 2008; Mittler et al., 2012). Plants begin to deplete carbohydrate reserves as heat stress remains to reduce net photosynthesis resulting in growth decrease. During HT stress, photosynthesis is inhibited due to a decrease in RuBisCo activity (Allakhverdiev et al., 2008). HT inhibits the photosystem II (PS II) functioning by reducing electron transport removing foreign proteins and releasing magnesium and calcium ions from the binding site (Mittler et al., 2012; Wahid et al., 2007; Zlatev and Lidon, 2012). Under HT, excess generation of singlet oxygen causes damage to proteins of D1 and D2 (Yoshioka et al., 2006). Heat stress inhibits starch produced through photosynthetic assimilation causing cereals or legume seed growth to be hampered (Stone and Nicolas, 1994). Heat stress has long been known to be particularly sensitive to reproductive development This is the primary reason for plant output loss (Thakur et al., 2010). In comparison to the pistil or the female gametophyte, the gametophyte in males is more susceptible to HT (Hedhly, 2011). Successful fertilization is severely impeded in the flowers of tomatoes due to reduced viability of pollen and anthers which further impedes embryo development (Barnabás et al., 2008). Pollen viability was diminished as soluble carbohydrates found in pollen grains and anther were reduced by high temperatures (Ismail and Hall, 1999). Another development is impeded in barley and Arabidopsis in heat stress (30°C–35°C) due to cell expansion seizure and aberrant development of mitochondria, chloroplasts, and vacuoles (Sakata et al., 2010). Leaf area and photosynthesis decrease as water use efficiency declines throughout wheat reproductive development. Grain weight and kernel sugar content as a result dropped (Shah and Paulsen, 2003) affecting the nutritional quality of flour (Hedhly et al., 2009). HT lowers water movement and uptake of nutrients by affecting xylem and phloem loading (Taiz and Zeiger, 2006). The transpiration rate of tomato plants was dramatically raised by high temperatures (4 days, 38°C) (Cheng et al., 2009). Throughout the 21 century, rising temperatures may cause yield reductions of 2.5%–10% in certain agronomic species (Hatfield et al., 2011). Increases of 1°C in growth periods have been anticipated to result in a 6.7%–10% drop in the yield of rice (Peng et al., 2004; Lyman et al., 2013) and a 10% drop in wheat output (You et al., 2009). Temperatures exceeding 35°C reduce the vitality of maize pollen (Dupuis and Dumas, 1990). In tropical humid climates, high temperatures 30°C or more reduce the production of seed in soybean crops (Lindsey and Thomson, 2012) and also in common bean plants (Porch, 2006). When the daytime temperature climbed by 30°C–35°C, reduced seed set on soybean plants that are female-fertile but male-sterile (Wiebbecke et al., 2012). Citrus sinensis fruit dropped when temperatures rose above 30°C (Cole and McCloud, 1985). According to Prasad et al. (2011), HT (31/18°C and 34/22°C, day/night) significantly decreased grain yield in wheat by reducing grain numbers by 56% and individual grain weight by 25%. This was in comparison to the control temperature of 24/14°C. A 54%–64% decline in sorghum seed set was associated with poor fertilization, smaller panicles, and damaged floral primordia (Jain et al., 2007).

Freezing stress is defined as damage brought on by the formation of ice crystals inside plant tissues while chilling stress is defined as damage brought on by low temperatures without the formation of ice crystals. The ability of different plant species to withstand cold or freezing temperatures varies. By interfering with different aspects of plant growth and development, the cold reduces crop yield (Sanghera et al., 2011). Early plant growth is negatively impacted by chilling stress. Rice seed germination requires a temperature range of 20°C–35°C. Rice seeds do not germinate below 10°C, according to Yoshida (1981), who also stated that this is the minimum critical temperature. For *B. napus*, 50% germination required 3 days at 8°C, but 13 days at 2°C (Angadi et al., 2000). Compared to temperatures below 20°C, *T*riticum *aestivum* plants exhibited the highest seed germination and vigor index at 20°C–30°C (Buriro et al., 2011). According to Krzyzanowski and Delouche (2011), Gossypium hirsutum germination was hindered and delayed at low temperatures (LT) of less than 20°C. Temperatures below 8°C–10°C significantly decreased *T. aestivum* germination (Zabihi-e-Mahmoodabad et al., 2011). According to Nahar et al. (2012), cold temperatures during the vegetative stage of a plant limit the growth of seedlings and cause abnormal phenotypes such as stem discolouration, yellowing or whitening of leaf, white patches on leaf, wilting and reduced tillering. When mungbean seedlings were exposed to a low temperature (6°C), their development and dry weight reduced. These seedlings' phenotypic appearance was also abnormal (Nahar et al., 2015b). Nahar et al. reported that rice displayed symptoms of damage from cold (11°C) (2009). Limited development, foliage chlorosis, an unequal amount of tillers and indications of deformed and discolored grain were common in rice plants afflicted by chill. Lower temperatures have an impact on the growth duration, accumulation of biomass, harvested index, total amount of seeds and weight of soybeans, according to Calviño et al. (2003). Meiosis, tapetal hypertrophy and the growth process of male gametophytes are all affected under cooling stress. In cold climates, pollen tubes distort, anther protein deteriorates, and pollen grains become twisted and small, rendering the pollen sterile. chill-induced suppression of male and female gametophyte growth, mostly because of smaller reproductive organs and abnormal embryo sac structures, which prevent fertilization and allow fruit and seed to continue developing (Jiang et al., 2002). According to studies by Jiang et al. (2002) and Thakur et al. (2010), rice and other grain crops impacted by chilling exhibit delayed blooming, bud abscission, sterile flowers, aborted embryos, and empty grain. Low temperatures (between 2°C and 5.5°C) decreased the amount of floral buds in Simmondsia chinensis (Nelson et al., 1993). Chinese cabbage has early blooming when the temperature is lowered (Kalisz and Cebula, 2001). Rice genotypes under 11°C freezing stress altered the mode of panicle initiation, emergence, delayed heading, and sterile and malformed spikelets (Nahar et al., 2017). Different canola genotypes displayed delayed blossoming, delayed flowering, and a higher rate of pod abortion and pollen ejection under freezing stress (Miller et al., 2001). Freezing temperatures cause damage to precipitation, membrane permeability, protein denaturation and alteration of protein-lipid structure (Wang and Li, 2006). The membrane undergoes a solid gel phase transformation as a result of the chilling stress. Ion leakage, unequal anion/cation exchange, disruption of anions and cations homeostasis and disruption of osmosis and diffusion processes are the fates of chill-affected plants (Farooq et al., 2009a,b). Photosynthesis is inhibited by cold temperatures because the thylakoid membrane and chloroplast structure are disrupted. The chilling temperature impedes carbon cycle metabolism, stomatal conductance and electron transport. Chill-induced damage is thought to be primarily directed at the PS II. PS II's quantum efficiency is greatly reduced as the temperature is lowered. The photophosphorylation is confined and RuBisCO regeneration is restricted by the chilling temperature (Allen and Ort, 2001). After 14 days at LT (15/10°C), the total chlorophyll content of rice was reduced by 50% (Aghaee et al., 2011). In comparison to control seedlings, the total chlorophyll (a + b) of mungbean plants dropped by 32 and 38%, respectively, after 2 and 3 days of exposure to a chilling temperature of 6°C (Nahar et al., 2015b). According to Yordanova and Popova (2007), wheat plants that were chilled to 3°C for 48 or 72 hours showed reduced rates of $CO_2$ assimilation, transpiration and chlorosis. Cell freezing components, water and solutes can also cause dehydration stress (Yadav, 2010). Ice formation begins in intracellular areas at temperatures below 0°C which is also the primary cause of solute transport obstruction (Thomashow, 1999; Yadav, 2010). Dehydration and osmotic stress are caused by compromising the cold temperature of the root system which lowers nutrient and water intake (Chinnusamy et al., 2007). Chill-induced dehydration stress has been linked to root system damage and stomatal closure (Yadav, 2010). Chill-induced damage was seen in cucumber root cortical cells. The endoplasmic reticulum was also injured by the cold temperature which increased cytoplasm density (Lee et al., 2002). Chill temperature causes reduced hydraulic conductivity of root, turgor potentials and leaf water in sensitive plants (Aroca et al., 2003). Plant stress metabolic activities are altered during chill-M-induced dehydration including the buildup of isozymes and enzymes (Hurry et al., 1994). Low temperature-induced yield drop is a common occurrence in plants because of its detrimental effects on the vegetative and reproductive stages (Nahar et al., 2017; Kalbarczyk, 2009; Riaz-ud-din et al., 2010). Chilling temperature in rice decreased flower abortion, pollen and ovule sterility, skewed fertilization, poor seed filling, and seed setting (Thakur et al., 2010). In temperate zones, cooling the temperature lowered yield by 30%–40%, according to another study (Andaya and Mackill, 2003). Because of the chilling stress imposed by late sowing, yields of BRRI dhan31 and BRRI dhan46 rice types were reduced by 16% and 37%, respectively (Nahar et al., 2017). In *B. rapa* and *B. napus*, low temperatures caused abnormal seed shape. Low temperatures caused red-brown pigmentation, folded seed, white patches, shriveled seed, and white reticulation by reducing the diameter and causing white reticulation, shriveled seed and white patches, red-brown pigmentation and folded seed (Angadi et al., 2000). Rasolohery et al. (2008) found that the weight of seed in Glycine *max* crops dropped by 5% when uncovered to a cool temperature of 23/13°C (night/day).

With rapid industrialization, the population of the world is growing every day. As a result, a large quantity of harmful metals is released into the environment (Sarma et al., 2012). Increased hazardous metal contamination is growing in importance in the contemporary world (Sun et al., 2005). Clear losses in farm productivity are pretty more widespread as a result of toxic metal poisoning in the soil, posing a major hazard to human health and animals (Sharma and Dubey, 2007; Sharma and Dietz, 2009). Toxic metals can quickly influence organisms' all development processes of energy due to their high reactivity. Thus, one of the most significant abiotic stressors having a negative influence on plant and animal health is metal toxicity (Maksymiec, 2007). Plants may absorb too many dangerous metals, which could lead to a range of physiological and biochemical reactions and could interfere with normal growth by interfering with absorption, translocation, or synthesis processes (Hasanuzzaman and Fujita, 2012a,b). Toxic metal/metalloids severely damaged the plant from germination to seed production. As the germination of seeds marks the beginning of plant life, they are very susceptible to a wide range of environmental factors. *Salicornia brachiata* seed germination was negatively impacted by varying dosages of Cd, Ni, and As (Sharma et al., 2011). Furthermore, a number of studies found that, with the exception of Cd, which interfered with both processes at the same dose, heavy metals (Hg, Pb, Cu, Zn, etc.) had a stronger effect on seedling growth than on seed germination (Li et al., 2005). A wide range of dangerous metals and metalloids can induce oxidative stress in plants. Reactive oxygen species (ROS) are the primary plant response to metal stress. Many metals can directly produce ROS through Haber-Weiss reactions. Moreover, the indirect effects of heavy metal toxicity in plants may give rise to ROS overproduction and oxidative stress (Wojtaszek, 1997; Mithöfer et al., 2004). The most detrimental effect of heavy metal stress on cells is the direct induction of bio-membrane deterioration by lipid peroxidation. Hasanuzzaman et al. reported that 48 hours of Cd stress (1 mM $CdCl_2$) increased the levels of malondialdehyde (MDA) by 134% and $H_2O_2$ by 60% in rapeseed seedlings (2012b). Numerous investigations have revealed that plants' antioxidant enzyme activity is altered by harmful metals and/or metalloids. By interfering with enzymatic catalase (CAT) and non-enzymatic (AsA and GSH), ascorbate peroxidase (APX), glutathione reductase, superoxide dismutase (SOD) and $H_2O_2$ content, Cd stress in mungbean plants decreased growth, damaged chloroplasts, decreased leaf RWC, changed proline (Pro), and increased oxidative damage (lipid peroxidation, $H_2O_2$ content, $O_2$ generation rate). Similar physiological alterations were seen in wheat plants under arsenic (As) stress and rice plants under Cd stress (Rahman et al., 2016; Hasanuzzaman and Fujita, 2013).

## 5.4 Role of Zn for alleviating abiotic stresses of plants

In addition, zinc is essential for plant growth and development and for reducing abiotic stress on plants. Thus, to maximize plant uptake of zinc and to

**FIGURE 5.4** Multiple benefits of Zn fertilization on crop and soil.

minimize zinc shortage while also promoting plant stress tolerance, zinc levels at the root-soil interface should be raised (Fig. 5.4).

### 5.4.1 Soil salinity

Salinity is an important environmental stress that significantly hampered agricultural productivity globally by reducing plant growth and development (Ashraf and McNeilly, 2004; Ahanger and Agarwal, 2017). Almost 20% of fertile land on earth is affected by high salinity, which in most cases accelerates urbanization and industrialization (Gupta and Huang, 2014). Salinity stress inhibits the growth of plants by decreasing cell division and enlargement, decreasing photosynthetic efficacy, modifying metabolic processes and also causing osmotic stress, ion toxicity, oxidative stress, other physiological diseases and genotoxicity (Ibrahim et al., 2012; Wani et al., 2013; Yan et al., 2013; Anjum et al., 2015; Ahmad et al., 2016; Hussein et al., 2017).

Overproduction of reactive oxygen species (ROS) brought on by salinity stress damages lipids, DNA and proteins which can result in cell death (Ahmad et al., 2010; Gill and Tuteja, 2010; Anjum et al., 2015). Plants have inherent oxidative defense mechanisms (enzymatic and non-enzymatic antioxidants) to counteract ROS generated by salt. The detrimental effects of NaCl on growth and biomass yield may have been caused by osmotic stress and insufficient absorption of essential components (Hashem et al., 2014). To extract the maximum economic yield from stressed crops, many strategies are being employed. For example, adding enough mineral macro and micronutrient supplements reduces the harmful effects of stressors, promoting growth and increased productivity (Ahanger et al., 2015; Ahmad et al., 2015; Siddiqui et al., 2015; Ahanger and Agarwal, 2017). Zn treatment dramatically reduces detrimental effects of salt stress by absorption and partitioning of necessary mineral elements (Weisany et al., 2014). Zinc is a crucial structural motif for DNA binding and it is necessary for enzyme activity, protein stability and membrane stabilization (Tavallali et al., 2009). Zinc is required as a cofactor for more than 300 enzymes and related proteins for the metabolism of glucose, nucleic acids and proteins synthesis and also cell division (Osredkar and Sustar, 2011; Singh et al., 2015). Moreover, Zn forms the potent antioxidant Cu/Zn-SOD, which might help to manage stress (Ahmad et al., 2010). By boosting the synthesis of indole acetic acid, or natural auxin, plants can grow faster (Ali and Mahmoud, 2013), accumulate phospholipids (Jiang et al., 2014), maintain the structural integrity of their membranes (Weisany et al., 2014), enhance protein synthesis (Ebrahimian and Bybordi, 2011) and scavenge free oxygen radicals (Weisany et al., 2011; Siddiqui et al., 2015; Yousuf et al., 2016a,b). Zn supplementation helped to keep *B*rassica *juncea*'s tissue water content stable even in saline circumstances (Tavallali et al., 2009). Zn finger $C_2H_2$-type proteins increase salt tolerance by elevating osmotic adjustment substance concentrations (Xu et al., 2008; Sun et al., 2010) and increasing the number of soluble sugars and free proline in rice, as well as up-regulating a number of genes in sweet potatoes that are involved in the production of osmotic chemicals to increase salt tolerance (Wang et al., 2016).

## 5.4.2 Drought

These days, the severe abiotic stress of drought has a disastrous effect on agricultural production (Qados, 2011). Climate change, rising atmospheric temperatures and erratic rainfall are making it increasingly common (Whitmore, 2000). Drought stress significantly reduces plant development, albeit the amount of growth decrease varies depending on the gradation of osmotic stress, plant stage and variety (Fahad et al., 2017). Drought stress lowers crop productivity because it affects gas exchange rate, leaf hydration status, and water uptake capacity (Saud et al., 2016). Moreover, it induces withering by lowering stomatal conductance, which raises temperature of leaf (Farooq et al., 2017; Sehgal et al., 2017). Additional consequences of water scarcity include impairments in plant photosynthetic efficiency and membrane permeability (Awasthi et al., 2014; Rahbarian et al., 2011; Samarah et al., 2009; Gunes et al., 2006). By turning on their antioxidant defense mechanism, which scavenges reactive oxygen species (ROS) and causes a variety of physiochemical processes, plants mitigate the effects of drought (Jubany-Mar et al., 2010). ROS damages macromolecules including proteins and lipids as well as plant cell membranes (Wu et al., 2014). Plant drought tolerance ability is determined by a variety of physiological and biochemical traits, including membrane permeability, relative water content (RWC), osmotic adjustment, stomatal conductance, electrolytic leakage, structural catalytic and co-catalytic components, phytohormones and other characteristics (Bajji et al., 2002; Sánchez-Blanco et al., 2002; Liu et al., 2000; DaCosta and Huang, 2006; Sharma et al., 1995). Drought stress inhibited seed germination and seedling emergence during the early stages of plant growth (Sedghi et al., 2013). Moreover, plumules get shorter and lose dry weight as a result. Zn (priming) boosts plumule length and dry weight during drought conditions by promoting the synthesis of IAA and $GA_3$ (Cakmak, 2008). Zinc promotes the preservation of WUE, antioxidant activity, photosynthetic efficiency, and membrane permeability (Karam et al., 2007; Bagci et al., 2007; Babaeian et al., 2010). Zn treatment leads to increased growth and output under dry conditions by increasing leaf area and chlorophyll content, photosynthetic pigments and stomatal conductance (Karim and Rahman, 2015). Zinc increases carotenoid and relative water content (RWC), which lowers electrolyte loss and water consumption in dry conditions Khan et al. (2016). One of the main processes linked to plants' ability to retain water and maintain higher turgor potential during droughts is osmotic adjustment (DaCosta and Huang, 2006). Zinc is a structural catalytic and co-catalytic component of about 300 enzymes, including aldolase, fructose 1, 6 bisphosphatase, carboxypeptidase, alcohol dehydrogenase, and carbonic anhydrase (Ahanger et al., 2016). Reactive oxygen species (ROS) overproduction during drought stress damages and kills cell membranes (Wu et al., 2015). According to Tsonev and Cebola Lidon (2012), zinc binds to histidine and cysteine to prevent the buildup of reactive oxygen species (ROS) and shield the cell membrane. Additionally, it enhances the activity of the antioxidant system, assisting in the reduction of drought-induced oxidative stress (Wu et al., 2015; Thounaojam et al., 2014). Zn supplementation raises the levels of ascorbate (ASC), total flavonoid content (TFC), and total phenolic content (TPC) under drought stress, which shields cells from reactive oxygen species (ROS) (Tavallali et al., 2010; Song et al., 2015).

### 5.4.3 Waterlogging

Although plants need enough water to thrive correctly, waterlogging stress can occur very fast when the soil becomes saturated or even super-saturated. Reduced root respiration and the buildup of harmful chemicals during waterlogging stress negatively impact on both vegetative and reproductive stages, resulting severe yield loss (Hirabayashi et al., 2013; Xu et al., 2014; Herzog et al., 2016; Tian et al., 2019; Ding et al., 2020; Zhou et al., 2020). Waterlogging causes the stomata of leaves to close, and leaf senescence, yellowing, and chlorophyll degradation reduce the leaves' capacity to absorb light, which lowers the photosynthetic rate (Kuai et al., 2014; Yan et al., 2018). Because waterlogging eliminates air from soil pores, it prevents gas exchange between soil and atmosphere. However, water has a far slower rate of oxygen diffusion than air. Because of this, there is very little oxygen available in wet soil, which lowers root respiration, root activity and energy levels (van Veen et al., 2014). On the other hand, prolonged waterlogging and anaerobic respiration cause the buildup of harmful metabolites such as lactic acid, ethanol, and aldehydes in addition to a rise in ROS, especially hydrogen peroxide, which ultimately causes plant senescence and cell death

(Xu et al., 2014; Zhang et al., 2017). Although plants suffer greatly when they are wet, they may still adapt to the harm that comes from this kind of environmental stress by using a range of strategies (Fukao et al., 2011; Xu et al., 2016; Doupis et al., 2017; Yin et al., 2019). Certain plants have morphological changes that mitigate the effects of waterlogging-induced root respiratory damages from energy metabolism. Examples of morphological alteration was the development of adventitious roots (ARs) or other aeration tissues, the quick apical meristematic tissue extension, barriers to radial oxygen loss (ROL) and the creation of air films in the upper cuticle (Hattori et al., 2009; Pedersen et al., 2009; Yamauchi et al., 2017; Qi et al., 2019). The entire life cycle of plants is regulated by endogenous plant hormones and proper hormone balance is necessary for healthy physiological metabolism, growth and development (Bartoli et al., 2013; Miransari and Smith, 2014; Wang et al., 2020). The plant modifies the response to waterlogging and the balance of plant hormone synthesis and transport through intricate interactions. Plant hormones, play a critical role in the process of waterlogging tolerance (Benschop et al., 2006; Wu et al., 2019; Yamauchi et al., 2020).

Abscisic acid (ABA) is an essential stomata regulator that modifies guard cell size and, in turn, controls the water potential of plants. Consequently, ABA is thought to be an important hormone in responses to water stress (Zhu, 2016; He et al., 2018). Zn stimulates enzymes in plants that are involved in the metabolism of glucose, protein synthesis, cellular membrane integrity, regulation of auxin production, and pollen formation (Marschner, 1995). Zinc is required for the production of auxin, an essential growth hormone and tryptophan, a precursor to IAA (Alloway, 2004; Brennan, 2005).

### 5.4.4 High temperature

Worldwide warming is the term used to describe a rise of the average worldwide surface temperature brought on by either natural or man-made (Wani et al., 2010). The heat thus represents the biggest danger to food security in an area where the second most important food grain is wheat and the population is growing quickly (Indexmundi, 2011). Rising temperatures could result in yield reductions of 2.5%–10% for a range of agronomic crops during the 21st century (Hatfield et al., 2011). High temperatures have an effect on the morphological and physiochemical processes that plants go through. Some of the main effects include sunburning branches and stems, scorching aerial plant parts, leaf death, abscission and senescence, inhibiting both shoot and root development and also lowering yield (Ismail and Hall, 1999; Wahid et al., 2007). High nighttime temperatures decrease crop productivity by reducing sugar and starch content, respiration rate (Mohammed and Tarpley, 2009b), photosynthetic function (Loka and Oosterhuis, 2010; Turnbull et al., 2002), suppressing the development of floral buds (Ahmed and Hall, 1993), increasing male sterility and reducing pollen viability and changing crop maturity (Mohammed and Tarpley, 2009a). Temperatures below 10°C or above 25°C (the ideal range is 12°C–25°C), according to a number of research findings, affect the phenology, growth, and development of the current varieties of Bangladeshi wheat and ultimately lower their yield (Hakim et al., 2012; Hossain et al., 2009a,b; 2011, 2012; Nahar et al., 2010; Rahman et al., 2009). Plants react differently to heat stress in different ways, but one of the most frequent ways that damage occurs is when changes in membrane structure occur, leading to increased membrane leakiness and function loss (Sayed, 2003; Wahid et al., 2007). Under heat stress Reactive oxygen species (ROS) are produced more frequently in plants due to damage to their membranes (Liu and Huang, 2000; Barnabás et al., 2008).

Under heat stressors, adequate Zn can help to maintain membrane integrity and photosynthetic activity (Cakmak, 2000). Adequate Zn helps rice, maize and leafy vegetables to maintain photosynthetic activity and plant production (Coolong et al., 2004; Hafeez et al., 2013; Sarwar et al., 2017). Thermo tolerance is linked to Zn-dependent physiological processes such as continuous enzyme activity, membrane integrity preservation and photosynthetic activity. As a result, adequate Zn nutrition may help plants resist the adverse outcomes led by heat stress. Natural Zn deficiency, on the other hand, is common around the globe (Zheng, 1994; Razzaq et al., 2013). High Zn nutrition can counteract the detrimental effects of brief heat stress on wheat kernel growth and chloroplast function, as per Peck and McDonald (2010). By binding sulfhydryl groups and phospholipids, zinc ions have been identified to stabilize and protect cellular membranes. The ability to detoxify ROS produced during heat stress is a critical mechanism for preventing membrane damage (Almeselmani et al., 2006; Cui et al., 2006). One of the many enzymatic and non-enzymatic antioxidants that shield cells from oxidative damage is zinc copper SOD (Cakmak, 2000).

### 5.4.5 Chilling

One of the most challenging environment for higher plants is the combination of dryness and low temperature, or chilling stress (Theocharis et al., 2012; Zhou et al., 2017). There are two kinds of low-temperature (LT) stress: freezing stress and chilling stress (Mickelbart et al., 2015; Guo et al., 2017; Liu and Zhou, 2018; Shi et al., 2018). It affects the growth and development of the plant. Tropical and subtropical plants are less able to tolerate low temperatures and are more vulnerable to chilling stress. Conversely, temperate plants may withstand freezing temperatures through a process called "cold adaptation," which occurs after a period of exposure to non-freezing environments (Chinnusamy et al., 2007). To survive in such stress condition, plants need to modify their cell behavior and activity, especially cell membrane stability and the structure of protein with biological activity during LT stress (Chen et al., 2018). As a result of LT, rice's chlorophyll concentration also decreased. Rice treated to low temperatures changed the amount of chloroplasts, the shape of the grana and structures of lamellar within the chloroplasts (Sun et al., 2017). This is because of aberrant chloroplast growth during and after low-temperature stress (Cui et al., 2019). A comprehensive comprehension of the mechanisms, the problem of cold stress injury in plants may be resolved if underlying cold damage is present (Yang et al., 2013; Cheng et al., 2014; Sui, 2015; Pareek et al., 2017).

At the molecular level, freezing stress usually follows dehydration stress. As ice crystals form, there will be a rise in electrolyte leakage and change in membrane lipid phase. As the freezing process advances, osmotic pressures produce cellular dryness, which promotes the formation of intracellular ice crystals. Plant cells can be ruptured by ice crystals, which can lead to cytosol leakage and ultimately plant death (Zhang et al., 2011; Sun et al., 2019; Demidchik et al., 2014). The production of reactive oxygen species (ROS) by NADPH oxidase and the breakdown of antioxidative defense systems also have an impact on chilling-induced cell damage (Shen et al., 2000; Aroca et al., 2005; Wang et al., 2005). It was discovered that citrus trees grown in Zn-deficient environments were more vulnerable to freezing temperatures and wintertime peroxidative damage (Cakmak et al., 1995).

Plants with increased cold resistance are benefited by $C_2H_2$ Zn finger proteins, which directly affect genes related to cold. For example, to regulate cold acclimation, ZAT12 controls the expression of 15 genes that are inhibited by cold and 9 genes that are induced by cold. Additionally, ZAT12 decreases the expression of CBF genes, which play a negative role for ZAT12 in plant adaptation to cold stress (Vogel et al., 2005). Plant cold resistance can be enhanced by $C_2H_2$ Zn finger proteins through an increase in osmotic chemical amounts. ZFP182 increases the expression of OsP5CS and OsLEA3, as well as the build-up of osmoprotectants, hence improving the cold tolerance of rice overexpression lines (Huang et al., 2012). By rising proline and soluble content and decreasing membrane lipid peroxidation, GmZF1 enhanced cold resistance in transgenic Arabidopsis under cold stress (Yu et al., 2014). Eventually, the ABA signaling pathway regulates low-temperature stress via numerous $C_2H_2$-type Zn finger proteins. Through ABA-dependent signaling pathways, SCOF-1 improves plant cold tolerance (Kim et al., 2001).

### 5.4.6 Toxic metals

Human activities such as mining and industry continuously increased heavy metal concentrations in agricultural soils (Ikenaka et al., 2010). According to Mahar et al. (2016) and Klang-Westin and Eriksson (2003), improper use of pesticides and fertilizers, solid waste, gas deposition, and sewage irrigation can all lead to heavy metal pollution. Two types of metals are found in soils: nonessential elements which have unknown biological and physiological functions (Cd, Sb, Cr, Pb, As, Co, Ag, Se, and Hg) and essential micronutrients (Fe, Mn, Zn, Cu, Mg, Mo, and Ni) for normal plant growth (Schutzendubel and Polle, 2002; Tangahu et al., 2011; Zhou et al., 2014). Their toxicity is determined by the target organism's bioavailability and sensitivity (Rascio and Navari-Izzo, 2011).

Heavy metal accumulation in plants has been shown to have deleterious effects on photosynthetic pigment biosynthesis, disrupt photosynthesis and modify the chloroplast membrane (Ventrella et al., 2009). It has long been established that heavy metals impair crop productivity, stomatal conductance, photosynthetic rate, plant development and transpiration rate (Yusuf et al., 2012). To defend themselves against the harmful effects of heavy metals, plants have evolved a variety of defense mechanisms (Latowski et al., 2011).

There are two types of Cr: hexavalent $Cr^{6+}$ and trivalent $Cr^{3+}$. The former is less toxic than the latter. Plants absorb and transfer Cr via carrier ions such as iron or sulfate rather than being absorbed directly by plants (Singh et al., 2013; Gajalakshmi et al., 2012). Cr poisoning prevents plant roots from dividing and lengthening, shortening their overall length (Shanker et al., 2005). Aluminum is a metal that is known as a plant development inhibitor especially in acid soil with pH values of 5 or 5.5 where the most phytotoxic form of aluminum ($Al^{3+}$) predominates (Liu et al., 2014a,b). Al toxicity primarily affects plant roots, where an accumulation of Al can quickly restrict root growth within minutes or hours (Ma et al., 2001). It can cause lateral roots to get thicker and turn brown instead of white (Mossor-Pietraszewska, 2001). Al toxicity also results in changes to the enzymatic regulation of sugar phosphorylation and a decrease in root respiration (Rout et al., 2001). The plant type, age, temperature, and amount of light all affect how poisonous manganese is (González et al., 1998; Kavvadias and Miller, 1999; Soceanu et al., 2005). Some indicators of the disease include wrinkled leaves (Reddy, 2006), darkening of the leaf veins on older foliage (Schubert, 1992), brown spots and chlorosis on older leaves (Dragišić Maksimović et al., 2012) and black colored dots on stems (Vitosh et al., 1994). Lower $CO_2$ assimilation but unchanged levels of chlorophyll (Chl) have been associated with Mn poisoning in citrus grandis seedlings (Li et al., 2010). The level of amount representing nickel toxicity in plants varies from 25 to 246 g/g dry weight (DW) of plant tissue, depending on the plant species and cultivars (Iyaka, 2011). Overexposure to nickel inhibits the absorption of several cations, such as $Fe^{2+}$ and $Zn^{2+}$, which causes a deficiency in either element and the development of chlorosis in plants (Khan and Khan, 2010). A surplus of nickel has a severe effect on seed germination and growth of seedlings because it inhibits the action of some enzymes like as amylase and protease and stops the breakdown of food storage in germination seeds (Aydinalp and Marinova, 2009; Sethy and Ghosh, 2013). Although copper can be found in soils in several states, plants mostly absorb copper as $Cu^{2+}$ (Maksymiec,

1998). Healthy plants may absorb 20–30 g/g DW of copper, with soil concentrations typically ranging from 2 to 250 g/g (Azooz et al., 2012). Conversely, the phyto-availability of copper is strongly correlated with soil pH, and it rises with decreasing pH (Sheldon and Menzies, 2005). Additionally, the nutritional status of the plant, the concentration of $Cu^{2+}$ in the soil, the duration of exposure, and the species' genotype all affect how much copper is absorbed by plants and how harmful it is (Nicholls and Mal, 2003).

One of the most important nutrients for maintaining the homeostasis of heavy metals in plant cells is zinc. Therefore, by limiting their availability, metal toxicity is decreased (Appenroth, 2010). Zn and other micronutrients accelerate the growth and development of plants (Hänsch and Mendel, 2009). It is crucial to many different metabolic activities. Zinc activates a large number of enzymes involved in the metabolism of lipids, nucleic acids and also proteins (Bonnet et al., 2000). For abiotic tolerance in a range of plants, zinc is crucial. Plants have been demonstrated to absorb more zinc when zinc is present, reducing copper toxicity (Hafeez et al., 2013). Zinc lowers oxidative stress and Cd toxicity in rice (Hassan et al., 2005) and wheat (Hassan et al., 2005; Zhao et al., 2011). Numerous investigations have shown that zinc (Zn) lowers the production of free radicals that harm cells and aids in the preservation of cell membrane integrity in plant roots (Cakmak, 2000).

Zinc was applied to rice cultivars to reducing oxidative stress and growth inhibition caused by Cd. Cd reduced biomass, rate of photosynthesis and plant development, but it increased enzyme activity and malondialdehyde levels. Following Zn treatment, plant growth, biomass, photosynthetic rate, and chlorophyll content all increased, whereas antioxidant enzyme activity and malondialdehyde concentration fell (Hassan et al., 2005). Zn reduces ROS levels and mitigates Cu-induced stress (Aravind and Prasad, 2005; Upadhyay and Panda, 2010). Zn enhanced rice seedlings' antioxidant efficiency while they were under Cu stress (Aravind and Prasad, 2005; Upadhyay and Panda, 2010). Zn treatment raised glutathione levels under Cd stress by lowering Cd toxicity. Additionally, it was discovered that zinc decreased the uptake and accumulation of Cd in the plants *Chara australis* (R. Br.) and *Calendula officinalis* L. (Clabeaux et al., 2013; Moustakas et al., 2011).

## 5.5 Physiological mechanisms of Zn during abiotic stress

Trace elements are minutely necessary elements in plants and other biota to ensure normal growth, development, and biological processes. Conversely, elevated concentrations of certain trace elements—often referred to as "heavy metals"—such as cadmium (Cd), lead (Pb), and mercury (Hg) may cause toxicity and harm to plants instead of serving any known biological purpose in plants (Daud et al., 2009; Ali et al., 2014; Khan et al., 2014). According to Khan et al. (2018), plants those are exposed to abiotic stress condition benefit from the trace element zinc.

Reactive oxygen species (ROS) are typically produced in response to abiotic stress. However, in certain plant species, oxidative stress can be effected by an increase in reactive oxygen species generation in unfavorable climatic conditions (Smirnoff, 1993; Mullineaux & Karpinski, 2002; Miller et al., 2010, Boguszewska and Zagdańska, 2012). $C_2H_2$ Zn finger proteins are expressed more frequently in the presence of excess ROS, which aids in keeping plant ROS levels steady. On plant growth and development proecess, this is known to as a stabilizing impact (Gadjev et al., 2006; Choudhury et al., 2017).

Zinc is one of the micro-elements that is most frequently used in biological systems. Large field crops frequently have deficiencies in this nutrient. Growth retardation, chlorosis, necrosis, and delayed seed development are signs of zinc deficiency. Consequently, Zn fortification of primary crops has been the focus of the current study (Poletti et al., 2004; Lu et al., 2008). Zinc plays several biological roles in plants, including stabilizing cell membranes, producing chlorophyll, stabilizing and synthesizing proteins, activating enzymes, controlling auxin, and more. It is present in more than 1200 proteins and has a role in transcription, protein-protein interactions, RNA synthesis and editing, and DNA regulation. According to Yang et al. (2009), a large number of regulatory proteins containing a Zn-finger motif is essential for tolerance to abiotic stress.

Zinc is an essential plant nutrient that works in a variety of enzyme systems as a structural, functional, or regulatory cofactor or as a metal component (Marschner, 1995). Antioxidant enzymes, namely catalase (CAT) and superoxide dismutase (SOD), are widely recognized for serving as a cell's first guide of defense against reactive oxygen species (ROS) (Alscher et al., 2002). In well-watered maize leaves, Wang and Jin (2007) found that Zn treatment increased SOD activity; however, during a drought, there was no discernible difference between Zn-treated and non-Zn-treated plants. Zn soil treatment increased SOD transcription in flag leaves of wheat independent of the soil water regime. The varying growth phases and amounts of Zn applied were the main causes of the variations in SOD activity and SOD gene expression responses to Zn administration. According to Gao et al. (2009), increased $Zn^{2+}$ under low-temperature stress boosted both SOD gene expression and SOD activity in seedling leaves of cucumber. Thus, increasing $Zn^{2+}$ may enhance the ability to scavenge reactive oxygen species (ROS) by inducing the expression of SOD genes and elevating SOD activity. Plant photosynthetic characteristics were enhanced by soil and foliar Zn treatment under water stress, leading to improvements in SPAD and Fv/Fm. Similar to this, adding zinc to the soil increased drought resistance in wheat and chickpea by promoting SOD and POD activity (Khan et al., 2003; Peleg et al., 2008, Yavas and Unay, 2016).

Zinc is an essential plant nutrient that promotes plant growth. Under unfavorable circumstances such as salt stress, zinc therapy has been shown to boost chlorophyll a and b content, photosynthetic rate, and plant development (Hassan et al., 2005; Tavallali et al., 2010). In tomato plants afflicted by cadmium poisoning, zinc treatment restored several lipid types, including galactolipids, neutral lipids, total lipids, and phospholipids (Ammar et al., 2015). Zn enhanced the production of ascorbic acid and tocopherol in green chilli to scavenge free radical species (Manas et al., 2014). It is commonly known that zinc, through Cu-Zn SOD enzymes, helps plants such as sugar beets (Hajheidari et al., 2005), periwinkle (Jaleel et al., 2007), and oilseed rape (Abedi and Pakniyat, 2010) to lessen the effects of drought stress. Comparably, it has been shown that Cu-Zn SOD protects against salt stress in rice (Tanaka et al., 1999), citrus (Gueta-Dahan et al., 1997), tobacco (Wang et al., 2003), and lentils (Bandeoğlu et al., 2004). Two Cu-Zn SODs genes were increased in a variety of tomato seedling organs throughout plant growth and light-induced stress (Perl-Treves and Galun, 1991). Applying Mn, Zn, and B foliarly to winter wheat enhanced its agronomic characteristics and lessened the adverse effects of late-season water shortage (Karim et al., 2012). Certain low molecular weight antioxidant active chemicals, like phenolics, GSH, and ASC, may aid plants that are suffering from drought (Rice-Evans et al., 1997; Chen et al., 2003). According to Cakmak's (2000) theory, zinc shields chloroplasts from reactive oxygen species (ROS)-induced photooxidative damage. Zn treatment raised ASC and GSH levels in wheat flag leaves during drought conditions. After Zn and boron were added, tobacco showed an increase in ASC and GSH (Jiang et al., 2000). Secondary metabolites called TPC and TFC shield cells from ROS. Treatment of the soil with zinc could increase TPC and TFC in wheat flag leaves under drought stress. In a similar vein, berry total phenol and flavonoid accumulation were enhanced by Zn treatments (Song et al., 2015). Tavallali et al. (2010) found that application of Zn in soil resulted in a significant increase in ASC content and phenolics in pistachio leaves compared to salt stress alone. The amount of antioxidants increased by an enhancement in antioxidant production that Zn was credited with. Zinc has the ability to affect GSH levels by controlling its synthesis, maintaining the GSH's reactive cysteine residue, or ensuring that associated enzymes function as intended (Foyer and Halliwell, 1976). Application of zinc may increase antioxidant content to resist damage from reactive oxygen species (ROS) brought on by drought, supporting earlier research suggesting micronutrients may be important for drought stress tolerance by shielding membranes from oxidative damage (Cakmak, 2000; Ducic and Polle, 2005).

Superoxide dismutase (SOD), the primary scavenger in the detoxification of active oxygen species in plants, was developed by Irwin Fridovich and Joe McCord in 1969 (McCord and Fridovich, 1969). By accelerating its dismutation, which lowers one $O_2$ molecule to $H_2O_2$ and another to $O_2$, SODs deplete $O_2$. The subcellular localization of the enzyme may be the cause of SOD function specialization (Mittler, 2002). SODs are metalloproteins that fall into three groups according to the metal cofactor they contain iron (Fe-SOD), manganese (Mn-SOD) and copper/zinc (Cu/Zn-SOD), all of which are found in distinct cellular compartments. Copper serves as a redox catalyst for around 30 enzymes involved in several metabolic pathways (Harrison et al., 1999). Plastocyanin, cytochrome c oxidase, and Cu-Zn SODs are the three primary Cu proteins found in plant cells. Photosynthesis requires plastocyanin, which is present in chloroplasts and acts as an electron carrier between PSI and PSII. The mitochondrial transmembrane protein cytochrome c oxidase is also involved in electron transfer, which results in the synthesis of ATP and H2O. Cu-Zn SODs are classified according to their location as either cytosolic or chloroplastic. In higher plants, the cytosolic fraction and chloroplasts contain Cu/Zn-SOD. While cytosolic Cu-Zn SOD is present in the nucleus and apoplastic regions of thylakoids, its primary location is in the stroma of these organelles. According to Khan et al. (2018), Cu-Zn SODs help reduce ROS in plants. The Cu-Zn SOD response varies in plant species that are tolerant and sensitive. Most earlier studies have discovered that tolerant cultivars exhibited increased Cu-Zn SOD activity. Cu-Zn SOD activity rose in seedlings of a salt-tolerant cultivar of *Setaria italica* when they were subjected to salinity stress (Sreenivasulu et al., 2000). Cu-Zn SOD activity was also higher in salt-tolerant pea chloroplasts than in sensitive kinds. In 1995, Hernandez et al. (1995) Research on tolerance mechanisms in plants under stress could be facilitated by using transgenic plants that overexpress the Cu-Zn SOD genes. In previous work, the Cu-Zn SOD gene from *Avicennia marina* was inserted into *Oryza sativa* to increase rice tolerance to MV and salt stresses (Prashanth et al., 2008). Comparably, transgenic tobacco's resistance to stress brought on by low temperatures and a shortage of water was enhanced by increased expression of Cu-Zn SOD genes (Faize et al., 2011). Moreover, downregulating microRNAs in *Arabidopsis thaliana* enhanced the production of Cu-Zn SOD genes in response to high light and MV stressors (Sunkar et al., 2006).

The ASC−GSH cycle and phenolic biosynthesis are components of the non-enzymatic antioxidative system. The ASC−GSH cycle involves the enzymes MDHAR, DHAR, GS, and APX. Liu et al. (2012) suggest that plant stress tolerance may be more significantly influenced by APX and MDAR. Zn treatment increased the transcription of GS, APX, DHAR, MDHAR, and DHAR in plants under drought stress. Zn supplementation increased pistachio APX activity under salt stress (Tavallali et al., 2010). Two important enzymes in the early phases of the flavonoid biosynthesis pathway are PAL and CHS. It was indentified that exogenous silicon increased the amount of PAL and CHS transcripts in rice and wheat (Fleck et al., 2011; Ma et al., 2016a,b). Zinc may be capable to change the expression of genes by taking up a position on a transcription factor that accelerates transcription (Cousins, 1998). Wheat's resilience to drought may be improved by zinc treatment, which may increase the transcription of genes involved in plant antioxidant defense.

Antioxidase genes involved in ROS scavenging during abiotic stress can be influenced by $C_2H_2$ zinc finger proteins (Liu et al., 2017). Ascorbate peroxidase 1, or Apx1, is an essential $H_2O_2$ scavenger in plants (Rizhsky et al., 2004). ZAT12 and ZAT7, two TF genes, express more when $H_2O_2$ levels rise in the loss-of-function mutant apx1. Comparing with the wild-type plants, ZAT12-deficient plants are more sensitive to high $H_2O_2$ levels (Rizhsky et al., 2004; Davletova et al., 2005). APX1, APX2, and FeSOD1 are significantly increased in ZAT10 and ZAT12 transgenic lines under high-light stress but downregulated in zat10 loss-of-function mutants (Mittler et al., 2006). According to Huang et al. (2009b), DST binds to the promoter region of the peroxidase24 precursor, controlling its production and preventing $H_2O_2$ accumulation in plants. Overexpressing ZFP245 and ZFP179 causes a significant increase in SOD and POD activity in rice plants, which increases resistance to abiotic stress and ROS at the seedling stage (Huang et al., 2009a; Sun et al., 2010). Under oxidative stress and water, overexpressing ZFP36 reduces ROS damage by upregulating SOD and APX activity, while ZFP36-RNAi lines show significant ROS damage as a result of decreased SOD and APX activity (Zhang et al., 2014). Antioxidative stress responses are mediated by $C_2H_2$-type Zn finger proteins via the ABA and MAPK signaling pathways. ZFP36 is notably impacted by ABA-produced increases in $H_2O_2$ and OsMPK activity. ZFP36 also increases the expression of the genes MAPK and NADPH oxidase in the ABA signaling pathway. According to Zhang et al. (2014), ZFP36 controls the interaction between NADPH oxidase, $H_2O_2$, and MAPK in the ABA signaling pathway.

Application of Zn fertilizer has been suggested as a viable method of increasing grain yield and Zn content. Plant roots, on the other hand, rely heavily on soil moisture for micronutrients. According to Wang and Jin (2007), maize plants should be fertilized with Zn and given enough water. In comparison to non-Zn treated plants, Zn application boosted grain production and Zn concentration of grain in the studies whether in AW supply or drought stress. Under drought conditions, Zn application enhanced the grain production of wheat as well as Zn content of grain (Karim et al., 2012). The plant's increased uptake of zinc from the soil or leaves was reflected in its increased grain yield and zinc content.

Zinc fertilizer has the potential to increase grain yield and zinc content in grains, independent of soil water supply circumstances. Moreover, in SD conditions, Application of Zn fertilizer had a higher effect on accumulating Zn content in grain than AW supply. Zinc therapy may reduce oxidative stress in wheat by transcriptionally modulating many defense mechanisms, including antioxidant enzymes, the ASC–GSH cycle, and flavonoid secondary metabolism (Ma et al., 2017). Zinc supplementation has the potential to mitigate the effects of drought stress by increasing the generation of reactive species. The physiochemical processes underlying application of Zn and plant resistance to abiotic stress are better-understood thanks to this review.

## 5.6 Zn-associated genes in response to abiotic stress tolerance of plants

The growth and productivity of crop plants are greatly impacted by a variety of abiotic stresses, including salinity, drought, cold, availability of water (less or excess), extremes temperature (freezing, cold, or high), metal/metalloids, light intensity and also nutrient stress. These stresses essentially pose a serious threat to global agriculture by inhibiting crop plants from expressing their full genetic potential and resulting in significant yield reduction on a global scale (Gill et al., 2014; Zhou et al., 2018; Wang et al., 2019; Han et al., 2020). To cope with the stresses resulting from constantly changing complex and multigenic traits, plants have evolved sophisticated regulatory mechanisms. These mechanisms include morphological and physiochemical modifications that are regulated by transcriptional genetic control (Pastori and Foyer, 2002; Gill et al., 2016; Wang et al., 2019). To maintain plant survival in different challenging environmental in the form of various abiotic stresses, perception of stress signals and their transmission are crucial steps in the process (Gill et al., 2016).

It has been determined that the zinc-finger protein family, which is primarily composed of the *N*-terminal A20 and the C-terminal AN1 zinc-finger domain, is an important gene family for plant defense against environmental stresses (Liu et al., 2015a,b,c; Liu et al., 2014; Wang et al., 2016; Zhai et al., 2016; Zhou et al., 2018). The amount and location of cysteine (C) and/or histidine (H) residues in zinc finger proteins determine which subclasses they belong to: $C_2H_2$ (TFIIIA), $C_8$ (Steroid-thyroid receptor), $C_6$ (GAL4), $C_3HC_4$ (RING finger), $C_2HC$ (Retroviral nucleocapsid), $C_2HC_5$ (LIM domain), $C_4$ (GATA-1), $C_3H$ (Nup 475), $C_4HC_3$ (Retroviral nucleocapsid), $C_4HC_3$ (Requium) (Berg and Shi, 1996). Most of these subclasses of Zn finger proteins are composed of $C_2H_2$ Zn finger proteins (Cys2/His2), which have been the subject of the most research (Liu et al., 2015a,b,c; Muthamilarasan et al., 2014; Lyu and Cao, 2018). Low-temperature tolerance is linked to SCOF-1, a soybean cold-inducible $C_2H_2$ Zn finger transcription factor (Kim et al., 2001). Overexpression of SCOF-1 enhanced cold tolerance in transgenic tobacco and Arabidopsis plants (Wang et al., 2019). Kim et al. (2016) found that transgenic potato plants with enhanced SCOF-1 expression under the promoter of the oxidative stress-inducible SWPA2 gene were more resistant to freezing and cold stress. Plant resistance to low temperatures can be enhanced by ZAT12 derived from Arabidopsis (Vogel et al., 2005). DST is a drought-responsive gene in rice (Huang et al., 2009b), while ZFP179 and ZFP182 were identified from rice and functionally described as a novel salt-responsive gene (Kodaira et al., 2011; Sun et al., 2010; Huang et al., 2007). In Arabidopsis, STZ and AZF1, 2, and 3 can enhance salinity and low-temperature tolerance (Sakamoto et al., 2000). Salinity stress in Arabidopsis is linked to the Zn-associated gene TaZNF (Ma et al., 2016a,b). In rice, the genes ZFP245 and RZF5, 71 are responsive to drought stress and salt stress, respectively (Guo et al., 2007; Huang et al., 2009a; Huang et al., 2012). ZAT18 is the cause of Arabidopsis's drought stress (Yin et al., 2017). The IbZFP1 gene in sweet potatoes is associated with drought stress and salinity (Wang et al., 2016). According to Iida et al. (2000) RHL41 gene can help Arabidopsis withstand light exposure. BnLATE can increase mustard's resistance to silique shattering (Tao et al., 2017).

## 5.7 Conclusion

Food security, sustainable agriculture, ecological intensification, quality food production, conserving the environment by reducing natural resources destroying practices of agriculture, introducing organic farm culture and creating a congenial environment for human and plant's perpetuation are the long-awaited demand from the masses to the agricultural scientists. Population explosion, agricultural intensification, competition for increasing quantities of food production from the limited natural resources and abiotic pressures brought on by climate change are the main forces which have been acting against sustainable agriculture. Abiotic stresses are now common phenomena in almost all parts of the world in reducing tolerance of crop plants failing to provide food, fiber, fuel and residence to the people. Increasing crops' ability to withstand abiotic stress should be viewed as the panacea for sustaining agriculture and the environment in the age of global warming but plant physiology is altered or changed under the different stressors. Understanding the mitigating mechanisms of different plants in contrast to abiotic stress and developing tolerance-boosting techniques need to be emphasized more in the research activities of crop scientists. However, ensuring the required nutrition of plants in an optimum time is prime for increasing tolerance to abiotic stress where the essentiality of trace elements for crops is neglected in most of the farm culture, consequently, leaving plants unable to fight against stresses. Among the trace elements, Zn has special importance as it is involved in almost all parts of plants as a structural component and presents in nearly all metabolic reactions, oxidative stress minimization and abiotic stress tolerance maximization. On the other hand, extensive researches are needed to discover as well as apply the function of Zn in increasing the abiotic stress tolerance of plants.

## References

Abedi, T., Pakniyat, H., 2010. Antioxidant enzymes change in response to drought stress in ten cultivars of oilseed rape (*Brassica napus* L.). Czech Journal of Genetics and Plant Breeding 46 (1), 27–34.

Aghaee, A., Moradi, F., Zare-Maivan, H., Zarinkamar, F., Irandoost, H.P., Sharifi, P., 2011. Physiological responses of two rice (*Oryza sativa* L.) genotypes to chilling stress at seedling stage. African Journal of Biotechnology 10 (39), 7617–7621.

Ahanger, M.A., Agarwal, R.M., 2017. Salinity stress induced alterations in antioxidant metabolism and nitrogen assimilation in wheat (*Triticum aestivum* L) as influenced by potassium supplementation. Plant Physiology and Biochemistry 115, 449–460.

Ahanger, M.A., Agarwal, R.M., Tomar, N.S., Shrivastava, M., 2015. Potassium induces positive changes in nitrogen metabolism and antioxidant system of oat (*Avena sativa* L cultivar Kent). Journal of plant interactions 10 (1), 211–223.

Ahanger, M.A., Morad-Talab, N., Abd-Allah, E.F., Ahmad, P., Hajiboland, R., 2016. Plant growth under drought stress: significance of mineral nutrients. Water Stress and Crop Plants: A Sustainable Approach 2.

Ahmad, P., Jaleel, C.A., Salem, M.A., Nabi, G., Sharma, S., 2010. Roles of enzymatic and nonenzymatic antioxidants in plants during abiotic stress. Critical Reviews in Biotechnology 30 (3), 161–175.

Ahmad, P., Jamsheed, S., Hameed, A., Rasool, S., Sharma, I., Azooz, M.M., et al., 2014. Drought stress induced oxidative damage and antioxidants in plants. Oxidative Damage to Plants. Academic Press, pp. 345–367.

Ahmad, P., Sarwat, M., Bhat, N.A., Wani, M.R., Kazi, A.G., Tran, L.S.P., 2015. Alleviation of cadmium toxicity in *Brassica juncea* L. (Czern. & Coss.) by calcium application involves various physiological and biochemical strategies. PLoS One 10 (1), e0114571.

Ahmad, P., Abdel Latef, A.A., Hashem, A., Abd_Allah, E.F., Gucel, S., Tran, L.S.P., 2016. Nitric oxide mitigates salt stress by regulating levels of osmolytes and antioxidant enzymes in chickpea. Frontiers in Plant Science 7, 347.

Ahmed, F.E., Hall, A.E., 1993. Heat injury during early floral bud development in cowpea. Crop Science 33 (4), 764–767.

Ahmed, M., Asif, M., Goyal, A., 2012. Silicon the non-essential beneficial plant nutrient to enhanced drought tolerance in wheat. Crop Plant 31–48.

Akman, Z., 2009. Comparison of high temperature tolerance in maize, rice and sorghum seeds by plant growth regulators. Journal of Animal and Veterinary Advances 8 (2), 358–361.

Aktaş, H., Abak, K., Öztürk, L., Çakmak, İ., 2007. The effect of zinc on growth and shoot concentrations of sodium and potassium in pepper plants under salinity stress. Turkish Journal of Agriculture and Forestry 30 (6), 407–412.

Allen, D.J., Ort, D.R., 2001. Impacts of chilling temperatures on photosynthesis in warm-climate plants. Trends in Plant Science 6 (1), 36–42.

Ali, E.A., Mahmoud, A.M., 2013. Effect of foliar spray by different salicylic acid and zinc concentrations on seed yield and yield components of mungbean in sandy soil. Asian Journal of Crop Science 5 (1), 33.

Ali, B., Qian, P., Jin, R., Ali, S., Khan, M., Aziz, R., et al., 2014. Physiological and ultra-structural changes in *Brassica napus* seedlings induced by cadmium stress. Biologia Plantarum 58 (1), 131–138.

Allakhverdiev, S.I., Kreslavski, V.D., Klimov, V.V., Los, D.A., Carpentier, R., Mohanty, P., 2008. Heat stress: an overview of molecular responses in photosynthesis. Photosynthesis Research 98 (1), 541–550.

Alloway, B.J., 2004. Zinc in soils and crop nutrition. International Zinc Association Communications. IZA Publications, Brussels.

Almeselmani, M., Deshmukh, P.S., Sairam, R.K., Kushwaha, S.R., Singh, T.P., 2006. Protective role of antioxidant enzymes under high temperature stress. Plant Science 171 (3), 382–388.

Alscher, R.G., Erturk, N., Heath, L.S., 2002. Role of superoxide dismutases (SODs) in controlling oxidative stress in plants. Journal of Experimental Botany 53 (372), 1331–1341.

Amirjani, M.R., 2011. Effect of salinity stress on growth, sugar content, pigments and enzyme activity of rice. International Journal of Botany 7 (1), 73–81.

Ammar, W.B., Zarrouk, M., Nouairi, I., 2015. Zinc alleviates cadmium effects on growth, membrane lipid biosynthesis and peroxidation in *Solanum lycopersicum* leaves. Biologia (Lahore, Pakistan) 70 (2), 198–207.

Andaya, V., Mackill, D., 2003. QTLs conferring cold tolerance at the booting stage of rice using recombinant inbred lines from a japonica × indica cross. Theoretical and Applied Genetics 106 (6), 1084–1090.

Angadi, S.V., Cutforth, H.W., McConkey, B.G., 2000. Seeding management to reduce temperature stress in Brassica species. In Soils and Crops Workshop.

Anjum, N.A., Sofo, A., Scopa, A., Roychoudhury, A., Gill, S.S., Iqbal, M., et al., 2015. Lipids and proteins—major targets of oxidative modifications in abiotic stressed plants. Environmental Science and Pollution Research 22 (6), 4099–4121.

Appenroth, K.J., 2010. Definition of "heavy metals" and their role in biological systems. Soil Heavy Metals. Springer, Berlin, Heidelberg, pp. 19–29.

Aravind, P., Prasad, M.N.V., 2005. Modulation of cadmium-induced oxidative stress in *Ceratophyllum demersum* by zinc involves ascorbate–glutathione cycle and glutathione metabolism. Plant Physiology and Biochemistry 43 (2), 107–116.

Arbona, V., Hossain, Z., López-Climent, M.F., Pérez-Clemente, R.M., Gómez-Cadenas, A., 2008. Antioxidant enzymatic activity is linked to waterlogging stress tolerance in citrus. Physiologia Plantarum 132 (4), 452–466.

Aroca, R., Vernieri, P., Irigoyen, J.J., Sánchez-Dıaz, M., Tognoni, F., Pardossi, A., 2003. Involvement of abscisic acid in leaf and root of maize (*Zea mays* L.) in avoiding chilling-induced water stress. Plant Science 165 (3), 671–679.

Aroca, R., Amodeo, G., Fernández-Illescas, S., Herman, E.M., Chaumont, F., Chrispeels, M.J., 2005. The role of aquaporins and membrane damage in chilling and hydrogen peroxide induced changes in the hydraulic conductance of maize roots. Plant Physiology 137 (1), 341–353.

Ashraf, M.H.P.J.C., Harris, P.J., 2013. Photosynthesis under stressful environments: an overview. Photosynthetica 51 (2), 163–190.

Ashraf, M., McNeilly, T., 2004. Salinity tolerance in Brassica oilseeds. Critical Reviews in Plant Sciences 23 (2), 157–174.

Awasthi, R., Kaushal, N., Vadez, V., Turner, N.C., Berger, J., Siddique, K.H., et al., 2014. Individual and combined effects of transient drought and heat stress on carbon assimilation and seed filling in chickpea. Functional Plant Biology 41 (11), 1148–1167.

Aydinalp, C., Marinova, S., 2009. The effects of heavy metals on seed germination and plant growth on alfalfa plant (*Medicago sativa*). Bulgarian Journal of Agricultural Science 15 (4), 347–350.

Azooz, M.M., Abou-Elhamd, M.F., Al-Fredan, M.A., 2012. Biphasic effect of copper on growth, proline, lipid peroxidation and antioxidant enzyme activities of wheat ('*Triticum aestivum*'cv. Hasaawi) at early growing stage. Australian Journal of Crop Science 6 (4), 688.

Babaeian, M., Heidari, M., Ghanbari, A., 2010. Effect of water stress and foliar micronutrient application on physiological characteristics and nutrient uptake in sunflower (*Helianthus annus* L.). Iranian Journal of crop sciences 12 (4).

Bagci, S.A., Ekiz, H., Yilmaz, A., Cakmak, I., 2007. Effects of zinc deficiency and drought on grain yield of field-grown wheat cultivars in Central Anatolia. Journal of Agronomy and Crop Science 193 (3), 198–206.

Bailey-Serres, J., Colmer, T.D., 2014. Plant tolerance of flooding stress–recent advances. Plant, Cell & Environment 37 (10), 2211–2215. Available from: https://doi.org/10.1111/pce.12420.

Bajji, M., Kinet, J.M., Lutts, S., 2002. The use of the electrolyte leakage method for assessing cell membrane stability as a water stress tolerance test in durum wheat. Plant Growth Regulation 36 (1), 61–70.

Bandeoğlu, E., Eyidoğan, F., Yücel, M., Öktem, H.A., 2004. Antioxidant responses of shoots and roots of lentil to NaCl-salinity stress. Plant Growth Regulation 42 (1), 69–77.

Banerjee, S., Dey, N., Adak, M.K., 2015. Assessment of some biomarkers under submergence stress in some rice cultivars varying in responses. American Journal of Plant Sciences 6 (01), 84.

Barnabás, B., Jäger, K., Fehér, A., 2008. The effect of drought and heat stress on reproductive processes in cereals. Plant, Cell & Environment 31 (1), 11–38.

Bartoli, C.G., Casalongué, C.A., Simontacchi, M., Marquez-Garcia, B., Foyer, C.H., 2013. Interactions between hormone and redox signalling pathways in the control of growth and cross tolerance to stress. Environmental and Experimental Botany 94, 73–88.

Bates, B., Kundzewicz, Z., & Wu, S. 2008. Climate change and water. Intergovernmental Panel on Climate Change Secretariat, Geneva.

Benschop, J.J., Bou, J., Peeters, A.J., Wagemaker, N., Gühl, K., Ward, D., et al., 2006. Long-term submergence-induced elongation in *Rumex palustris* requires abscisic acid-dependent biosynthesis of gibberellin1. Plant Physiology 141 (4), 1644–1652.

Berg, J.M., Shi, Y., 1996. The galvanization of biology: a growing appreciation for the roles of zinc. Science (New York, N.Y.) 271 (5252), 1081–1085.

Bhatt, R., Hossain, A., Sharma, P., 2020. Zinc biofortification as an innovative technology to alleviate the zinc deficiency in human health: a review. Open Agriculture 5 (1), 176–186. Available from: https://doi.org/10.1515/opag-2020-0018.

Blaha, G., Stelzl, U., Spahn, C.M., Agrawal, R.K., Frank, J., Nierhaus, K.H., 2000. Preparation of functional ribosomal complexes and effect of buffer conditions on tRNA positions observed by cryoelectron microscopy.

Boguszewska, D., Zagdańska, B., 2012. ROS as signaling molecules and enzymes of plant response to unfavorable environmental conditions. Oxidative Stress–Molecular Mechanisms and Biological Effects. InTech, Rijeka, Croatia, pp. 341–362.

Bonnet, M., Camares, O., Veisseire, P., 2000. Effects of zinc and influence of *Acremonium lolii* on growth parameters, chlorophyll a fluorescence and antioxidant enzyme activities of ryegrass (*Lolium perenne* L. cv Apollo). Journal of Experimental Botany 51 (346), 945–953.

Bordi, A., 2010. The influence of salt stress on seed germination, growth and yield of canola cultivars. Notulae Botanicae Horti Agrobotanici 38, 128–133.

Bray, E.A., 2000. Response to abiotic stress. Biochemistry and Molecular Biology of Plants 1158–1203.

Brennan, R.F., 2005. Zinc application and its availability to plants (Doctoral dissertation, Murdoch University).

Budak, H., Kantar, M., Yucebilgili Kurtoglu, K., 2013. Drought tolerance in modern and wild wheat. The Scientific World Journal 2013.

Buriro, M., Oad, F.C., Keerio, M.I., Tunio, S., Gandahi, A.W., Hassan, S.W.U., et al., 2011. Wheat seed germination under the influence of temperature regimes. Sarhad Journal of Agriculture 27 (4), 539–543.

Cakmak, I., 2000. Tansley Review No. 111: possible roles of zinc in protecting plant cells from damage by reactive oxygen species. New Phytologist 146 (2), 185–205.

Cakmak, I., 2008. Enrichment of cereal grains with zinc: agronomic or genetic biofortification? Plant and Soil 302 (1), 1–17.

Cakmak, I., Atli, M., Kaya, R., Evliya, H., Marschner, H., 1995. Association of high light and zinc deficiency in cold-induced leaf chlorosis in grapefruit and mandarin trees. Journal of Plant Physiology 146 (3), 355–360.

Calviño, P.A., Sadras, V.O., Andrade, F.H., 2003. Development, growth and yield of late-sown soybean in the southern Pampas. European Journal of Agronomy 19 (2), 265–275.

Camejo, D., Rodríguez, P., Morales, M.A., Dell'Amico, J.M., Torrecillas, A., Alarcón, J.J., 2005. High temperature effects on photosynthetic activity of two tomato cultivars with different heat susceptibility. Journal of Plant Physiology 162 (3), 281–289.

Chen, Z., Young, T.E., Ling, J., Chang, S.C., Gallie, D.R., 2003. Increasing vitamin C content of plants through enhanced ascorbate recycling. Proceedings of the National Academy of Sciences 100 (6), 3525–3530.

Chen, L., Zhao, Y., Xu, S., Zhang, Z., Xu, Y., Zhang, J., et al., 2018. Os MADS 57 together with Os TB 1 coordinates transcription of its target Os WRKY 94 and D14 to switch its organogenesis to defense for cold adaptation in rice. New Phytologist 218 (1), 219–231.

Cheng, L., Zou, Y., Ding, S., Zhang, J., Yu, X., Cao, J., et al., 2009. Polyamine accumulation in transgenic tomato enhances the tolerance to high temperature stress. Journal of Integrative Plant Biology 51 (5), 489–499.

Cheng, S., Yang, Z., Wang, M., Song, J., Sui, N., Fan, H., 2014. Salinity improves chilling resistance in *Suaeda salsa*. Acta Physiologiae Plantarum 36 (7), 1823–1830.

Chinnusamy, V., Schumaker, K., Zhu, J.K., 2004. Molecular genetic perspectives on cross-talk and specificity in abiotic stress signalling in plants. Journal of Experimental Botany 55 (395), 225–236.

Chinnusamy, V., Zhu, J., Zhu, J.K., 2007. Cold stress regulation of gene expression in plants. Trends in Plant Science 12 (10), 444–451.

Choudhury, F.K., Rivero, R.M., Blumwald, E., Mittler, R., 2017. Reactive oxygen species, abiotic stress and stress combination. The Plant Journal 90 (5), 856–867.

Ciscar, J.C., 2012. The impacts of climate change in Europe (the PESETA research project). Climatic Change 112 (1), 1–6.

Clabeaux, B.L., Navarro, D.A., Aga, D.S., Bisson, M.A., 2013. Combined effects of cadmium and zinc on growth, tolerance, and metal accumulation in *Chara australis* and enhanced phytoextraction using EDTA. Ecotoxicology and Environmental Safety 98, 236–243.

Cole, P.J., McCloud, P.I., 1985. Salinity and climatic effects on the yields of citrus. Australian Journal of Experimental Agriculture 25 (3), 711–717.

Condon, A.G., Richards, R.A., Rebetzke, G.J., Farquhar, G.D., 2004. Breeding for high water-use efficiency. Journal of Experimental Botany 55 (407), 2447–2460.

Coolong, T.W., Randle, W.M., Toler, H.D., Sams, C.E., 2004. Zinc availability in hydroponic culture influences glucosinolate concentrations in *Brassica rapa*. HortScience: A Publication of the American Society for Horticultural Science 39 (1), 84–86.

Cousins, R.J., 1998. A role of zinc in the regulation of gene expression. Proceedings of the Nutrition Society 57 (2), 307–311.

Cui, L., Li, J., Fan, Y., Xu, S., Zhang, Z., 2006. High temperature effects on photosynthesis, PSII functionality and antioxidant activity of two *Festuca arundinacea* cultivars with different heat susceptibility. Botanical Studies 47 (1), 61–69.

Cui, X., Wang, Y., Wu, J., Han, X., Gu, X., Lu, T., et al., 2019. The RNA editing factor DUA 1 is crucial to chloroplast development at low temperature in rice. New Phytologist 221 (2), 834–849.

Cvikrová, M., Gemperlová, L., Martincová, O., Vanková, R., 2013. Effect of drought and combined drought and heat stress on polyamine metabolism in proline-over-producing tobacco plants. Plant Physiology and Biochemistry 73, 7–15.

DaCosta, M., Huang, B., 2006. Osmotic adjustment associated with variation in bentgrass tolerance to drought stress. Journal of the American Society for Horticultural Science 131 (3), 338–344.

Damanik, R.I., Maziah, M., Ismail, M.R., Ahmad, S., 2010. Responses of the antioxidative enzymes in Malaysian rice (*Oryza sativa* L.) cultivars under submergence condition. Acta Physiologiae Plantarum 32 (4), 739–747.

Dantas, B.F., Ribeiro, L.D.S., Aragão, C.A., 2007. Germination, initial growth and cotyledon protein content of bean cultivars under salinity stress. Revista Brasileira de Sementes 29 (2), 106–110.

Das, S., Jahiruddin, M., Islam, M.R., Al Mahmud, A., Hossain, A., Laing, A.M., 2020. Zinc biofortification in the grains of two wheat (*Triticum aestivum* L.) varieties through fertilization. Acta Agrobotanica 73 (1), 1–13. Available from: https://doi.org/10.5586/aa.7312.

Daud, M.K., Sun, Y., Dawood, M., Hayat, Y., Variath, M.T., Wu, Y.X., et al., 2009. Cadmium-induced functional and ultrastructural alterations in roots of two transgenic cotton cultivars. Journal of Hazardous Materials 161 (1), 463–473.

Davletova, S., Rizhsky, L., Liang, H., Shengqiang, Z., Oliver, D.J., Coutu, J., et al., 2005. Cytosolic ascorbate peroxidase 1 is a central component of the reactive oxygen gene network of Arabidopsis. The Plant Cell 17 (1), 268–281.

Demidchik, V., Straltsova, D., Medvedev, S.S., Pozhvanov, G.A., Sokolik, A., Yurin, V., 2014. Stress-induced electrolyte leakage: the role of K + -permeable channels and involvement in programmed cell death and metabolic adjustment. Journal of Experimental Botany 65 (5), 1259–1270.

Ding, J., Liang, P., Wu, P., Zhu, M., Li, C., Zhu, X., et al., 2020. Effects of waterlogging on grain yield and associated traits of historic wheat cultivars in the middle and lower reaches of the Yangtze River, China. Field Crops Research 246, 107695.

Doupis, G., Kavroulakis, N., Psarras, G., Papadakis, I.E., 2017. Growth, photosynthetic performance and antioxidative response of 'Hass' and 'Fuerte'avocado (*Persea americana* Mill.) plants grown under high soil moisture. Photosynthetica 55 (4), 655–663.

Dragišić Maksimović, J., Mojović, M., Maksimović, V., Römheld, V., Nikolic, M., 2012. Silicon ameliorates manganese toxicity in cucumber by decreasing hydroxyl radical accumulation in the leaf apoplast. Journal of Experimental Botany 63 (7), 2411–2420.

Ducic, T., Polle, A., 2005. Transport and detoxification of manganese and copper in plants. Brazilian Journal of Plant Physiology 17 (1), 103–112.

Dupuis, I., Dumas, C., 1990. Influence of temperature stress on in vitro fertilization and heat shock protein synthesis in maize (*Zea mays* L.) reproductive tissues. Plant Physiology 94 (2), 665–670.

Ebrahimian, E., Bybordi, A., 2011. Exogenous silicium and zinc increase antioxidant enzyme activity and alleviate salt stress in leaves of sunflower. Journal of Food, Agriculture & Environment 9 (1), 422–427.

Ella, E.S., Kawano, N., Ito, O., 2003. Importance of active oxygen-scavenging system in the recovery of rice seedlings after submergence. Plant Science 165 (1), 85–93.

Fahad, S., Bajwa, A.A., Nazir, U., Anjum, S.A., Farooq, A., Zohaib, A., et al., 2017. Crop production under drought and heat stress: plant responses and management options. Frontiers in plant science 8, 1147.

Faize, M., Burgos, L., Faize, L., Piqueras, A., Nicolas, E., Barba-Espin, G., et al., 2011. Involvement of cytosolic ascorbate peroxidase and Cu/Zn-superoxide dismutase for improved tolerance against drought stress. Journal of Experimental Botany 62 (8), 2599–2613.

Farooq, M., Aziz, T., Wahid, A., Lee, D.J., Siddique, K.H., 2009a. Chilling tolerance in maize: agronomic and physiological approaches. Crop and Pasture Science 60 (6), 501–516.

Farooq, M., Wahid, A., Kobayashi, N.S.M.A., Fujita, D.B.S.M.A., Basra, S.M.A., 2009b. Plant drought stress: effects, mechanisms and management. Sustainable Agriculture 153–188.

Farooq, M., Gogoi, N., Barthakur, S., Baroowa, B., Bharadwaj, N., Alghamdi, S.S., et al., 2017. Drought stress in grain legumes during reproduction and grain filling. Journal of Agronomy and Crop Science 203 (2), 81–102.

Fisarakis, I., Chartzoulakis, K., Stavrakas, D., 2001. Response of Sultana vines (*V. vinifera* L.) on six rootstocks to NaCl salinity exposure and recovery. Agricultural Water Management 51 (1), 13–27.

Fleck, A.T., Nye, T., Repenning, C., Stahl, F., Zahn, M., Schenk, M.K., 2011. Silicon enhances suberization and lignification in roots of rice (*Oryza sativa*). Journal of Experimental Botany 62 (6), 2001–2011.

Foyer, C.H., Halliwell, B., 1976. The presence of glutathione and glutathione reductase in chloroplasts: a proposed role in ascorbic acid metabolism. Planta 133 (1), 21–25.

Fu, X.Z., Xing, F., Wang, N.Q., Peng, L.Z., Chun, C.P., Cao, L., et al., 2014. Exogenous spermine pretreatment confers tolerance to combined high-temperature and drought stress in vitro in trifoliate orange seedlings via modulation of antioxidative capacity and expression of stress-related genes. Biotechnology & Biotechnological Equipment 28 (2), 192–198.

Fukao, T., Yeung, E., Bailey-Serres, J., 2011. The submergence tolerance regulator SUB1A mediates crosstalk between submergence and drought tolerance in rice. Plant Cell 23, 412–427.

Gadjev, I., Vanderauwera, S., Gechev, T.S., Laloi, C., Minkov, I.N., Shulaev, V., et al., 2006. Transcriptomic footprints disclose specificity of reactive oxygen species signaling in Arabidopsis. Plant Physiology 141 (2), 436–445.

Gajalakshmi, S., Iswarya, V., Ashwini, R., Divya, G., Mythili, S., Sathiavelu, A., 2012. Evaluation of heavy metals in medicinal plants growing in Vellore District. European Journal of Experimental Biology 2 (5), 1457–1461.

Gao, J.J., Tao, L.I., Yu, X.C., 2009. Gene expression and activities of SOD in cucumber seedlings were related with concentrations of Mn2 + , Cu2 + , or Zn2 + under low temperature stress. Agricultural Sciences in China 8 (6), 678–684.

García-Sánchez, F., Syvertsen, J.P., Gimeno, V., Botía, P., Perez-Perez, J.G., 2007. Responses to flooding and drought stress by two citrus rootstock seedlings with different water-use efficiency. Physiologia Plantarum 130 (4), 532–542.

Gill, S.S., Tuteja, N., 2010. Reactive oxygen species and antioxidant machinery in abiotic stress tolerance in crop plants. Plant Physiology and Biochemistry 48 (12), 909–930.

Gill, S.S., Gill, R., Tuteja, R., Tuteja, N., 2014. Genetic engineering of crops: a ray of hope for enhanced food security. Plant Signaling & Behavior 9 (3), e28545.

Gill, S.S., Anjum, N.A., Gill, R., Tuteja, N., 2016. Abiotic stress signaling in plants—an overview. Abiotic Stress Response in Plants Edn .

Gomes-Filho, E., Lima, C.R.F.M., Costa, J.H., da Silva, A.C.M., Lima, M.D.G.S., de Lacerda, C.F., et al., 2008. Cowpea ribonuclease: properties and effect of NaCl-salinity on its activation during seed germination and seedling establishment. Plant Cell Reports 27 (1), 147–157.

González, A., Steffen, K.L., Lynch, J.P., 1998. Light and excess manganese: implications for oxidative stress in common bean. Plant Physiology 118 (2), 493–504.

Gueta-Dahan, Y., Yaniv, Z., Zilinskas, B.A., Ben-Hayyim, G., 1997. Salt and oxidative stress: similar and specific responses and their relation to salt tolerance in citrus. Planta 203 (4), 460–469.

Gunes, A., Cicek, N., Inal, A., Alpaslan, M., Eraslan, F., Guneri, E., et al., 2006. Genotypic response of chickpea (*Cicer arietinum* L.) cultivars to drought stress implemented at pre-and post-anthesis stages and its relations with nutrient uptake and efficiency. Plant Soil and Environment 52 (8), 368.

Guo, S.Q., Huang, J., Jiang, Y., Zhang, H.S., 2007. Cloning and characterization of RZF71 encoding a C2H2-type zinc finger protein from rice. Yi chuan = Hereditas 29 (5), 607–613.

Guo, Y.H., Yu, Y.P., Wang, D., Wu, C.A., Yang, G.D., Huang, J.G., et al., 2009. GhZFP1, a novel CCCH-type zinc finger protein from cotton, enhances salt stress tolerance and fungal disease resistance in transgenic tobacco by interacting with GZIRD21A and GZIPR5. New Phytologist 183 (1), 62–75.

Guo, X., Zhang, L., Zhu, J., Liu, H., Wang, A., 2017. Cloning and characterization of SiDHN, a novel dehydrin gene from Saussurea involucrata Kar. et Kir. that enhances cold and drought tolerance in tobacco. Plant Science 256, 160–169.

Gupta, B., Huang, B., 2014. Mechanism of salinity tolerance in plants: physiological, biochemical, and molecular characterization. International Journal of Genomics 2014.

Hafeez, B., Khanif, Y.M., Saleem, M., 2013. Role of zinc in plant nutrition-a review. Journal of Experimental Agriculture International 374–391.

Hajheidari, M., Abdollahian-Noghabi, M., Askari, H., Heidari, M., Sadeghian, S.Y., Ober, E.S., et al., 2005. Proteome analysis of sugar beet leaves under drought stress. Proteomics 5 (4), 950–960.

Hakim, M.A., Hossain, A., da Silva, J.A.T., Zvolinsky, V.P., Khan, M.M., 2012. Protein and starch content of 20 wheat (*Triticum aestivum* L.) genotypes exposed to high temperature under late sowing conditions. Journal of Scientific Research 4 (2), 477.

Han, G., Lu, C., Guo, J., Qiao, Z., Sui, N., Qiu, N., et al., 2020. C2H2 zinc finger proteins: master regulators of abiotic stress responses in plants. Frontiers in Plant Science 11, 115.

Hänsch, R., Mendel, R.R., 2009. Physiological functions of mineral micronutrients (cu, Zn, Mn, Fe, Ni, Mo, B, cl). Current Opinion in Plant Biology 12 (3), 259–266.

Harrison, M.D., Jones, C.E., Dameron, C.T., 1999. Copper chaperones: function, structure and copper-binding properties. JBIC Journal of Biological Inorganic Chemistry 4 (2), 145–153.

Hartmann, H.T., Kester, D.E., 1975. Plant Propagation: Principles and Practices. Prentice-Hall.

Hasanuzzaman, M., Fujita, M., 2012a. Heavy metals in the environment: current status, toxic effects on plants and possible phytoremediation. Phytotechnologies: Remediation of Environmental Contaminants. CRC Press, Boca Raton, pp. 7–73.

Hasanuzzaman, M., Fujita, M., 2012b. Selenium and plants' health: the physiological role of selenium. Selenium: Sources, Functions and Health Effects 101–122.

Hasanuzzaman, M., Fujita, M., 2013. Exogenous sodium nitroprusside alleviates arsenic-induced oxidative stress in wheat (*Triticum aestivum* L.) seedlings by enhancing antioxidant defense and glyoxalase system. Ecotoxicology (London, England) 22 (3), 584–596.

Hasanuzzaman, M., Fujita, M., Islam, M.N., Ahamed, K.U., Nahar, K., 2009. Performance of four irrigated rice varieties under different levels of salinity stress. International Journal of Integrative Biology 6 (2), 85–90.

Hasanuzzaman, M., Hossain, M.A., Fujita, M., 2011a. Nitric oxide modulates antioxidant defense and the methylglyoxal detoxification system and reduces salinity-induced damage of wheat seedlings. Plant Biotechnology Reports 5 (4), 353–365.

Hasanuzzaman, M., Hossain, M.A., Fujita, M., 2011b. Selenium-induced up-regulation of the antioxidant defense and methylglyoxal detoxification system reduces salinity-induced damage in rapeseed seedlings. Biological Trace Element Research 143 (3), 1704–1721.

Hasanuzzaman, M., Hossain, M.A., da Silva, J.A.T., Fujita, M., 2012a. Plant response and tolerance to abiotic oxidative stress: antioxidant defense is a key factor. Crop Stress and Its Management: Perspectives and Strategies. Springer, Dordrecht, pp. 261–315.

Hasanuzzaman, M., Hossain, M.A., Fujita, M., 2012b. Exogenous selenium pretreatment protects rapeseed seedlings from cadmium-induced oxidative stress by upregulating antioxidant defense and methylglyoxal detoxification systems. Biological Trace Element Research 149 (2), 248–261.

Hasanuzzaman, M., Gill, S.S., Fujita, M., 2013a. Physiological role of nitric oxide in plants grown under adverse environmental conditions. Plant Acclimation to Environmental Stress. Springer, New York, NY, pp. 269–322.

Hasanuzzaman, M., Nahar, K., Fujita, M., 2013b. Extreme temperature responses, oxidative stress and antioxidant defense in plants. Abiotic Stress-Plant Responses and Applications in Agriculture 13, 169–205.

Hasanuzzaman, M., Nahar, K., Fujita, M., 2013c. Plant response to salt stress and role of exogenous protectants to mitigate salt-induced damages. Ecophysiology and Responses of Plants Under Salt Stress. Springer, New York, NY, pp. 25–87.

Hasanuzzaman, M., Nahar, K., Fujita, M., Ahmad, P., Chandna, R., Prasad, M.N.V., et al., 2013d. Enhancing plant productivity under salt stress: relevance of poly-omics. Salt Stress in Plants 113–156.

Hasanuzzaman, M., Alam, M., Rahman, A., Hasanuzzaman, M., Nahar, K., Fujita, M., 2014a. Exogenous proline and glycine betaine mediated upregulation of antioxidant defense and glyoxalase systems provides better protection against salt-induced oxidative stress in two rice (*Oryza sativa* L.) varieties. BioMed Research International 2014.

Hasanuzzaman, M., Nahar, K., Fujita, M., 2014b. Silicon and selenium: two vital trace elements that confer abiotic stress tolerance to plants. Emerging Technologies and Management of Crop Stress Tolerance. Academic press, pp. 377–422.

Hasanuzzaman, M., Nahar, K., Alam, M.M., Bhowmik, P.C., Hossain, M.A., Rahman, M.M., et al., 2014c. Potential use of halophytes to remediate saline soils. BioMed research international 2014 (1), 589341.

Hasanuzzaman, M., Nahar, K., Fujita, M., 2015. Arsenic toxicity in plants and possible remediation. Soil Remediation and Plants: Prospects and Challenges 433–501.

Hasanuzzaman, M., Nahar, K., Rahman, A., Mahmud, J.A., Hossain, M.S., Fujita, M., 2016. Soybean production and environmental stresses. In: Miransari, M. (Ed.), Environmental Stresses in Soybean Production: Soybean Production, Vol. 2. Elsevier, New York, pp. 61–102.

Hashem, A., Abd_Allah, E.F., Alqarawi, A.A., Al Huqail, Egamberdieva, D., 2014. Alleviation of abiotic salt stress in Ochradenus baccatus (Del.) by Trichoderma hamatum (Bonord.) Bainier. Journal of Plant Interactions 9 (1), 857–868.

Hassan, M.J., Zhang, G., Wu, F., Wei, K., Chen, Z., 2005. Zinc alleviates growth inhibition and oxidative stress caused by cadmium in rice. Journal of Plant Nutrition and Soil Science 168 (2), 255–261.

Hatfield, J.L., Boote, K.J., Kimball, B.A., Ziska, L.H., Izaurralde, R.C., Ort, D., et al., 2011. Climate impacts on agriculture: implications for crop production. Agronomy Journal 103 (2), 351–370.

Hattori, Y., Nagai, K., Furukawa, S., Song, X.-J., Kawano, R., Sakakibara, H., et al., 2009. The ethylene response factors SNORKEL1 and SNORKEL2 allow rice to adapt to deep water. Nature 460, 1026–1030. Available from: https://doi.org/10.1038/nature08258.

He, F., Wang, H.L., Li, H.G., Su, Y., Li, S., Yang, Y., et al., 2018. Pe CHYR 1, a ubiquitin E3 ligase from *Populus euphratica*, enhances drought tolerance via ABA-induced stomatal closure by ROS production in Populus. Plant Biotechnology Journal 16 (8), 1514–1528.

Hedhly, A., 2011. Sensitivity of flowering plant gametophytes to temperature fluctuations. Environmental and Experimental Botany 74, 9–16.

Hedhly, A., Hormaza, J.I., Herrero, M., 2009. Global warming and sexual plant reproduction. Trends in Plant Science 14 (1), 30–36.

Hernandez, J.A., Olmos, E., Corpas, F.J., Sevilla, F., Del Rio, L.A., 1995. Salt-induced oxidative stress in chloroplasts of pea plants. Plant Science 105 (2), 151–167.

Herzog, M., Striker, G.G., Colmer, T.D., Pedersen, O., 2016. Mechanisms of waterlogging tolerance in wheat–a review of root and shoot physiology. Plant, Cell & Environment 39 (5), 1068–1086.

Hirabayashi, Y., Mahendran, R., Koirala, S., Konoshima, L., Yamazaki, D., Watanabe, S., et al., 2013. Global flood risk under climate change. Nature Climate Change 3, 816–821. Available from: https://doi.org/10.1038/nclimate1911.

Hossain, Z., López-Climent, M.F., Arbona, V., Pérez-Clemente, R.M., Gómez-Cadenas, A., 2009a. Modulation of the antioxidant system in citrus under waterlogging and subsequent drainage. Journal of Plant Physiology 166, 1391–1404.

Hossain, A., Sarker, M.A.Z., Saifuzzaman, M., Akhter, M.M., Mandal, M.S.N., 2009b. Effect of sowing dates on yield of wheat varieties and lines developed since 1998. Bangladesh Journal of Progressive Science and Technology 7 (1), 5–8.

Hossain, A., Sarker, M.A.Z., Hakin, M.A., Lozovskaya, M.V., Zvolinsky, V.P., 2011. Effect of temperature on yield and some agronomic characters of spring wheat (*Triticum aestivum* L.) genotypes. International Journal of Agricultural Research, Innovation and Technology (IJARIT) 1 (2355-2020-1489), 44–54.

Hossain, A., da Silva, J.A.T., Lozovskaya, M.V., Zvolinsky, V.P., 2012. The effect of high temperature stress on the phenology, growth and yield of five wheat (*Triticum aestivum* L.) Varieties. Asian and Australasian Journal of Plant Science and Biotechnology 6 (1), 14–23.

Hossain, M.M., Liu, X., Qi, X., Lam, H.M., Zhang, J., 2014. Differences between soybean genotypes in physiological response to sequential soil drying and rewetting. The Crop Journal 366–380.

Huang, J., Yang, X., Wang, M.M., Tang, H.J., Ding, L.Y., Shen, Y., et al., 2007. A novel rice C2H2-type zinc finger protein lacking DLN-box/EAR-motif plays a role in salt tolerance. Biochimica et Biophysica Acta (BBA)-Gene Structure and Expression 1769 (4), 220–227.

Huang, X.Y., Chao, D.Y., Gao, J.P., Zhu, M.Z., Shi, M., Lin, H.X., 2009a. A previously unknown zinc finger protein, DST, regulates drought and salt tolerance in rice via stomatal aperture control. Genes & Development 23 (15), 1805–1817.

Huang, J., Sun, S.J., Xu, D.Q., Yang, X., Bao, Y.M., Wang, Z.F., et al., 2009b. Increased tolerance of rice to cold, drought and oxidative stresses mediated by the overexpression of a gene that encodes the zinc finger protein ZFP245. Biochemical and Biophysical Research Communications 389 (3), 556–561.

Huang, S., Clark, R.J., Zhu, L., 2007. Highly sensitive fluorescent probes for zinc ion based on triazolyl-containing tetradentate coordination motifs. Organic Letters 9 (24), 4999–5002.

Huang, J., Sun, S., Xu, D., Lan, H., Sun, H., Wang, Z., et al., 2012. A TFIIIA-type zinc finger protein confers multiple abiotic stress tolerances in transgenic rice (*Oryza sativa* L.). Plant Molecular Biology 80 (3), 337–350.

Hurry, V.M., Malmberg, G., Gardestrom, P., Oquist, G., 1994. Effects of a short-term shift to low temperature and of long-term cold hardening on photosynthesis and ribulose-1, 5-bisphosphate carboxylase/oxygenase and sucrose phosphate synthase activity in leaves of winter rye (*Secale cereale* L.). Plant Physiology 106 (3), 983–990.

Hussein, M., Embiale, A., Husen, A., Aref, I.M., Iqbal, M., 2017. Salinity-induced modulation of plant growth and photosynthetic parameters in faba bean (*Vicia faba*) cultivars. Pakistan Journal of Botany 49 (3), 867–877.

Ibrahim, M., Balakrishnan, R., Shamsudeen, S., Adam, F., Bhawani, S., 2012. A concise review of the natural existance, synthesis, properties, and applications of syringaldehyde. BioResources 7 (3), 4377–4399.

Iida, A., Kazuoka, T., Torikai, S., Kikuchi, H., Oeda, K., 2000. A zinc finger protein RHL41 mediates the light acclimatization response in Arabidopsis. The Plant Journal 24 (2), 191–203.

Ikenaka, Y., Nakayama, S.M., Muzandu, K., Choongo, K., Teraoka, H., Mizuno, N., et al., 2010. Heavy metal contamination of soil and sediment in Zambia. African Journal of Environmental Science and Technology 4 (11), 729–739.

Indexmundi, 2011. Bangladesh wheat production by year: market year, production (1000MT) and growth rate (%). Available online: http://www.indexmundi.com/agriculture/?

Ismail, A.M., Hall, A.E., 1999. Reproductive-stage heat tolerance, leaf membrane thermostability and plant morphology in cowpea. Crop Science 39 (6), 1762–1768.

Iyaka, Y.A., 2011. Nickel in soils: a review of its distribution and impacts. Scientific Research and Essays 6 (33), 6774–6777.

Jackson, M.B., Ram, P.C., 2003. Physiological and molecular basis of susceptibility and tolerance of rice plants to complete submergence. Annals of Botany 91 (2), 227–241.

Jain, M., Prasad, P.V., Boote, K.J., Hartwell, A.L., Chourey, P.S., 2007. Effects of season-long high temperature growth conditions on sugar-to-starch metabolism in developing microspores of grain sorghum (*Sorghum bicolor* L. Moench). Planta 227 (1), 67–79.

Jaleel, C.A., Manivannan, P., Sankar, B., Kishorekumar, A., Gopi, R., Somasundaram, R., et al., 2007. Induction of drought stress tolerance by ketoconazole in *Catharanthus roseus* is mediated by enhanced antioxidant potentials and secondary metabolite accumulation. Colloids and Surfaces B: Biointerfaces 60 (2), 201–206.

James, R.A., von Caemmerer, S., Condon, A.T., Zwart, A.B., Munns, R., 2008. Genetic variation in tolerance to the osmotic stress componentof salinity stress in durum wheat. Functional Plant Biology 35 (2), 111–123.

Jiang, L., Tang, Y., Zhang, R., Wang, W., 2000. Effect of fertilizer activating agent on the ascorbate-glutathione cycle and potassium content in tobacco leaves. Journal of Nanjing Agricultural University 23 (4), 13–16.

Jiang, Q.W., Kiyoharu, O., Ryozo, I., 2002. Two novel mitogen-activated protein signaling components, OsMEK1 and OsMAP1, are involved in a moderate low-temperature signaling pathway in Rice. Plant Physiology 129, 1880–1891.

Jiang, W., Sun, X.H., Xu, H.L., Mantri, N., Lu, H.F., 2014. Optimal concentration of zinc sulfate in foliar spray to alleviate salinity stress in *Glycine soja*.

Jubany-Mar, T., Munn Bosch, S., Alegre, L., 2010. Redox regulation of water stress responses in field-grown plants. Role of hydrogen peroxide and ascorbate. Plant Physiology Biochemistry 48, 351–358.

Kalbarczyk, R., 2009. Potential reduction in cucumber yield (*Cucumis sativus* L.) in Poland caused by unfavourable thermal conditions of soil. Acta Scientiarum Polonorum Hortorum Cultus 8, 45–58.

Karam, F., Lahoud, R., Masaad, R., Kabalan, R., Breidi, J., Chalita, C., et al., 2007. Evapotranspiration, seed yield and water use efficiency of drip irrigated sunflower under full and deficit irrigation conditions. Agricultural Water Management 90 (3), 213–223.

Karim, M.R., Zhang, Y.Q., Zhao, R.R., Chen, X.P., Zhang, F.S., Zou, C.Q., 2012. Alleviation of drought stress in winter wheat by late foliar application of zinc, boron, and manganese. Journal of Plant Nutrition and Soil Science 175 (1), 142–151.

Kaveh, H., Nemati, H., Farsi, M., Jartoodeh, S.V., 2011. How salinity affect germination and emergence of tomato lines. Journal of Biological & Environmental Sciences 5 (15), 159–163.

Kavvadias, V.A., Miller, H.G., 1999. Manganese and calcium nutrition of *Pinus sylvestris* and *Pinus nigra* from two different origins I. Manganese. Forestry 72 (1), 35–45.

Khan, M.R., Khan, M.M., 2010. Effect of varying concentration of nickel and cobalt on the plant growth and yield of chickpea. Australian Journal of Basic and Applied Sciences 4 (6), 1036–1046.

Khan, H.R., McDonald, G.K., Rengel, Z., 2003. Zn fertilization improves water use efficiency, grain yield and seed Zn content in chickpea. Plant and Soil 249 (2), 389–400.

Khan, M.A., Ḵẖān, M.A., Weber, D.J. (Eds.), 2006. Ecophysiology of High Salinity Tolerant Plants, Vol. 40. Springer Science & Business Media.

Khan, M., Khan, M.D., Basharat, A.L.I., Muhammad, N., Shuijin, Z.H.U., 2014. Differential physiological and ultrastructural responses of cottonseeds under Pb toxicity. Polish Journal of Environmental Studies 23 (6), 2063–2070.

Khan, R., Gul, S., Hamayun, M., Shah, M., Sayyed, A., Ismail, H., et al., 2016. Effect of foliar application of zinc and manganese on growth and some biochemical constituents of *Brassica junceae* grown under water stress. Journal of Agriculture and Environmental Sciences 16, 984–997.

Khan, M., Ahmad, R., Khan, M.D., Rizwan, M., Ali, S., Khan, M.J., et al., 2018. Trace elements in abiotic stress tolerance. Plant Nutrients and Abiotic Stress Tolerance. Springer, Singapore, pp. 137–151.

Kim, J.C., Lee, S.H., Cheong, Y.H., Yoo, C.M., Lee, S.I., Chun, H.J., et al., 2001. A novel cold-inducible zinc finger protein from soybean, SCOF-1, enhances cold tolerance in transgenic plants. The Plant Journal: For Cell and Molecular Biology 25, 247–259. Available from: https://doi.org/10.1046/j.1365-313x.2001.00947.x.

Kim, Y.H., Kim, M.D., Park, S.C., Jeong, J.C., Kwak, S.S., Lee, H.S., 2016. Transgenic potato plants expressing the cold-inducible transcription factor SCOF-1 display enhanced tolerance to freezing stress. Plant Breeding 135 (4), 513–518.

Klang-Westin, E., Eriksson, J., 2003. Potential of Salix as phytoextractor for Cd on moderately contaminated soils. Plant and Soil 249 (1), 127–137.

Kodaira, K.S., Qin, F., Tran, L.S.P., Maruyama, K., Kidokoro, S., Fujita, Y., et al., 2011. Arabidopsis Cys2/His2 zinc-finger proteins AZF1 and AZF2 negatively regulate abscisic acid-repressive and auxin-inducible genes under abiotic stress conditions. Plant Physiology 157 (2), 742–756.

Krzyzanowski, F.C., Delouche, J.C., 2011. Germination of cotton seed in relation to temperature. Revista Brasileira de Sementes 33 (3), 543–548.

Kuai, J., Liu, Z., Wang, Y., Meng, Y., Chen, B., Zhao, W., et al., 2014. Waterlogging during flowering and boll forming stages affects sucrose metabolism in the leaves subtending the cotton boll and its relationship with boll weight. Plant Science 223, 79–98.

Kumar, P., Pal, M., Joshi, R., Sairam, R.K., 2013. Yield, growth and physiological responses of mung bean [*Vigna radiata* (L.) Wilczek] genotypes to waterlogging at vegetative stage. Physiology and Molecular Biology of Plants 19 (2), 209–220.

Kumutha, D., Ezhilmathi, K., Sairam, R.K., Srivastava, G.C., Deshmukh, P.S., Meena, R.C., 2009. Waterlogging induced oxidative stress and antioxidant activity in pigeonpea genotypes. Biologia Plantarum 53 (1), 75–84.

Kutman, U.B., Yildiz, B., Ozturk, L., Cakmak, I., 2010. Biofortification of durum wheat with zinc through soil and foliar applications of nitrogen. Cereal Chemistry 87, 1–9.

Latowski, D., Kuczyńska, P., Strzałka, K., 2011. Xanthophyll cycle–a mechanism protecting plants against oxidative stress. Redox Report 16 (2), 78–90.

Lee, H., Guo, Y., Ohta, M., Xiong, L., Stevenson, B., Zhu, J.K., 2002. LOS2, a genetic locus required for cold-responsive gene transcription encodes a bi-functional enolase. The EMBO Journal 21 (11), 2692–2702.

Leisner, C.P., Cousins, A.B., Offermann, S., Okita, T.W., Edwards, G.E., 2010. The effects of salinity on photosynthesis and growth of the single-cell C 4 species *Bienertia sinuspersici* (Chenopodiaceae). Photosynthesis Research 106 (3), 201–214.

Li, W., Khan, M.A., Yamaguchi, S., Kamiya, Y., 2005. Effects of heavy metals on seed germination and early seedling growth of *Arabidopsis thaliana*. Plant Growth Regulation 46 (1), 45–50.

Li, Q., Chen, L.S., Jiang, H.X., Tang, N., Yang, L.T., Lin, Z.H., et al., 2010. Effects of manganese-excess on CO2 assimilation, ribulose-1, 5-bisphosphate carboxylase/oxygenase, carbohydrates and photosynthetic electron transport of leaves, and antioxidant systems of leaves and roots in Citrus grandis seedlings. BMC Plant Biology 10 (1), 1–16.

Li, J., Cang, Z., Jiao, F., Bai, X., Zhang, D., Zhai, R., 2017. Influence of drought stress on photosynthetic characteristics and protective enzymes of potato at seedling stage. Journal of the Saudi Society of Agricultural Sciences 16 (1), 82–88.

Lindey, L., Thomson, P., 2012. High temperature effects on corn and soybean. C.O.R.N Newsletter 2012, 23–26.

Liu, X., Huang, B., 2000. Heat stress injury in relation to membrane lipid peroxidation in creeping bentgrass. Crop Science 40 (2), 503–510.

Liu, Y., Zhou, J., 2018. MAPping kinase regulation of ICE1 in freezing tolerance. Trends in Plant Science 23 (2), 91–93.

Liu, Q., Zhao, N., Yamaguch-Shinozaki, K., Shinozaki, K., 2000. Regulatory role of DREB transcription factors in plant drought, salt and cold tolerance. Chinese Science Bulletin 45, 970–975.

Liu, S., Liu, S., Wang, M., Wei, T., Meng, C., Wang, M., Xia, G., 2014. A wheat SIMILAR TO RCD-ONE gene enhances seedling growth and abiotic stress resistance by modulating redox homeostasis and maintaining genomic integrity. The Plant Cell 26 (1), 164–180.

Liu, Y.J., Yuan, Y., Liu, Y.Y., Liu, Y., Fu, J.J., Zheng, J., Wang, G.Y., 2012. Gene families of maize glutathione–ascorbate redox cycle respond differently to abiotic stresses. Journal of Plant Physiology 169 (2), 183–192.

Liu, D., He, S., Zhai, H., Wang, L., Zhao, Y., Wang, B., et al., 2014a. Overexpression of IbP5CR enhances salt tolerance in transgenic sweetpotato. Plant Cell, Tissue and Organ Culture (PCTOC) 117 (1), 1–16.

Liu, J., Piñeros, M.A., Kochian, L.V., 2014b. The role of aluminum sensing and signaling in plant aluminum resistance. Journal of Integrative Plant Biology 56 (3), 221–230.

Liu, D., He, S., Song, X., Zhai, H., Liu, N., Zhang, D., et al., 2015a. IbSIMT1, a novel salt-induced methyltransferase gene from Ipomoea batatas, is involved in salt tolerance. Plant Cell, Tissue and Organ Culture (PCTOC) 120 (2), 701–715.

Liu, Y., Ji, X., Nie, X., Qu, M., Zheng, L., Tan, Z., et al., 2015b. Arabidopsis Atb HLH 112 regulates the expression of genes involved in abiotic stress tolerance by binding to their E-box and GCG-box motifs. New Phytologist 207 (3), 692–709.

Liu, S.J., Xu, H.H., Wang, W.Q., Li, N., Wang, W.P., Møller, I.M., et al., 2015c. A proteomic analysis of rice seed germination as affected by high temperature and ABA treatment. Physiologia Plantarum 154 (1), 142–161.

Liu, D., Yang, L., Luo, M., Wu, Q., Liu, S., Liu, Y., 2017. Molecular cloning and characterization of PtrZPT2-1, a ZPT2 family gene encoding a Cys2/His2-type zinc finger protein from trifoliate orange (*Poncirus trifoliata* (L.) Raf.) that enhances plant tolerance to multiple abiotic stresses. Plant Science 263, 66–78.

Loka, D.A., Oosterhuis, D.M., 2010. Effect of high night temperatures on cotton respiration, ATP levels and carbohydrate content. Environmental and Experimental Botany 68 (3), 258–263.

Lombardi, T., Lupi, B., 2006. Effect of salinity on the germination and growth of *Hordeum secalinum* Schreber (Poaceae) in relation to the seeds after-ripening time. Atti della Società Toscana di Scienze Naturali, Memorie, Serie B 113, 37–42.

Long, S.P., Ort, D.R., 2010. More than taking the heat: crops and global change. Current Opinion in Plant Biology 13 (3), 240–247.

Lu, K., Li, L., Zheng, X., Zhang, Z., Mou, T., Hu, Z., 2008. Quantitative trait loci controlling Cu, Ca, Zn, Mn and Fe content in rice grains. Journal of Genetics 87 (3), 305.

Luo, H.H., Zhang, Y.L., Zhang, W.F., 2016. Effects of water stress and rewatering on photosynthesis, root activity, and yield of cotton with drip irrigation under mulch. Photosynthetica 54 (1), 65–73.

Lyman, N.B., Jagadish, K.S., Nalley, L.L., Dixon, B.L., Siebenmorgen, T., 2013. Neglecting rice milling yield and quality underestimates economic losses from high-temperature stress. PLoS One 8 (8), e72157.

Lyu, T., Cao, J., 2018. Cys2/His2 zinc-finger proteins in transcriptional regulation of flower development. International Journal of Molecular Sciences 19 (9), 2589.

Kalisz, A., Cebula, S., 2001. Direct plant covering and soil mulching in the spring production of some Chinese cabbage cultivars. Growth and yielding Folia Horticulturae 3–12.

Karim, M.R., Rahman, M.A., 2015. Drought risk management for increased cereal production in Asian least developed countries. Weather and Climate Extremes 24–35.

Ma, J.F., Ryan, P.R., Delhaize, E., 2001. Aluminium tolerance in plants and the complexing role of organic acids. Trends in Plant Science 6 (6), 273–278.

Ma, X., Liang, W., Gu, P., Huang, Z., 2016a. Salt tolerance function of the novel C2H2-type zinc finger protein TaZNF in wheat. Plant Physiology and Biochemistry 106, 129–140.

Ma, D., Sun, D., Wang, C., Qin, H., Ding, H., Li, Y., et al., 2016b. Silicon application alleviates drought stress in wheat through transcriptional regulation of multiple antioxidant defense pathways. Journal of Plant Growth Regulation 35 (1), 1–10.

Ma, D., Sun, D., Wang, C., Ding, H., Qin, H., Hou, J., et al., 2017. Physiological responses and yield of wheat plants in zinc-mediated alleviation of drought stress. Frontiers in Plant Science 8, 860.

Mahajan, S., Tuteja, N., 2005. Cold, salinity and drought stresses: an overview. Archives of Biochemistry and Biophysics 444 (2), 139–158.

Mahar, A., Wang, P., Ali, A., Awasthi, M.K., Lahori, A.H., Wang, Q., et al., 2016. Challenges and opportunities in the phytoremediation of heavy metals contaminated soils: a review. Ecotoxicology and Environmental Safety 126, 111–121.

Maksymiec, W., 1998. Effect of copper on cellular processes in higher plants. Photosynthetica 34 (3), 321–342.

Maksymiec, W., 2007. Signaling responses in plants to heavy metal stress. Acta Physiologiae Plantarum 29 (3), 177.

Malakouti, M.J., 2000. Balanced Nutrition of Wheat: An Approach Towards Self-Sufficiency and Enhancement of National Health. "A compilation of papers.". Ministry of Agriculture, Karaj, Iran, p. 544.

Malakouti, M.J., 2007. Zinc is a neglected element in the life cycle of plants. Middle Eastern and Russian Journal of Plant Science and Biotechnology 1 (1), 1–12.

Malakouti, M.J., Malakouti, A., Bybordi, I., Khamesi, E., 2006. Zinc (Zn) is the neglected element in the life cycle of plant, animal and human health (9th Edn). Technical bulletin No. 475. Sana Publication Co., Ministry of Jihad-eAgriculture, Tehran, Iran, p. 12.

Malakouti, M.J., 1998. Increasing grain yield and community's health through the use of zinc sulfate in wheat fields. Journal of Soil and Water Sciences 12, 34–43.

Malesh, A.A., Mengistu, D.K., Aberra, D.A., 2016. Linking agriculture with health through genetic and agronomic biofortification. Agricultural Sciences 7, 295–307.

Manas, D., Chakravarty, A., Pal, S., Bhattacharya, A., 2014. Influence of foliar applications of chelator and micronutrients on antioxidants in green chilli. International Journal of Nutrition and Metabolism 6 (2), 18–27.

Marcińska, I., Czyczyło-Mysza, I., Skrzypek, E., Filek, M., Grzesiak, S., Grzesiak, M.T., et al., 2013. Impact of osmotic stress on physiological and biochemical characteristics in drought-susceptible and drought-resistant wheat genotypes. Acta Physiologiae Plantarum 35 (2), 451–461.

Marschner, H., 1995. Mineral Nutrition of Higher Plants, second ed. Academic Press, New York, pp. 15–22.

McCord, J.M., Fridovich, I., 1969. Superoxide dismutase: an enzymic function for erythrocuprein (hemocuprein). Journal of Biological Chemistry 244 (22), 6049–6055.

Mickelbart, M.V., Hasegawa, P.M., Bailey-Serres, J., 2015. Genetic mechanisms of abiotic stress tolerance that translate to crop yield stability. Nature Reviews. Genetics 16 (4), 237–251.

Miller, P., Lanier, W., Brandt, S., 2001. Using growing degree days to predict plant stages, Ag/ Extension Communications Coordinator, Communications Services, 59717. Montana State University-Bozeman, Bozeman, MO, pp. 994–2721, 406.

Miller, G.A.D., Suzuki, N., Ciftci-Yilmaz, S.U.L.T.A.N., Mittler, R.O.N., 2010. Reactive oxygen species homeostasis and signalling during drought and salinity stresses. Plant, Cell & Environment 33 (4), 453–467.

Miransari, M., Smith, D.L., 2014. Plant hormones and seed germination. Environmental and Experimental Botany 99, 110–121.

Mithöfer, A., Schulze, B., Boland, W., 2004. Biotic and heavy metal stress response in plants: evidence for common signals. FEBS Letters 566 (1-3), 1–5.

Mittler, R., 2002. Oxidative stress, antioxidants and stress tolerance. Trends in Plant Science 7 (9), 405–410.

Mittler, R., Blumwald, E., 2010. Genetic engineering for modern agriculture: challenges and perspectives. Annual Review of Plant Biology 61, 443–462.

Mittler, R., Kim, Y., Song, L., Coutu, J., Coutu, A., Ciftci-Yilmaz, S., et al., 2006. Gain-and loss-of-function mutations in Zat10 enhance the tolerance of plants to abiotic stress. FEBS Letters 580 (28-29), 6537–6542.

Mittler, R., Finka, A., Goloubinoff, P., 2012. How do plants feel the heat? Trends in Biochemical Sciences 37 (3), 118–125.

Mohammed, A.R., Tarpley, L., 2009a. High nighttime temperatures affect rice productivity through altered pollen germination and spikelet fertility. Agricultural and Forest Meteorology 149 (6-7), 999–1008.

Mohammed, A.R., Tarpley, L., 2009b. Impact of high nighttime temperature on respiration, membrane stability, antioxidant capacity, and yield of rice plants. Crop Science 49 (1), 313–322.

Monjezi, F., Vazin, F., Hassanzadehdelouei, M., 2013. Effects of iron and zinc spray on yield and yield components of wheat (*Triticum aestivum* L.) in drought stress. Cercetari Agronomice in Moldova 46 (1), 23–32.

Morison, J.I.L., Baker, N.R., Mullineaux, P.M., Davies, W.J., 2008. Improving water use in crop production. Philosophical Transactions of the Royal Society B: Biological Sciences 363 (1491), 639–658.

Mossor-Pietraszewska, T., 2001. Effect of aluminium on plant growth and metabolism. Acta Biochimica Polonica 48 (3), 673–686.

Moussa, H.R., Abdel-Aziz, S.M., 2008. Comparative response of drought tolerant and drought sensitive maize genotypes to water stress. Australian Journal of Crop Science 1 (1), 31–36.

Moustakas, N.K., Akoumianaki-Ioannidou, A., Barouchas, P.E., 2011. The effects of cadmium and zinc interactions on the concentration of cadmium and zinc in pot marigold ('*Calendula officinalis*' L.). Australian Journal of Crop Science 5 (3), 277.

Mullineaux, P., Karpinski, S., 2002. Signal transduction in response to excess light: getting out of the chloroplast. Current Opinion in Plant Biology 5 (1), 43–48.

Munns, R., 2002. Comparative physiology of salt and water stress. Plant, Cell & Environment 25 (2), 239–250.

Munns, R., Tester, M., 2008. Mechanisms of salinity tolerance. Annual Review of Plant Biology 59, 651–681.

Muthamilarasan, M., Bonthala, V.S., Mishra, A.K., Khandelwal, R., Khan, Y., Roy, R., et al., 2014. C 2 H 2 type of zinc finger transcription factors in foxtail millet define response to abiotic stresses. Functional & Integrative Genomics 14 (3), 531–543.

Nahar, K., Hasanuzzaman, M., 2009. Germination, growth, nodulation and yield performance of three mungbean varieties under different levels of salinity stress. Green Farming 2 (12), 825–829.

Nahar, K., Ahamed, K.U., Fujita, M., 2010. Phenological variation and its relation with yield in several wheat (*Triticum aestivum* L.) cultivars under normal and late sowing mediated heat stress condition. Notulae Scientia Biologicae 2 (3), 51–56.

Nahar, K., Biswas, J.K., Shamsuzzaman, A.M.M., 2012. Cold Stress Tolerance in Rice Plant: Screening of Genotypes Based on Morphophysiological Traits. LAP Lambert Academic, Berlin.

Nahar, K., Hasanuzzaman, M., Ahamed, K.U., Hakeem, K.R., Ozturk, M., Fujita, M., 2015a. Plant responses and tolerance to high temperature stress: role of exogenous phytoprotectants. Crop Production and Global Environmental Issues. Springer, Cham, pp. 385–435.

Nahar, K., Hasanuzzaman, M., Alam, M., Fujita, M., 2015b. Exogenous spermidine alleviates low temperature injury in mung bean (*Vigna radiata* L.) seedlings by modulating ascorbate-glutathione and glyoxalase pathway. International Journal of Molecular Sciences 16 (12), 30117–30132.

Nahar, K., Hasanuzzaman, M., Alam, M.M., Rahman, A., Mahmud, J.A., Suzuki, T., et al., 2017. Insights into spermine-induced combined high temperature and drought tolerance in mung bean: osmoregulation and roles of antioxidant and glyoxalase system. Protoplasma 254 (1), 445–460.

Nelson, J.M., Palzkill, D.A., Bartels, P.G., 1993. Irrigation cut-off date affects growth, frost damage, and yield of jojoba. Journal of the American Society for Horticultural Science 118 (6), 731–735.

Nicholls, A.M., Mal, T.K., 2003. Effects of lead and copper exposure on growth of an invasive weed, *Lythrum salicaria* L. (Purple Loosestrife).

Nishiuchi, S., Yamauchi, T., Takahashi, H., Kotula, L., Nakazono, M., 2012. Mechanisms for coping with submergence and waterlogging in rice. Rice 5 (1), 1–14.

Niu, G., Rodriguez, D.S., Wang, Y.T., 2006. Impact of drought and temperature on growth and leaf gas exchange of six bedding plant species under greenhouse conditions. HortScience: A Publication of the American Society for Horticultural Science 41 (6), 1408–1411.

Obidiegwu, J.E., Bryan, G.J., Jones, H.G., Prashar, A., 2015. Coping with drought: stress and adaptive responses in potato and perspectives for improvement. Frontiers in Plant Science 6, 542.

Oliver, M.A., 1997. Soil and human health: a review. European Journal of Soil Science 48 (4), 573–592.

Osredkar, J., Sustar, N., 2011. Copper and zinc, biological role and significance of copper/zinc imbalance. Journal of Clinical Toxicology Open Access 3 (2161), 0495.

Othman, Y., Al-Karaki, G., Al-Tawaha, A.R., Al-Horani, A., 2006. Variation in germination and ion uptake in barley genotypes under salinity conditions. World Journal of Agricultural Sciences 2 (1), 11–15.

Pandey, G.K., 2015. Elucidation of Abiotic Stress Signaling in Plants: Functional Genomics Perspectives, Vol. 2. Springer, New York.

Pareek, A., Khurana, A., K Sharma, A., Kumar, R., 2017. An overview of signaling regulons during cold stress tolerance in plants. Current Genomics 18 (6), 498–511.

Pastori, G.M., Foyer, C.H., 2002. Common components, networks, and pathways of cross-tolerance to stress. The central role of "redox" and abscisic acid-mediated controls. Plant Physiology 129 (2), 460–468.

Peacock, J.M., Miller, W.B., Matsuda, K., Robinson, D.L., 1990. Role of heat girdling in early seedling death of sorghum. Crop Science 30 (1), 138–143.

Pearson, J.N., Rengel, Z., 1994. Distribution and remobilization of Zn and Mn during grain development in wheat. Journal of Experimental Botany 281, 1829–1835.

Peck, A.W., McDonald, G.K., 2010. Adequate zinc nutrition alleviates the adverse effects of heat stress in bread wheat. Plant and soil 355–374.

Pedersen, O., Rich, S.M., Colmer, T.D., 2009. Surviving floods: leaf gas films improve O2 and CO2 exchange, root aeration, and growth of completely submerged rice. The Plant Journal 58 (1), 147–156.

Peleg, Z., Saranga, Y., Yazici, A., Fahima, T., Ozturk, L., Cakmak, I., 2008. Grain zinc, iron and protein concentrations and zinc-efficiency in wild emmer wheat under contrasting irrigation regimes. Plant and Soil 306 (1), 57–67.

Peng, S., Huang, J., Sheehy, J.E., Laza, R.C., Visperas, R.M., Zhong, X., et al., 2004. Rice yields decline with higher night temperature from global warming. Proceedings of the National Academy of Sciences 101 (27), 9971–9975.

Perl-Treves, R., Galun, E., 1991. The tomato Cu, Zn superoxide dismutase genes are developmentally regulated and respond to light and stress. Plant Molecular Biology 17 (4), 745–760.

Pirovano, L., Morgutti, S., Espen, L., Cocucci, S.M., 1997. Differences in the reactivation process in thermosensitive seeds of *Phacelia tanacetifolia* with germination inhibited by high temperature (30 C). Physiologia Plantarum 99 (2), 211–220.

Poletti, S., Gruissem, W., Sautter, C., 2004. The nutritional fortification of cereals. Current Opinion Biotechnology 15, 162–165.

Porch, T.G., 2006. Application of stress indices for heat tolerance screening of common bean. Journal of Agronomy and Crop Science 192 (5), 390–394.

Praharaj, S., Skalicky, M., Maitra, S., Bhadra, P., Shankar, T., Brestic, M., et al., 2021. Zinc biofortification in food crops could alleviate the zinc malnutrition in human health. Molecules (Basel, Switzerland) 26 (12), 3509. Available from: https://doi.org/10.3390/molecules26123509.

Prasad, P.V.V., Pisipati, S.R., Momčilović, I., Ristic, Z., 2011. Independent and combined effects of high temperature and drought stress during grain filling on plant yield and chloroplast EF-Tu expression in spring wheat. Journal of Agronomy and Crop Science 197 (6), 430–441.

Prashanth, S.R., Sadhasivam, V., Parida, A., 2008. Over expression of cytosolic copper/zinc superoxide dismutase from a mangrove plant *Avicennia marina* in indica rice var Pusa Basmati-1 confers abiotic stress tolerance. Transgenic Research 17 (2), 281–291.

Qados, A.M.A., 2011. Effect of salt stress on plant growth and metabolism of bean plant *Vicia faba* (L.). Journal of the Saudi Society of Agricultural Sciences 10 (1), 7–15.

Qi, X., Li, Q., Ma, X., Qian, C., Wang, H., Ren, N., et al., 2019. Waterlogging-induced adventitious root formation in cucumber is regulated by ethylene and auxin through reactive oxygen species signalling. Plant, Cell & Environment 42 (5), 1458–1470.

Rahbarian, R., Khavari-Nejad, R., Ganjeali, A., Bagheri, A., Najafi, F., 2011. Drought stress effects on photosynthesis, chlorophyll fluorescence and water relations in tolerant and susceptible chickpea (*Cicer arietinum* L.) genotypes. Acta Biologica Cracoviensia s. Botanica .

Rahman, S., Matsumuro, T., Miyake, H., Takeoka, Y., 2000. Salinity-induced ultrastructural alterations in leaf cells of rice (*Oryza sativa* L.). Plant Production Science 3 (4), 422–429.

Rahman, M.M., Hossain, A.K.B.A.R., Hakim, M.A., Kabir, M.R., Shah, M.M.R., 2009. Performance of wheat genotypes under optimum and late sowing condition. Int. J. Sustain. Crop Prod 4 (6), 34–39.

Rahman, A., Mostofa, M.G., Nahar, K., Hasanuzzaman, M., Fujita, M., 2016. Exogenous calcium alleviates cadmium-induced oxidative stress in rice (*Oryza sativa* L.) seedlings by regulating the antioxidant defense and glyoxalase systems. Brazilian Journal of Botany 39 (2), 393–407.

Ram, P.C., Singh, A.K., Singh, B.B., Singh, V.K., Singh, H.P., Setter, T.L., et al., 1999. Environmental characterization of floodwater in eastern India: relevance to submergence tolerance of lowland rice. Experimental Agriculture 35 (2), 141–152.

Rascio, N., Navari-Izzo, F., 2011. Heavy metal hyperaccumulating plants: how and why do they do it? And what makes them so interesting? Plant Science 180 (2), 169–181.

Rasolohery, C.A., Berger, M., Lygin, A.V., Lozovaya, V.V., Nelson, R.L., Dayde, J., 2008. Effect of temperature and water availability during late maturation of the soybean seed on germ and cotyledon isoflavone content and composition. Journal of the Science of Food and Agriculture 88 (2), 218–228.

Raziuddin, F., Hassan, G., Akmal, M., Shah, S.S., Fida, M., Shafi, M., Bakht, J., Zhou, W., 2011. Effects of cadmium and salinity on growth and photosynthesis parameters of Brassica species. Pakistan Journal of Botany 43 (1), 333–340.

Razzaq, K., Khan, A.S., Malik, A.U., Shahid, M., Ullah, S., 2013. Foliar application of zinc influences the leaf mineral status, vegetative and reproductive growth, yield and fruit quality of 'Kinnow' mandarin. Journal of Plant Nutrition 36 (10), 1479–1495.

Reddy, K.J., 2006. Nutrient stress. In: Madhava, K., Raghavendra, A., Janardhan Reddy, K. (Eds.), Physiology and Molecular Biology of Stress Tolerance in Plants. Springer Netherlands, Dordrecht, pp. 187–217. Available from: https://doi.org/10.1007/1-4020-4225-6_7.

Ren, B., Zhang, J., Li, X., Fan, X., Dong, S., Liu, P., et al., 2014. Effects of waterlogging on the yield and growth of summer maize under field conditions. Canadian Journal of Plant Science 94 (1), 23–31.

Riaz-ud-Din, M.S., Ahmad, N., Hussain, M., Rehman, A.U., 2010. Effect of temperature on development and grain formation in spring wheat. Pakistan Journal of Botany 42 (2), 899–906.

Rice-Evans, C.A., Miller, N.J., Paganga, G., 1997. Antioxidant properties of phenolic compounds. Trends Plant Science 2, 152–159.

Rilkey, G.J., 1981. Effects of high temperature on protein synthesis during germination of maize (*Zea mays* L.). Planta 151 (1), 75–80.

Rizhsky, L., Davletova, S., Liang, H., Mittler, R., 2004. The zinc finger protein Zat12 is required for cytosolic ascorbate peroxidase 1 expression during oxidative stress in Arabidopsis. Journal of Biological Chemistry 279 (12), 11736–11743.

Robertson, D., Zhang, H., Palta, J.A., Colmer, T., Turner, N.C., 2009. Waterlogging affects the growth, development of tillers, and yield of wheat through a severe, but transient, N deficiency. Crop and Pasture Science 60 (6), 578–586.

Rout, G., Samantaray, S., Das, P., 2001. Aluminium toxicity in plants: a review. Agronomie 21 (1), 3–21.

Sairam, R.K., Dharmar, K., Lekshmy, S., Chinnusamy, V., 2011. Expression of antioxidant defense genes in mung bean (*Vigna radiata* L.) roots under water-logging is associated with hypoxia tolerance. Acta Physiologiae Plantarum 33 (3), 735–744.

Sakamoto, H., Araki, T., Meshi, T., Iwabuchi, M., 2000. Expression of a subset of the Arabidopsis Cys2/His2-type zinc-finger protein gene family under water stress. Gene 248 (1–2), 23–32.

Sakata, T., Oshino, T., Miura, S., Tomabechi, M., Tsunaga, Y., Higashitani, N., et al., 2010. Auxins reverse plant male sterility caused by high temperatures. Proceedings of the National Academy of Sciences 107 (19), 8569–8574.

Samarah, N.H., Alqudah, A.M., Amayreh, J.A., McAndrews, G.M., 2009. The effect of late-terminal drought stress on yield components of four barley cultivars. Journal of Agronomy and Crop Science 195 (6), 427–441.

Sánchez-Blanco, M.J., Rodrıguez, P., Morales, M.A., Ortuño, M.F., Torrecillas, A., 2002. Comparative growth and water relations of *Cistus albidus* and *Cistus monspeliensis* plants during water deficit conditions and recovery. Plant Science 162 (1), 107–113.

Sanghera, G.S., Wani, S.H., Hussain, W., Singh, N.B., 2011. Engineering cold stress tolerance in crop plants. Current Genomics 12 (1), 30.

Sapeta, H., Costa, J.M., Lourenco, T., Maroco, J., Van der Linde, P., Oliveira, M.M., 2013. Drought stress response in *Jatropha curcas*: growth and physiology. Environmental and Experimental Botany 85, 76–84.

Sarma, H., Deka, S., Deka, H., Saikia, R.R., 2012. Accumulation of heavy metals in selected medicinal plants. In: Whitacre, D. (Ed.), Reviews of Environmental Contamination and Toxicology, 214. Springer, New York, NY. Available from: https://doi.org/10.1007/978-1-4614-0668-6_4.

Sarwar, S., Rafique, E., Gill, S.M., Khan, M.Z., 2017. Improved productivity and zinc content for maize grain by different zinc fertilization techniques in calcareous soils. Journal of Plant Nutrition 40 (3), 417–426.

Saud, S., Yajun, C., Fahad, S., Hussain, S., Na, L., Xin, L., et al., 2016. Silicate application increases the photosynthesis and its associated metabolic activities in Kentucky bluegrass under drought stress and post-drought recovery. Environmental Science and Pollution Research 23 (17), 17647–17655.

Sayed, O.H., 2003. Chlorophyll fluorescence as a tool in cereal crop research. Photosynthetica 41 (3), 321–330.

Schubert, T.S., 1992. Manganese Toxicity of Plants in Florida, Vol. 353. Florida Department of Agriculture & Consumer Services, Division of Plant Industry.

Schutzendubel, A., Polle, A., 2002. Plant responses to abiotic stresses: heavy metal-induced oxidative stress and protection by mycorrhization. Journal of Experimental Botany 53 (372), 1351–1365.

Sedghi, M., Hadi, M., Toluie, S.G., 2013. Effect of nano zinc oxide on the germination parameters of soybean seeds under drought stress. Annales of West University of Timisoara 16 (2), 73. Series of Biology.

Sehgal, A., Sita, K., Kumar, J., Kumar, S., Singh, S., Siddique, K.H., et al., 2017. Effects of drought, heat and their interaction on the growth, yield and photosynthetic function of lentil (*Lens culinaris* Medikus) genotypes varying in heat and drought sensitivity. Frontiers in Plant Science 8, 1776.

Sethy, S.K., Ghosh, S., 2013. Effect of heavy metals on germination of seeds. Journal of Natural Science, Biology, and Medicine 4 (2), 272.

Setter, T.L., Waters, I., 2003. Review of prospects for germplasm improvement for waterlogging tolerance in wheat, barley and oats. Plant and Soil 253 (1), 1–34.

Shah, N.H., Paulsen, G.M., 2003. Interaction of drought and high temperature on photosynthesis and grain-filling of wheat. Plant and Soil 257 (1), 219–226.

Shahbaz, M., Ashraf, M., 2013. Improving salinity tolerance in cereals. Critical Reviews in Plant Sciences 32 (4), 237–249.

Shanker, A.K., Cervantes, C., Loza-Tavera, H., Avudainayagam, S., 2005. Chromium toxicity in plants. Environment International 31 (5), 739–753.

Sharma, S.S., Dietz, K.J., 2009. The relationship between metal toxicity and cellular redox imbalance. Trends in Plant Science 14 (1), 43–50.

Sharma, P., Dubey, R.S., 2007. Involvement of oxidative stress and role of antioxidative defense system in growing rice seedlings exposed to toxic concentrations of aluminum. Plant Cell Reports 26 (11), 2027–2038.

Sharma, P.N., Tripathi, A., Bisht, S.S., 1995. Zinc requirement for stomatal opening in cauliflower. Plant Physiology 107 (3), 751–756.

Sharma, A., Gontia-Mishra, I., Srivastava, A.K., 2011. Toxicity of heavy metals on germination and seedling growth of *Salicornia brachiata*. Journal of Phytology 3 (9).

Sheldon, A.R., Menzies, N.W., 2005. The effect of copper toxicity on the growth and root morphology of Rhodes grass (*Chloris gayana* Knuth.) in resin buffered solution culture. Plant and Soil 278 (1), 341–349.

Shen, W., Nada, K., Tachibana, S., 2000. Involvement of polyamines in the chilling tolerance of cucumber cultivars. Plant Physiology 124 (1), 431–440.

Shi, Y., Ding, Y., Yang, S., 2018. Molecular regulation of CBF signaling in cold acclimation. Trends in Plant Science 23 (7), 623–637.

Siddiqui, S.N., Umar, S., Iqbal, M., 2015. Zinc-induced modulation of some biochemical parameters in a high-and a low-zinc-accumulating genotype of *Cicer arietinum* L. grown under Zn-deficient condition. Protoplasma 252 (5), 1335–1345.

Singh, H.P., Mahajan, P., Kaur, S., Batish, D.R., Kohli, R.K., 2013. Chromium toxicity and tolerance in plants. Environmental Chemistry Letters 11 (3), 229–254.

Singh, S., Parihar, P., Singh, R., Singh, V.P., Prasad, S.M., 2015. Heavy metal tolerance in plants: role of transcriptomics, proteomics, metabolomics, and ionomics. Frontiers in Plant Science 6, 1143.

Smirnoff, N., 1993. Tansley Review No. 52. The role of active oxygen in the response of plants to water deficit and desiccation. New Phytologist 27–58.

Soceanu, A., Magearu, V., Popescu, V., Matei, N., 2005. Accumulation of manganese and IRON in citrus fruits. Analele Universitatii Bucuresti: Chimie 14 (1-2), 173–177.

Song, C.Z., Liu, M.Y., Meng, J.F., Chi, M., Xi, Z.M., Zhang, Z.W., 2015. Promoting effect of foliage sprayed zinc sulfate on accumulation of sugar and phenolics in berries of *Vitis vinifera* cv. Merlot growing on zinc deficient soil. Molecules (Basel, Switzerland) 20 (2), 2536–2554.

Sreenivasulu, N., Grimm, B., Wobus, U., Weschke, W., 2000. Differential response of antioxidant compounds to salinity stress in salt-tolerant and salt-sensitive seedlings of foxtail millet (*Setaria italica*). Physiologia Plantarum 109 (4), 435–442.

Steffens, B., Geske, T., Sauter, M., 2011. Aerenchyma formation in the rice stem and its promotion by H2O2. New Phytologist 190 (2), 369–378.

Stone, P.J., Nicolas, M.E., 1994. Wheat cultivars vary widely in their responses of grain yield and quality to short periods of post-anthesis heat stress. [Workshop paper]. Australian Journal of Plant Physiology (Australia).

Sudhir, P., Murthy, S.D.S., 2004. Effects of salt stress on basic processes of photosynthesis. Photosynthetica 42 (2), 481–486.

Sui, N., 2015. Photoinhibition of *Suaeda salsa* to chilling stress is related to energy dissipation and water-water cycle. Photosynthetica 53 (2), 207–212.

Sun, Q., Wang, X.R., Ding, S.M., Yuan, X.F., 2005. Effects of exogenous organic chelators on phytochelatins production and its relationship with cadmium toxicity in wheat (*Triticum aestivum* L.) under cadmium stress. Chemosphere 60 (1), 22–31.

Sun, S.J., Guo, S.Q., Yang, X., Bao, Y.M., Tang, H.J., Sun, H., et al., 2010. Functional analysis of a novel Cys2/His2-type zinc finger protein involved in salt tolerance in rice. Journal of Experimental Botany 61 (10), 2807–2818.

Sun, J., Zheng, T., Yu, J., Wu, T., Wang, X., Chen, G., et al., 2017. TSV, a putative plastidic oxidoreductase, protects rice chloroplasts from cold stress during development by interacting with plastidic thioredoxin Z. New Phytologist 215 (1), 240–255.

Sun, X., Zhu, Z., Zhang, L., Fang, L., Zhang, J., Wang, Q., et al., 2019. Overexpression of ethylene response factors VaERF080 and VaERF087 from Vitis amurensis enhances cold tolerance in Arabidopsis. Scientia Horticulturae 243, 320–326.

Sunkar, R., Kapoor, A., Zhu, J.K., 2006. Posttranscriptional induction of two Cu/Zn superoxide dismutase genes in Arabidopsis is mediated by downregulation of miR398 and important for oxidative stress tolerance. The Plant Cell 18 (8), 2051–2065.

Taiz, L., Zeiger, E., 2006. Stress physiology. In: Taiz, L., Zeiger, E. (Eds.), Plant Physiology, fifth ed. Sinauer Associates, Sunderland, pp. 671–681.

Tanaka, Y., Hibino, T., Hayashi, Y., Tanaka, A., Kishitani, S., Takabe, T., et al., 1999. Salt tolerance of transgenic rice overexpressing yeast mitochondrial Mn-SOD in chloroplasts. Plant Science 148 (2), 131–138.

Tangahu, B.V., Sheikh Abdullah, S.R., Basri, H., Idris, M., Anuar, N., Mukhlisin, M., 2011. A review on heavy metals (As, Pb, and Hg) uptake by plants through phytoremediation. International Journal of Chemical Engineering 2011.

Tao, Z., Huang, Y., Zhang, L., Wang, X., Liu, G., Wang, H., 2017. BnLATE, a Cys2/His2-type zinc-finger protein, enhances silique shattering resistance by negatively regulating lignin accumulation in the silique walls of *Brassica napus*. PLoS One 12 (1), e0168046. Available from: https://doi.org/10.1371/journal.pone.01680.

Tavakkoli, E., Fatehi, F., Coventry, S., Rengasamy, P., McDonald, G.K., 2011. Additive effects of Na + and Cl−ions on barley growth under salinity stress. Journal of Experimental Botany 62 (6), 2189–2203.

Tavallali, V., Rahemi, M., Maftoun, M., Panahi, B., Karimi, S., Ramezanian, A., et al., 2009. Zinc influence and salt stress on photosynthesis, water relations, and carbonic anhydrase activity in pistachio. Scientia Horticulturae 123 (2), 272–279.

Tavallali, V., Rahemi, M., Eshghi, S., Kholdebarin, B., Ramezanian, A., 2010. Zinc alleviates salt stress and increases antioxidant enzyme activity in the leaves of pistachio (*Pistacia vera* L.'Badami') seedlings. Turkish Journal of Agriculture and Forestry 34 (4), 349–359.

Thakur, P., Kumar, S., Malik, J.A., Berger, J.D., Nayyar, H., 2010. Cold stress effects on reproductive development in grain crops: an overview. Environmental and Experimental Botany 67 (3), 429–443.

Thameur, A., Lachiheb, B., Ferchichi, A., 2012. Drought effect on growth, gas exchange and yield, in two strains of local barley Ardhaoui, under water deficit conditions in southern Tunisia. Journal of Environmental Management 113, 495–500.

Theocharis, A., Clément, C., Barka, E.A., 2012. Physiological and molecular changes in plants grown at low temperatures. Planta 235 (6), 1091–1105.

Thomashow, M.F., 1999. Plant cold acclimation: freezing tolerance genes and regulatory mechanisms. Annual Review of Plant Biology 50 (1), 571–599.

Thounaojam, T.C., Panda, P., Choudhury, S., Patra, H.K., Panda, S.K., 2014. Zinc ameliorates copper-induced oxidative stress in developing rice (*Oryza sativa* L.) seedlings. Protoplasma 251 (1), 61–69.

Tian, L., Li, J., Bi, W., Zuo, S., Li, L., Li, W., et al., 2019. Effects of waterlogging stress at different growth stages on the photosynthetic characteristics and grain yield of spring maize (*Zea mays* L.) under field conditions. Agricultural Water Management 218, 250–258.

Tiryaki, I., Keles, H., 2012. Reversal of the inhibitory effect of light and high temperature on germination of *Phacelia tanacetifolia* seeds by melatonin. Journal of Pineal Research 52 (3), 332–339.

Tsonev, T., Cebola Lidon, F.J., 2012. Zinc in plants-an overview. Emirates Journal of Food & Agriculture (EJFA) 24 (4).

Turnbull, M.H., Murthy, R., Griffin, K.L., 2002. The relative impacts of daytime and night-time warming on photosynthetic capacity in *Populus deltoides*. Plant, Cell & Environment 25 (12), 1729–1737.

Upadhyay, R., Panda, S.K., 2010. Zinc reduces copper toxicity induced oxidative stress by promoting antioxidant defense in freshly grown aquatic duckweed *Spirodela polyrhiza* L. Journal of Hazardous Materials 175 (1-3), 1081–1084.

van Veen, H., Akman, M., Jamar, D.C., Vreugdenhil, D., Kooiker, M., van Tienderen, P., et al., 2014. Group VII Ethylene Response Factor diversification and regulation in four species from flood-prone environments. Plant, Cell & Environment 37 (10), 2421–2432.

Ventrella, A., Catucci, L., Piletska, E., Piletsky, S., Agostiano, A., 2009. Interactions between heavy metals and photosynthetic materials studied by optical techniques. Bioelectrochemistry (Amsterdam, Netherlands) 77 (1), 19–25.

Vitosh, M.L., Warncke, D.D., Lucas, R.E., 1994. Secondary and micronutrients for vegetables and field crops (Boron). Extension bulletin E-486, Department of Crop and Soil Science, Michigan State University.

Vogel, J.T., Zarka, D.G., Van Buskirk, H.A., Fowler, S.G., Thomashow, M.F., 2005. Roles of the CBF2 and ZAT12 transcription factors in configuring the low temperature transcriptome of Arabidopsis. The Plant Journal 41 (2), 195–211.

Wahid, A., Gelani, S., Ashraf, M., Foolad, M.R., 2007. Heat tolerance in plants: an overview. Environmental and Experimental Botany 61 (3), 199–223.

Wang, H., Jin, J., 2007. Effects of zinc deficiency and drought on plant growth and metabolism of reactive oxygen species in maize (Zea mays L.). Agricultural Sciences in China 6 (8), 988–995. Available from: https://doi.org/10.1016/S1671-2927(07)60138-2.

Wang, L.J., Li, S.H., 2006. Salicylic acid-induced heat or cold tolerance in relation to Ca2 + homeostasis and antioxidant systems in young grape plants. Plant Science 170 (4), 685–694.

Wang, W., Vinocur, B., Altman, A., 2003. Plant responses to drought, salinity and extreme temperatures: towards genetic engineering for stress tolerance. Planta 218 (1), 1–14.

Wang, Y., Wisniewski, M., Meilan, R., Cui, M., Webb, R., Fuchigami, L., 2005. Overexpression of cytosolic ascorbate peroxidase in tomato confers tolerance to chilling and salt stress. Journal of the American Society for Horticultural Science 130 (2), 167–173.

Wang, F., Tong, W., Zhu, H., Kong, W., Peng, R., Liu, Q., et al., 2016. A novel Cys 2/His 2 zinc finger protein gene from sweetpotato, IbZFP1, is involved in salt and drought tolerance in transgenic Arabidopsis. Planta 243 (3), 783–797.

Wang, K., Ding, Y., Cai, C., Chen, Z., Zhu, C., 2019. The role of C2H2 zinc finger proteins in plant responses to abiotic stresses. Physiologia Plantarum 165 (4), 690–700.

Wang, X., Li, M., Jannasch, A.H., Jiang, Y., 2020. Submergence stress alters fructan and hormone metabolism and gene expression in perennial ryegrass with contrasting growth habits. Environmental and Experimental Botany 179, 104202. Available from: https://doi.org/10.1016/j.envexpbot.2020.104202.

Wani, S.A., Mahdi, S.S., Samoon, S.A., Hassan, G.I., Dar, S.A., 2010. Climate change-its impact on agriculture. Journal of Phytology 2 (10), 82–86.

Wani, A.S., Ahmad, A., Hayat, S., Fariduddin, Q., 2013. Salt-induced modulation in growth, photosynthesis and antioxidant system in two varieties of *Brassica juncea*. Saudi Journal of Biological Sciences 20 (2), 183–193.

Weisany, W., Sohrabi, Y., Heidari, G., Siosemardeh, A., Ghassemi-Golezani, K., 2011. Physiological responses of soybean ('*Glycine max*'L.) To zinc application under salinity stress. Australian Journal of Crop Science 5 (11), 1441.

Weisany, W., Sohrabi, Y., Heidari, G., Siosemardeh, A., Badakhshan, H., 2014. Effects of zinc application on growth, absorption and distribution of mineral nutrients under salinity stress in soybean (*Glycine max* L.). Journal of Plant Nutrition 37 (14), 2255–2269.

Welch, R.M., Graham, R.D., 2004. Breeding for micronutrients in staple food crops from a human nutrition perspective. Journal of Experimental Botany 55 (396), 353–364.

Whitmore, T.C., 2000. The case of tropical rain forests. The sustainable development of forests: aspirations and the reality. Naturzale-Cuadernos de Ciencias Naturales (15), 13–15.

Wiebbecke, C.E., Graham, M.A., Cianzio, S.R., Palmer, R.G., 2012. Day temperature influences the male-sterile locus ms9 in soybean. Crop Science 52 (4), 1503–1510.

Wojtaszek, P., 1997. Oxidative burst: an early plant response to pathogen infection. Biochemical Journal 322 (3), 681–692.

Wu, F., Zhang, G., 2002. Alleviation of cadmium-toxicity by application of zinc and ascorbic acid in barley. Journal of Plant Nutrition 25 (12), 2745–2761.

Wu, S., Hu, C., Tan, Q., Nie, Z., Sun, X., 2014. Effects of molybdenum on water utilization, antioxidative defense system and osmotic-adjustment ability in winter wheat (*Triticum aestivum*) under drought stress. Plant Physiology and Biochemistry 83, 365–374.

Wu, S., Hu, C., Tan, Q., Li, L., Shi, K., Zheng, Y., et al., 2015. Drought stress tolerance mediated by zinc-induced antioxidative defense and osmotic adjustment in cotton (*Gossypium hirsutum*). Acta Physiologiae Plantarum 37 (8), 1–9.

Wu, H., Chen, H., Zhang, Y., Zhang, Y., Zhu, D., Xiang, J., 2019. Effects of 1-aminocyclopropane-1-carboxylate and paclobutrazol on the endogenous hormones of two contrasting rice varieties under submergence stress. Plant Growth Regulation 87 (1), 109–121.

Xu, D.Q., Huang, J., Guo, S.Q., Yang, X., Bao, Y.M., Tang, H.J., et al., 2008. Overexpression of a TFIIIA-type zinc finger protein gene ZFP252 enhances drought and salt tolerance in rice (*Oryza sativa* L.). FEBS Letters 582, 1037–1043. Available from: https://doi.org/10.1016/j.febslet.2008.02.052.

Xu, X., Wang, H., Qi, X., Xu, Q., Chen, X., 2014. Waterlogging-induced increase in fermentation and related gene expression in the root of cucumber (*Cucumis sativus* L.). Scientia Horticulturae 179, 388–395. Available from: https://doi.org/10.1016/j.scienta.2014.10.001.

Xu, X., Ji, J., Ma, X., Xu, Q., Qi, X., Chen, X., 2016. Comparative proteomic analysis provides insight into the key proteins involved in cucumber (*Cucumis sativus* L.) adventitious root emergence under waterlogging stress. Frontiers in Plant Science 7, 1515.

Yadav, S.K., 2010. Cold stress tolerance mechanisms in plants. A review. Agronomy for Sustainable Development 30 (3), 515–527.

Yamaguchi, T., Blumwald, E., 2005. Developing salt-tolerant crop plants: challenges and opportunities. Trends in Plant Science 10 (12), 615–620.

Yamauchi, T., Yoshioka, M., Fukazawa, A., Mori, H., Nishizawa, N.K., Tsutsumi, N., et al., 2017. An NADPH oxidase RBOH functions in rice roots during lysigenous aerenchyma formation under oxygen-deficient conditions. The Plant Cell 29 (4), 775–790.

Yamauchi, T., Tanaka, A., Tsutsumi, N., Inukai, Y., Nakazono, M., 2020. A role for auxin in ethylene-dependent inducible aerenchyma formation in rice roots. Plants (Basel) 9, 610.

Yan, K., Shao, H., Shao, C., Chen, P., Zhao, S., Brestic, M., et al., 2013. Physiological adaptive mechanisms of plants grown in saline soil and implications for sustainable saline agriculture in coastal zone. Acta Physiologiae Plantarum 35 (10), 2867–2878.

Yan, K., Zhao, S., Cui, M., Han, G., Wen, P., 2018. Vulnerability of photosynthesis and photosystem I in Jerusalem artichoke (*Helianthus tuberosus* L.) exposed to waterlogging. Plant Physiology and Biochemistry 125, 239–246.

Yang, Z., Wu, Y., Li, Y., Ling, H.Q., Chu, C., 2009. OsMT1a, a type 1 metallothionein, plays the pivotal role in zinc homeostasis and drought tolerance in rice. Plant Molecular Biology 70 (1-2), 219–229.

Yang, J.C., Li, M., Xie, X.Z., Han, G.L., Sui, N., Wang, B.S., 2013. Deficiency of phytochrome B alleviates chilling-induced photoinhibition in rice. American Journal of Botany 100 (9), 1860–1870.

Yavas, I., Unay, A., 2016. Effects of zinc and salicylic acid on wheat under drought stress. Journal of Animal and Plant Sciences 26 (4), 1012–1101.

Yin, M., Wang, Y., Zhang, L., Li, J., Quan, W., Yang, L., et al., 2017. The Arabidopsis Cys2/His2 zinc finger transcription factor ZAT18 is a positive regulator of plant tolerance to drought stress. Journal of Experimental Botany 68 (11), 2991–3005. Available from: https://doi.org/10.1093/jxb/erx157.

Yin, D., Sun, D., Han, Z., Ni, D., Norris, A., Jiang, C.Z., 2019. PhERF2, an ethylene-responsive element binding factor, plays an essential role in waterlogging tolerance of petunia. Horticulture Research 6 (1), 1–11.

Yordanova, R., Popova, L., 2007. Effect of exogenous treatment with salicylic acid on photosynthetic activity and antioxidant capacity of chilled wheat plants. General and Applied Plant Physiology 33 (3-4), 155–170.

Yoshida, S., 1981. Physiological analysis of rice yield. Fundamentals of Rice Crop Science. International Rice Research Institute, Los Banos, pp. 231–251.

Yoshioka, M., Uchida, S., Mori, H., Komayama, K., Ohira, S., Morita, N., et al., 2006. Quality control of photosystem II: cleavage of reaction center D1 protein in spinach thylakoids by FtsH protease under moderate heat stress. Journal of Biological Chemistry 281 (31), 21660–21669.

You, L., Rosegrant, M.W., Wood, S., Sun, D., 2009. Impact of growing season temperature on wheat productivity in China. Agricultural and Forest Meteorology 149 (6-7), 1009–1014.

Yousuf, P.Y., Ahmad, A., Aref, I.M., Ozturk, M., Ganie, A.H., Iqbal, M., 2016a. Salt-stress-responsive chloroplast proteins in *Brassica juncea* genotypes with contrasting salt tolerance and their quantitative PCR analysis. Protoplasma 253 (6), 1565–1575.

Yousuf, P.Y., Ahmad, A., Ganie, A.H., Iqbal, M., 2016b. Salt stress-induced modulations in the shoot proteome of *Brassica juncea* genotypes. Environmental Science and Pollution Research 23 (3), 2391–2401.

Yu, G.H., Jiang, L.L., Ma, X.F., Xu, Z.S., Liu, M.M., Shan, S.G., et al., 2014. A soybean C2H2-type zinc finger gene GmZF1 enhanced cold tolerance in transgenic Arabidopsis. PLoS One 9, e109399. Available from: https://doi.org/10.1371/journal.pone.0109399.

Yusuf, M., Fariduddin, Q., Varshney, P., Ahmad, A., 2012. Salicylic acid minimizes nickel and/or salinity-induced toxicity in Indian mustard (*Brassica juncea*) through an improved antioxidant system. Environmental Science and Pollution Research 19 (1), 8–18.

Zabihi-e-Mahmoodabad, R., Jamaati-e-Somarin, S., Khayatnezhad, M., Gholamin, R., 2011. Effect of cold stress on germination and growth of wheat cultivars. Advances in Environmental Biology 5 (1), 94–97.

Zaman, Q.U., Aslam, Z., Yaseen, M., Ihsan, M.Z., Khaliq, A., Fahad, S., et al., 2018. Zinc biofortification in rice: leveraging agriculture to moderate hidden hunger in developing countries. Archives of Agronomy and Soil Science 64 (2), 147–161.

Zhai, H., Wang, F., Si, Z., Huo, J., Xing, L., An, Y., et al., 2016. A myo-inositol-1-phosphate synthase gene, Ib MIPS 1, enhances salt and drought tolerance and stem nematode resistance in transgenic sweet potato. Plant Biotechnology Journal 14 (2), 592–602.

Zhang, J.H., Huang, W.D., Liu, Y.P., Pan, Q.H., 2005. Effects of temperature acclimation pretreatment on the ultrastructure of mesophyll cells in young grape plants (*Vitis vinifera* L. cv. Jingxiu) under cross-temperature stresses. Journal of Integrative Plant Biology 47 (8), 959–970.

Zhang, B., Liu, W., Chang, S.X., Anyia, A.O., 2010. Water-deficit and high temperature affected water use efficiency and arabinoxylan concentration in spring wheat. Journal of Cereal Science 52 (2), 263–269.

Zhang, F., Jiang, Y., Bai, L., Zhang, L., Chen, L., Li, H., et al., 2011. The ICE-CBF-COR pathway in cold acclimation and AFPs in plants. Middle East Journal of Scientific Research 8 (2), 493–498.

Zhang, H., Liu, Y., Wen, F., Yao, D., Wang, L., Guo, J., et al., 2014. A novel rice C2H2-type zinc finger protein, ZFP36, is a key player involved in abscisic acid-induced antioxidant defence and oxidative stress tolerance in rice. Journal of Experimental Botany 65 (20), 5795–5809.

Zhang, P., Lyu, D., Jia, L., He, J., Qin, S., 2017. Physiological and de novo transcriptome analysis of the fermentation mechanism of *Cerasus sachalinensis* roots in response to short-term waterlogging. BMC Genomics 18 (1), 1–14.

Zhao, A.Q., Tian, X.H., Lu, W.H., Gale, W.J., Lu, X.C., Cao, Y.X., 2011. Effect of zinc on cadmium toxicity in winter wheat. Journal of Plant Nutrition 34 (9), 1372–1385.

Zheng, L., 1994. Regularities of content and distribution of zinc in soils of China. Scientia Agricutura Sinica 1.

Zhou, B., Yao, W., Wang, S., Wang, X., Jiang, T., 2014. The metallothionein gene, TaMT3, from Tamarix androssowii confers Cd2 + tolerance in tobacco. International Journal of Molecular Sciences 15 (6), 10398–10409.

Zhou, M., Chen, H., Wei, D., Ma, H., Lin, J., 2017. Arabidopsis CBF3 and DELLAs positively regulate each other in response to low temperature. Scientific Reports 7 (1), 1–13.

Zhou, Y., Zeng, L., Chen, R., Wang, Y., Song, J., 2018. Genome-wide identification and characterization of stress-associated protein (SAP) gene family encoding A20/AN1 zinc-finger proteins in *Medicago truncatula*. Archives of Biological Sciences 70 (1), 087–098.

Zhou, W., Chen, F., Meng, Y., Chandrasekaran, U., Luo, X., Yang, W., et al., 2020. Plant waterlogging/flooding stress responses: from seed germination to maturation. Plant Physiology and Biochemistry: PPB / Societe Francaise de Physiologie Vegetale 148, 228–236. Available from: https://doi.org/10.1016/j.plaphy.2020.01.020.

Zhu, J.K., 2016. Abiotic stress signaling and responses in plants. Cell 167, 313–324. Available from: https://doi.org/10.1016/j.cell.2016.08.029.

Zlatev, Z., Lidon, F.C., 2012. An overview on drought induced changes in plant growth, water relationsand photosynthesis. Emirates Journal of Food and Agriculture 57–72.

Chapter 6

# Roles of zinc in alleviating environmental stress on plant photosynthesis: challenges and future outlook

Mohamed Salah Sheteiwy[1,2], Ahmed El-Sawah[3], Zaid Ulhassan[4], Sajad Hussain[5], Hiba Shaghaleh[6], Izabela Jośko[7], Yousef Alhaj Hamoud[8], Ali Raza Khan[9], Hamada Abdelgawad[10] and Weijun Zhou[4]

*[1]Department of Applied Biology, Faculty of Science, University of Sharjah, Sharjah, United Arab Emirates, [2]Department of Agronomy, Faculty of Agriculture, Mansoura University, Mansoura, Egypt, [3]Department of Agricultural Microbiology, Faculty of Agriculture, Mansoura University, Mansoura, Egypt, [4]Institute of Crop Science, Department of Agronomy, Ministry of Agriculture and Rural Affairs Key Lab of Spectroscopy Sensing, Zhejiang University, Hangzhou, P.R. China, [5]National Research Center of Intercropping, The Islamic University of Bahawalpur, Pakistan, [6]Key Lab of Integrated Regulation and Resource Development on Shallow Lakes, Ministry of Education, Hohai University, Nanjing, P.R. China, [7]Institute of Plant Genetics, Breeding and Biotechnology, Faculty of Agrobioengineering, University of Life Sciences, Lublin, Poland, [8]The National Key Laboratory of Water Disaster Prevention, College of Hydrology and Water Resources, Hohai University, Nanjing, P.R. China, [9]School of Emergency Management, Institute of Environmental Health and Ecological Security, School of Environment and Safety Engineering, Jiangsu Province Engineering Research Center of Green Technology and Contingency Management for Emerging Pollutants, Jiangsu University, Zhenjiang, P.R. China, [10]Department of Botany and Microbiology, Faculty of Science, Beni-Suef University, Beni-Suef, Egypt11 Department of Biology, University of Antwerp, Antwerp, Belgium*

## 6.1 Introduction

Due to the expected increase in escalating world population which is expected to be 9 billion by 2050. To ensure human and animal food security, the yield of the most strategic crops will be required to be increased by more than 70% during the next three decades (FAO, 2009; Tilman et al., 2011). However, there are great challenges to achieve this target according to the climate change scenario and vulnerabilities in the form of temperature,

**Zinc in Plants. DOI: https://doi.org/10.1016/B978-0-323-91314-0.00014-4**

drought, and flood stressors (Abdoulaye et al., 2019). It is expected that a 50% reduction in yield production will be achieved by 2050 as a result of these abiotic stress factors (Bierbaum et al., 2007).

Photosynthesis is a complicated process that is split into light reactions and photosynthetic Calvin cycle reactions, in which $CO_2$ fixation occurs in the stroma of chloroplasts (Renger, 2007), resulting in biosphere primary production (Baslam et al., 2020). The atmospheric $CO_2$ is the main substrate of photosynthesis, which would contribute to higher yields of the major strategic crops. There are nearly 6500–7000 billion tons of $CO_2$ in the atmosphere. From this, photosynthesis fixes more than 100 billion tons per year (Baslam et al., 2020). The amount of atmospheric $CO_2$ absorbed by plants through photosynthesis is enormous (Craggs, 2016); there is an evidence that increasing atmospheric $CO_2$ levels may increase photosynthesis (Hussain et al., 2021b). Depending on the crop plant and climatic circumstances, agricultural productivity is projected to rise by up to 30% (Long et al., 2006; Sanz-Saez et al., 2017). Therefore, several evidences have provided strategies for improving the photosynthetic process aiming to increase crop production and mitigate the environmental impacts. However, photosynthesis is challenged by environmental changes that hampered the crop productivity.

The functionality of photosynthesis process is critically affected by the environmental stress factors such as low to high temperatures, drought, salinity, and heavy metals/metalloids due to their detrimental consequences on pigments, photosystems, electron transport, and gas exchange (Ali et al., 2018c; Sheteiwy et al., 2018a and b; Ulhassan et al., 2019c; Mwamba et al., 2020; Sharma et al., 2020; Nazir et al., 2020; Alotaibi et al., 2021; Hussain et al., 2021b; Yang et al., 2021); moreover, photophosphorylation have negative impacts on photosynthetic metabolic activities and caused damages to the ultra-structures of organelles as well as thylakoid membranes (Lobell et al., 2014; Sieber et al., 2016). The photosynthetic efficiency of plants is severely affected by the abiotic stresses due to their detrimental consequences. Among them, drought is the most significant abiotic factor that decreases agricultural yield (Lobell et al., 2014; Sheteiwy et al., 2018b, 2020) due to photosynthetic mediated limitations executed by stomatal or nonstomatal traits (Kohzuma et al., 2009; Dahal et al., 2014). However, researchers have extensively investigated to boost yields along with improved plant responses to abiotic stresses and less attention has been given to photosynthesis. To attain this goal, new tools such as genomics and modeling methodologies have opened the road for enhancing photosynthesis to enhance agricultural production under various climate change scenarios (Long et al., 2015; Ort et al., 2015). Moreover, the interaction with critical elements such as Zn and its releasing approaches to soils have the potential to affect photosynthesis via diverse mechanisms. Here, our key target is to achieve the optimized status of the photosynthesis process under different levels of Zn under severe environmental scenarios. Zinc also plays great roles in plant metabolic processes to activate the several enzymes that participate in protein synthesis, photosynthesis, and auxin synthesis, as well as in nitrogen, carbohydrate, nucleic acid, and lipid metabolisms (Borowiak et al., 2017; Sheteiwy et al., 2021c). Additionally, Zn aids in the biosynthesis of indole-3-acetic acid (IAA) and modules physiological and metabolic processes (Fang et al., 2008). In this chapter, we have overviewed the interaction and/or beneficial impacts of Zn against severe environmental stress factors by targeting the stress alleviation effects on the photosynthesis process in plants.

## 6.2 Photosynthetic acclimation and potential roles of zinc for plant adaptation to abiotic stress factors

### 6.2.1 Salinity stress

The negative effects of salinity on arable land are increasing day after day due to the dramatic increase in saline areas, especially under the current climate changes (Qadir et al., 2014). Climate change leads to a loss in the agricultural area, eliminating 1.5 million hectares of arable land per year (Munns and Tester, 2006, 2008). Globally, salt stress affects approximately 34 Mha of arable land, resulting in severe agricultural production reductions (Florke et al., 2019). Plant growth is inhibited by salinity, and this inhibition is reliant on the plant's sensitivity to salinity as well as the salt level in the environment (Munns and Tester, 2008; Ning et al., 2019; Abbas et al., 2020). This inhibition is induced by different constraints. First, the ability of salt to inhibit the plants to uptake water and nutrient from the soil solution. This is regarded as an osmotic component of salt stress. Second, introducing unwarranted quantities of salt into the transpiration stream causes cell damage in transpiring leaves (Shabala and Munns, 2017). Several studies have reported on the regulation of photosynthetic under salt stress (Koyro 2006; Wei et al., 2006) and found that photosynthesis was generally hampered by salinity (Sudhir and Murthy 2004; Koyro 2006; Munns et al., 2006) through stomatal and nonstomatal factors (Naumann et al., 2007). In maize, chlorophyll fluorescence, rubisco activity, and chlorophyll contents were minimized by high salt stress leading to chloroplast damage (Yan and Tong, 2018). Salinity stress inhibited the activity of photosynthetic attributes, mainly PSII and *Fv/Fm* in the leaves of mustard plants (Wani et al., 2019), while in lisianthus cultivars, reduced the levels of gaseous exchange parameters (Ashrafi and Rezaei Nejad, 2017). Another research found that salt stress impaired the photosynthesis process by closing the leaf stomata and decreased the carbon metabolism (Çiçek et al., 2018). Salinity displayed significant impacts on the photosynthetic pigments, $CO_2$ assimilation as well as the photosynthetic efficiency of PSII, which might trigger the photoinhibition (Tarchoune et al., 2012). Thus, photosynthetic pigments and chlorophyll fluorescence may be considered simple and rapid indicators that can be used for varieties screening for their salinity tolerance (Demiral and Türkan 2006; Tarchoune et al., 2012). In contrast, some studies reported that *Chl a* fluorescence (*Fv/Fm*) is not an ideal tool as a salt stress marker (Debez et al., 2008; Hichem et al., 2009). While the effects of salt on plants become more acute in arid and semiarid locations, a lack of or an excess of Zn can also impede plant productivity in these areas, posing a major challenge to human nutrition (Tolay, 2021).

The rate of $CO_2$ assimilation is considered the most variable in the whole plant that can reflect its response to salinity (Pan et al., 2020). In fact, salinity imposes stomatal and nonstomatal limitations that can affect the photosynthesis processes (Munns et al., 2006; Chaves et al., 2009). The limitation of gas exchange parameters causes inhibitory effects on stomatal conductance and reduces the stomatal aperture (Pan et al., 2020). Salinity affects stomatal conductance through perturbed water relations and ABA synthesis (Fricke et al., 2004). These signals of stomatal are regulated by root via transferring ROS and $Ca^{2+}$ signals (Gilroy et al., 2014; Evans et al., 2016; Shabala et al., 2016). On the other hand, the nonstomatal limitation of photosynthesis occurred as a result of higher salt accumulation, which in turn can active the mesophyll tissues and lead to limited $CO_2$ assimilation, by influencing the mainly photosynthetic activities in chloroplasts (Pan et al., 2020). Also, salinity can inhibit the photosynthetic efficiency by increasing nonphotochemical associated quenching (Wang et al., 2007). Under salinity stress, the higher accumulation of $Na^+$ in cytosolic also disturbs the $K^+$ and $Ca^{2+}$ homeostasis in leaf mesophyll (Percey et al., 2016) which could affect the photosynthesis and light signal transduction (Pan et al., 2020).

The application of Zn can minimize the salt-induced toxic effects of stress by improving biomass, photosynthetic pigments, K uptake, and nutrient acquisition traits (Tolay, 2021). In rice, antioxidant enzyme activities were improved by Zn-applications that improved the salt tolerance (Nadeem et al., 2020). The improvement of these enzymes could scavenge the reactive oxygen species (ROS) under salinity stress and reduce the degradation of lipids, proteins, and nucleic acids, thus improving the photosynthesis process (Hichem et al., 2009). Baghalian et al. (2008) observed a direct correlation between chlorophyll content and Zn concentration under combined treatment of EC and Zn, reporting that Zn contributes to the improvement of photosynthesis, subsidizing the metabolic processes leading to an increase of biomass. Zn deficiency reduces auxin, leading to a reduction in leaf size due to roasting and leaf curling under this condition (Broadley et al., 2007). Reduction in auxin levels would be critical, especially under the combination of both environmental conditions and Zn deficiency, which ultimately causes inhibition in shoot growth (Mroue et al., 2018). Pervious investigations by Nadeem et al. (2020) showed that Zn decreased ABA levels and alleviated the salt stress by increasing the $K^+/Na^+$ ratio and RWC. Zn can also affect the integrity and the permeability of the cell membrane and thus can reduce the excessive Na uptake under saline conditions (Aktaş and Abak, 2006). Therefore, Zn deficiency led to high leakage of the cell membranes under salinity stress. Moreover, Zn effectively decreased the Na uptake and improved the $Na^+/K^+$ ratio in plants under salt stress (Saleh et al., 2009; Nadeem et al., 2020). The combined applications of salt stress and Zn deficiency on plant growth are significant and warrant further investigation, since higher accumulations of toxic $Na^+$ and $Cl^-$ ions were observed under these conditions (Tolay, 2021). Also, Mateos-Naranjo et al. (2008) investigated the effect of Zn on *S. dens* flora growth and photosynthetic apparatus in a glasshouse experiment, but they have not found the interaction with salt. However, further research into the underlying biochemical and genetic mechanisms involved in Zn-induced salinity tolerance is required (Tolay, 2021).

Zn deficiency is a worldwide problem and could be compounded in alkaline calcareous soils as well as in the submergence condition (Nadeem et al., 2020). In this context, plants grown under these conditions, such as rice, may encounter Zn deficiency more than other crop species due to the submerged conditions, in which Zn becomes sulfide under anaerobic conditions and thus becomes unavailable for rice plant (Nadeem et al., 2020). Moreover, under this condition, Zn in the form of $ZnCO_3$ also becomes unavailable when reacting with $CO_2$ (Johnson-Beebout et al., 2009). Additionally, Zn deficiency can also result from crop intensification which could increase Zn removal leading to a depletion of the soil Zn retention (Nadeem et al., 2020). Thus, a sufficient supply of different micronutrients, such as Zn, to plants may improve the growth of the plants under both regular as well as adverse environmental conditions.

### 6.2.2 Drought stress

Globally, drought stress is the most influential abiotic stress, affecting plant development and crop yield, particularly in arid and semiarid regions (Arough et al., 2016). Drought stress led to several biochemical and physiological changes as a response of plants under stress conditions (Tas and Tas, 2007; Sheteiwy et al., 2021a, b). Drought stress can directly affect the photosynthetic process, by disrupting the water balance and thylakoid electron transport, reducing the carbon cycle, and controlling the $CO_2$ assimilation (Allen and Ort, 2001). At the subcellular level, drought-increased lipid peroxidation could affect the chloroplast ultrastructure (Hu et al., 2018). The oxidative stress induced by drought stress, contributes to increasing ROS levels and causes severe damage to photosynthesis (Goodman and Newton, 2005). Drought stress often induces oxidative stress producing free radical causing an inhibition to the photosynthetic machinery and membrane proteins (Chen et al., 2004, Wang and Huang, 2004). Induction of reactive oxygen species under drought stress can cause serious damage to chloroplasts.

Therefore, plants have to activate their oxidative defense system to scavenge ROS by inducing the antioxidative enzymes which are considered important plant adaptations to drought stress (Sairam and Saxena, 2000).

Moreover, drought stress causes great damage to the stroma lamellae and grana of the chloroplast and reduces of accumulation of starch granules (Xu et al., 2009). Drought results in lipid peroxidation of the cell membrane system due to the higher accumulation of ROS (Sun et al., 2019). There was a contradiction in the reports regarding the pigments under drought stress, for example, drought stress augmented the senescence that was associated with the decline in chlorophyll contents (Farooq et al., 2014). However, Pavia et al. (2019) reported that there was no significant difference in the levels of light-harvesting pigments including chlorophylls and carotenoids under drought conditions. On the other hand, other studies testified a maintenance or slight increase in chlorophylls under drought stress (Simova-Stoilova et al., 2009; Fahad et al., 2017). Drought inhibits photochemical events, decreases the activity of enzymes involved in the Calvin cycle, decreases the assimilation of $CO_2$, reduces the stomatal conductance, and impairs the quantum yield of photosystem II (PSII) that directly affects the photosynthesis process (Jaleel et al., 2009).

Drought has a significant impact on photosynthesis, which is one of the most important physiological processes (Pavia et al., 2019; Sheteiwy et al., 2020). The damage in photosynthetic apparatus under drought stress may result from the excessive energy absorption and generation of reactive oxygen species (Zargar et al., 2017; Pavia et al., 2019). One of the most significant effects of drought on plants is a reduction in photosynthesis. The decrease in photosynthesis often results from stomatal constraints (Farooq et al., 2009). The closure of stomata is the key factor in reducing the $CO_2$ assimilation and reduction in photosynthesis process under drought stress (Farooq et al., 2009; Abid et al., 2016). The limitations in nonstomatal traits may also happen from moderate to severe drought stress (Brito et al., 2018). Under such severe drought stress, photosynthesis is mainly restricted by stomatal restraints, reduction in maximum quantum efficiency of photosystem II, and $CO_2$ assimilating related enzyme activities (Abid et al., 2016). The study by Pavia et al. (2019) reported that the nonregulated energy dissipation has also increased under severe drought stress. Therefore, plants have to adapt to drought stress through different mechanisms and processes. In this context, drought stress enhanced the leaf thickness accompanied by an increase in intrinsic water use efficiency (Abid et al., 2016). Such adaptations may also decrease the photoinhibition in plants under drought conditions (Izanloo et al., 2008; Abid et al., 2016). Drought causes considerable losses to crop yields and this loss was found to be severe under Zn deficiency (Ekiz et al., 1998).

The nutritional status of plants plays a critical role in plant adaptation to drought stress. Zn being an important micronutrient is indispensable for most important physiological and biochemical processes in plants (Pavia et al., 2019). Zn deficiency affects plant metabolism and the balance between ROS production and scavenging; thus, Zn deficiency affects plant drought tolerance, and drought effects are greatly influenced by their Zn nutritional status (Ullah et al., 2019). Depending upon the plant species and level of Zn deficiency, Zn deficiency can reduce the net photosynthesis by 50%–70%. The reduction in the enzyme activity of carbonic anhydrase, bisphosphate carboxylase (RuBPC), chlorophyll, and stomata conditions that affect the $CO_2$ availability in plants caused a decrease in photosynthesis efficiency (Alloway, 2008). Zn-priming could help to re-establish the Zn levels in Zn-deficient seeds, however, the mechanism by which Zn-priming improves the growth and yield needs to be further investigated (Pavia et al., 2019). Increased Zn in plants can reduce the damage to stomata, by upholding the membrane's integrity (Cakmak, 2000). Other reports showed that Zn priming or foliar applications increased the yield and alleviated the inhibitory effects of drought in wheat plants (Karim et al., 2012; Ma et al., 2017). In addition, Harris et al. (2007) showed that priming seeds with Zn has increased endogenous Zn contents in seeds, contributing to a higher plant growth and yield. However, higher Zn concentrations result in damage to stomata morphology and narrow stomata openings (Sagardoy et al., 2010). Zn participated in ROS-detoxification and thus reduced the production of free radicals by SOD under drought stress (Cakmak and Engels, 1999). Zn may diminish the drought stress impacts at a cellular level by being involved in membrane integrity and protecting cells from damage caused by extra ROS accumulation (Fageria, 2009). Another study reported that Zn, being a component of the carbonate anhydrase enzyme, facilitated the possibilities of stomata opening to stabilize the $HCO_3$ contents in guard cells (Sharma et al., 1995). Application of Zn at booting to anthesis stages alleviated drought stress effects by increasing the photosynthesis rate, pollen viability, number of fertile spikes, grains per spike, and water-use efficiency (Karim et al., 2012). Nano-ZnO improved the activities of antioxidants such as SOD, CAT, and APX, which helped plants to hunt free radicals under drought (Sun et al., 2019). Among the antioxidative enzymes, the Zn-associated overexpression of genes contributed to improving the drought lenience in maize plants (Malan et al., 1990).

However, there is a contradiction in the reports on the role of Zn in alleviating drought stress. In this regard, Wang et al. (2009) confirmed that Zn supplementation did not improve the maize growth or net photosynthesis rate under drought stress. In contrast, Baðci et al. (2007) determined that durum wheat plants were sensitive to drought stress conditions in case of Zn deficiency stress. Also, Zn treatments were unable to alleviate drought stress-induced severe yield loss (Pavia et al., 2019). However, other reports showed that drought-induced nonregulated energy dissipation was reduced by Zn treatments, minimizing the photosynthetic damages caused by drought (Pavia et al., 2019).

## 6.3 Cold or low-temperature stress

The plants' exposure to environmental stress such as chilling or drought, or heavy metals/metalloids caused an imbalance between the production and detoxication of reactive oxygen species (ROS) promoting the generation of ROS and malondialdehyde (MDA), which caused membrane lipid peroxidation, destroyed the membrane structures and leads to cell death (Margutti et al., 2017; Ali et al., 2018c; Ulhassan et al., 2018, 2019a, b; Elkelish et al., 2020; Mwamba et al., 2020; Hussain et al., 2021a, b; Sheteiwy et al., 2021a; Yang et al., 2021). Cold, chilling, or low-temperature limits the plant tolerance capacity and productivity. In addition, disrupts the integrity of chloroplast membranes by negatively affecting the structures of thylakoid membranes, activities of photosynthetic enzymes, chlorophyll levels, and electron transport chains (Hussain et al., 2018; Banerjee and Roychoudhury, 2019). Cold-stress-induced damage to photosynthetic apparatus is illustrated by the disturbances in chloroplasts equilibrium, inhibition in the functioning of photosystem II (PSII) reaction centers, imbalance in antenna complexes, loss in light-harvesting chlorophyll pigments (Ensminger et al., 2006), suppression of the carbon fixation capacity, and resultant Rubisco degradation (Augustyniak et al., 2018). In this situation, plants adopt different strategies such as adjustments in the alterations of cellular metabolism and regulatory mechanisms to restore their metabolic level or fluxes (Shi and Chan, 2014; Dhingra, 2015). Seed priming with chemical agents or compatible solutes (Farooq et al., 2017) and applications of phytohormones (Yang et al., 2018) exhibited protection to photosynthetic apparatus against cold stress.

It has been well documented that excessive concentrations of Zn can be phytotoxic by modifying the morpho-physiological, metabolic, and ultrastructural attributes (Subba et al., 2014; Hamed et al., 2017; Kaya et al., 2018). At low doses, Zn displays many physiological functions such as regulation of plant growth and photosynthetic machinery (Tavanti et al., 2021) and stress resistance (Ma et al., 2017; ul Hassan et al., 2017; Salam et al., 2021). During the early growth of maize, the supplementation of Zn enhanced the cold tolerance by stimulating the chlorophyll status and minimizing the leaf oxidative damage (Bradáčová et al., 2016). The outcomes from recent investigations indicated that the optimal amount of Zn content enhanced the photosynthetic rates, stomatal conductance, $CO_2$ assimilation rates, photosynthetic efficiency (*Fv/Fm*), and status of photosystem II (PSII) that leads to stimulating the photosynthesis process under low-temperature (4°C) stress conditions. On the contrary, excessive Zn levels further accelerated the low-temperature stress by impairing the above-mentioned photosynthetic process in wheat plants (Kaznina et al., 2019). Thus, low Zn levels enhance the plant's capacity to tolerate low-temperature stress with improved photosynthesis and excessive Zn reduces the plant's ability to cold adaptation with impairments in photosynthesis. A zinc finger transcription factor,

*PeSTZ1*, was involved in improving the freezing tolerance and efficiency of photosynthetic parameters in poplar 84K hybrids (He et al., 2019). These findings open an economic option for chilling or cold stress prophylaxis in agricultural crops. Still, a thorough understanding (at the molecular level) of Zn-mediated improvement in the photosynthesis process and cold adaptation is required.

## 6.4 High-temperature stress

The rising global temperatures are causing major threats to plant productivity (Wahid et al., 2007). The elevated temperature levels caused damages to pre- and post-harvesting that ultimately negatively affected the plant growth, metabolism, and yield attributes (Vollenweider and Gunthardt-Goerg, 2005). The photosynthesis process in plants is extremely temperature sensitive, and it is frequently harmed before the visible signs of high temperatures appear on plants. The principal sources of injury at high temperatures have been postulated to be impaired photochemical processes, carbon metabolism in the stroma of chloroplasts, and PSII activity (Wise et al., 2004). A study on cotton plants showed that high temperatures exacerbated the damage to lipid membranes by increasing the malondialdehyde levels and decreasing the chlorophyll contents, gas exchange parameters, and water status (Sarwar et al., 2019). However, when combined with other abiotic stressors such as mineral nutrient deficiencies, high temperatures become a major threat to plants. Foliar Zn application was shown in a recent study on the *Pak chio* plant to protect plant growth from heat stress by increasing Zn concentration in leaves, thereby maintaining SOD activity and membrane stability and protecting photosynthesis from heat damage (Han et al., 2020). Ullah et al. (2019) found that Zn supplementation improves the photosynthesis of chickpea plants under heat stress by regulating the leaf PSII efficiency, improving water relations and free proline. In addition, zinc applications have been shown to improve cotton yield by 17% under the supraoptimal thermal regime, 12% under the suboptimal thermal regime, and 19% under the high-temperature of field studies compared to water-exposed plants under the same temperature conditions (Sarwar et al., 2019).

## 6.5 Elevated $CO_2$

The earth's climate is changing due to the effects of anthropogenic activities. Atmospheric $CO_2$, currently about 385 ppm may surpass 700 ppm by the end of 21$^{th}$ century (Long et al., 2006). This rise is primarily due to $CO_2$ emissions from anthropogenic sources. Climate change adversely affects a broad spectrum of life. Plants form the basis of life on earth, hence studying the impact of climate change on plants is essential. In general, plants exposed to elevated $CO_2$ ($eCO_2$) show an increase in biomass relative to those grown in ambient $CO_2$ (Albert et al., 2011; Ainsworth et al., 2008). Since carbon dioxide essentially is a plant food, which is used during photosynthesis, and converted into the energy-rich bonds of carbohydrates during photosynthesis, changes in the $CO_2$ concentrations affect photosynthesis and consequently alter the whole plant metabolism (Albert et al., 2011). A great deal of studies have been performed on the effects of $eCO_2$ on the physiology of plants, mostly addressing changes in photosynthesis, biomass production, and nutrient relations (Albert et al., 2011). In general, increased $CO_2$ levels have a fertilizing effect in $C_3$ plants, most likely due to the enhancement in photosynthesis rates and water use efficiency (Albert et al., 2011), which result in higher plant growth and productivity (AbdElgawad et al. 2014).

Climate change is a multifaceted phenomenon, which includes $CO_2$ rises and other environmental factors such as mineral deficiency. Therefore, predicting the consequences of $eCO_2$ on plant growth is highly complex. Moreover, increasing $CO_2$ results in a global Zn deficiency. For instance, it has been shown that $eCO_2$ significantly decreased Zn concentration in C3 and C4 grasses (Meyers et al., 2014). High $CO_2$ reduced plant concentrations of several elements, including Zn (Loladze 2002). Consistent with findings from the short experiments of Loladze (2002), $eCO_2$ significantly reduced Zn levels in long-term field studies (Hagedorn et al., 2002). Given the key role of Zn in optimum plant growth, its absence reduces a variety of physiological processes, including net photosynthesis (Salama et al., 2006). Its deficiency disrupts photosynthetic activity through the limitation of substrate availability required for carboxylation (Salama et al., 2006). High $CO_2$ ameliorated the effect of Zn deficiency on photosynthesis; conversely, adequate Zn and well-watered plants can get more $CO_2$ by improving stomatal conductance (Asif et al., 2017).

Most of the earlier studies were related to the specific effects of Zn on the Calvin cycle, photosystem activities, and plant mineral composition (Loneragan and Webb 1993). Zinc is an essential micronutrient that is also involved in upholding photosynthetic activity and integrity of membranes (Brown et al., 1993). Zn improved the photochemical reactions of the photosystems including ribulose-1,5-bisphosphate carboxylase/oxygenase activity and equilibrium between $CO_2$ and $O_2$. Moreover, the decrease in photochemical activities may be attributed to the modification of cellular structures and configuration of membranes associated with thylakoids. Zn restricted the stomatal conductance and thus $CO_2$ fixation (Khudsar et al., 2008), while mesophyll tissues are negatively affected in accordance with both stomata cell number and stomata size (Sagardoy et al., 2010). Interestingly, the zinc finger transcription factor overexpression altered photosynthesis that enhanced tolerance photoinhibition induced by light and $H_2O_2$ exposure, and it also increased ROS detoxification gene expression (Rossel et al., 2007).

Alternatively, Zn has been reported to have a toxic effect on plants, influencing various functions. In this context, photosynthesis may be inhibited by

Zn toxicity at various stages and through diverse methods, including a decrease in net photosynthesis rate (Mateos-Naranjo et al., 2008). In *Beta vulgaris*, high Zn concentrations reduce net photosynthetic and respiration rates (Sagardoy et al., 2010), and reactive oxygen species are produced, which reduce chlorophyll contents (Cui and Zhao, 2011). The inhibition of the photosynthetic machinery is linked to the structure of chlorophyll at the molecular level. For instance, $Mg^{2+}$ can be replaced with Zn (Von Wettstein et al., 1995), which weakens LHCII activity (Küpper et al., 2002). In this regard, Zn toxic concentrations inhibited the photochemical efficiency of PSII and/or PSI (Di et al., 2003). Zn affected LHCII, where *Chl b* is located, rather than *Chl a* RC (Bertrand and Poirier, 2005). Szalontai et al. (1999) reported the changes in membrane architecture as a result of reducing electron transport with Zn. The *Chl* fluorescence kinetics among plant species also revealed these inhibitory targets on photosynthetic electron transport. In this regard, $F_v$ and $F_m$ inhibition in plant chloroplasts treated with $Zn^{2+}$ was considered as evidence for PSII activity inhibition (Plekhanov and Chemeris 2003). The elevated levels of minimum *Chl* fluorescence in the dark (*Fo*) were linked to a decrease in *Fv/Fm* (Vaillant et al., 2005), suggesting inactivation of PSII, consistent with the enhanced lipid peroxidation in high Zn treated plants. Nevertheless, the observed decreases in PSII activity can be explained by the downregulation induced by a decreased demand for ATP and NADPH in the Calvin cycle (Subrahmanyam and Rathore, 2000). In high-Zn-treated plants, catalase activity was decreased, which may suggest a drop in photorespiratory activity. However, photosynthetic thylakoid electron transport rates (ETR) were altered under Zn excess (Monnet et al., 2001), indicating an increase in electron consumption diverted to photorespiration.

High Zn concentrations have been found to block xylem elements, resulting in decreased stomatal conductance and, as a result, a decrease in transpiration rate (Sagardoy et al., 2009). The increased mesophyll resistance to the diffusion of $CO_2$ has also been found under high Zn concentrations (Prasad and Strzalka, 1999). Photosynthesis and respiratory metabolism interact, with Zn inhibiting the function of the mitochondrial cytochrome bc1 complex (Link and von Jagow, 1995), as well as AOX activity (Affourtit and Anthony, 2004). Zn-triggered changes in respiration can arise due to metabolic impacts that may possibly diminish the net photosynthesis, regardless of the processes involved.

### 6.5.1 Heavy metals stress

Heavy metals (HMs) contamination is causing global threats to the productivity of agricultural crops. The ingestion of these HMs-contaminated crops governed hazards to human health. The overaccumulation of toxic HMs in soil-plant systems caused a reduction in plant growth and development (Ahmad et al., 2017; Ali et al., 2018a, 2018b; Aslam et al., 2021; Hamid et al., 2020, 2021; Salam et al., 2021; Ulhassan et al., 2019c, 2023). The exogenous supplementation of Zn ions has been reported to alleviate the HMs-induced toxic effects in different plant species (ul Hassan et al., 2017; Rizwan et al., 2019; Adil et al., 2020). Recent investigations confirmed that an exogenous supply of Zn mitigated the Cd toxicity in *juvenile cacao* by minimizing the Cd-mediated programmed cell death and improving the leaf gas exchange parameters such as net photosynthetic rate and stomatal conductance and photosynthetic efficiency (dos Santos et al., 2020). In Cd-exposed *Oryza sativa* seedlings, the applications of Zn alleviated the Cd toxicity by minimizing the damages to chlorophylls, carotenoids, photosystem II (PSII), maximum photochemical efficiency (*Fv/Fm*), cell viability and reducing the overaccumulation of ROS and MDA in plant tissues (Adil et al., 2020). A recent study revealed that nano-size Zn ions mitigated the cobalt-induced toxic effects on maize seedlings and improved the photosynthetic efficiency or net photosynthesis process by maintaining the status of *Chl a*, *Chl b*, SPAD values, photochemical efficiency (*Fv/Fm*), and gas exchange parameters (stomatal conductance, net photosynthetic rate, transpiration rate, intercellular $CO_2$ concentration), thus improved plant growth and biomass production (Salam et al., 2021). In this way, Zn may improve the activities of Calvin cycle enzymes, electron transport chains of chloroplast, PSII functionality by limiting singlet oxygen species, RuBisCo efficiency, and photosynthetic system. These findings suggested that Zn, mainly at lower concentrations, can be utilized for biofortification purposes and effectively alleviate the HMs-induced phytotoxic effects in hydroponic as well as soil conditions. Still, the involved molecular mechanisms related to Zn-mediated metal tolerance by improving the photosynthetic apparatus in different plants are not clear and should be explored in upcoming research

## 6.6 Challenges and future perspectives

The potential to improve photosynthesis for improving crop yields is a targeted strategy, aiming to meet food production for future challenges (Nölke and Schillberg, 2020). The current study discusses and highlights the potential role of Zn in enhancing plant photosynthesis under different abiotic stresses. In the present overview, we have highlighted the existing knowledge gaps, which should be focused in future studies. For example, the effects of Zn treatment on nutrient acquisition and its crosstalk with other essential nutrients under different abiotic stresses are little unstated. It would be useful to investigate Zn signaling under abiotic stress to elucidate whether Zn contributes directly or indirectly to increasing the endogenous hormones against stress conditions. Moreover, omics approaches are needed to detect responsive proteins of abiotic stress tolerance. Future study is also needed to estimate the influence of Zn on the translocation of osmolytes such as proline, glycine-betaine, polyamines, and polyphenols under environmental changes.

Moreover, more investigations should consider the effects of Zn on the expression of heat shock proteins and embryogenesis-rich proteins under severe environmental conditions (Hassan et al., 2020). The interaction of Zn with microorganisms at molecular levels can open new avenues for abiotic stress tolerance (Watts-Williams et al., 2017; Hassan et al., 2017a, b; El-Sawah et al., 2021). More studies also are needed to investigate and understand Zn transport forms in different plant tissues. In addition, further research into the biochemical and genetic pathways behind Zn-induced abiotic stress resistance is needed (Tolay, 2021). The recent advancements in genome editing and synthetic biology may pave the path for the simultaneous introduction of transgenes. In addition to this, the targeted editing of endogenous genes can permit the implementation of more aggressive crop yield improvement strategies. It is also important to study the genotypic variation in response to environmental changes under excessive or Zn deficiency conditions. The manipulation of photosynthetic components through overexpression/silencing and/or genome editing of the photosynthesis limiting process may open new avenues for minimizing photosynthesis bottleneck steps and maximizing crop production improvement (Hussain et al., 2021b). In this review zinc-solubilizing microorganisms are an environmentally acceptable method that may be employed as biofertilizers to solubilize insoluble zinc in soil and increase zinc biofortification, photosynthesis, plant growth, and productivity. Here, we can minimize our reliance on chemical fertilizers.

## Acknowledgments

This work was supported by the Science and Technology Department of Zhejiang Province (2023C02002), and Collaborative Innovation Center for Modern Crop Production co-sponsored by Province and Ministry (CIC-MCP). We thank Deli Sun from Agricultural Experiment Station of Zhejiang University for his assistance.

## References

Abbas, G., Amjad, M., Saqib, M., Murtaza, B., Asif Naeem, M., Shabbir, A., Murtaza, G., 2020. Soil sodicity is more detrimental than salinity for quinoa (*Chenopodium quinoa* Willd.): a multivariate comparison of physiological, biochemical and nutritional quality attributes. Journal of Agronomy and Crop Science 207, 59–73.

AbdElgawad, H., Darin, P., Gaurav, Z., Wim, V.D.E., Ivan, A.J., Han, A., 2014. Climate extreme effects on the chemical composition of temperate grassland species under ambient and elevated $CO_2$: a comparison of fructan and non-fructan accumulators. PLoS One 9 (3), e92044.

Abdoulaye, A.O., Lu, H., Zhu, Y., Hamoud, Y.A., Sheteiwy, M.S., 2019. The global trend of the net irrigation water requirement of maize from 1960 to 2050. Climate 7, 124.

Abid, M., Tian, Z., Ata-Ul-Karim, S.T., Wang, F., Liu, Y., Zahoor, R., Jiang, D., Dai, T., 2016. Adaptation to and recovery from drought stress at vegetative stages in wheat (*Triticum aestivum*) cultivars. Functional Plant Biology 43, 1159–1169.

Adil, M.F., Sehar, S., Han, Z., Lwalaba, J.L.W., Jilani, G., Zeng, F., Chen, Z.H., Shamsi, I.H., 2020. Zinc alleviates cadmium toxicity by modulating photosynthesis, ROS homeostasis, and cation flux kinetics in rice. Environmental Pollution 265, 114979.

Affourtit, C., Anthony, L.M., 2004. Purification of the plant alternative oxidase from *Arum maculatum*: measurement, stability and metal requirement. Biochimica et Biophysica Acta (BBA)-Bioenergetics 1608 no. 2–3, 181–189.

Ahmad, R., Ali, S., Hannan, F., Rizwan, M., Iqbal, M., Ulhassan, Z., Akram, N.A., Maqbool, S., Abbas, F., 2017. Promotive role of 5-aminolevulinic acid on chromium-induced morphological, photosynthetic, and oxidative changes in cauliflower (*Brassica oleracea* botrytis L.). Environmental Science and Pollution Research 24, 8814–8824.

Ainsworth, E.A., Leakey, A.D., Ort, D.R., Long, S.P., 2008. FACE-ing the facts: inconsistencies and interdependence among field, chamber and modeling studies of elevated [$CO_2$] impacts on crop yield and food supply. New Phytologist 179 (1), 5–9.

Aktaş, H., Abak, K., Öztürk, L., Cakmak, I., 2006. The effect of zinc on growth and shoot concentrations of sodium and potassium in pepper plants under salinity stress. Turkish Journal of Agriculture and Foresty 30, 407–412.

Albert, K.R., Mikkelsen, T.N., Michelsen, A., Ro-Poulsen, H., van der Linden, L., 2011. Interactive effects of drought, elevated $CO_2$ and warming on photosynthetic capacity and photosystem performance in temperate heath plants. Journal of Plant Physiology 168 (13), 1550–1561.

Ali, S., Jin, R., Gill, R.A., Mwamba, T.M., Zhang, N., Ulhassan, Z., Islam, F., Ali, S., Zhou, W. J., 2018a. Beryllium stress-induced modifications in antioxidant machinery and plant ultrastructure in the seedlings of black and yellow seeded oilseed rape. Biomedical Research International 1615968. Available from: https://doi.org/10.1155/2018/1615968.

Ali, S., Gill, R.A., Mwamba, T.M., Zhang, N., Lv, M.T., Ulhassan, Z., Islam, F., Zhou, W.J., 2018b. Differential cobalt-induced effects on plant growth, ultrastructural modifications, and antioxidative response among four *Brassica napus* L. cultivars. International Journal of Environmental Science and Technology 15, 2685–2700. Available from: https://doi.org/10.1007/s13762-017-1629-z.

Ali, S., Gill, R.A., Ulhassan, Z., Najeeb, U., Kanwar, M.K., Abid, M., Mwamba, T.M., Huang, Q., Zhou, W., 2018c. Insights on the responses of *Brassica napus* cultivars against the cobalt-stress as revealed by carbon assimilation, anatomical changes and secondary metabolites. Environmental and Experimental Botany 156, 183–196.

Allen, D.J., Ort, D.R., 2001. Impacts of chilling temperatures on photosynthesis in warm-climate plants. Trends in Plant Science 6, 36–42.

Alloway, B.J., 2008. Zinc in Soils and Crop Nutrition (Brussels: IZA and IFA), 137. Brussels, Belgium/Paris, France.

Alotaibi, M.O., Saleh, A.M., Sobrinho, R.L., Sheteiwy, M.S., El-Sawah, A.M., Mohammed, A. E., AbdElgawad, H., 2021. Arbuscular mycorrhizae mitigate aluminum toxicity and regulate proline metabolism in plants grown in acidic soil. Journal of Fungi 7 (7), 531.

Arough, Y.K., Sharifi, R.S., Sharifi, R.S., 2016. Bio fertilizers and zinc effects on some physiological parameters of triticale under water-limitation condition. Journal of Plant Interactions 1, 167–177.

Ashrafi, N., Rezaei Nejad, A., 2017. Lisianthus response to salinity stress. Photosynthetica 56, 487–494.

Aslam, M., Aslam, A., Sheraz, M., Ali, B., Ulhassan, Z., Najeeb, U., Zhou, W., Gill, R.A., 2021. Lead toxicity in cereals: mechanistic insight into toxicity, mode of action, and management. Frontiers in Plant Science 11, 2248.

Augustyniak, A., Perlikowski, D., Rapacz, M., Kościelniak, J., Kosmala, A., 2018. Insight into cellular proteome of *Lolium multiflorum/Festuca arundinacea* introgression forms to decipher crucial mechanisms of cold acclimation in forage grasses. Plant Science 272, 22–31.

Baðci, S.A., Ekiz, H., Yilmaz, A., Cakmak, I., 2007. Effect of Zn deficiency and drought on grain yield of field grown wheat cultivars in Central Anatolia. Journal of Agronomy and Crop Science 193, 198–206.

Baghalian, K., Haghiry, A., Naghavi, M.R., Mohammadi, A., 2008. ). Effect of saline irrigation water on agronomical and phytochemical characters of chamomile (*Matricaria recutita* L.). Scientia Horticulturae 116, 437–441.

Banerjee, A., Roychoudhury, A., 2019. Cold stress and photosynthesis. Photosynthesis, Productivity and Environmental Stress 27–37.

Baslam, M., Mitsui, T., Hodges, M., Priesack, E., Herritt, M.T., Aranjuelo, I., Sanz-Sáez, A., 2020. Photosynthesis in a changing global climate: scaling up and scaling down in crops. Frontiers in Plant Science 11, 882.

Bertrand, M., Poirier, I., 2005. Photosynthetic organisms and excess of metals. Photosynthetica 43 (3), 345–353.

Bierbaum, R.M., Holdren, J.P., MacCracken, M.C., Moss, R.H., Raven, P.H., 2007. Confronting Climate Change: Avoiding the Unmanageable and Managing the Unavoidable. The United Nations Foundation, Washington, DC.

Bradáčová, K., Weber, N.F., Morad-Talab, N., Asim, M., Imran, M., Weinmann, M., Neumann, G., 2016. Micronutrients (Zn/Mn), seaweed extracts, and plant growth-promoting bacteria as cold-stress protectants in maize. Chemical and Biological Technologies in Agriculture 3 (1), 1–10.

Brito, C., Dinis, L.-T., Meijón, M., Ferreira, H., Pinto, G., Moutinho-Pereira, J., Correia, C., 2018. Salicylic acid modulates olive tree physiological and growth responses to drought and rewatering events in a dose dependent manner. Journal of Plant Physiology 230, 21–32.

Broadley, M., White, P.J., Hammond, J.P., Zelko, I., Lux, A., 2007. Zinc in plants. New Phytologist 173, 677–702.

Brown, P.H., Cakmak, I., Zhang, Q., 1993. Form and function of zinc plants. In: Robson, A.D. (Ed.), Zinc in Soils and Plants: Proceedings of the International Symposium on 'Zinc in Soils and Plants' Held at the University of Western Australia, 27–28 September, 1993. Springer, Netherlands, Dordrecht, pp. 93–106.

Cakmak, I., 2000. Possible roles of zinc in protecting plant cells from damage by reactive oxygen species. New Phytologist 146, 185–205.

Cakmak, I., Engels, C., 1999. Role of mineral nutrients in photosynthesis and yield formation. In: Rengel, Z. (Ed.), Mineral Nutrition of Crops. Haworth Press, New York, pp. 141–168.

Chaves, M., Flexas, J., Pinheiro, C., 2009. Photosynthesis under drought and salt stress: regulation mechanisms from whole plant to cell. Annals of Botany 103, 551–560.

Chen, K.M., Gong, H.J., Chen, G.C., Wang, S.M., Zhang, C.L., 2004. Gradual drought under field conditions influences glutathione metabolism, redox balance and energy supply in spring wheat. Journal of Plant Growth Regulation 23, 20–28.

Çiçek, N., Oukarroum, A., Strasser, R.J., Schansker, G., 2018. Salt stress effects on the photosynthetic electron transport chain in two chickpea lines differing in their salt stress tolerance. Photosynthesis Research 136 (3), 291–301.

Craggs, G.J.P., 2016. Photosynthesis and its role in climate change and soil regeneration. Future Analysis Directions 1–6.

Cui, Y., Zhao, N., 2011. Oxidative stress and change in plant metabolism of maize (*Zea mays* L.) growing in contaminated soil with elemental sulfur and toxic effect of zinc. Plant Soil and Environment 57, 34–39.

Dahal, K., Wang, J., Martyn, G.D., Rahimy, F., Vanlerberghe, G.C., 2014. Mitochondrial alternative oxidase maintains respiration and preserves photosynthetic capacity during moderate drought in *Nicotiana tabacum*. Plant Physiology 166, 1560–1574.

Debez, A., Koyro, H.W., Grignon, C., Abdelly, C., Huczermeyer, B., 2008. Relationship between the photosynthetic activity and the performance of *Cakile maritima* after long-term salt treatment. Physiologia Plantarum 133, 373–385.

Demiral, T., Türkan, I., 2006. Exogenous glycinebetaine affects growth and proline accumulation and retards senescence in two rice cultivars under NaCl stress. Environmental and Experimental Botany 56, 72–79.

Dhingra, M., 2015. Physiological responses and tolerance mechanisms of low temperature stress in plants. International Journal of Advanced Research 3, 637–646.

Di, B., Tognetti, D., Sebastiani, L.R., Vitagliano, C., 2003. Responses of *Populus deltoides* × *Populus nigra* (Populus × euramericana) clone I-214 to high zinc concentrations. New Phytologist 159 (2), 443–452.

dos Santos, M.L.S., de Almeida, A.A.F., da Silva, N.M., Oliveira, B.R.M., Silva, J.V.S., Junior, J.O.S., Ahnert, D., Baligar, V.C., 2020. Mitigation of cadmium toxicity by zinc in juvenile cacao: physiological, biochemical, molecular and micromorphological responses. Environmental and Experimental Botany 179, 104201.

Ekiz, H., Bagci, S.A., Kıral, A.S., Eker, S., Gültekin, I., Alkan, A., Cakmak, I., 1998. Effects of zinc fertilization and irrigation on grain yield and zinc concentration of various cereals grown in zinc-deficient calcareous soils. Journal of Plant Nutrition 21, 2245–2256.

Elkelish, A., Qari, S.H., Mazrou, Y.S., Abdelaal, K.A., Hafez, Y.M., Abu-Elsaoud, A.M., et al., 2020. Exogenous ascorbic acid induced chilling tolerance in tomato plants through modulating metabolism, osmolytes, antioxidants, and transcriptional regulation of catalase and heat shock proteins. Plants 9, 431.

El-Sawah, A.M., El-Keblawy, A., Ali, D.F.I., Ibrahim, H.M., El-Sheikh, M.A., Sharma, A., Hamoud, Y.A., Shaghaleh, H., Brestič, M., Skalicky, M., Xiong, Y., Sheteiwy, M., 2021. Arbuscular mycorrhizal fungi and plant growth-promoting rhizobacteria enhance soil key enzymes, plant growth, seed yield, and qualitative attributes of guar (*Cyamopsis tetragonoloba* L.). Agriculture 11, 194.

Ensminger, I., Busch, F., Huner, N.P., 2006. Photostasis and cold acclimation: sensing low temperature through photosynthesis. Physiologia Plantarum 126 (1), 28–44.

Evans, M.J., Choi, W.G., Gilroy, S., Morris, R.J., 2016. A ROS-assisted calcium wave dependent on the AtRBOHD NADPH oxidase and TPC1 cation channel propagates the systemic response to salt stress. Plant Physiology 171, 1771–1784.

Fageria, N.K., 2009. The Use of Nutrients in Crop Plants. CRC Press, Boca Raton London, p. 448.

Fahad, S., Bajwa, A.A., Nazir, U., Anjum, S.A., Farooq, A., Zohaib, A., Sadia, S., Nasim, W., Adkins, S., Saud, S., Ihsan, M.Z., Alharby, H., Wu, C., Wang, D., Huang, J., 2017. Crop production under drought and heat stress: plant responses and management options. Frontiers in Plant Science 8.

Fang, Y., Wang, L., Xin, Z., Zhao, L., An, X., Hu, Q., 2008. Effect of foliar application of zinc, selenium, and iron fertilizers on nutrients concentration and yield of rice grain in China. Journal of Agricultural and Food Chemistry 56, 2079–2084.

FAO, 2009. Feeding the World in 2050. World Agricultural Summit on Food Security. FAO, Rome.

Farooq, M., Hussain, M., Siddique, K.H.M., 2014. Drought stress in wheat during flowering and grain-filling periods. Critical Reviews in Plant Sciences 33, 331–349.

Farooq, M., Wahid, A., Kobayashi, N., Fujita, D., Basra, S.M.A., 2009. Plant drought stress: effects, mechanisms and management. Agronomy for Sustainable Development 29, 185–212.

Farooq, M., Hussain, M., Nawaz, A., Lee, D.-J., Alghamdi, S.S., Siddique, K.H., 2017. Seed priming improves chilling tolerance in chickpea by modulating germination metabolism, trehalose accumulation and carbon assimilation. Plant Physiology and Biochemistry 111, 274–283.

Florke, M., Barlund, I., van Vliet, M.T.H., Bouwman, A.F., Wada, Y., 2019. Analysing trade-offs between SDGs related to water quality using salinity as a marker. Current Opinion in Environmental Sustainability 36, 96–104.

Fricke, W., Akhiyarova, G., Veselov, D., Kudoyarova, G., 2004. Rapid and tissue-specific changes in ABA and in growth rate response to salinity in barley leaves. Journal of Experimental Botany 55, 1115–1123.

Gilroy, S., Suzuki, N., Miller, G., Choi, W.G., Toyota, M., Devireddy, A.R., Mittler, R., 2014. A tidal wave of signals: calcium and ROS at the forefront of rapid systemic signaling. Trends in Plant Science 19, 623–630.

Goodman, B.A., Newton, A.C., 2005. Effects of drought stress and its sudden relief on free radical processes in barley. Journal of the Science of Food and Agriculture 85, 47–53.

Hagedorn, F., Landolt, W., Tarjan, D., Egli, P., Bucher, J.B., 2002. Elevated $CO_2$ influences nutrient availability in young beech-spruce communities on two soil types. Oecologia 132 (1), 109–117.

Hamed, S.M., Zinta, G., Klöck, G., Asard, H., Selim, S., AbdElgawad, H., 2017. Zinc-induced differential oxidative stress and antioxidant responses in *Chlorella sorokiniana* and *Scenedesmus acuminatus*. Ecotoxicology and Environmental Safety 140, 256–263.

Hamid, Y., Tang, L., Hussain, B., Usman, M., Liu, L., Cao, X., Ulhassan, Z., Khan, M.B., Yang, X., 2020. Cadmium mobility in three contaminated soils amended with different additives as evaluated by dynamic flow-through experiments. Chemosphere 261, 127763. Available from: https://doi.org/10.1016/j.chemosphere.2020.127763.

Hamid, Y., Tang, L., Hussain, B., Usman, M., Liu, L., Ulhassan, Z., He, Z., Yang, X., 2021. Sepiolite clay: a review of its applications to immobilize toxic metals in contaminated soils and its implications in soil-plant system. Environmental Technology & Innovation 101598.

Han, W., Huang, L., Owojori, O.J., 2020. Foliar application of zinc alleviates the heat stress of pakchoi (*Brassica chinensis* L.). Journal of Plant Nutrition 43 (2), 194–213.

Harris, D., Rashid, A., Miraj, G., Arif, M., Shah, H., 2007. 'On-farm' seed priming with zinc sulphate solution—a cost-effective way to increase the maize yields of resource-poor farmers. Field Crop Research: A Journal of Science and its Applications 102, 119–127.

Hassan, Z., Ali, S., Ahmad, R., Rizwan, M., Ab-bas, F., Yasmeen, T., Iqbal, M., 2017a. Biochemical and molecular responses of oilseed crops to heavy metal stress. Oilseed Crops: Yield and Adaptations under Environmental Stress. John Wiley & Sons, pp. 236–248. Available from: https://doi.org/10.1002/9781119048800.ch13.

Hassan, Z., Ali, S., Rizwan, M., Ibrahim, M., Nafees, M., Waseem, M., 2017b. Role of bioremediation agents (bacteria, fungi, and algae) in alleviating heavy metal toxicity. In: Kumar, V., Kumar, M., Sharma, S., Prasad, R. (Eds.), Probiotics in Agroecosystem. Springer, Singapore. Available from: 10.1007/978-981-10-4059-7_27.

Hassan, M.U., Aamer, M., Chattha, M.U., Haiying, T., Shahzad, B., Barbanti, L., Nawaz, M., Rasheed, A., Afzal, A., Liu, Y., Guoqin, H., 2020. The critical role of zinc in plants facing the drought stress. Agriculture 10, 0396.

He, F., Li, H.G., Wang, J.J., Su, Y., Wang, H.L., Feng, C.H., Yang, Y., Niu, M.X., Liu, C., Yin, W., Xia, X., 2019. Pe STZ 1, a C2H2-type zinc finger transcription factor from *Populus euphratica*, enhances freezing tolerance through modulation of ROS scavenging by directly regulating Pe APX 2. Plant Biotechnology Journal 17 (11), 2169–2183.

Hichem, H., Naceur, E., Mounir, D., 2009. Effects of salt stress on photosynthesis. PSII photochemistry and thermal energy dissipation in leaves of two corn (*Zea mays* L.) varieties. Photosynthesis 47, 517–526.

Hu, W., Tian, S.B., Di, Q., Duan, S.H., Dai, K., 2018. Effects of exogenous calcium on mesophyll cell ultrastructure, gas exchange, and photosystem II in tobacco (*Nicotiana tabacum* Linn.) under drought stress. Photosynthetica 56, 1204–1211.

Hussain, H.A., Hussain, S., Khaliq, A., Ashraf, U., Anjum, S.A., Men, S., et al., 2018. Chilling and drought stresses in crop plants: implications, cross talk, and potential management opportunities. Frontiers in Plant Science 9.

Hussain, S., Mumtaz, M., Manzoor, S., Shuxian, L., Ahmed, I., Skalicky, M., Brestic, M., Rastogi, A., Ulhassan, Z., Shafiq, I., Allakhverdiev, S.I., 2021a. Foliar application of silicon improves growth of soybean by enhancing carbon metabolism under shading conditions. Plant Physiology and Biochemistry 159, 43–52.

Hussain, S., Ulhassan, Z., Brestic, M., Zivcak, M., Zhou, W., Allakhverdiev, S.I., Yang, X., Safdar, M.E., Yang, W., Liu, W., 2021b. Photosynthesis research under climate change. Photosynthesis Research . Available from: https://doi.org/10.1007/s11120-021-00861-z.

Izanloo, A., Condon, A.G., Langridge, P., Tester, M., Schnurbusch, T., 2008. Different mechanisms of adaptation to cyclic water stress in two South Australian bread wheat cultivars. Journal of Experimental Botany 59, 3327–3346.

Jaleel, C.A., Manivannan, P., Wahid, A., Farooq, M., Al-Juburi, H.J., Somasundaram, R., Panneerselvam, R., 2009. Drought stress in plants: a review on morphological characteristics and pigments composition. International Journal of Agriculture and Biology 11, 100–105.

Johnson-Beebout, S.E., Lauren, J.G., Duxbury, J.M., 2009. Immobilization of zinc fertilizer in flooded soils monitored by adapted DTPA soil test. Communications in Soil Science and Plant Analysis 40, 1842–1861.

Karim, M.R., Zhang, Y.-Q., Zhao, R.-R., Chen, X.-P., Zhang, F.-S., Zou, C.-Q., 2012. Alleviation of drought stress in winter wheat by late foliar application of zinc, boron, and manganese. Journal of Plant Nutrition and Soil Science 175, 142–151.

Kaya, C., Ashraf, M., Akram, N.A., 2018. Hydrogen sulfide regulates the levels of key metabolites and antioxidant defense system to counteract oxidative stress in pepper (*Capsicum annuum* L.) plants exposed to high zinc regime. Environmental Science and Pollution Research 25 (13), 12612–12618.

Kaznina, N.M., Batova, Y.V., Laidinen, G.F., Sherudilo, E.G., Titov, A.F., 2019. Low-temperature adaptation of winter wheat seedlings under excessive zinc content in the root medium. Russian Journal of Plant Physiology 66 (5), 763–770.

Khudsar, T., Arshi, A., Siddiqi, T.O., Mahmooduzzafar, Iqbal, M., 2008. Zinc-induced changes in growth characters, foliar properties, and Zn-accumulation capacity of pigeon pea at different stages of plant growth. Journal of Plant Nutrition 31 (2), 281–306.

Kohzuma, K., Cruz, J.A., Akashi, K., Hoshiyasu, S., Munekage, Y.N., Yokota, A., Kramer, D. M., 2009. The long-term responses of the photosynthetic proton circuit to drought. Plant, Cell & Environment 32, 209–219.

Koyro, H.W., 2006. ). Effect of salinity on growth, photosynthesis water relations and solute composition of the potential cash crop halophyte *Plantago coronopus* (L.). Environmental and Experimental Botany 56, 136–146.

Küpper, H., Šetlík, I., Spiller, M., Küpper, F.C., Prášil, O., 2002. Heavy metal-induced inhibition of photosynthesis: targets of in vivo heavy metal chlorophyll formation1. Journal of Phycology 38 (3), 429–441.

Link, T.A., von Jagow, G., 1995. Zinc ions inhibit the Qp center of bovine heart mitochondrial bc1 complex by blocking a protonatable group (*). Journal of Biological Chemistry 270 (42), 25001–25006.

Lobell, D.B., Roberts, M.J., Schlenker, W., Braun, N., Little, B.B., Rejesus, R.M., 2014. Greater sensitivity to drought accompanies maize yield increase in the U.S. Midwest. Science (New York, N.Y.) 344, 516–519.

Loladze, I., 2002. Rising atmospheric $CO_2$ and human nutrition: toward globally imbalanced plant stoichiometry? Trends in Ecology & Evolution 17 (10), 457–461.

Loneragan, J.F., Webb, M.J., 1993. Interactions between zinc and other nutrients affecting the growth of plants. Zinc in Soils and Plants. Springer, Dordrecht, pp. 119–134.

Long, S.P., Marshall-Colon, A., Zhu, X.G., 2015. Meeting the global food demand of the future by engineering crop photosynthesis and yield potential. Cell 161, 56–66.

Long, S.P., Zhu, X.G., Naidu, S.L., Ort, D.R., 2006. Can improvement in photosynthesis increase crop yields? Plant, Cell & Environment 29, 315–330.

Ma, D., Sun, D., Wang, C., Ding, H., Qin, H., Hou, J., Huang, X., Xie, Y., Guo, T., 2017. Physiological responses and yield of wheat plants in zinc-mediated alleviation of drought stress. Frontiers in Plant Science 8, 860.

Malan, C., Greyling, M.M., Gressel, J., 1990. Correlation between CuZn superoxide dismutase and glutathione reductase, and environmental and xenobiotic stress tolerance in maize inbred. Plant Science 69, 157–166.

Margutti, M.P., Reyna, M., Meringer, M.V., Racagni, G.E., Villasuso, A.L., 2017. Lipid signalling mediated by PLD/PA modulates proline and $H_2O_2$ levels in barley seedlings exposed to short-and long-term chilling stress. Plant Physiology and Biochemistry 113, 149–160.

Mateos-Naranjo, E., Redondo-Gómez, S., Cambrollé, J., Luque, T., Figueroa, M.E., 2008. Growth and photosynthetic responses to zinc stress of an invasive cordgrass, *Spartina densiflora*. Plant Biology 10 (6), 754–762.

Meyer, E., Aspinwall, M.J., Lowry, D.B., Palacio-Mejía, J.D., Logan, T.L., Fay, P.A., Juenger, T.E., 2014. ). Integrating transcriptional, metabolomic, and physiological responses to drought stress and recovery in switchgrass (*Panicum virgatum* L.). BMC Genomics 15 (1), 1–15.

Monnet, F., Nathalie, V., Philippe, V., Alain, C., Huguette, S., Adnane, H., 2001. Relationship between PSII activity, $CO_2$ fixation, and Zn, Mn and Mg contents of *Lolium perenne* under zinc stress. Journal of Plant Physiology 158 (9), 1137–1144.

Mroue, S., Simeunovic, A., Robert, H.S., 2018. Auxin production as an integrator of environmental cues for developmental growth regulation. Journal of Experimental Botany 69, 201–212.

Munns, R., James, R.A., Lauchli, A., 2006. Approaches to increasing the salt tolerance of wheat and other cereals. Journal of Experimental Botany 57, 1025–1043.

Munns, R., James, R.A., Laüchli, A., 2006. Approaching to increasing the salt tolerance of wheat and other cereals. Journal of Experimental Botany 57, 1025–1043.

Munns, R., Tester, M., 2008. Mechanisms of salinity tolerance. Annual Review Plant Biology 59, 651–681.

Mwamba, T.M., Islam, F., Ali, B., Lwalaba, J.L.W., Gill, R.A., Zhang, F., Farooq, M.A., Ali, S., Ulhassan, Z., Huang, Q., Zhou, W., 2020. Comparative metabolomic responses of low-and high-cadmium accumulating genotypes reveal the cadmium adaptive mechanism in *Brassica napus*. Chemosphere 250, 126308.

Nadeem, F., Azhar, M., Anwar-ul-Haq, M., Sabir, M., Samreen, T., Tufail, A., Umair, H., Awan, M., Juan, W., 2020. Comparative response of two rice (*Oryza sativa* L.) cultivars to applied zinc and manganese for mitigation of salt stress. Journal of Soil Science and Plant Nutrition 20, 2059–2072.

Naumann, J.C., Young, D.R., Anderson, J.E., 2007. Linking leaf chlorophyll fluorescence properties to physiological responses for detection of salt and drought stress in coastal plant species. Physiologia Plantarum 131, 422–433.

Nazir, M.M., Ulhassan, Z., Zeeshan, M., Ali, S., Gill, M.B., 2020. Toxic metals/metalloids accumulation, tolerance, and homeostasis in brassica oilseed species. In: Hasanuzzaman, M. (Ed.), The Plant Family Brassicaceae. Springer, Singapore. Available from: https://doi.org/10.1007/978-981-15-6345-4_13.

Ning, L., Du, W., Song, H., Shao, H., Qi, W., Sheteiwy, M.S., Yu, D., 2019. Identification of responsive miRNAs involved in combination stresses of phosphate starvation and salt stress in soybean root. Environmental and Experimental Botany 167, 103823.

Nölke, G., Schillberg, S., 2020. Strategies to enhance photosynthesis for the improvement of crop yields. In: Kumar, A., et al., (Eds.), Climate Change, Photosynthesis and Advanced Biofuels.

Ort, D.R., Merchant, S.S., Alric, J., Barkan, A., Blankenship, R.E., Bock, R., Croce, R., Hanson, M. R., Hibberd, J.M., Long, S.P., Moore, T.A., Moroney, J., Niyogi, K.K., Parry, M.A.J., Peralta-Yahya, P.P., Prince, R.C., Redding, K.E., Spalding, M.H., Wijk, K.J.V., Vermaas, W.F.J., Caemmerer, S.V., Weber, A.P.M., Yeates, T.O., Yuan, J.S., Zhu, X.G., 2015. Redesigning photosynthesis to sustainably meet global food and bioenergy demand. Proceedings of the National Academy of Sinces of the United States of America 112, 8529–8536.

Pan, T., Liu, M., Kreslavski, V.D., Zharmukhamedov, S.K., Nie, C., Yu, M., Kuznetsov, V.V., Allakhverdiev, S.I., Shabala, S., 2020. Non-stomatal limitation of photosynthesis by soil salinity. Critical Reviews in Environmental Science and Technology 1–36.

Pavia, I., Roque, J., Rocha, L., Ferreira, H., Castro, C., Carvalho, A., Silva, E., Brito, C., Gonçalves, A., Lima-Brito, J., Correia, C., 2019. Zinc priming and foliar application enhances photoprotection mechanisms in drought-stressed wheat plants during anthesis. Plant Physiology and Biochemistry 140, 27–42.

Percey, W.J., McMinn, A., Bose, J., Breadmore, M.C., Guijt, R.M., Shabala, S., 2016. Salinity effects on chloroplast PSII performance in glycophytes and halophytes1. Functional Plant Biology 43, 1003–1015.

Plekhanov, S.E., Chemeris, Y.K., 2003. Early toxic effects of zinc, cobalt, and cadmium on photosynthetic activity of the green alga *Chlorella pyrenoidosa* Chick S-39. Biology Bulletin of the Russian Academy of Sciences 30 (5), 506–511.

Prasad, M.N.V., Strzalka, K., 1999. Impact of heavy metals on photosynthesis. In: Prasad, M.N.V., Hagemeyer, J. (Eds.), Heavy Metal Stress in Plants. Springer-Verlag, Berlin, pp. 117–138.

Qadir, M., Quillerou, E., Nangia, V., Murtaza, G., Singh, M., Thomas, R.J., Drechsel, P., Noble, A.D., 2014. Economics of salt-induced land degradation and restoration. Natural Resources Forum 38, 282–295.

Renger, G., 2007. Primary processes of photosynthesis: principles and apparatus, 2 parts. Publishers. Royal Society of Chemistry, Cambridge.

Rizwan, M., Ali, S., ur Rehman, M.Z., Maqbool, A., 2019. A critical review on the effects of zinc at toxic levels of cadmium in plants. Environmental Science and Pollution Research 26 (7), 6279–6289.

Rossel, J.B., Wilson, P.B., Hussain, D., Woo, N.S., Gordon, M.J., Mewett, O.P., Pogson, B.J., 2007. Systemic and intracellular responses to photooxidative stress in Arabidopsis. The Plant Cell 19 (12), 4091–4110.

Sagardoy, R.U.T.H., Morales, F.E.R.M.Í.N., López-Millán, A.F., Abadía, A.N.U.N.C.I.A.C.I.Ó.N., Abadía, J.A.V.I.E.R., 2009. Effects of zinc toxicity on sugar beet (*Beta vulgaris* L.) plants grown in hydroponics. Plant Biology 11 (3), 339–350.

Sagardoy, R., Vázquez, S., Florez-Sarasa, I.D., Albacete, A., Ribas-Carbó, M.J., Abadía, F.J., Morales, F., 2010. Stomatal and mesophyll conductances to $CO_2$ are the main limitations to photosynthesis in sugar beet (*Beta vulgaris*) plants grown with excess zinc. New Phytologist 187, 145–158.

Sairam, R.S., Saxena, D.C., 2000. Oxidative stress and antioxidants in wheat genotypes: possible mechanism of water stress tolerance. Journal of Agronomy and Crop Science 184, 55–61.

Salam, A., Khan, A.R., Liu, L., Yang, S., Azhar, W., Ulhassan, Z., Zeeshan, M., Wu, J., Fan, X., Gan, Y., 2021. Seed priming with zinc oxide nanoparticles downplayed ultrastructural damage and improved photosynthetic apparatus in maize under cobalt stress. Journal of Hazardous Materials 127021.

Salama, Z.A., El-Fouly, M.M., Lazova, G., Popova, L.P., 2006. Carboxylating enzymes and carbonic anhydrase functions were suppressed by zinc deficiency in maize and chickpea plants. Acta Physiologia Plantarum 28, 445–451.

Saleh, J., Maftoun, M., Safarzadeh, S., Gholami, A., 2009. Growth, mineral composition, and biochemical changes of broad bean as affected by sodium chloride and zinc levels and sources. Communications in Soil Science and Plant Analysis 40, 3046–3060.

Sanz-Saez, A., Koester, R.P., Rosenthal, D.M., Montes, C.M., Ort, D.R., Ainsworth, E.A., 2017. Leaf and canopy scale drivers of genotypic variation in soybean response to elevated carbon dioxide concentration. Global Change Biology 23, 3908–3920.

Sarwar, M., Saleem, M.F., Ullah, N., Ali, S., Rizwan, M., Shahid, M.R., Alyemeni, M.N., Alamri, S.A., Ahmad, P., 2019. Role of mineral nutrition in alleviation of heat stress in cotton plants grown in glasshouse and field conditions. Scientific Reports 9, 13022.

Shabala, S., Munns, R., 2017. Salinity stress: Physiological constraints and adaptive mechanisms. In: Shabala, S. (Ed.), Plant Stress Physiology. CABI, Wallingford, UK, pp. 24–63.

Shabala, S., White, R.G., Djordjevic, M.A., Ruan, Y.L., Mathesius, U., 2016. Root-to-shoot signalling: integration of diverse molecules, pathways and functions. Functional Plant Biology 43, 87–104.

Sharma, A., Kumar, V., Shahzad, B., Ramakrishnan, M., Sidhu, G.P.S., Bali, A.S., Handa, N., Kapoor, D., Yadav, P., Khanna, K., Bakshi, P., Rehman, Abdul, Kohli, S.K., Khan, E.A., Parihar, R.D., Yuan, H., Thukral, A.K., Bhardwaj, R., Zheng, B., 2020. Photosynthetic response of plants under different abiotic stresses: a review. Journal of Plant Growth Regulation 39, 509–531.

Sharma, P., Tripathi, A., Bisht, S.S., 1995. Zinc requirement for stomatal opening in cauliflower. Plant Physiology 107, 751–756.

Sheteiwy, M.S., Ali, D.F.I., Xiong, Y.C., Brestic, M., Skalicky, M., Alhaj Hamoud, Y., Ulhassan, Z., Shaghaleh, H., AbdElgawad, H., Farooq, M., Sharma, A., El-Sawah, A.M., 2021a. Physiological and biochemical responses of soybean plants inoculated with Arbuscular mycorrhizal fungi and Bradyrhizobium under drought stress. BMC Plant Biology 21, 195. Available from: https://doi.org/10.1186/s12870-021-02949-z.

Sheteiwy, M.S., AbdElgawad, H., Xiong, Y.C., Macovei, A., Brestic, M., Skalicky, M., Shaghaleh, H., Alhaj Hamoud, Y., El-Sawah, A.M., 2021b. Inoculation with *Bacillus amyloliquefaciens* and mycorrhiza confers tolerance to drought stress and improve seed yield and quality of soybean plant. Physiologia Plantarum 5, 1–17.

Sheteiwy, M.S., Gong, D., Gao, Y., Pan, R., Hu, J., Guan, Y., 2018a. Priming with methyl jasmonate alleviates polyethylene glycol-induced osmotic stress in rice seeds by regulating the seed metabolic profile. Environmental and Experimental Botany 153, 236–248.

Sheteiwy, M.S., An, J., Yin, M., Jia, X., Guan, Y., He, F., Hu, J., 2018b. Cold plasma treatment and exogenous salicylic acid priming enhances salinity tolerance of *Oryza sativa* seedlings. Protoplasma 256, 79–99.

Sheteiwy, M.S., Shaghaleh, H., Alhaj Hamoud, Y., Holford, P., Shao, H., Qi, W., Hashmi, M.Z., Guan, Y., Hu, J., 2021c. Zinc oxide nanoparticles: potential effects on crop production, soil properties, antibacterial activity, food processing and quality. Environmental Science and Pollution Research 7, 1–25.

Sheteiwy, M.S., Shao, H., Qi, W., Daly, P., Sharma, A., Shaghaleh, H., Hamoud, Y.A., El-Esawi, M.A., Pan, R., Wan, Q., Lu, H., 2020. Seed priming and foliar application with jasmonic acid enhance salinity stress tolerance of soybean (*Glycine max* L.) seedlings. Journal of the Science of Food and Agriculture 1–15.

Shi, H., Chan, Z., 2014. Improvement of plant abiotic stress tolerance through modulation of the polyamine pathway. Journal of Integrative Plant Biology 56, 114–121.

Sieber, M.H., Thomsen, M.B., Spradling, A.C., 2016. Electron transport chain remodeling by GSK3 during oogenesis connects nutrient state to reproduction. Cell 164, 420–432.

Simova-Stoilova, L., Demirevska, K., Petrova, T., Tsenov, N., Feller, U., 2009. Antioxidative protection and proteolytic activity in tolerant and sensitive wheat (*Triticum aestivum* L.) varieties subjected to long-term field drought. Plant Growth Regulation 58, 107–117.

Subba, P., Mukhopadhyay, M., Mahato, S.K., Bhutia, K.D., Mondal, T.K., Ghosh, S.K., 2014. Zinc stress induces physiological, ultra-structural and biochemical changes in mandarin orange (Citrus reticulata Blanco) seedlings. Physiology and Molecular Biology of Plants 20 (4), 461–473.

Subrahmanyam, D., Rathore, V.S., 2000. Influence of manganese toxicity on photosynthesis in ricebean (*Vigna umbellata*) seedlings. Photosynthetica 38 (3), 449–453.

Sudhir, P., Murthy, S.D.S., 2004. Effects of salt stress on basic processes of photosynthesis. Photosynthesis 42, 481–486.

Sun, L., Song, F., Guo, J., Zhu, X., Liu, S., Liu, F., Li, X., 2019. Nano-ZnO-induced drought tolerance is associated with melatonin synthesis and metabolism in maize. International Journal of Molecular Science 21, 782.

Szalontai, B., Horváth, L.I., Debreczeny, M., Droppa, M., Horváth, G., 1999. Molecular rearrangements of thylakoids after heavy metal poisoning, as seen by Fourier transform infrared (FTIR) and electron spin resonance (ESR) spectroscopy. Photosynthesis Research 61 (3), 241–252.

Tarchoune, I., Degl'Innocenti, E., Kaddour, R., Guidi, L., Lachaa, M., Navari-Izzo, F., Ouerghi, Z., 2012. Effects of NaCl or $Na_2SO_4$ salinity on plant growth, ion content and photosynthetic activity in *Ocimum basilicum* L. Acta Physiologiae Plantarum 34, 607–615.

Tas, S., Tas, B., 2007. Some physiological responses of drought stress in wheat genotypes with different ploidity in Turkey. World Journal of Agriculture and Soil Science 3, 178–183.

Tavanti, T.R., de Melo, A.A.R., Moreira, L.D.K., Sanchez, D.E.J., dos Santos Silva, R., da Silva, R.M., Dos Reis, A.R., 2021. Micronutrient fertilization enhances ROS scavenging system for alleviation of abiotic stresses in plants. Plant Physiology and Biochemistry .

Tilman, D., Balzer, C., Hill, J., Befort, B.L., 2011. Global food demand and the sustainable intensification of agriculture. Proceedings of the National Academy of Sciences of the United States of America 108, 20260–20264.

Tolay, I., 2021. The impact of different zinc (Zn) levels on growth and nutrient uptake of Basil (*Ocimum basilicum* L.) grown under salinity stress. PLoS One 16 (2), e0246493.
ul Hassan, Z., Ali, S., Rizwan, M., Hussain, A., Akbar, Z., Rasool, N., Abbas, F., 2017. Role of zinc in alleviating heavy metal stress. In: Naeem, A.G.M., et al., (Eds.), Essential Plant Nutrients. Springer International Publishing. Available from: 10.1007/978-3-319-58841-4_14.
Ulhassan, Z., Ali, S., Gill, R.A., Mwamba, T.M., Abid, M., Li, L., Zhang, N., Zhou, W., 2018. Comparative orchestrating response of four oilseed rape (*Brassica napus*) cultivars against the selenium stress as revealed by physio-chemical, ultrastructural and molecular profiling. Ecotoxicology and Environmental Safety 161, 634–647.
Ulhassan, Z., Gill, R.A., Ali, S., Mwamba, T.M., Ali, B., Wang, J., Huang, Q., Aziz, R., Zhou, W., 2019a. Dual behavior of selenium: Insights into physio-biochemical, anatomical and molecular analyses of four *Brassica napus* cultivars. Chemosphere 225, 329–341.
Ulhassan, Z., Huang, Q., Gill, R.A., Ali, S., Mwamba, T.M., Ali, B., Hina, F., Zhou, W., 2019b. Protective mechanisms of melatonin against selenium toxicity in *Brassica napus*: insights into physiological traits, thiol biosynthesis, and antioxidant machinery. BMC Plant Biology 19 (507), 2019. Available from: https://doi.org/10.1186/s12870-019-2110-6.
Ulhassan, Z., Gill, R.A., Huang, H., Ali, S., Mwamba, T.M., Ali, B., Huang, Q., Hamid, Y., Khan, A.R., Wang, J., Zhou, W., 2019c. Selenium mitigates the chromium toxicity in *Brassicca napus* L. by ameliorating nutrients uptake, amino acids metabolism and antioxidant defense system. Plant Physiology and Biochemistry 145, 142–152. Available from: https://doi.org/10.1016/j.plaphy.2019.10.035.
Ulhassan, Z, Yang, S, He, D, Khan, A.R, Salam, A, Azhar, W, Muhammad, S, Ali, S, Hamid, Y, Khan, I, Sheteiwy, M.S, Zhou, W, 2023. Seed priming with nano-silica effectively ameliorates chromium toxicity in Brassica napus. Journal of Hazardous Materials 458, 131906. Available from: https://doi.org/10.1016/j.jhazmat.2023.131906In this issue.
Ullah, A., Romdhane, L., Rehman, A., Farooq, M., 2019. Adequate zinc nutrition improves the tolerance against drought and heat stresses in chickpea. Plant Physiology and Biochemistry 143, 11–18.
Vaillant, N., Monnet, F., Hitmi, A., Sallanon, H., Coudret, A., 2005. Comparative study of responses in four Datura species to a zinc stress. Chemosphere 59, 1005–1013.
Vollenweider, P., Gunthardt-Goerg, M.S., 2005. Diagnosis of abiotic and biotic stress factors using the visible symptoms in foliage. Environmental Pollution 137, 455–465.
Von Wettstein, D., Gough, S., Kannangara, C.G., 1995. Chlorophyll biosynthesis. The Plant Cell 7 (7), 1039.
Wahid, A., Gelani, S., Ashraf, M., Foolad, M., 2007. Heat tolerance in plants: an overview. Environmental and Experimental Botany 61 (3), 199–223.
Wang, H., Liu, R.L., Jin, J.Y., 2009. Effects of zinc and soil moisture on photosynthetic rate and chlorophyll fluorescence parameters of maize. Biologia Plantarum 53, 191–194.
Wang, R., Chen, S., Deng, L., Fritz, E., H€uttermann, A., Polle, A., 2007. Leaf photosynthesis, fluorescence response to salinity and the relevance to chloroplast salt compartmentation and anti-oxidative stress in two poplars. Trees 21 (5), 581–591.
Wang, Z.L., Huang, B.R., 2004. Physiological recovery of Kentucky bluegrass from simultaneous drought and heat stress. Crop Science 44, 1729–1736.
Wani, A.S., Ahmad, A., Hayat, S., Tahir, I., 2019. Epibrassinolide and proline alleviate the photosynthetic and yield inhibition under salt stress by acting on antioxidant system in mustard. Plant Physiology and Biochem, (Paris) 135, 385–394.

Watts-Williams, S.J., Tyerman, S.D., Cavagnaro, T.R., 2017. The dual benefit of arbuscular mycorrhizal fungi under soil zinc deficiency and toxicity: linking plant physiology and gene expression. Plant and Soil 420, 375–388.

Wei, Y., Xu, X., Tao, H., Wang, P., 2006. Growth performance and physiological response in the halophyte *Lycium barbarum* grown at salt-affected soil. Annals of Applied Biology 149, 263–269.

Wise, R.R., Olson, A.J., Schrader, S.M., Sharkey, T.D., 2004. Electron transport is the functional limitation of photosynthesis in field-grown Pima cotton plants at high temperature. Plant, Cell & Environment 27, 717–724.

Xu, Z.Z., Zhou, G.S., Shimizu, H., 2009. Effects of soil drought with nocturnal warming on leaf stomatal traits and mesophyll cell ultrastructure of a perennial grass. Crop Science 49, 1843–1851.

Yan, Lu, H.X., Tong, S., 2018. Effects of arbuscular mycorrhizal fungi on photosynthesis and chlorophyll fluorescence of maize seedlings under salt stress. Emirates Journal of Food Agriculture 30, 199–204.

Yang, S., Ulhassan, Z., Shah, A.M., Khan, A.R., Azhar, W., Hamid, Y., Hussain, S., Sheteiwy, M.S., Salam, A., Zhou, W., 2021. Salicylic acid underpins silicon in ameliorating chromium toxicity in rice by modulating antioxidant defense, ion homeostasis and cellular ultrastructure. Plant Physiology and Biochemistry 166, 1001–1013.

Yang, X.L., Xu, H., Li, D., Gao, X., Li, T.L., Wang, R., 2018. Effect of melatonin priming on photosynthetic capacity of tomato leaves under low-temperature stress. Photosynthetica 56 (3), 884–892.

Zargar, S.M., Gupta, N., Nazir, M., Mahajan, R., Malik, F.A., Sofi, N.R., Shikari, A.B., Salgotra, R.K., 2017. Impact of drought on photosynthesis: molecular perspective. Plant Gene 11, 154–159.

Chapter 7

# ZIP proteins related to zinc metabolism in plants

**Dalila Jacqueline Escudero-Almanza[1], Dámaris Leopoldina Ojeda-Barrios[1], Oscar Cruz-Alvarez[1], Ofelia Adriana Hernández-Rodríguez[1] and Yuridia Ortiz-Rivera[2]**
[1]*Universidad Autónoma de Chihuahua, Facultad de Ciencias Agrotecnológicas, Chihuahua, Mexico,* [2]*Universidad Autónoma de Ciudad Juárez Chihuahua, Mexico*

## 7.1 Introduction

The growth and development of plants and their ability to respond to changes in the environment are highly dependent on information from their genome. This gene expression process requires the transcription and translation of messenger RNA genes where transcription factors (TFs) play a key role (Turchi et al., 2013). TFs are proteins capable of specifically binding to short DNA sequences (cis-elements) located in gene promoters, and of interacting with the preinitiation of the transcription complex to induce or inhibit the activity of the RNA polymerase II enzyme (Ma et al., 2020).

The large number of TFs in plants suggests a high degree of complexity in the regulation of gene expression. Furthermore, the activities of TFs can be regulated through many post-translational modifications that often occur in response to the activation of a specific signal transduction pathway. For example, phosphorylation of the TF by a specific protein kinase can modify DNA binding or act as a signal for its degradation (Buchanan et al., 2015). The activities of TFs are regulated by changes in their within-cellular location; while the storage of a TF in the cytoplasm renders it inactive, the induction of its import to the nucleus makes it active (Buchanan et al., 2015). The ZIP genes encode a family of TFs that have the same name, the ZIP family, which are specific to plants and are involved in several biological processes such as the response to stress, and auxin transportation, among others (Wei et al., 2019).

In another way, recent findings link the concentration of zinc with the activity of these TFs involved in the regulation of ZIP genes. For example, the results of the expression and mapping of PvZIP12 encodes proteins

**Zinc in Plants. DOI: https://doi.org/10.1016/B978-0-323-91314-0.00015-6**
 

associated with the increase in the concentration of zinc (Zn) in seeds. Likewise, understanding the expression of PvZIP12 has helped to unravel the role of ZIP genes in the uptake, distribution, and accumulation of Zn in *Phaseolus vulgaris* L. (Astudillo et al., 2013). The main problem in understanding gene regulation is identifying the genes that control each FT and the Cis elements with which it binds (Chew et al., 2013). To address this problem, several experimental approaches have been developed to locate the binding site of an FT in the genome (Buchanan et al., 2015).

The characterization of these valuable resources provides a view of their evolutionary development and a valuable basis for the establishment of plant genetic improvement programs (Huang et al., 2019). On the other hand, one of the factors involved in $Zn^{2+}$ homeostasis is ZIP transporters, which are regulated to provide a suitable amount of Zn depending on the type of cell and its stage of development (Claus and Chavarría-Krauser, 2012). Likewise, gene expression is increased for some ZIP transporters (tZIP1, OsZIP1, ZmZIP3, OsZIP4, OsZIP5, and HvZIP7) under low-Zn condition to facilitate greater uptake of this micronutrient. Furthermore, the genetic modification of crops with ZIP genes could be a useful tool to correct the low-Zn condition. Therefore, the structural and functional understanding of the ZIP transporter in plants is essential (Ajeesh Krishna et al., 2020). That is why the present review aims to document the relationship and interaction of Zn with HD-ZIP proteins in physiological and biochemical processes such as enzymatic catalysis, and synthesis of proteins and hormones.

## 7.2 Transcription factors and regulation of ZIP genes

All living organisms need to maintain optimal concentrations of nutrients, including Zn, to carry out their cellular functions. Plants have homeostatic mechanisms to maintain Zn levels in a certain concentration range to avoid low-Zn or high-Zn conditions (toxicity) (Zlobin et al., 2020). Zn homeostasis involves a series of physiological processes such as uptake, storage, mobility, complex formation, and detoxification (Ajeesh Krishna et al., 2020). One of the most important factors involved in Zn homeostasis is ZIP transporters, proteins that are responsible for regulating the optimal concentration of Zn (Claus and Chavarría-Krauser, 2012).

Few information is available on the regulation of plant ZIP transporters and the mechanism of Zn homeostasis. Likewise, the mechanism by which the plant detects and transfers the signal of low-Zn condition has not been spell fully understood. Furthermore, it has been discovered that some TFs are essential for the regulation of genes involved in the maintenance of Zn homeostasis (Ajeesh Krishna et al., 2020). In Arabidopsis, TFs have been identified as crucial in the adaptive response to low-Zn conditions involved in the regulation of many physiological processes, including responses to biotic and abiotic stress. These FT belong to the group of ZIP proteins, which are also involved in the positive regulation of ZIP transporters in Arabidopsis during low-Zn condition (Assunção et al., 2013). The FT bZIP19 and bZIP23 are the essential regulators of genes involved in the homeostasis of low-Zn condition. In plants, the bZIP19 and bZIP23 proteins are single (monomers) under suitable range of Zn. The low-Zn condition leads to activation (binding) of bZIP9 and bZIP23 (dimerization) which induces expression of ZIP genes. dimer bZIP19 and bZIP23 bind to specific elements sensitive to low-Zn condition present in the promoter region of the genes involved (Lilay et al., 2019; Ajeesh Krishna et al., 2020). Currently, several ZIP genes have been identified in different plant species (Table 7.1).

## 7.3 The HD-ZIP family

The HD-ZIP family are TFs that are generally composed of two domains (the DNA-binding domain and homeodomain), with different functions (Roodbarkelari and Groot, 2017). The DNA-binding domain is the one that is responsible for the recognition and binding to a cis-element (located in the

**TABLE 7.1** ZIP genes associated with low-Zn condition in several crops.

| Scientific name | ZIP genes | Reference |
|---|---|---|
| *Arabidopsis thaliana* L. | *AtZIP1-AtZIP4* | Wei et al. (2019) |
| *Poncirus trifoliate* L. | *PtZIP1-PtZIP3, PtZIP5-PtZIP7, PtZIP9 and PtZIP10- PtZIP14* | Fu et al. (2017), Ma et al. (2020) |
| *Setaria italica* L. | *SiZIP1-SiZIP7* | Chai et al. (2018) |
| *Glycine max* L. | *GmZIP1, GmZIP4, GmZIP4, GmZIP6, GmZIP10 y GmZIP11* | Chen et al. (2014) |
| *Oryza sativa* L. | *OsZIP1- OsZIP16* | Bang et al. (2019) |
| *Phaseolus vulgaris* L. | *PvZIP1- PvZIP18* | Astudillo et al. (2013) |
| *Zea mays* L. | *ZmZIP1- ZmZIP12* | Tiong et al. (2015) |
| *Triticum aestivum* L. | *TaZIP1-TaZIP3, TaZIP5-TaZIP7, TaZIP9-TaZIP11, TaZIP13, TaZIP14 y TaZIP16* | Evens et al. (2017) |
| *Citrus sinensis* L. | *CsZIP1-CsZIP4* | Xing et al. (2016) |
| *Triticum durum* L. | *TdZIP1, TdZIP3, TdZIP7 TdZIP10* y *TdZIP15* | Deshpande et al. (2018) |
| *Triticum turgidum* L. *var. dicoccoides* | *TdZIP1* | Evens et al. (2017) |

DNA promoter). Moreover, the other domain is involved in transcription and determines the activity of the protein. A domain is a specific region of the protein that binds to the DNA of a gene and encodes a TF of other genes, determining the appearance of one structure or another (Barabasz et al., 2019).

These TFs are an amino acid sequence or protein domain called the homeodomain. These can adhere to segments of DNA to regulate the expression of another gene involved in development (Buchanan et al., 2015; Sessa et al., 2018). The HD-ZIP family of proteins takes its name from the initials in English of the first two proteins identified: ZRT (zinc-regulated transporter) and IRT (iron-regulated transporter), with the letter P added at the end to refer to the proteins (Lawson et al., 2017; Roodbarkelari and Groot, 2017). Otherwise, HD is an abbreviation for homeodomain (Sessa et al., 2018). The proposed structure for ZIP proteins is like that of all Zn transporter proteins. Their amino and carboxyl-terminal ends are outside the plasma membrane, they have transmembrane domains 355–490 amino acids long (Gainza-Cortés et al., 2012). TFs also often have a localization sequence, which allows proteins to enter the nucleus after being synthesized in the cytoplasm.

To date, several types of domains have been identified and their structures have been clarified at the atomic level, thus providing detailed information on how they interact with DNA (Gong et al., 2019). For example, zinc fingers (a type of transcription factor belonging to the HD-ZIP family) are characterized by a domain that carries a set of projections that bind to the DNA double helix. The projections are created by coordinating a zinc ion with four amino acids (usually two histidines and two cysteines). Another structural domain frequently found in TFs is that of the leucine zipper (bZIP). Proteins with this domain form dimers that bind to a recognition site in DNA. Each bZIP monomer contains an α-helix in which every seventh amino acid residue is a leucine. In the dimer, the leucine residues are linked together and arranged in a structural manner like a fastener or zipper. The propellers are wound into a coil, which holds the dimer together and protects its main region (Fig. 7.1) (Buchanan et al., 2015).

A domain is a specific region of the protein that binds to the DNA of a gene and encodes TFs from other genes, determining the appearance of one structure or another. These TFs are an amino acid sequence or protein domain called a homeodomain. These can adhere to segments of DNA to regulate the expression of another gene involved in development. In addition to the above, Buchanan et al. (2015) mentioned that the name HD-ZIP is due to how the leucine is arranged in the form of a zipper, which also refers to its name from the English word "zips." Leucine zipper homeodomain proteins (HD-ZIP) are TFs that are present only in plants, the most studied so far is that of Arabidopsis thaliana where the HD-ZIP are encoded by more than 25 genes. Likewise, all families consist of a homeodomain (HD) and the Leucine zipper (bZIP).

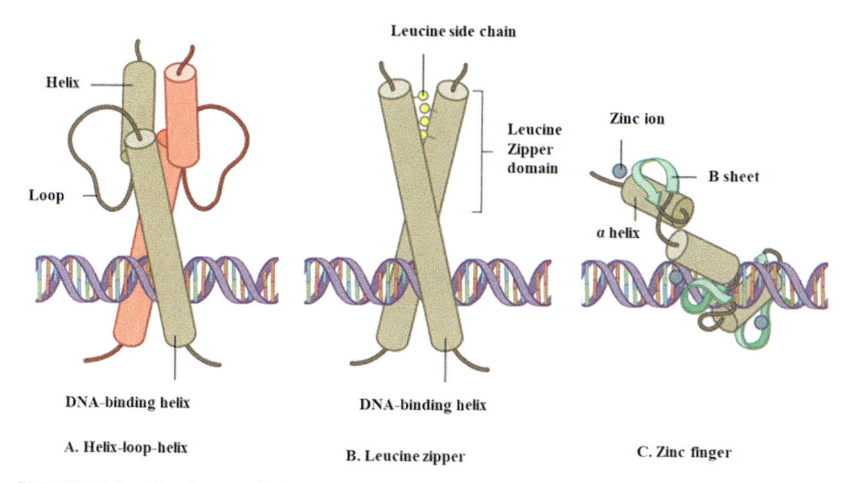


**FIGURE 7.1** The three main DNA-binding domains. *From: Buchanan, B.B., Gruissem, W., & Jones, R.L. 2015. Biochemistry & Molecular Biology of Plants (2nd Edition). John Wiley & Sons, Ltd., Oxford, USA, pp. 151–188.*

The HD domain is a sequence of approximately 60 amino acids that folds into a three-helix structure involved in DNA binding through interaction with the recognition helix, while the ZIP domain participates in heterodimerization. HD-ZIPs are classified into four groups or subfamilies (I, II, III, and IV). This classification is based on the following characteristics: (1) The conservation of the HD-ZIP domain that determines the specificities of DNA binding; (2) protein structure; (3) additional domains; and (4) functions. The distinctive structural features of each subfamily are described below.

## 7.4 Subfamily HD-ZIP I

The proteins of this subfamily are approximately 35 kDa and exhibit a highly conserved HD and a less conserved LZ (Leucine zipper) (Fig. 7.2). In the same way, protein sequences begin with an amino end and end with a carboxyl end, likewise their length is less than or equal to 300 amino acids (Zou et al., 2017; Sessa et al., 2018; Yue et al., 2018). It has also been observed in this subfamily that the N-terminal of the flexible arm plays an important role in DNA-protein interaction, increasing its affinity without changing the specificity (Turchi et al., 2015). In an in vitro description of the complete sequence of HD-ZIP I proteins, the binding sites were observed, and it was determined that the proteins form dimers that recognize the pseudopalindromic sequence CAAT (A/T) ATTG (Chen et al., 2014). Likewise, the inability of HD-ZIP proteins to bind to DNA in its monomeric form like

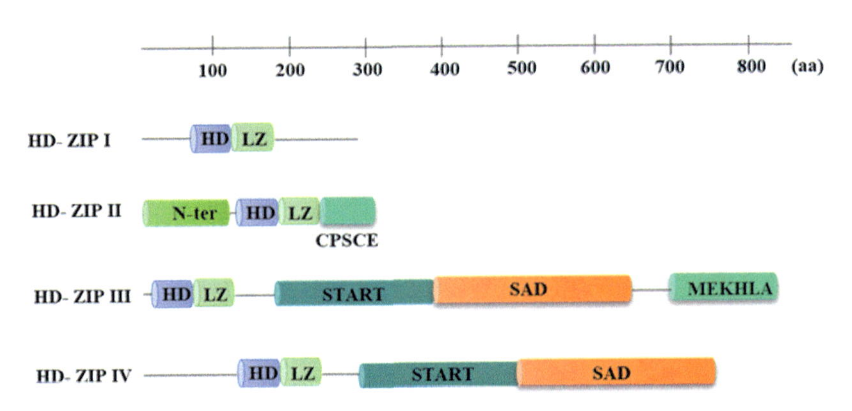


**FIGURE 7.2** wSchematic representation of the protein domains that each HD-ZIP family possesses. *Adaptation from Sessa, G., Carabelli, M., Possenti, M., Morelli, G., & Ruberti, I. 2018. Multiple links between HD-Zip proteins and hormone networks. International Journal of Molecular Sciences 19(12). doi: 10.3390/ijms19124047(Available at CC BY 4.0).*

most HD proteins is determined by the composition of the loop located between helices I and II (Yue et al., 2018).

## 7.5 Subfamily HD-ZIP II

Members of the HD-ZIP II subfamily are similar in size to those of subfamily I and possess two introns within the HB sequence. They show high conservation in the HD and LZ domains (Miyashima et al., 2019). They also present conservation in the additional domain: CPSCE (which takes its name from the five conserved amino acids Cys, Pro, Ser, Cys, and Glu in the one-letter code) (Roodbarkelari and Groot, 2017). Its length is like the members of subfamily I with approximately 300 amino acids (Fig. 7.2). (Sessa et al., 2018). The CPSCE domain participates in the detection of the oxidation-reduction state in the cell. In *vitro* tests have shown that, in an oxidizing environment, these TFs form high molecular weight multimers linked through Cys–Cys bridges intermolecularly.

This type of macromolecule is probably not transported to the nucleus to play its role, which suggests that the redox state of the cell could regulate the activity of HD-ZIP II proteins (Chew et al., 2013). The proteins encoded by this subfamily have the form of dimers that are characterized by the pseudopalindromic sequence CAAT (C/G) ATTG, differing from the members of subfamily I, only in the recognized nucleotide located in the center of the central site (Zou et al., 2017). The central nucleotide-binding specificity of the pseudopalindromic sequence appears to be conferred in part by amino acids 46 and 56 of helix III (Ala and Trp in HD-ZIP Glu and Thr in HD-ZIP II), along with a different spatial orientation of the Arg55 conserved in both proteins where Arg55 would be solely responsible for the interaction (Yue et al., 2018).

## 7.6 Subfamily HD-ZIP III

The members of the HD-ZIP III subfamily are the longest in the ZIP family with more than 800 amino acids. Similar to the other subfamilies, the structure consists of an HD domain and an LZ domain with four additional amino acids between the domains (Fig. 7.2) (Sessa et al., 2018).

Likewise, in this subfamily, more than half of the amino acids are conserved and exhibit a START domain (steroidogenic acute regulatory protein-related lipid-transfer domain) followed by the SAD domain (domain adjacent to START). The difference between groups III and IV depends mainly on a specific MEKHLA domain located at the C-terminal that responds to the redox mechanism of oxygen and light. This domain shares a significant similarity with the domain that takes its name from the three proteins where it is found: Per (circadian period regulatory protein), Arnt (nuclear receptor translocator protein Ah), and Sim (single-minded protein), which regulates the development of the central nervous system of *Drosophila*. The PAS domain is found in many proteins in all kingdoms of life and is involved in the detection of light, oxygen, and redox potential (Zhao et al., 2018). In Fig. 7.2, the characteristic domains of the family are shown schematically.

## 7.7 Subfamily HD-ZIP IV

The proteins encoded by the genes of this family also have the sequence of domains of the HD-ZIP-START-SAD type (Fig. 7.2), like those of HD-ZIP III (Li et al., 2019b). Moreover, their binding and dimerization structures are like those observed in subfamilies I and II, whose distinctive features include the presence of a loop in the middle of the LZ domain and it lacks the MEKHLA domain (Chew et al., 2013). This group of proteins has also been named HD-ZIP GL2 or simply GL2 family after its founding member, the GLABRA2 protein from *Arabidopsis tahitiana*. The proteins belonging to this subfamily show a binding preference for alternative sequences such as the following: CATT (A/T) AATG, which were identified through studies carried out in a member of the sunflower (*Helianthus annuus* L.) through the expression of the HAHR1 gene (Perotti et al., 2021). The sequences directed by the HD-ZIP IV proteins are characterized by a TAAA core (Yue et al., 2018).

## 7.8 Physiological mechanisms involving HD-ZIP proteins

The expression of ZIP genes is regulated by external factors such as drought, extreme temperatures, osmotic stresses, and lighting conditions

(Gong et al., 2019; Wei et al., 2019). Likewise, its response is based on the tissue and organs of the plant. In addition to being linked to regulation by several HD-ZIP proteins that act as FTs. The role of HD-ZIP as TFs is related to development events in response to these environmental conditions, especially those in which abiotic factors generate stress (Li et al., 2019a; Lilay et al., 2019). The proteins of the HD-ZIP family are involved in several physiological processes (transport and cellular homeostasis of Zn) (Xing et al., 2016; Zlobin et al., 2020; Lilay et al., 2021). Table 7.2 describes a series of physiological processes in which each HD-ZIP subfamily is involved. The regulation of the expression of ZIP genes depends on the plant tissue, the concentration of minerals in the cellular medium,

**TABLE 7.2 Physiological events involving genes encoding HD-ZIP proteins.**

| Subfamily | Physiological response | References |
|---|---|---|
| HD-ZIP I | Response to abiotic stress | Gong et al. (2019) |
| | Response to abscisic acid | Gonzalez-Grandio et al. (2017) |
| | Response to light | Gong et al. (2019) |
| | Desethiolation | Sessa et al. (2018) |
| HD-ZIP II | Apical embryo pattern | Sarvari et al. (2017) |
| | Lateral polarity of the organ | Sessa et al. (2018) |
| | Gynoecium development | Wei et al. (2019) |
| | Response to auxins | Turchi et al. (2015) |
| | Auxin transportation | Turchi et al. (2015) |
| | Cytokinin signaling | Roodbarkelari and Groot (2017) |
| HD-ZIP III | Embryogenesis | Miyashima et al. (2019) |
| | Development of the vascular system | Wei et al. (2019) |
| | Auxin transport | Turchi et al. (2015) |
| | Meristematic regulation | Mandel et al. (2016) |
| HD-ZIP IV | Epidermal cell differentiation | Galletti and Ingram (2015) |
| | Root development | Perotti et al. (2021) |
| | Trichome formation | Zhang et al. (2021) |
| | Anthocyanin build-up Flavonoid biosynthesis | Huang et al. (2019), Wang et al. (2020) |

and the plant species (Sessa et al., 2018). In accordance with the above, it is important to mention some important aspects about the balance of Zn in the plant, which are discussed below.

## 7.9 Zinc homeostasis

Heavy metals are present in the soil because of anthropogenic activities or its natural components. Different metals that are part of minerals can be found, such as magnesium (Mg), nickel (Ni), and Zn (Rodríguez-Jiménez et al., 2016). Plant cells require optimal levels of $Zn^{2+}$ for normal physiological functions (structural and function of the biological membrane, protein synthesis, gene expression and regulation, and defense against disease) (Capella et al., 2016; Lira-Morales et al., 2019).

However, the control mechanism of Zn homeostasis is still unknown (Sadeghzadeh, 2013). $Zn^{2+}$ is necessary in low concentrations (0.5–2 μM) in the soil for normal plant function (Ajeesh Krishna et al., 2020). In another way, for the uptake of Zn, several factors must be considered since the availability of micronutrients is a function of the "form" in which it is found in the soil that determines its "mobility" toward the roots of the plants (Ojeda-Barrios et al., 2014). In addition, it has been observed that in alkaline soils with a pH ranging between 7 and 8.6, Zn is bound by the bicarbonate ion that forms zinc carbonate ($ZnCO_3$), which is very insoluble (Gupta et al., 2016). Otherwise, other factors, such as low organic matter content and high iron content, limit the use of Zn (Gupta et al., 2016).

The Zn in the soil may not be available to the plant. For this, Fig. 7.3 summarizes and schematizes the availability of Zn in the soil, an essential part of Zn homeostasis. This could be the reason why the application of Zn in the soil is not effective and is limited only to noncalcareous soils (Ajeesh Krishna et al., 2020). Zn is taken in from the soil solution as a divalent cation ($Zn^{2+}$), transport through the plant plasma membrane is driven by an electrochemical proton gradient generated by the plasma membrane after being uptake, Zn is transported through the xylem, where it is binding by several small molecules including organic acids such as citrate, malate, and nicotinamide (Gupta et al., 2016).

When the supply of Zn is high, much of this nutrient is also complexed in the cell with organic acids (malate and citrate), amino acids (histidine, phytates, and metallothionein), and the rest is stored in the vacuole (Kambe et al., 2015). The Zn uptake is mediated by the membrane potential and the transporters facilitate its diffusion through the cell membrane to the cytoplasm (Lira-Morales et al., 2019). Zn, when moving long distances, follows a simplistic path before reaching the xylem through active transport (Olsen and Palmgren, 2014).

In addition, Zn is an essential catalytic component of more than 300 metalloenzymes, including alkaline phosphatase, alcohol dehydrogenase,

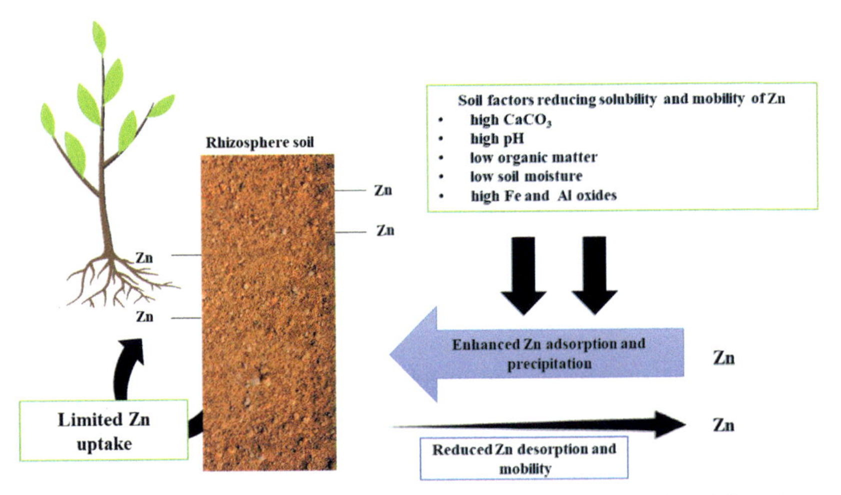


**FIGURE 7.3** Schematic representation of zinc availability in the soil. *Adapted from Gupta, N., Ram, H., & Kumar, B. 2016. Mechanism of zinc absorption in plants: uptake, transport, translocation, and accumulation. Reviews in Environmental Science and Bio/Technology. doi: 10.1007/s11157-016-9390-1.*

Cu-Zn superoxide dismutase, and carbonic anhydrase. One of the main characteristics of this cation ($Zn^{2+}$) is its chemical stability because it neither oxidizes nor reduces, and in the cells, it tends to form more stable tetrahedral complexes (Escudero-Almanza et al., 2012; Castillo-Gonzalez et al., 2018). Therefore, it is easier to provide stability in the domains of transcriptionally regulated proteins, such as the HD-ZIP proteins (Olsen and Palmgren, 2014).

## 7.10 Membrane transporters and zinc homeostasis

Higher plants modulate the processes that contribute to Zn homeostasis according to external supply and internal requirements to keep Zn concentrations within physiological limits. (Lilay et al., 2019). This process begins with Zn input and uptake from the soil in the roots. Subsequently, Zn is uptake through the plasma membrane of the root cells predominantly as a free ion (Fig. 7.4). In the cytoplasm of plant cells, Zn is attached to low-weight molecules to avoid precipitation (Fu et al., 2017).

Once in the root symplastic-flux, Zn can be immobilized through transport to vacuoles, or it can be transported according to the Zn-dependent processes and the cellular requirements (Fig. 7.4). Vacuoles are the main site for the storage and detoxification of excess Zn and a source Zn under low-Zn condition (Lira-Morales et al., 2019). $Zn^{2+}$ cations are strong Lewis acids and readily complex with many organic molecules. To have binding control

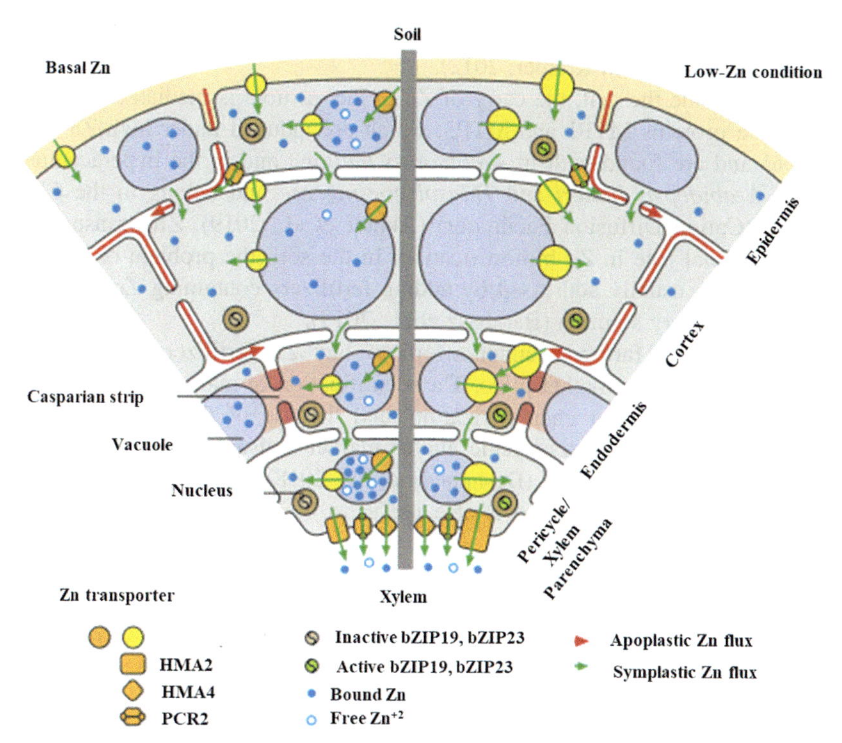


**FIGURE 7.4** Regulation of Zn in put by membrane transporters. *Adapted from Sinclair, S.A., & Krämer, U. 2012. The zinc homeostasis network of land plants. Biochimica et Biophysica Acta: Molecular Cell Research 1823(9): 1553–1567. doi: 10.1016/j.bbamcr.2012.05.016.*

of $Zn^{2+}$ to other compounds, Zn is present frequently in its bound form in the cytoplasm and other cell compartments (Sinclair and Krämer, 2012).

Otherwise, Mendoza-Tafolla et al. (2019) associate Zn with it as a growth promoter due to its activity as a precursor in the auxins output. Auxins promote cell division and elongation; moreover, they influence the production of indoleacetic acid, which acts as a hormonal phytostimulant. The sizes of the transporter symbols depict the protein levels from transcription and the number of Zn symbols are related to the Zn concentrations at different places within the root (Lira-Morales et al., 2019). Responses to low-Zn conditions include increased sympalastic-flux or root uptake, enhanced Zn remobilization from vacuoles, and increased Zn concentration in the xylem (Barabasz et al., 2019).

The model also shows that the active bZIP19/bZIP23 complex improves the transcription of ZIP1, ZIP3, ZIP4, ZIP5, ZIP9, ZIP10, ZIP12, IRT3, and NAS2. The transcription levels of NAS4 HMA2 are also known to be enhanced with low-Zn conditions, but the underlying mechanism remains unknown (Akhtar et al., 2019). The rendering of Zn concentrations in the root is based on fluorescence images in studies carried out in *Arabidopsis thaliana* (Sinclair and Krämer, 2012).

Once inside the cell, the carry of Zn to the vacuole is mediated by metal tolerance proteins MTP1 and MTP3, which are induced under high-Zn conditions and are found both in *Arabidopsis thaliana* and in the hyperaccumulators *Arabidopsis halleri* and *Thlaspi goesingense* and belong to the CDF family (Cation Diffusion Facilitator) (Akhtar et al., 2019). ZIP transporters play a crucial role in Zn biofortification. In the soil, the problem of low-Zn conditions is usually addressed by adding fertilizers containing Zn, but it is only a temporary solution (Barabasz et al., 2019).

Furthermore, farmers cannot afford to buy Zn fertilizers permanently because of the high market price. The identification of main genes related to ZIP transporters, their characterization, and the use of this information to develop crops through transgenic and/or marker-assisted selection can help develop Zn-efficient crops (Deshpande et al., 2018). The genetic modification of crops with ZIP genes is also useful to improve crops related to the efficiency of Zn use.

Therefore, the structural and functional understanding of the plant ZIP transporter is essential (Ajeesh Krishna et al., 2020). In some genomic studies on *Arabidopsis thaliana* and yeast *Saccharomyces cerevisiae*, several ZIP proteins (ZIP1, ZIP3, and ZIP4) have been identified with a key role in the uptake and transport of Zn in the plant (Chaudhuri et al., 2020; Becares et al., 2021). The carboxyl and amino terminals of ZIP proteins exist on the outer surface of the cell membrane for efficient Zn transport. The ZIP1 and ZIP3 proteins are expressed in roots in case of Zn deficiency.

An increase in Zn uptake was observed by *Arabidopsis halleri* and *Thlaspi caerulescens* (Zn hyperaccumulator mutants), which could be linked to the overexpression of ZIP proteins located in the plasma membrane (Akhtar et al., 2019). As reviewed, zinc homeostasis is overly complex and multifactorial, the plant physiology, and soil-composition affect the uptake, transportation, and storage of Zn. The possible molecular mechanisms involved in those processes are summarized in Fig. 7.5.

## 7.11 Mechanism of Interaction between Zn and the ZIP family

Zn is essential for the functioning of several enzymes (carbonic anhydrase, superoxide dismutase, and so on) and transcription factors such as bZIP19-bZIP23 (Lilay et al., 2019). Zn transporters tightly regulate Zn homeostasis (Roodbarkelari and Groot, 2017). Plants contain many genes that respond to Zn-concentration, which are specifically expressed under Zn-deficient conditions to ensure the coordination of assimilation pathways and meet physiological requirements (Sessa et al., 2018).

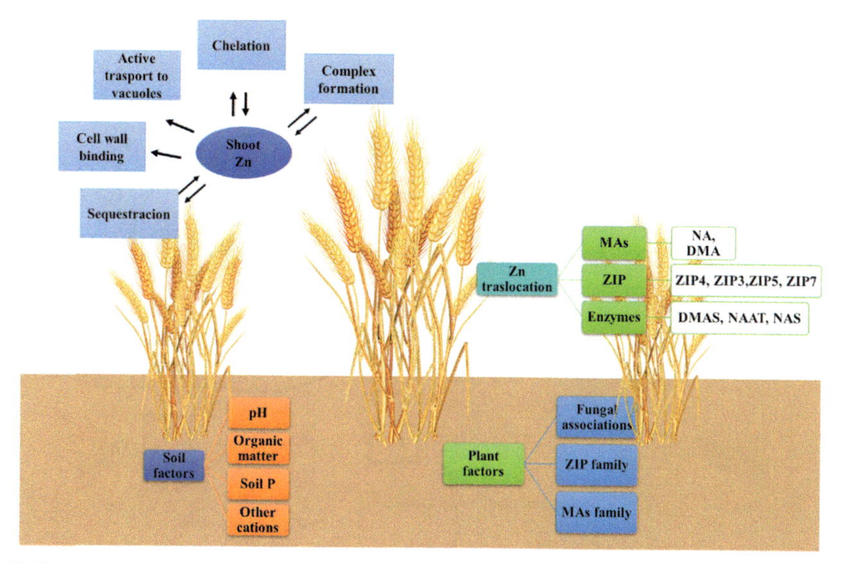


**FIGURE 7.5** Factors involved in Zn homeostasis. *Adapted form Akhtar, M., Yousaf, S., Sarwar, N., & Hussain, S. 2019. Zinc biofortification of cereals—role of phosphorus and other impediments in alkaline calcareous soils. Environmental Geochemistry and Health 41(5): 2365–2379. doi: 10.1007/s10653-019-00279-6.*

Likewise, it is important to study different conditions of variation on Zn-concentration that are useful in genetic improvement (Marín-Montes et al., 2020). Several years of research have led to a better understanding of the interaction between Zn and the secondary structure of this HD-ZIP protein family (Capella et al., 2016; Ajeesh Krishna et al., 2020). This concludes that the expression of several members of the ZIP family is regulated by TFs of the bZIP family. Members of the bZIP family contain two structural regions, the first comprises a specific DNA binding site and a domain that interacts with the cis-regulatory elements of the recognition sequence, and the second is a leucine dimerization region (Assunção et al., 2013) (Fig. 7.6).

Currently, there is a proposed mechanism for the regulation of HD-ZIP proteins, which is illustrated in Fig. 7.7 and described further. Under normal within-cellular Zn conditions, the bZIP19 and bZIP23 proteins remain separate (I). Under low-Zn conditions, bZIP19 and bZIP23 bind to form a dimer to regulate ZIP (II) gene transcription. Later, bZIP19 and bZIP23 recognize the cis-element ZDRE (Zn deficiency response element) present in some genes that are members of the ZIP (III) family. Followed by the induction of transcription of a range of ZIP genes (IV), which leads to the production of transmembrane proteins (V), which are capable of Zn cellular uptake to maintain within-cellular homeostasis (Deshpande et al., 2018).

The ZDRE element is present in some of the ZIP genes of *Arabidopsis thaliana* (Lira-Morales et al., 2019; Zlobin et al., 2020). The CysHis-rich

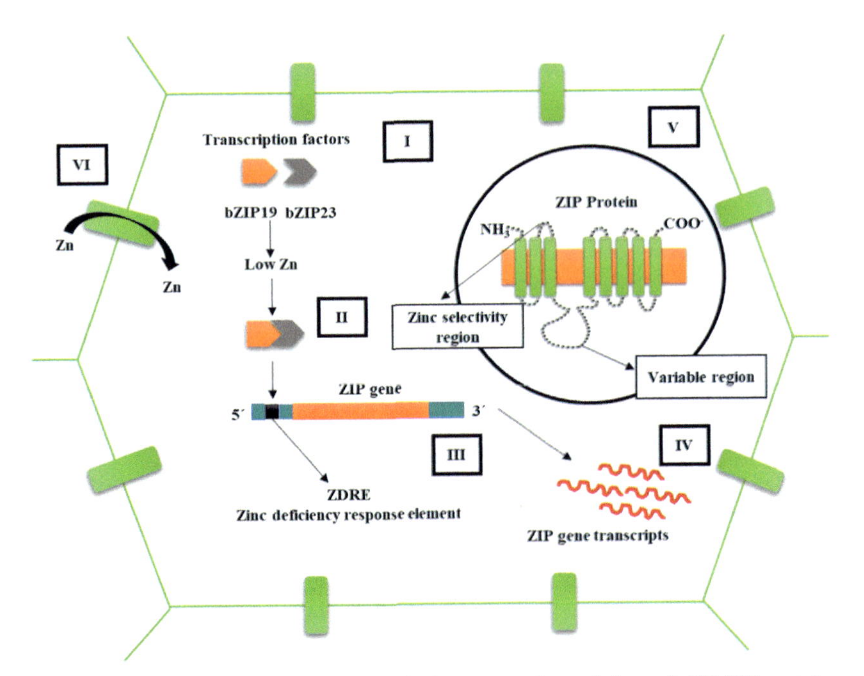


**FIGURE 7.6** Schematic representation of the proposed regulation of HD-ZIP proteins. *Adaptation from: Lira-Morales, J.D., Varela-Bojórquez, N., Montoya-Rojo, M.B., & Sañudo-Barajas, J.A. 2019. The role of zip proteins in zinc assimilation and distribution in plants: current challenges. Czech Journal of Genetics and Plant Breeding 55(2): 45–54. doi: 10.17221/54/2018-CJGPB.(Available at CC BY-NC 4.0 licence)*

domain of bZIP19 and bZIP23 is thought to act as a zinc sensor, thus playing a role in its Zn-dependent regulatory function (Barabasz et al., 2019). A hypothetical function of the bZIP19/23 Zn sensor would require a reversible Zn junction. Therefore, free cytosolic $Zn^{2+}$ can act through binding to the Cys-His domain as a signal for the activity of bZIP19 and bZIP23 and for the modulation of cell Zn homeostasis (Assunção et al., 2013).

Under normal-Zn condition, binding of Zn to the domain would render the TF nonfunctional through an effect on its conformation, its DNA-binding activity, or movement to the nucleus. All strategies before mentioned are involved in the posttranslational regulation of bZIP TF. Following Zn deficiency, coordinated Zn "release" from the Cys-His motif from the bZIPs would allow the transcription factor to become functional. Likewise, the proposal is stated as to how the cellular concentration of Zn can act as a Zn-sensor for the transcription factors bZIP19 and bZIP23 (Assunção et al., 2013; Lilay et al., 2019).

There is no distinction in this scheme between bZIP19 and bZIP23. Under low-Zn condition, the release of coordinated Zn from the Cys-His domain of bZIPs allows the TF to be functional. Possibly there is a reservoir of bZIP19

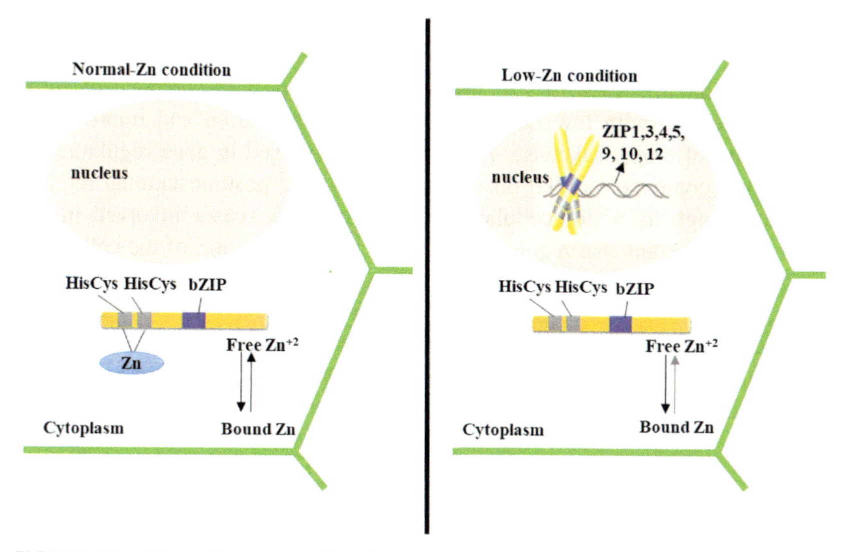


**FIGURE 7.7** Schematic representation of the proposal that some HD-ZIP proteins act as a low-Zn condition sensor. *Adaptation from Assunção, A.G.L., Persson, D.P., Husted, S., Schjørring, J.K., Alexander, R.D., et al. 2013. Model of how plants sense zinc deficiency. Metallomics 5(9): 1110–1116. doi: 10.1039/c3mt00070b.*

and bZZIP23 in the cytosol where Zn-sensor activity and TF activation take place (Tiong et al., 2015). Activation of the TF in the nucleus cannot be excluded. Fig. 7.7 shows the previous proposal, where the bZIP domain of the TF is represented in purple and contains the basic domains of the leucine zipper. HisCys domains are represented in gray (Lilay et al., 2019, 2021).

## 7.12 Perspectives and future research

To complete the missing links within the Zn homeostasis in the plant, there are still some challenges to face to clarify the entire mechanism of Zn regulation in the plant. Among these, the identification of the Zn-transporter sequence regions that may be associated with a higher affinity for Zn, as well as the determination of specific regions of the TF and other cis-elements of the ZIP genes that have the main influence on transcription under low-Zn, normal-Zn, and high-Zn conditions.

In addition to the above, the identification and characterization of ZIP proteins and related TF, in many crops would help to better understand Zn. Likewise, it is crucial to identify the specific role of each ZIP protein in different organs and tissues, as well as its association with a specific gene. Zn-transporters must carrier Zn from the xylem of the leaf to the phloem of the developing seeds and finally it is assimilated into the seeds with the help of ZIP genes.

Moreover, the identification and characterization of ZIP proteins and related TF, in many crops would help to better understand Zn homeostasis. In addition, a broader view of ZIP transporters in crops would reduce the problems associated with low-Zn soils through biofortification and improve the profitability of crops. Likewise, Zn is not only involved in gene regulation as a structural component of TF but could also regulate posttranslational modifications through its within-cellular concentration as a sensor involved in the synthesis of proteins that regulate its passage into the interior. of the cell.

However, more evidence is needed to support the mechanisms discussed above, as well as the identification of transcription factors associated with a specific gene. Therefore, future research is suggested to fully elucidate Zn homeostasis in the plant and its relationship with HD-ZIP protein homeostasis. In addition, a broader view of ZIP transporters in crops would reduce the problems associated with low-Zn soils through biofortification and improve the profitability of crops.

Likewise, Zn is not only involved in gene regulation as a structural component of TF but could also regulate posttranslational modifications through its within-cellular concentration as a Zn-sensor involved in the synthesis of proteins that regulate its within-cellular flux. However, more evidence is needed to support the mechanisms discussed above, as well as the identification of transcription factors associated with a specific gene. Therefore, future research is suggested to fully unravel the Zn homeostasis in the plant and its relationship with HD-ZIP proteins.

## References

Ajeesh Krishna, T.P., Maharajan, T., Victor Roch, G., Ignacimuthu, S., Antony Ceasar, S., 2020. Structure, function, regulation and phylogenetic relationship of ZIP family transporters of plants. Frontiers in Plant Science 11, 1–18. Available from: https://doi.org/10.3389/fpls.2020.00662. May.

Akhtar, M., Yousaf, S., Sarwar, N., Hussain, S., 2019. Zinc biofortification of cereals—role of phosphorus and other impediments in alkaline calcareous soils. Environmental Geochemistry and Health 41 (5), 2365–2379. Available from: https://doi.org/10.1007/s10653-019-00279-6.

Assunção, A.G.L., Persson, D.P., Husted, S., Schjørring, J.K., Alexander, R.D., et al., 2013. Model of how plants sense zinc deficiency. Metallomics 5 (9), 1110–1116. Available from: https://doi.org/10.1039/c3mt00070b.

Astudillo, C., Fernandez, A.C., Blair, M.W., Cichy, K.A., 2013. The *Phaseolus vulgaris* ZIP gene family: identification, characterization, mapping, and gene expression. Frontiers in Plant Science 4, 1–15. Available from: https://doi.org/10.3389/fpls.2013.00286. JUL.

Bang, S.W., Lee, D.K., Jung, H., Chung, P.J., Kim, Y.S., et al., 2019. Overexpression of OsTF1L, a rice HD-Zip transcription factor, promotes lignin biosynthesis and stomatal closure that improves drought tolerance. Plant Biotechnology Journal . Available from: https://doi.org/10.1111/pbi.12951.

Barabasz, A., Palusińska, M., Papierniak, A., Kendziorek, M., Kozak, K., et al., 2019. Functional analysis of NtZIP4B and zn status-dependent expression pattern of tobacco ZIP genes. Frontiers in Plant Science 9, 1–18. Available from: https://doi.org/10.3389/fpls.2018.01984. January.

Becares, E.R., Pedersen, P.A., Gourdon, P., Gotfryd, K., 2021. Overproduction of human Zip (SLC39) zinc transporters in *Saccharomyces cerevisiae* for biophysical characterization. Cells . Available from: https://doi.org/10.3390/cells10020213.

Buchanan, B.B., Gruissem, W., Jones, R.L., 2015. Biochemistry & Molecular Biology of Plants, 2nd Edition John Wiley & Sons, Ltd., Oxford, USA, pp. 151–188, By.

Capella, M., Ribone, P.A., Arce, A.L., & Chan, R.L. 2016. Homeodomain-Leucine Zipper Transcription Factors: Structural Features of These Proteins, Unique to Plants. Plant Transcription Factors: Evolutionary, Structural and Functional Aspects.

Castillo-Gonzalez, J., Ojeda-Barrios, D., Hernandez-Rodriguez, A., González-Franco, A.C., Robles-Hernández, L., et al., 2018. Zinc metaloenzimes in plants. Interciencia 43 (4), 242–248. Available from: https://www.interciencia.net/wp-content/uploads/2018/05/242-6131-OJEDA-43-04.pdf.

Chai, W., Si, W., Ji, W., Qin, Q., Zhao, M., et al., 2018. Genome-wide investigation and expression profiling of HD-Zip transcription factors in foxtail millet (*Setaria italica* L.). BioMed Research International 2018. Available from: https://doi.org/10.1155/2018/8457614.

Chaudhuri, A., Das, S., Das, B., 2020. Localization elements and zip codes in the intracellular transport and localization of messenger RNAs in *Saccharomyces cerevisiae*. Wiley Interdisciplinary Reviews. RNA . Available from: https://doi.org/10.1002/wrna.1591.

Chen, X., Chen, Z., Zhao, H., Zhao, Y., Cheng, B., et al., 2014. Genome-wide analysis of soybean HD-Zip gene family and expression profiling under salinity and drought treatments. PLoS One 9 (2). Available from: https://doi.org/10.1371/journal.pone.0087156.

Chew, W., Hrmova, M., Lopato, S., 2013. Role of homeodomain leucine zipper (HD-Zip) iv transcription factors in plant development and plant protection from deleterious environmental factors. International Journal of Molecular Sciences 14 (4), 8122–8147. Available from: https://doi.org/10.3390/ijms14048122.

Claus, J., Chavarría-Krauser, A., 2012. Modeling regulation of zinc uptake via ZIP transporters in yeast and plant roots. PLoS One 7 (6). Available from: https://doi.org/10.1371/journal.pone.0037193.

Deshpande, P., Dapkekar, A., Oak, M., Paknikar, K., Rajwade, J., 2018. Nanocarrier-mediated foliar zinc fertilization influences expression of metal homeostasis related genes in flag leaves and enhances gluten content in durum wheat. PLoS One . Available from: https://doi.org/10.1371/journal.pone.0191035.

Escudero-Almanza, D.J., Ojeda-Barrios, D.L., Hernández-Rodríguez, O.A., Chávez, E.S., Ruíz-Anchondo, T., et al., 2012. Carbonic anhydrase and zinc in plant physiology. Chilean Journal of Agricultural Research 72 (1). Available from: https://doi.org/10.4067/S0718-58392012000100022.

Evens, N.P., Buchner, P., Williams, L.E., Hawkesford, M.J., 2017. The role of ZIP transporters and group F bZIP transcription factors in the Zn-deficiency response of wheat (*Triticum aestivum*). The Plant Journal: for Cell and Molecular Biology . Available from: https://doi.org/10.1111/tpj.13655.

Fu, X.Z., Zhou, X., Xing, F., Ling, L.L., Chun, C.P., et al., 2017. Genome-wide identification, cloning and functional analysis of the zinc/iron-regulated transporter-like protein (ZIP) gene family in trifoliate orange (*Poncirus trifoliata* L. Raf.). Frontiers in Plant Science . Available from: https://doi.org/10.3389/fpls.2017.00588.

Gainza-Cortés, F., Pérez-Dïaz, R., Pérez-Castro, R., Tapia, J., Casaretto, J.A., González, S., et al., 2012. Characterization of a putative grapevine Zn transporter, VvZIP3, suggests its involvement in early reproductive development in *Vitis vinifera* L. BMC Plant Biology 12 (1), 111. 10.1186/1471-2229-12-111.

Galletti, R., Ingram, G.C., 2015. Communication is key: reducing DEK1 activity reveals a link between cell-cell contacts and epidermal cell differentiation status. Communicative & Integrative Biology . Available from: https://doi.org/10.1080/19420889.2015.1059979.

Gong, S., Ding, Y., Hu, S., Ding, L., Chen, Z., et al., 2019. The role of HD-Zip class I transcription factors in plant response to abiotic stresses. Physiologia Plantarum 167 (4), 516–525. Available from: https://doi.org/10.1111/ppl.12965.

Gonzalez-Grandio, E., Pajoro, A., Franco-Zorrilla, J.M., Tarancon, C., Immink, R.G.H., et al., 2017. Abscisic acid signaling is controlled by a branched 1/HD-ZIP i cascade in Arabidopsis axillary buds. Proceedings of the National Academy of Sciences of the United States of America 114 (2), E245–E254. Available from: https://doi.org/10.1073/pnas.1613199114.

Gupta, N., Ram, H., Kumar, B., 2016. Mechanism of zinc absorption in plants: uptake, transport, translocation, and accumulation. Reviews in Environmental Science and Bio/Technology . Available from: https://doi.org/10.1007/s11157-016-9390-1.

Huang, Y., Xiao, L., Zhang, Z., Zhang, R., Wang, Z., et al., 2019. The genomes of pecan and Chinese hickory provide insights into Carya evolution and nut nutrition. Gigascience 8 (5), 1–17. Available from: https://doi.org/10.1093/gigascience/giz036.

Kambe, T., Tsuji, T., Hashimoto, A., Itsumura, N., 2015. The physiological, biochemical, and molecular roles of zinc transporters in zinc homeostasis and metabolism. Physiological Reviews . Available from: https://doi.org/10.1152/physrev.00035.2014.

Lawson, R., Maret, W., Hogstrand, C., 2017. Expression of the ZIP/SLC39A transporters in ß-cells: a systematic review and integration of multiple datasets. BMC Genomics 18 (1), 1–13. Available from: https://doi.org/10.1186/s12864-017-4119-2.

Li, W., Dong, J., Cao, M., Gao, X., Wang, D., et al., 2019a. Genome-wide identification and characterization of HD-ZIP genes in potato. Gene . Available from: https://doi.org/10.1016/j.gene.2019.02.024.

Li, L., Zheng, T., Zhuo, X., Li, S., Qiu, L., et al., 2019b. Genome-wide identification, characterization and expression analysis of the HD-Zip gene family in the stem development of the woody plant Prunus mume. PeerJ 2019 (8). Available from: https://doi.org/10.7717/peerj.7499.

Lilay, G.H., Castro, P.H., Campilho, A., Assunção, A.G.L., 2019. The arabidopsis bZIP19 and bZIP23 activity requires zinc deficiency—insight on regulation from complementation lines. Frontiers in Plant Science 9, 1–13. Available from: https://doi.org/10.3389/fpls.2018.01955. January.

Lilay, G.H., Persson, D.P., Castro, P.H., Liao, F., Alexander, R.D., et al., 2021. Arabidopsis bZIP19 and bZIP23 act as zinc sensors to control plant zinc status. Nature Plants . Available from: https://doi.org/10.1038/s41477-021-00856-7.

Lira-Morales, J.D., Varela-Bojórquez, N., Montoya-Rojo, M.B., Sañudo-Barajas, J.A., 2019. The role of zip proteins in zinc assimilation and distribution in plants: current challenges. Czech Journal of Genetics and Plant Breeding 55 (2), 45–54. Available from: https://doi.org/10.17221/54/2018-CJGPB.

Ma, Y.J., Li, P.T., Sun, L.M., Zhou, H., Zeng, R.F., et al., 2020. HD-ZIP I transcription factor (PtHB13) negatively regulates citrus flowering through binding to FLOWERING LOCUS C promoter. Plants 9 (1), 1–14. Available from: https://doi.org/10.3390/plants9010114.

Mandel, T., Candela, H., Landau, U., Asis, L., Zelinger, E., et al., 2016. Differential regulation of meristem size, morphology and organization by the ERECTA, CLAVATA and class III HD-ZIP pathways. Development . Available from: https://doi.org/10.1242/dev.129973.

Marín-Montes, I.M., Lobato-Ortiz, R., Carrillo-Castañeda, G., Rodríguez-Pérez, J.E., Jesús García-Zavala, J., et al., 2020. Genetic parameters of an interspecific cross between *S. lycopersicum* L. and *S. habrochaites* Knapp & Spooner. Revista Chapingo, Serie Horticultura . Available from: https://doi.org/10.5154/R.RCHSH.2020.01.003.

Mendoza-Tafolla, R.O., Juarez-Lopez, P., Ontiveros-Capurata, R.E., Sandoval-Villa, M., Alia-Tejacal, I., et al., 2019. Estimating nitrogen and chlorophyll status of romaine lettuce using SPAD and at LEAF readings. Notulae Botanicae Horti Agrobotanici Cluj-Napoca . Available from: https://doi.org/10.15835/nbha47311589.

Miyashima, S., Roszak, P., Sevilem, I., Toyokura, K., Blob, B., et al., 2019. Mobile PEAR transcription factors integrate positional cues to prime cambial growth. Nature . Available from: https://doi.org/10.1038/s41586-018-0839-y.

Ojeda-Barrios, D.L., Perea-Portillo, E., Hernández-Rodríguez, O.A., Ávila-Quezada, G., Abadía, J., et al., 2014. Foliar fertilization with zinc in pecan trees. HortScience: A Publication of the American Society for Horticultural Science 49 (5), 562–566. Available from: https://doi.org/10.21273/hortsci.49.5.562.

Olsen, L.I., Palmgren, M.G., 2014. Many rivers to cross: the journey of zinc from soil to seed. Frontiers in Plant Science . Available from: https://doi.org/10.3389/fpls.2014.00030.

Perotti, M.F., Arce, A.L., Chan, R.L., 2021. The underground life of homeodomain-leucine zipper transcription factors. Journal of Experimental Botany . Available from: https://doi.org/10.1093/jxb/erab112.

Rodríguez-Jiménez, T.J., Ojeda-Barrios, D.L., Blanco-Macías, F., Valdez-Cepeda, R.D., Parra-Quezada, R., 2016. Urease and nickel in plant physiology. Revista Chapingo, Serie Horticultura 22 (6), 69–81. Available from: https://doi.org/10.5154/r.rchsh.2014.11.051.

Roodbarkelari, F., Groot, E.P., 2017. Regulatory function of homeodomain-leucine zipper (HD-ZIP) family proteins during embryogenesis. The New Phytologist 213 (1), 95–104. Available from: https://doi.org/10.1111/nph.14132.

Sadeghzadeh, B., 2013. A review of zinc nutrition and plant breeding. Journal of Soil Science and Plant Nutrition 13 (4), 907–927. Available from: https://doi.org/10.4067/S0718-95162013005000072.

Sarvari, M., Darvishzadeh, R., Najafzadeh, R., Maleki, H.H., 2017. Physio-biochemical and enzymatic responses of sunflower to drought stress. Journal of Plant Physiology and Breeding 7 (1), 105–119.

Sessa, G., Carabelli, M., Possenti, M., Morelli, G., Ruberti, I., 2018. Multiple links between HD-Zip proteins and hormone networks. International Journal of Molecular Sciences 19 (12). Available from: https://doi.org/10.3390/ijms19124047.

Sinclair, S.A., Krämer, U., 2012. The zinc homeostasis network of land plants. Biochimica et Biophysica Acta: Molecular Cell Research 1823 (9), 1553–1567. Available from: https://doi.org/10.1016/j.bbamcr.2012.05.016.

Tiong, J., Mcdonald, G., Genc, Y., Shirley, N., Langridge, P., et al., 2015. Increased expression of six ZIP family genes by zinc (Zn) deficiency is associated with enhanced uptake and root-to-shoot translocation of Zn in barley (*Hordeum vulgare*). The New Phytologist 207 (4), 1097–1109. Available from: https://doi.org/10.1111/nph.13413.

Turchi, L., Carabelli, M., Ruzza, V., Possenti, M., Sassi, M., et al., 2013. Arabidopsis HD-Zip II transcription factors control apical embryo development and meristem function. Development . Available from: https://doi.org/10.1242/dev.092833.

Turchi, L., Baima, S., Morelli, G., Ruberti, I., 2015. Interplay of HD-Zip II and III transcription factors in auxin-regulated plant development. Journal of Experimental Botany .

Wang, Z., Wang, S., Xiao, Y., Li, Z., Wu, M., et al., 2020. Functional characterization of a HD-ZIP IV transcription factor NtHDG2 in regulating flavonols biosynthesis in *Nicotiana tabacum*. Plant Physiology and Biochemistry: PPB/Societe Francaise de Physiologie Vegetale 146, 259–268. Available from: https://doi.org/10.1016/j.plaphy.2019.11.033. November 2019.

Wei, M., Liu, A., Zhang, Y., Zhou, Y., Li, D., et al., 2019. Genome-wide characterization and expression analysis of the HD-Zip gene family in response to drought and salinity stresses in sesame. BMC Genomics 20 (1), 1–13. Available from: https://doi.org/10.1186/s12864-019-6091-5.

Xing, F., Fu, X.Z., Wang, N.Q., Xi, J.L., Huang, Y., et al., 2016. Physiological changes and expression characteristics of ZIP family genes under zinc deficiency in navel orange (*Citrus sinensis.* ). Journal of Integrative Agriculture 15 (4), 803–811. Available from: https://doi.org/10.1016/S2095-3119(15)61276-X.

Yue, H., Shu, D., Wang, M., Xing, G., Zhan, H., et al., 2018. Genome-wide identification and expression analysis of the hd-zip gene family in wheat (*Triticum aestivum* L.). Genes (Basel) 9 (2). Available from: https://doi.org/10.3390/genes9020070.

Zhang, L., Lv, D., Pan, J., Zhang, K., Wen, H., et al., 2021. A SNP of HD-ZIP I transcription factor leads to distortion of trichome morphology in cucumber (*Cucumis sativus* L.). BMC Plant Biology . Available from: https://doi.org/10.1186/s12870-021-02955-1.

Zhao, Y., Upadhyay, S., Lin, X., 2018. PAS domain protein pas3 interacts with the chromatin modifier bre1 in regulating cryptococcal morphogenesis. MBio . Available from: https://doi.org/10.1128/mBio.02135-18.

Zlobin, I.E., Pashkovskiy, P.P., Kartashov, A.V., Nosov, A.V., Fomenkov, A.A., et al., 2020. The relationship between cellular Zn status and regulation of Zn homeostasis genes in plant cells. Environmental and Experimental Botany . Available from: https://doi.org/10.1016/j.envexpbot.2020.104104.

Zou, X., Ratti, B.A., O'Brien, J.G., Lautenschlager, S.O., Gius, D.R., et al., 2017. Manganese superoxide dismutase (SOD2): is there a center in the universe of mitochondrial redox signaling? Journal of Bioenergetics and Biomembranes 49 (4), 325–333. Available from: https://doi.org/10.1007/s10863-017-9718-8.

Chapter 8

# Zinc and plant hormones: an updated review

Isha Madaan[1,2], Pooja Sharma[3,4], Arun Dev Singh[3], Shalini Dhiman[3], Jaspreet Kour[3], Pardeep Kumar[3], Gurvarinder Kaur[1], Indu Sharma[5], Vandana Gautam[3], Rupinder Kaur[6], Ashutosh Sharma[7], Geetika Sirhindi[1] and Renu Bhardwaj[3]

[1]*Department of Botany, Punjabi University, Patiala, Punjab, India,* [2]*Science Department, Government College of Education, Jalandhar, Punjab, India,* [3]*Department of Botanical and Environmental Sciences, Guru Nanak Dev University, Amritsar, Punjab, India,* [4]*Department of Microbiology, DAV University, Jalandhar, Punjab, India,* [5]*University Institute of Sciences, Sant Baba Bhag Singh University, Jalandhar, Punjab, India,* [6]*Department of Biotechnology, DAV College, Amritsar, Punjab, India,* [7]*Faculty of Agricultural Sciences, DAV University, Jalandhar, Punjab, India*

## 8.1 Introduction

Heavy metals (HMs) concentrations in soil ecosystems are of greater concern to mankind, as well as to the flora and fauna including microbial populations. These concentrations develop environmental constraints act toxic to plants and further hamper the growth, yield, and developmental processes in plants (Dhalaria et al., 2020; Khan et al., 2021). Out of the 90 metal elements that exist naturally in the environment, 53 are considered to be HMs because of their high densities (5 g/cm$^3$) in the natural environment. Some of these HMs are known as micronutrients and are of significant importance to plants. However, they show negative impacts at higher conc. toward growth and metabolic activities in plants (Singh et al., 2018; Arif et al., 2016). In addition to the natural composition of HMs, anthropogenic activities, industrialization, and modern agricultural practices are also contributing to the rapid accumulation and intake of these pollutants inside the plant system (Thakur et al., 2016). These pollutants mainly include Zn (zinc), Hg (mercury), Cd (cadmium), Pb (lead), and Al (aluminum), etc. which enter through the roots system and further disturb the balance in water uptake, mineral uptake, photosynthetic contents and ion distributions inside the plans tissues (Ghori et al., 2019; Sofo et al., 2017). These pollutants directly or indirectly find their entrances into the edible parts of the crop plants and finally penetrate into the food chains.

**Zinc in Plants. DOI: https://doi.org/10.1016/B978-0-323-91314-0.00016-8**

Zn is among the most frequently available elements in industrial polluted areas and is highly toxic to plant tissues when present in excess thus arresting their normal biochemical and physiological functions and accumulating in different cellular compartments (Sofo et al., 2017; Cuypers et al., 2013). It is also one of the essential micronutrients to all kinds of living entities. They are crucial in the case of plants and contribute to cellular processes and as a cofactor in different proteins, metallo-enzymes, and transcription factors. Also, Zn plays an important role in biological events like photosynthesis, carbohydrates, lipids and nucleic acid metabolism, and oxidative stress management. However, Zn has its essential role as a cofactors in many of the cell macromolecules and plant growth at its normal concentration of 30–200 μg/g (DW) whereas a concentration of 400 μg/g (DW) in any of the plant's tissues is highly toxic to crop plants (Brokbartold et al., 2012; Li et al., 2020). Zn being an essential micronutrient is utilized by plants for the normal functioning of primary and secondary metabolic processes. Among the various transition heavy metals, Zn ranks second in abundance after iron in living beings. Zn is present naturally in the soil as a result of various processes such as pedogenesis, leaching, and volcanic eruptions. However, industrialization and many other anthropogenic activities also act as additional sources of Zn to the environment. Due to the versatile roles of Zn in plants, it is up taken by plant roots through Zn transporters and cation channels from the soil and water. Apart from the interplay of Zn with various primary and secondary metabolites within the plants, it is also known to show cross-talk with phytohormones produced in minute amounts in various parts of plants. These phytohormones act as plant growth regulators and play a role in controlling various signaling mechanisms of cell division, differentiation, and hence overall plant development. Zn acts as a positive regulator of various signaling mechanisms controlled by certain phytohormones viz. auxins, gibberellins (GA), and cytokinin. However, other phytohormones have been reported to mitigate the noxious effects of excessive Zn present in the plant's vicinity. Zn forms an important structural component of receptor proteins of classic phytohormones, i.e., auxins and GA. Hence, this makes Zn an enhancer for the transcription of proteins involved in hormone signaling pathways.

Zn is suitable for plants at a particular threshold value. Its presence below this value may lead to deficiency and above it may cause toxicity. Zn is involved in the regulation of several plant development and signaling processes owing to its structural, functional, and catalytic properties. It forms an important constituent of several enzymes such as carbonic anhydrase, and aldolase, which are involved in carbon and nucleic acid metabolism. Deficiency of Zn in growing habitats has posed deleterious effects on the yield and quality of crop plants. On the other hand, the presence of excessive amounts of Zn causes toxicity and again proves deteriorating for the plants (Hassan et al., 2020).

In order to prevent themselves from the toxic effects of Zn, plants adopt various strategies that downregulate the uptake of Zn, constrain their accumulation within roots, and prevent their transport to the upper plant parts. Nowadays, eco-friendly plant growth regulators (PGRs) are also being given to plants at the seed or seedling level that help them to maintain normal metabolism under different kinds of stresses. This chapter mainly focusses on the dynamics of Zn within the various organs of different plants and its interplay with classic and recent phytohormones/signaling molecules produced intracellularly, as well as given exogenously in different sections discussed ahead.

## 8.2 Zinc dynamics in plants

Zn is naturally released in soils from minerals rich in carbonates, phosphates, sulfates, sulfides, and zinc oxides, which are released from the pedogenetic processes of parent rocks. It is also released from various biotic and atmospheric processes. Besides its natural occurrence, various industrial and anthropogenic activities, smelting, mining, sewage sludge, and extensive use of Zn fertilizers add up Zn in the environment (Mateos-Naranjo et al., 2014). Zn occurs in water-soluble cation $Zn^{2+}$, adsorbed to clayey colloids, humins, and hydroxides of iron and aluminum, and forms insoluble complexes and minerals (Montalvo et al., 2016; Sturikova et al., 2018). It is predominantly found in the colloidal fraction in the form of neutral sulfate ($ZnSO_4$) and phosphate ($ZnHPO_4$) (Sadeghzadeh and Rengel, 2011). The bioavailability of Zn is influenced by many factors such as pH, soil organic matter, rhizospheric microbial communities, root exudates, and chemical and mineral composition (Kwon et al., 2017; González-Alcaraz and van Gestel, 2015). The distribution of zinc in soil is influenced by the pH, which impacts the mobility of zinc ions. At pH more than 7.5, the mobility of Zn in plants decreases which is due to the altered form of zinc in the solution. At pH 7.7 $Zn^{2+}$ prevails and at pH above 9–11, the neutral Zn $(OH)_2$ predominates in the soil solution (Cao et al., 2004). Root exudates consisting of low molecular weight compounds like organic acids, amino acids, phenolic compounds, sugars, and high molecular weight compounds such as polysaccharides and proteins secreted by the root cells in the rhizosphere modify the soil properties such as pH, bioavailability, mobility, and microbial activity (Hou et al., 2016; Chen et al., 2017).

Plant roots take up the Zn in the form of $Zn^{2+}$ ions or complexes with organic acid chelates from the soil solution (Palmgren et al., 2008) and translocated to the above-ground part of the plant via the xylem. Zn is transferred from the soil at the root by the heavy metal transport protein belonging to the ZIP family (Zinc Regulatory Transporter—Iron Regulatory Transporter-like protein) (Sturikova et al., 2018). Zn is an essential micronutrient that is required for plant growth and development (Fig. 8.1). It possesses both positive and noxious effects on plant cells. The deficiency of Zn deteriorates and

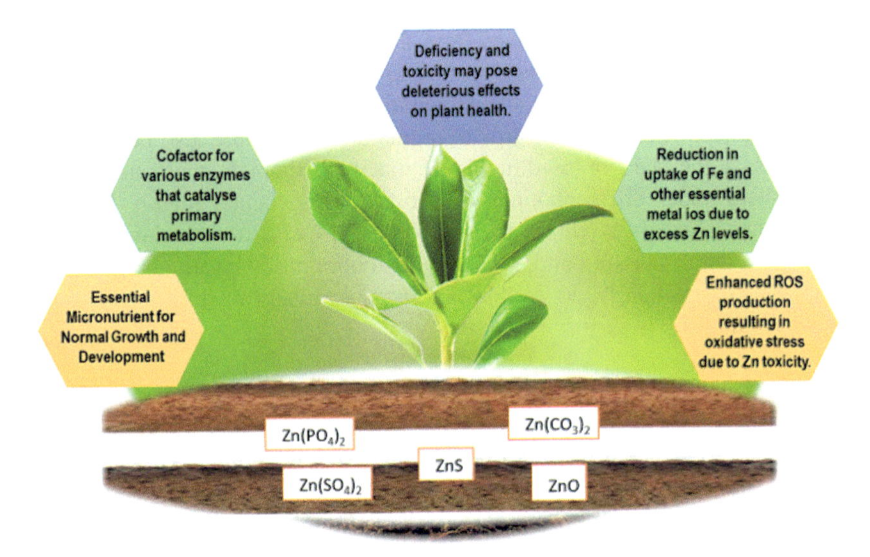


**FIGURE 8.1** Zn dynamics in plants.

diminishes the yield and quality of crops while the application of Zn helps to improve the quality and production of crops (Chattha et al., 2017; Mousavi et al., 2007). It acts as a cofactor for various enzymes such as aldolase, carboxypeptidase, and Zn-superoxide dismutase (Tsonev and Cebola Lidon, 2012; Balafrej et al., 2020). It is also an important constituent of lipids and proteins, and also serves as a cofactor for auxins; it also plays a crucial role in nucleic acid metabolism in plants (Mengel et al., 2001). Its deficiency disrupts the activity of enzymes which results in the inhibition of photosynthesis. Zn is a cofactor of carbonic anhydrase that enhances $CO_2$ concentration in the chloroplast, and therefore, upturns the carboxylation proficiency of the Rubisco enzyme (Salama et al., 2006).

Zn ions are the component of the transcription factor, called "zinc fingers" that regulate the proliferation and differentiation of cells (Palmer and Guerinot, 2009). In addition, Zn also plays an imperative role in the development of chloroplasts and repair of the photosystem II by "recyclation" of the D1 protein impaired by SPP peptidase activity or light radiation (Hänsch and Mendel, 2009).

Excess amount of Zn shows phytotoxicity and results in the suppression of cell division and elongation, which leads to reduced biomass production (Tsonev and Cebola Lidon, 2012). Zn toxicity occurs as a result of competition by excess Zn ions in the cell for binding sites intended for other important biologically active ions (Kramer, 2010). Its toxicity also causes leaf chlorosis due to reduced uptake of iron ions (Ren et al., 1993). Zn causes a cessation of the general metabolism in the roots because of the inhibition of aerobic respiration (Sagardoy et al., 2010). Higher accumulation of Zn in the roots leads to the generation of reactive oxygen species like hydrogen peroxide and superoxide anion which results in oxidative damage (Anwaar et al., 2015; Feigl et al., 2015).

## 8.3 Classic phytohormones and zinc

The excessive amount of Zn has now become a concern to the living world. There are reports which clearly say that anthropogenic activities—including incinerators, smelters, and disposal practices of industrial and municipal wastes—are a challenge to the world's environment (Li et al., 2020). Zn toxicity in plants leads to symptoms such as leaf chlorosis, necrosis, stunted growth, arrests of primary root growth, and other reproductive developments. In plants, the primary organs targeted by Zn toxicity includes "primary roots (PR)" which promotes stunted and retarded root system and hence the impaired hydraulic conductivity in plants (Bazihizina et al., 2014; Wang et al., 2015; Khare et al., 2017). Plants show adaptations to the HMs concentration to an extent through pumping action of these elements from cytosol into the apoplasm and vacuoles and by chelating action with suitable ligands. Also, plants adopt to high Zn toxicity by reallocation of these elements in different tissues and limit their entry inside the roots and epidermal cells (Wang and Yang, 2021). Vassilev et al. (2011) found that bean seedlings grown under high Zn concentration are found to have a decrease in their fresh biomass, net photosynthetic rate, electron transport chain, transpiration, and stomatal conductivity. In *Arabidopsis thaliana*, seedlings grown in Ms-medium when treated with 50 μM $ZnSO_4$ solution for the 12 constant days show retardation in root growth and mass up to 75% and accumulate Zn concentration up to 240 μg/g DW in leaf tissues. Excessive amounts of Zn ions follow the displacement of $Fe^{2+}$ ions from its protein binding sites and cause alterations in the functioning of these proteins thus higher Zn concentration leads to Fe deficiency in plants (Wintz et al., 2003). Xu et al. (2010) also reported that *Solanum nigrum* grown under a high Zn toxic environment were found to have retarded growth in primary roots whereas induced the formation of lateral root growth by activating the NO-induced ROS signal transduction pathway.

These multiple adaptations to Zn stress are integrated results of all the environmental cues and the endogenous phytohormones and their signaling cascade triggered by the signaling molecules to reshape the growth activities and counter these HM stressors. However, phytohormones such as IAA, GA, CK, BRs, SA, ET, and ABA are introduced as an efficient and eco-friendly strategy during HMs stress. These phytohormones are synthesized by plants in very low concentrations and further show their regulatory mechanisms in various growth and developmental pathways as well as to counter the abiotic stress factors (Rhaman et al., 2021; Wani et al., 2016). Auxin (IAA) has well-documented ameliorative roles under metalloids and metal stress conditions and further helps the plants to enhance their biomass and yield (Zhang and Friml, 2020). Das and Sarkar (2018) also reported that auxin synthesizing PGPRs (*Acinetobacter lwoffii* strain-RJB-2) inoculations to *Vigna radiata* plants grown under As (arsenic) stress helps the plants in maintaining the normal physiological and morphological activities. Mabrouk et al. (2019) found that another plant hormone salicylic acid (SA) 0.2 mM application to fenugreek seedlings grown under Zn and As stress results in an increase in the contents of proteins, amino acids, and soluble sugars in cotyledons and radicles and further ameliorates the metalloids toxicity. Therefore, phytohormones are considered endogenous regulators and scavengers that activates a plethora of signaling cascades to counter these stressor molecules. Synergistic interactions of Zn with different classic phytohormones are discussed in the following sections.

### 8.3.1 Auxins

Involvement in controlling the plant tropism towards light and gravity led to the recognition of auxins as phytohormones which later were discovered to regulate a wide array of plant growth and development responses like embryogenesis, development of shoot, rhizogenesis, vascular differentiation, development of fruit, and many more (Miransari et al., 2014; Estelle, 2011). Auxin shows its pivotal role in the growth and developmental activities of plants as well as acts against abiotic stressors. IAA (i.e., indole 3-acetic acid) mediates the growth-related processes in plants such as controlling cell differentiation, cell division, and roots initiation, and triggers responses against low nutrient availability. Auxin under phosphate (P) deficiency modulates the root architecture to assimilate and uptake more phosphate. Thus, auxin gradients in root tips modulate the root architectures which regulate the elongation of primary roots (PR) and further initiate the development of lateral roots (LR) (Rayle and Cleland, 1992; Al-Ghazi et al., 2003; Nan et al., 2014; Shi et al., 2015a, b). Auxin transport mainly occurs through carrier proteins present on the plasma membranes. AUX-1 is an auxin influx carrier protein and further helps in auxin gradient establishments and promotion of the primary root growth and LR initiation (Shi et al., 2015a, b; De Smet et al., 2007). However, there are reports which signify the active roles of auxin signaling under HMs stimuli (Mroue et al., 2018; Angulo-Bejarano et al., 2021). Yuan and Huang (2016) reported that in *Arbidopsis*, Cd stress lowers or inhibits the expression of PIN—proteins related genes (PIN1, PIN 3, PIN7) which however minimizes the auxin conc in roots. Furthermore, in another case, Cu toxicity not only affects the PR growth through the repression of PIN 1 genes but also interferes with the biosynthesis of auxin by transcriptional repression of YUCCA and TAA-1 genes (Yuan et al., 2013; Song et al., 2017). Likewise, higher Zn concentration in *Arabidopsis thaliana* severely affects the physiological and morphological activities in seedlings as well as in mature plants. Zhang et al. (2018a,b) found that high Zn concentration in *Arabidopsis thaliana* limits the expression of PIN-4 proteins and therefore affects the auxin concentration and transportation, whereas Sofo et al. (2017) reported the same after finding the activity of AUX-1 and LAX-3 gene expressions. It is also reported that excess of Zn ions retards the activity of nitrogenase enzymes and compromises the activity of the Calvin cycle and electron transport chains. In the case of *Brassica juncea* seedlings when applied with 10 mM Zn concentration for the respective 10 days are found to have a significant increase in their antioxidative enzyme activities such as SOD, POD, CAT, DHAR, MDHAR, and GR in comparison to their control plants (Sidhu, 2016).

In other case, Ramakrishna and Rao (2013) found that Zn toxicity in *Raphanus sativus* promotes the activity of MDHAR and APX by 32% and 40.4% in comparison to the control plants whereas lowers the activity of GR and DHAR by 52.7% and 20.7%. Auxin levels in *Arabidopsis thaliana* seedlings are influenced accordingly with the variable Zn conc. (Zn50, Zn100, and Zn350). Zn50 and Zn100 concentration promotes the transcriptional levels of YUC3 and YUC11 genes in the roots of Arabidopsis seedlings whereas Zn350 concentration are found to significantly downregulate the activity of auxin biosynthetic genes such as YUC3, YUC11, and TAA1 in root tips. However exogenous application of IAA (5 nM) to Arabidopsis seedlings under high Zn concentration are found to have an elongation in the PR length and further shows an increase in the auxin concentration inside the root tips was observed (Wang and Yang, 2021). Furthermore, a decrease in the IAA and GA1 levels was observed in plants, i.e., *Zea mays*, *Avena sativa*, and *Hordeum vulgare* grown under Zn deficient environment (Suge et al., 1986; Sekimoto et al., 1997). Also, in the case of *Lactuca sativa* grown under Zn-deficient environments, a decrease in the levels of IAA and GA3 was observed as Zn is required for the synthesis of tryptophan an auxin precursor molecule (Navarro-León et al., 2016). Thus, Zn is an essential element for the auxin biosynthesis and auxin plays an ameliorative role under conditions of Zn toxicity in plants.

### 8.3.2 Gibberellins

GA are considered a huge group of tetracyclic diterpenoid carboxylic acids (Sponsel and Hedden, 2010). In the whole life cycle of a plant, GA are involved in the stimulation of plant growth by the enhancement of the cell division and the elongation of cell. It also helps in the promotion of the transition phases from seed dormancy to adult growth phase and from developmental to reproductive phase (Griffiths et al., 2006). GA are biosynthesized in plastids through the methylerythritol phosphate pathway with the help of trans-geranyl-geranyl-diphosphate (Kasahara et al., 2002). The GA signaling includes the biosynthesis of hormones, perception, transduction of signals, and inactivation. All the steps are regulated by either the various signals which may also include the abiotic stress (Hedden and Thomas, 2012).

For the proper growth of the plant (both vegetative and reproductive), there must be balanced homeostasis of the phytohormones including GA (Yao et al., 2018). The role of GA in the development of male and female reproductive organs in *Arabidopsis thaliana* and *Oryza sativa* has been studied by Hu et al., 2008; Rieu et al., 2008; Sakata et al., 2014. They found that GA is involved in the development of male reproductive organs whereas the development of female reproductive organs is not affected by the low or insufficient concentration of GA in plants. Another report by Aya et al., 2009 stated that due to GA insensitive mutants or due to the deficiency of GA, several species of plant display the abnormal growth of anther.

It has been observed that in angiosperms, the biosynthesis of GA is regulated by environmental as well as endogenous factors. Mostly, this regulation is exerted at the level of transcription (Hedden and Thomas, 2012). During the biosynthesis of GA, the environmental and endogenous factors target the 2-OGD genes which increase or decrease the GA levels (Fukazawa et al., 2014). There is another class of genes as Swollen Anther Wall 1 (SAW1) encoding the CCCH-type Zinc finger class of proteins. This gene is responsible for the development of anther and GA biosynthesis. Some CCCH-type zinc finger proteins known to bind DNA are also responsible for their role in the regulation of the expression of genes by acting as transcription factors., whereas some of these proteins are involved in RNA metabolism (Zhang et al., 2018a,b).

A study conducted by Wang et al. (2018a,b) to study the role of CCCH-type zinc finger proteins in the GA homeostasis and the development of anther has concluded that this protein binds directly to the promoter region of OSGA2oox3 (GA synthesizing gene) for the induction of anther specific expression. There was another study conducted by Wang et al., 2014 on the role of the zinc finger transcription factor gene on the biosynthesis of GA. He concluded that $BBX_{24}$ in *Chrysanthemum morifolium* integrates the synthesis of GA and also modulates flowering and stress tolerance in plants by modulating GA biosynthesis. Another study on the $ZFP_{1}85$ gene that encodes $A_{20}/AN_{1}$ type zinc finger protein in *Oryza sativa* is responsible for the downregulation in the biosynthesis of GA and ABA (Zhang et al., 2016). From these reports, there seems to be a close relationship between zinc and the biosynthesis of GA (both in the upregulation and the downregulation of the GA biosynthesis).

Besides the various roles of GA in plant growth and development, it is also involved in the amelioration of heavy metal stress to regulate various plant processes (Javed et al., 2021). In many plants of *Vicia faba* treated with GA under heavy metal stress, the percentage of the chromosomal aberrations was reduced. It was later observed that after GA application in the stressed plants, there was more accumulation of proteins, soluble sugars, and

RNA (Mansour and Kamel, 2005). It was observed by Falkowska et al., (2011) that in Cd and Pd stressed *Chlorella vulgaris*, the application of GA increases the protein content and number of cells. There is another report indicating the concentration-dependent role of GA in pea plants stressed with Cr. It was observed by Gangwar and Singh (2011) that at lower concentrations, GA improved the defense mechanism of plants whereas at higher concentrations the results were different. Javed et al. (2021) found that in pea plants that were grown under Cu stress, GA application helped in strengthening the defense mechanism in plants. All these reports suggest the link of metal stress and protection with the help of exogenous GA application in dose-dependent manner. There was a report according to the study conducted by Atici et al., 2005. He found that when Zn was present at lower concentrations, there was an increase in endogenous GA concentration and it increased when Zn was present at higher concentrations.

### 8.3.3 Cytokinins

Cytokinin is essentially an important phytohormone needed for plant growth, development, cell cycle, cell division, and nutritional responses (Ullah et al., 2018). Moreover, plant senescence is also slowed down by cytokinins, which prevent the degradation of chlorophyll, proteins, nucleic acids (DNA and RNA), and other components in plants and re-distribute them (Liu et al., 2020). Cytokinin phytohormone is not only growth-promoting hormone but is also used for ameliorating an enormous number of biotic and abiotic stresses (Liu et al., 2020). All the plants naturally produced either isoprenoid cytokinins (isopentenyl adenine cytokinin and trans/cis/dihydro zeatin's) or aromatic cytokinins (ortho/meso topolins or benzyladenine) (Sakakibara, 2006; Liu et al., 2020). Endogenous applied synthetic cytokinins such as benzyl adenine, 6-benzylaminopurine, kinetin, and zeatin also regulate the plant's growth as well as various plant's developmental stages (Liu et al., 2020). Cytokinin improves zinc nutrient status mainly through strictly monitoring the transporter that helps in Zn uptake and chelators that are responsible for transporting Zn inside plant systems (Gao et al., 2019). The cytokinin oxidase/dehydrogenase (*CKX*) gene expressed during the cytokinin metabolism is also responsible for enhancing zinc nutrients and the production of *Oryza sativa* L. ssp. *japonica* (Gao et al., 2019). Cytokinin negatively regulates root growth. The increased *CKX* gene expression lowers the cytokinin level inside the root system and thus, further causes the enhancement in root growth. *CKX* transgenics of *Arabidopsis thaliana*, *Nicotiana tabacum*, and *Hordeum vulgare* increased calcium (Ca), phosphorus (P), sulfur (S), copper (Cu), iron (Fe), manganese (Mn), and zinc (Zn) in leaves (Ramireddy et al., 2018; Werner et al., 2010). Such increase in the uptake of macroelements and microelements depends upon the improved root system size in *CKX* transgenics plants and because of the enhanced expression of genes that encodes zinc, manganese, sulfate, and phosphate transporters (Werner et al., 2010). *CKX* genes transgenic barley grains also show an increase in the Zn element (Ramireddy et al., 2018). Therefore, *CKX* genes transgenic barley grains are quite good sources for human consumption due to the presence of Zn element.

Cytokinin metabolism is affected by Zn element concentration inside the plant system. At high concentrations of Zn, there was a decrease in the cytokinin content whereas at optimal concentration of Zn, cytokinin synthesis increased (Atici et al., 2005). Moreover, excess zinc concentration inside the plants causes stress by negatively regulating the plant metabolism (Fig. 8.2). However, cytokinin synthesis inside the plant system helps in the amelioration of Zn stress in many plants like expression in isopentyl transferase (ipt) genes increased the endogenous production of cytokinin content that improved the Zn stress tolerances inside the tobacco plants (Pavlíková et al., 2014). Cytokinin-like growth activity containing synthetic growth regulators, i.e., thidiazuron and kinetin reported to ameliorate the zinc and nickel toxicity in *Zea mays* L. seedling (Lukatkin et al., 2007).

## 8.4 Abscisic acid

The combined application of ABA and zinc is involved in the rise of ABA content which mitigates the negative effect of zinc in *Triticum aestivum* (Kosakivska et al., 2019). Excess Zn leads to slowed germination of cucumber and chickpea seeds in turn elevates the level of ABA (Atici et al., 2005; Wang et al., 2014). ABA content is also increased at all Zn concentrations in *Cicer arietinum* (Atici et al., 2005). However, high Zn concentration leads to a decline in Abscisic acid content in poplar roots and an increase in ABA content in poplar leaves (Shi et al., 2015a, b). Endogenous ABA concentrations gradually increase with the increase in Zn concentration.

Zn finger proteins play an important role in plant growth and development (Schnable et al., 2009). The combination of zinc finger proteins with

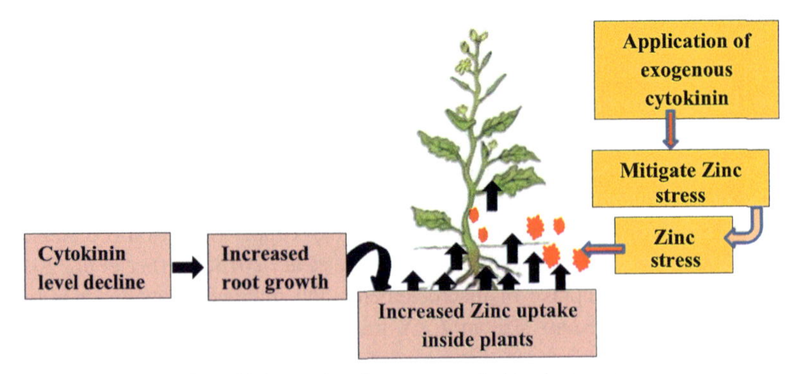


FIGURE 8.2 Effect of cytokinin on zinc element status inside plant system.

abscisic acid is involved in seed development and germination (Joseph et al., 2014). Various responses of abscisic acid and zinc finger proteins are shown in Table 8.1.

## 8.5 Gaseous phytohormones in Zn tolerance

The significance of gaseous signaling molecules such as nitric oxide (NO), carbon monoxide (CO), hydrogen gas ($H_2$), hydrogen sulfide ($H_2S$), methane ($CH_4$), and gaseous phytohormone—such as ethylene—has been appraised in abiotic stress management in plants (Jahan et al., 2021; Karle et al., 2021; Weisany et al., 2012). Ethylene ($CH_2 = CH_2$) is one of the simplest gaseous signaling molecules that was identified for the first time and has a crucial role in fruit ripening as well as in combating stress. Besides ethylene, nitric oxide (NO) is identified as the smallest diatomic gas and a signaling gaseous molecule that controls various events related to seed germination, senescence, photosynthesis, and oxidative stress management in plants (Jahan et al., 2021). Being small in size, NO (acts as both a signal molecule and a reactive nitrogen species) diffuses easily across the membranes/walls of plant cells and thus, interacts with various cellular components and may induce the expression of transcription genes via activating secondary messengers. It has been also reported to react with Zn in metalloproteins thereby resulting in metal nitrosylation (Martínez-Ruiz et al., 2011). This section discusses the interaction of Zn with gaseous phytohormones/signals in plant growth and development.

### 8.5.1 Ethylene and zinc

In *Lolium perenne*, the uptake of copper (Cu) and zinc (Zn) ions under the influence of ethylene diamine disuccinic acid and gibberellic acid was studied (Borker et al., 2020). It was observed that Zn accumulation in roots and its translocation to shoot resulted in significant damage to the roots under Zn toxicity. The treatment of ethylene diamine disuccinic acid alone or in combination with Zn, decreased the rate of transpiration whereas it increased the Zn translocation thereby, revealing its potential use in phytoextraction of metal ions (Cu and Zn) by *Lolium perenne*.

Another study by Huang et al. (2020) emphasized that the transcription factor, *Zinc finger protein 5* (*ZFP5*), regulates the transcription of *Arabidopsis EIN2 (ETHYLENE INSENSITIVE 2)* gene which controls ethylene signaling in response to the phosphate (P) and potassium (K) deficiency, thereby resulting in the development/elongation of root hairs. Both P and K are macronutrients whereas, in *Arabidopsis*, *ZFP5* is a regulator of root hair/trichome development/elongation. It was concluded that the expression of *ZFP5* was regulated by the deficiency of either P or K which stimulated ethylene to induce the development of root hair (Table 8.2). Recently, zinc oxide (ZnO) nanoparticles along with a phytohormone (24-epibrassinolide)

**TABLE 8.1** Role of abscisic acid and Zn finger proteins in different plants.

| S. no. | Zinc finger proteins interacting with ABA | Plant | Response | References |
|---|---|---|---|---|
| 1. | GhWIP2 | *Gerbera hybrida* | GhWIP2 promotes the ABA accumulation | Ren et al. (2018) |
| 2. | AtTZF4, AtTZF5, AtTZF6 | *Arabidopsis thaliana* | ABA accumulation | Bogamuwa and Jang (2014) |
| 3. | ZFP182 | *Oryza sativa* | ABA-induced antioxidant defense | Zhang et al. (2012) |
| 4. | SlZFP2 | *Solanum lycopersicum* | SlZFP2 transcriptional repressor to inhibit ABA production and fruit ripening | Weng et al. (2015) |
| 5. | LlZFHD4 | *Arabidopsis thaliana* | LlZFHD4 involve in cold and water stress tolerances by upregulating ABA genes | Yong et al. (2021) |
| 6. | ZFP3 | *Arabidopsis thaliana* | Inhibition of ABA concentrations promotes seed germination | Joseph et al. (2014) |
| 7. | ZmFLZ25 | *Zea mays* | ABA increase in seedlings | Chen et al. (2021) |
| 8. | ZFP185 | *Oryza sativa* | ZFP185 negative regulator in ABA signaling | Zhang et al. (2018a,b) |
| 9. | BcabaR1 | *Botrytis cinerea* | BcabaR1stimulate the production of ABA | Wang et al. (2018a,b) |
| 10. | SlNCED1 | *Solanum lycopersicum* | Increased ABA levels are due to SlNCED1 overexpression | Kai et al. (2019) |
| 11. | ZFP3 | *Arabidopsis thaliana* | ZFP3 negative regulator of ABA | Joseph et al. (2014) |
| 12. | AZF1 and AZF2 | *Arabidopsis thaliana* | Protein-repressed expression of ABA-repressive genes | Kodaira et al. (2011) |
| 13. | TaCHP | *Arabidopsis thaliana* | TaCHP downregulated by ABA | Li et al. (2010) |
| 14. | MdBBX10 | *Arabidopsis thaliana* | MdBBX10 enhanced abiotic stress tolerance by ABA | Liu et al. (2019) |

**TABLE 8.2** Stress-mediated response of gaseous phytohormone/molecules in relation with Zn.

| Gaseous phytohormone/ molecules | Deficiency/ stress | Mechanism of action | Response | References |
|---|---|---|---|---|
| Ethylene | P or K deficiency | *Zinc finger protein 5* (*ZFP5*) regulates the transcription of *Arabidopsis EIN2* (*ETHYLENE INSENSITIVE 2*) gene | Root hair/trichome development/elongation in *Arabidopsis* | Huang et al. (2020) |
| Ethylene | Zinc oxide (ZnO) nanoparticles | Ethylene-insensitive mutants (ethylene-signaling defective mutants i.e., *ETR1−3* and *EIN2−1*) significantly tolerated ZnO nanoparticles-induced stress as compared to wild-type plants | Ethylene-induced oxidative stress in *Arabidopsis* plants under ZnO nanoparticles induced stress | Khan et al. (2019) |
| Ethylene | Cd, Cu, and Zn stress | *TdSHN1* gene promoter regulated the expression of heavy metals induced *GUS* gene in transgenic tobacco lines. Activities of superoxide dismutase and catalase were increased; thus, less ROS were produced. Enhanced root growth, biomass accumulation, and chlorophyll levels were observed. | Ethylene-responsive transcription factor (*TdSHN1*), has been observed to confer heavy metals (cadmium, Zn, and copper) tolerance in transgenic tobacco plants, durum wheat, and yeast | Djemal and Khoudi (2021) |
| Ethephon application resulted in ethylene formation | Ni and Zn stress | Ethylene induced increased net rate of photosynthesis, altered activity of antioxidant enzymes (ascorbate peroxidase and glutathione reductase), and enhanced reduced glutathione levels. | Ethephon-induced formation of ethylene reduced the oxidative stress caused by Ni and Zn toxicity in mustard. | Khan and Khan (2014) |
| NaHS, as a source of $H_2S$ | Zn stress | $H_2S$ applications resulted in decreased accumulation of cytosolic Zn ions, downregulated Zn uptake, and gene expression of homeostasis-related genes, whereas upregulation of antioxidant enzymes and | $H_2S$ induced Zn stress amelioration in *S. nigrum* by regulating the antioxidant defense system and metallothioneins. | Liu et al. (2016) |

(*Continued*)

**TABLE 8.2** (Continued)

| Gaseous phytohormone/molecules | Deficiency/stress | Mechanism of action | Response | References |
|---|---|---|---|---|
| | | metallothioneins enhanced the chelation of free Zn ions. | | |
| Sodium nitroprusside as a source of nitric oxide (NO) | Zn stress | Both salicylic acid and NO decreased the contents of malondialdehyde, α-tocopherol, hydrogen peroxide ($H_2O_2$), phytochelatins, and proline under Zn stress. The levels of chlorophyll and the activity of antioxidant enzymes (catalase, ascorbate peroxidase, and guaiacol peroxidase) were increased under Zn stress with application of NO along with salicylic acid. | Zn-induced deleterious effects on the growth and development of safflower were reversed with the combined treatment of salicylic acid and NO. | Namdjoyan et al. (2018) |
| Ethephon-mediated production of ethylene | Zn stress | Ethylene resulted in increased activity of antioxidant enzymes, enhanced metabolism of proline and glyoxalase system, and altered nutrient homeostasis in Zn stressed mustard. | Exogenous application of ethylene promotes the Zn stress tolerance through modulation of defense mechanisms in mustard. | Khan et al. (2019b) |
| Exogenous treatment of NaHS, as a source of $H_2S$ | Zn stress | NaHS reduced Zn concentrations and increased iron (Fe) and nitrogen (N) in root and leaf and root. Also, levels of malondialdehyde, H2O2, and electrolyte leakage were decreased. Activities of antioxidant enzymes altered. | $H_2S$ application overcomes Zn-induced oxidative stress by regulating the antioxidant defense system in pepper. | Kaya et al. (2018) |

have been reported to control the stomatal movement, regulate the phenomenon of photosynthesis, scavenging of reactive oxygen species (ROS) and Copper (Cu)-stress protection in tomato (Faizan et al., 2021). It was further elaborated that Zn is released from these nanoparticles which regulate the growth and development in tomato plants under Cu metal stress. Zn is directly associated with the auxin synthesis which mediates the cell division and cell expansion in plants. Hence, Zn has the potential to affect various signaling pathways in plants. However, Khan and co-workers (2019) highlighted that over-exploitation of metallic nanoparticles such as ZnO has led to Zn/metal-accumulation in the soils which may enter into the food chain through crop plants. This study also revealed that ethylene-induced oxidative stress in the *Arabidopsis* plants subjected to ZnO nanoparticle toxicity.

In transgenic tobacco plants, durum wheat, and yeast, ethylene-responsive transcription factor (*TdSHN1*) has been observed to confer heavy metals (cadmium or Cd, Zn, and Cu) tolerance (Djemal and Khoudi, 2021). *TdSHN1* is a SHINE-type ethylene-responsive transcription factor and its expression was observed to be induced in response to amelioration of metal stress. *TdSHN1* gene promoter regulated the expression of heavy metals induced *GUS* gene in transgenic tobacco lines which resulted in increased activities of superoxide dismutase and catalase (ROS-scavenging antioxidant enzymes), thereby conferred the heavy metal stress tolerance. In bananas, two $C_2H_2$ zinc finger proteins, i.e., *MaC2H2−1* and *MaC2H2−2*, were identified, characterized, and observed to be localized in the nucleus (Han et al., 2016). Both of them have been reported as transcriptional repressors that mediate the regulation of ethylene production in bananas during fruit ripening. These $C_2H_2$ zinc finger transcriptional repressors (*MaC2H2−1* and *MaC2H2−2*) regulate the ethylene synthesis through the repression of ethylene biosynthetic genes.

In mustard (*Brassica juncea* L.), nickel (Ni) and Zn have been reported to induce oxidative stress and increase the activity of 1-aminocyclopropane carboxylic acid synthase (Khan and Khan, 2014). However, the supplementation of ethylene reversed the Ni- and Zn-induced photosynthetic inhibition by altering PS II activity, antioxidant metabolism, and photosynthetic nitrogen use efficiency and by enhancing the activity of ribulose-1,5-bisphosphate carboxylase. This was further confirmed by employing an ethylene action inhibitor, norbornadiene. Further study also emphasized that exogenous application of ethylene promotes the Zn stress tolerance through modulation of defense mechanisms in mustard (Khan et al., 2019b). It was reported that ethylene treatment resulted in the enhanced activity of antioxidant enzymes, reduced oxidative stress, restored rate of photosynthesis, enhanced proline metabolism, nutrient homeostasis, and glyoxalase system in Zn-stressed mustard plants.

The seedlings of a wetland halophyte, *Kosteletzkya pentacarpos* were exposed to Cd, Zn, combination of Cd and Zn in the presence/absence of salt (NaCl) to study the physiological response under metal and salt stress (Zhou et al., 2018).

The salt stress was observed to influence the interaction of metal ions and the synthesis of ethylene and polyamine in *K. pentacarpos*. Metal stress induced the ethylene synthesis but salinity stress was observed to reduce the same. Levels of putrescine (polyamine) were recorded to increase, whereas levels of spermine and spermidine were reported to decline under Cd stress. However, Zn had no significant impact on the synthesis of polyamines. Further, aminovynilglycine, an inhibitor of ethylene synthesis, was observed to restore the synthesis of polyamines (by decreasing putrescine; increasing spermidine and spermine contents), decreasing senescence, and increasing plant growth under metal stress. It has been reported that the same biosynthetic precursors are involved in the synthesis of both ethylene and polyamines (Kolbert et al., 2019). Moreover, the oxidation of polyamines can produce NO. It unfolds an indirect biosynthetic link as well as a ppossible interplay between ethylene and NO. Both ethylene and NO are required for various developmental stages of plants and stress management. The detailed underlying mechanism of interaction between ethylene and NO may be helpful for plant stress protection and management.

### 8.5.2 Nitric oxide and zinc

The sodium nitroprusside has been applied as a source of nitric oxide (NO), an endogenous gasotransmitter and its interactive effect with salicylic acid has been observed in the safflower (*Carthamus tinctorius* L.) under Zn stress (Namdjoyan et al., 2018). Both salicylic acid and NO (in the form of sodium nitroprusside) reversed the Zn-induced enhancement in the contents of malondialdehyde, α-tocopherol, hydrogen peroxide ($H_2O_2$), phytochelatins and proline, whereas the concentration of chlorophyll and the activity of antioxidant enzymes (catalase, ascorbate peroxidase, and guaiacol peroxidase) were increased with the application of salicylic acid and NO in safflower plants subjected to Zn toxicity. Another study also revealed that Zn toxicity has been reported to be alleviated by the treatment of NO and salicylic acid in *C. tinctorius* L. (Namdjoyan et al., 2017).

In *Salmonella enterica* serovar Typhimurium, NO has been observed to disrupt the homeostasis of Zn (Frawley et al., 2018). Several genes were identified that target the synthesis of zinc metalloproteins prerequisite for DNA replication and repair, proteins, and various metabolites. In *S. typhimurium*, the Zn-induced cytotoxic effects were ameliorated by ZntA and ZitB zinc efflux transporters. Exogenous applications of NO were observed to enhance yield, performance, increased concentration of Zn in roots, decreased level of Zn in shoots, and biochemical parameters in sunflowers subjected to Zn stress. (Akladious and Mohamed, 2017). The sodium nitroprusside was applied as a source for NO. Its application remarkably enhanced growth, the content of photosynthetic pigments, ascorbic acid, and glutathione under Zn stress. Also, the activities of antioxidant enzymes and the quality of the oil (with increased unsaturated fatty acids) were alleviated by NO treatments. In maize, NO was also reported to improve the high Zn tolerance (Kaya, 2016). Exogeneous foliar applications of NO have been reported to decrease the content of Zn in both leaf and root; and enhanced levels of nitrogen in both leaf and root in maize under Zn stress. It further reduced the membrane permeability in maize plants subjected to high Zn concentrations.

### 8.5.3 Hydrogen sulfide and zinc

Hydrogen sulfide ($H_2S$) is an endogenous gasotransmitter in plants that was previously considered to be a toxic gas, however, of late it has been identified as a gaseous molecule with a significant role in plant growth, development, and physiology (Banerjee et al., 2018). It also interacts with the signaling pathways of NO and $Ca^{2+}$ and when applied exogenously $H_2S$ has been reported to ameliorate abiotic stress in plants. In sunflower plants, potassium phosphate, zinc sulfate, and $H_2S$ have been reported to mitigate drought stress (Almeida et al., 2020). Application of $H_2S$ and zinc sulfate retained the water potential, decreased leaf malonaldehyde content, and hence reduced the rate of lipid peroxidation under water deficit. Besides this, the activity of antioxidant enzymes was altered, leaf osmotic potential was restored and an adequate amount of water in sunflower leaf tissues was also maintained by application of zinc sulfate and $H_2S$ under drought stress. Similarly, in soybeans, potassium phosphate, zinc sulfate, and $H_2S$ also ameliorated drought stress conditions by upregulating the activity of the antioxidant enzyme, increasing the accumulation of free amino acids, soluble sugars, prolines, and maintaining the osmotic potential (Batista et al., 2020).

In pepper (*Capsicum annuum* L.), $H_2S$ application has been reported to overcome Zn-induced oxidative stress by regulating the antioxidant defense system (Kaya et al., 2018). Zn-stress reduced chlorophyll, dry biomass, leaf maximum fluorescence, fruit yield, and relative water content, whereas increased the contents of free proline, endogenous hydrogen peroxide, malondialdehyde, and enhanced electrolyte leakage, activities of antioxidant enzymes. Exogenous treatment of NaHS, as a source of $H_2S$ significantly increased fruit yield, dry matter, chlorophyll, plant growth, water status, reduced Zn concentration, and proline under Zn-toxicity. $H_2S$ has also been reported to reduce Zn uptake and regulate the expression of antioxidant enzymes and metallothioneins in *Solanum nigrum* L. roots under Zn toxicity (Liu et al., 2016).

## 8.6 Plant steroids in Zn stress alleviation

Brassinosteroids (BRs) are the polyhydroxylated steroidal hormones ever isolated from plants (Zhabinskii et al., 2015). The structure of those plant hormones is comparable to animal steroidal hormones for example ecdysone or progesterone. The very first categorized brassinosteroid, brassinolide (BL), was discovered in rape pollen grains. To date, 70 compounds of BRs have been identified and isolated in plant kingdom (Bhardwaj et al., 2014). They regulate the various physiological processes under normal circumstances and play a crucial role in the activation of adoptive responses to abiotic stress conditions including the metal stress (Gautam et al., 2017).

In a research conducted through growth chamber experiments on *Raphanus sativus* L. plants, Ramakrishna and Rao (2013) reported that BRs (24-epibrassinolide) (EBL) and 28-homobrassinolide (HBL) application through leaves ease the symptoms of zinc metal toxicity. The outcome of this study indicated that BRs confronted the $Zn^{2+}$ stress and imparted better plant health and growth by elevating the enzymatic as well as nonenzymatic antioxidants which led to improved photosynthetic attributes and relative water content. Also, out of two tested BRs, EBL was concluded to be more efficient than HBL in the amelioration of $Zn^{2+}$ stress. The exogenous application of 24-Epibrassinolide upholds a high redox state of ascorbate and glutathione in Zinc stressed *Raphanus sativus* L. seedlings and amplified the plant development in terms of seedling length, fresh weight, and dry weight (Ramakrishna and Rao, 2013). The exogenous supplementation of Zn-stressed radish seedlings with 28-Homobrassinolide results in elevated glutathione redox homeostasis. HBL regulates the GSH levels by enhancing the activities of gamma-glutamylcysteine synthetase, glutathione synthetase, and glutathione reductase. The HBL carried out decontamination of Zn stress by maintaining glutathione peroxidase and glutathione-S-transferase activities. Also, the phytochelatins and cysteine levels were enhanced due to HBL treatment in Zn-stressed radish seedlings (Ramakrishna and Rao, 2013). In another study, Ramakrishna and Rao (2013) reported that EBL improves the Zn-induced stress conditions in radish seedlings by activating the antioxidative apparatus. Additionally, EBL supplementation improved the phenol and free proline levels in Zn-stressed seedlings. Brassinosteroids (24-epibrassinolide) have also been reported to ameliorate ZnO (zinc oxide) nanoparticles generated stress by convalescing the antioxidant capacity and redox homeostasis in *Solanum lycopersicum* seedlings (Li et al., 2016). EBL enhanced the activities of antioxidant enzymes, remarkably decreased the concentration of hydrogen peroxide and malondialdehyde, and increased the lengths of roots and shoots of plants under Zn stress. dos Santos et al., 2017 reported that in zinc-stressed soybean plants, 24-Epibrassinolide recovers the root anatomy and antioxidative enzymes. Supplementary, the gas exchange stimulation induced by this plant steroid was illustrated by constructive foliar anatomy and stomatal recital and enhanced $CO_2$ diffusion.

The results of the above studies inferred that plant steroidal hormones such as 24-epibrassinolide and 28-homobrassinolide alleviate the ill effects of Zn toxicity in different plant species by regulating the oxidative defense system. Fig. 8.3 depicts that the plant steroids control various antioxidant

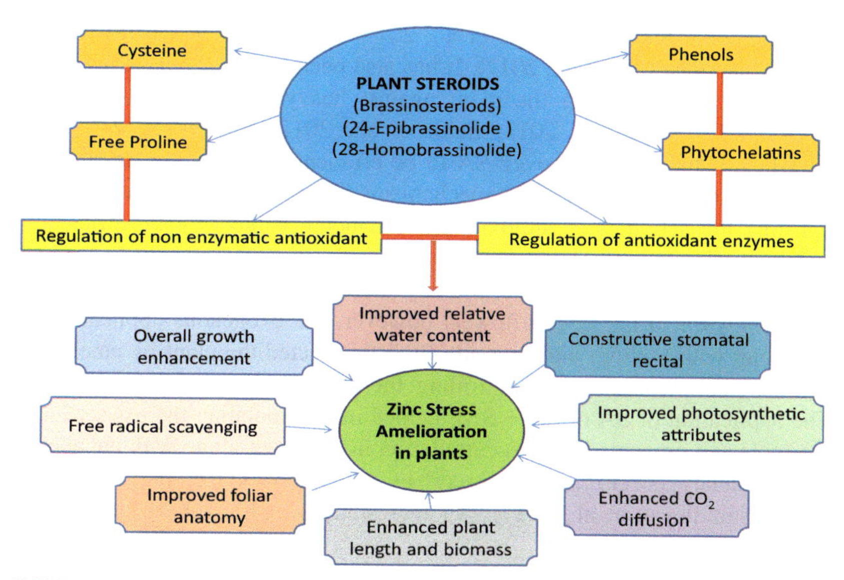


**FIGURE 8.3** Plant steroids mediated alleviation of the Zn toxicity in different plant species.

enzymes and nonenzymatic antioxidants to regulate the normal growth and metabolism of plants under Zn stress. Also, under the Zn metal oxidative stress, the plant steroids regulate the phenols and phytochelatins to encounter adverse effects on photosynthetic apparatus, foliar anatomy, stomatal structure, and other crucial functions.

## 8.7 Salicylic acid and Zn stress amelioration

Salicylic acid (SA), also known as *ortho*-hydroxybenzoic acid, is an essential phenolic compound that is known to regulate several fundamental processes like photosynthesis, nitrogen metabolism, mineral nutrition, plant water relations, stomatal functioning, flower and fruit ripening and quality, osmolyte contents, and antioxidant efficiency (Emamverdian et al., 2020; Ahmad et al., 2019). The process of photosynthesis is also regulated by SA as it modulates the functions of several essential enzymes like Ribulose-1,5-Bisphosphate Carboxylase/Oxygenase (RuBisCo) and carbonic anhydrase along with controlling stomatal movements and chloroplast structure (Kohli et al., 2017). It also mediates a wide array of growth responses in plants under stress conditions (Ahmad et al., 2019). SA is known to reinforce the antioxidant defense machinery of the plants when subjected to any stress conditions and also alter several physio-genomic mechanisms that work to maintain plant growth under environmental cues. SA has proven its efficacy in shielding plants against several known abiotic stress conditions (Abbaszadeh et al., 2020; Wang et al., 2018a,b; Sharma et al., 2017) and biotic attacks (Stella de Freitas et al., 2019; Deenamo et al., 2018; Yousif, 2018; Chandrasekhar et al., 2017). It has also been unraveled that SA plays a potent role in protecting the plant against heavy metal stress (Zanganeh et al., 2019; Lu et al., 2018; Saidi et al., 2017; Gondor et al., 2016). Significant studies have proved SA to be efficient in mitigating Zn stress in several plant species by various mechanisms like controlling oxidative stress, uplifting the activities of antioxidants, enhancing photosynthetic efficiency, and promoting growth (Table 8.3).

It has been reported that Zn toxicity led to a decrease in the biomass of *Carthamus tinctorius* L. (Safflower) however, the exogenous application of SA along with Sodium nitroprusside (SNP) protected the plant by ameliorating the plant biomass and controlling the reactive oxygen species (ROS). The application of SA or SA + SNP helped in accumulating the excess Zn in the roots and prevented their accumulation in the shoot along with enhancing the efficiency of enzymes of ascorbate-glutathione cycle and glyoxalase cycle (Namdjoyan et al., 2017). Another study in safflower by Namdjoyan

**TABLE 8.3 Salicylic acid-mediated protection against Zn stress in different plant species.**

| Plant species | SA treatment | Plant response | Reference |
|---|---|---|---|
| *Helianthus annuus* | 0.5 mmol/L SA | Improved seed germination and Zn translocation | Huang et al. (2021) |
| *Triticum aestivum* | 100 μM SA | Enhanced activity of antioxidants | Mazumder et al. (2021) |
| *Salvia officinalis* | 0.5- and 1-mM SA | Improved calcium, potassium, and phosphorus contents and glandular hair density | Es-sbihi et al. (2020) |
| *Trigonella foenum-graecum* | 0.2 mM SA | Enhanced levels of amino acids, proteins, and sugars and improved activity of hydrolytic enzymes | Mabrouk et al. (2019) |
| *Carthamus tinctorius* | 100 μM SA + 100 μM SNP | Decreased levels of ROS and enhanced synthesis of phytochelatins | Namdjoyan et al. (2018) |
| *Carthamus tinctorius* | 100 μM SA + 100 μM SNP | Enhanced efficiency of ascorbate-glutathione cycle and glyoxalase cycle enzymes | Namdjoyan et al. (2017) |
| *Zea mays* | 4.14 mg/L SA | Enhanced biomass of plant | Santos et al. (2017) |

et al. (2018) demonstrated the potential of SA in controlling the levels of Malondialdehyde (MDA) and hydrogen peroxide ($H_2O_2$) in Zn-stressed safflower. The combination of SA and SNP led to the enhanced production of phytochelatins and maintained the optimum levels of photosynthetic contents and osmolytes. Ameliorative potential of SA has also been documented in *Trigonella foenum-graecum* L. (fenugreek), where the heavy metal toxicity with Zn and arsenic (As) led to a decrease in the seedling growth and inhibited the enzyme activity of amylases along with reducing the contents of photosynthetic pigments, soluble sugars, and amino acids. However, pretreatment with SA proved to be beneficial in restoring the enzymatic activities of amylases and maintaining the levels of free amino acids, sugars, and proteins in fenugreek (Mabrouk et al., 2019).

Exposure of toxic concentrations of Zn led to a decrease in essential oil yield, mineral uptake, chlorophyll content, and growth of *Salvia officinalis* L. due to the accumulation of Zn in different parts of the plant. The spray of SA mitigated the toxic effects of Zn in the stressed plants by improving the growth and mineral status. The abnormalities in the glandular hair caused by the Zn stress were also mitigated by SA application which improved the peltate glandular trichome density (Es-sbihi et al., 2020). Huang et al. (2021) reported that Zn stress-associated decrease in seed germination was ameliorated on exogenous application of SA in sunflowers. SA application also regulated the zinc-transport-associated genes that helped in maintaining the levels of Zn in the plant body. Oxidative stress caused by Zn toxicity and reduced efficiency of antioxidants was restored by SA exogenous treatment in *Triticum aestivum* (Mazumder et al., 2021). The Zn stress reduced the growth parameters due to overproduction of ROS but SA treatment enhanced the activities of antioxidants like glutathione and ascorbic acid for

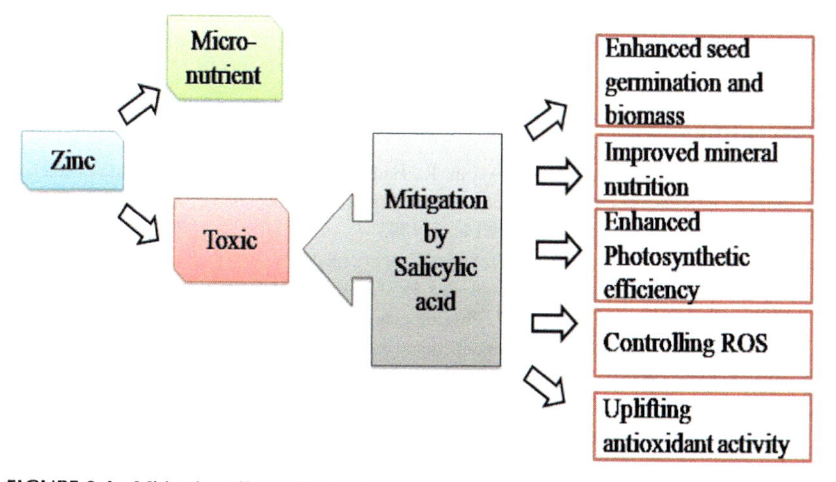


**FIGURE 8.4** Mitigating effects of aalicylic acid against Zn stress.

controlling the over-produced ROS. Santos et al., 2017 reported that Zn, when provided in optimum amounts, acts as a micronutrient and a growth enhancer. However, while increasing the amount of Zn supplemented to plants, it inhibits the growth of the *Zea mays* which was restored on SA supplementation. The mitigating potential of SA in enduring Zn toxicity has been well acknowledged in various plant species (Fig. 8.4). The exogenous application of SA opens up new ways for the emerging field of chemicogenetics that involves the use of plant growth regulators for managing stress conditions prevalent in the environment.

## 8.8 Conclusion

Plant nutrition plays an indispensable role in regulating their growth and development. Most of the nutrients required by plants are available in sufficient amounts in soil and water. However, deficiency and toxicity of certain nutrients may pose deleterious effects on plant health and deteriorate crop productivity. Zn being an essential micronutrient controls a vast array of physiological and biochemical processes within the plants. Being part of several transcription factors and proteins, Zn also shows crosstalk with signaling mechanisms induced by various phytohormones under different growth conditions. Nowadays, several anthropogenic activities have led to an enormous rise in Zn levels within the environment. Therefore, measures need to be taken to control the availability of Zn in soil and its entry into the plant system. Phytohormones being eco-friendly and nontoxic in nature can be utilized for the detoxification of different HMs including Zn in a variety of agricultural crop plants. Owing to their versatile roles, different phytohormones have the potential to mitigate the noxious effects of Zn toxicity in plants by inducing different signaling mechanisms. The components of these mechanisms are yet to be unveiled.

## References

Abbaszadeh, B., Layeghhaghighi, M., Azimi, R., Hadi, N., 2020. Improving water use efficiency through drought stress and using salicylic acid for proper production of *Rosmarinus officinalis* L. Industrial Crops and Products 144, 111893.

Ahmad, F., Singh, A., Kamal, A., 2019. Salicylic acid–mediated defense mechanisms to abiotic stress tolerance. Plant Signaling Molecules 355–369.

Akladious, S.A., Mohamed, H.I., 2017. Physiological role of exogenous nitric oxide in improving performance, yield and some biochemical aspects of sunflower plant under zinc stress. Acta Biologica Hungarica 68 (1), 101–114.

Almeida, G.M., Silva, A.A.D., Batista, P.F., Moura, L.M.D.F., Vital, R.G., Costa, A.C., 2020. Hydrogen sulfide, potassium phosphite and zinc sulfate as alleviators of drought stress in sunflower plants. Ciência e Agrotecnologia 44.

Al-Ghazi, Y., Muller, B., Pinloche, S., Tranbarger, T.J., Nacry, P., Rossignol, M., et al., 2003. Temporal responses of Arabidopsis root architecture to phosphate starvation: evidence for the involvement of auxin signalling. Plant, Cell & Environment 26 (7), 1053–1066.

Angulo-Bejarano, P.I., Puente-Rivera, J., Cruz-Ortega, R., 2021. Metal and metalloid toxicity in plants: an overview on molecular aspects. Plants 10 (4), 635.

Anwaar, S.A., Ali, S., Ali, S., Ishaque, W., Farid, M., Farooq, M.A., et al., 2015. Silicon (Si) alleviates cotton (*Gossypium hirsutum* L.) from zinc (Zn) toxicity stress by limiting Zn uptake and oxidative damage. Environmental Science and Pollution Research 22 (5), 3441–3450.

Arif, N., Yadav, V., Singh, S., Singh, S., Ahmad, P., Mishra, R.K., et al., 2016. Influence of high and low levels of plant-beneficial heavy metal ions on plant growth and development. Frontiers in Environmental Science 4, 69.

Atici, Ö., Ağar, G., Battal, P.E.Y.A.M.İ., 2005. Changes in phytohormone contents in chickpea seeds germinating under lead or zinc stress. Biologia Plantarum 49 (2), 215–222.

Aya, K., Ueguchi-Tanaka, M., Kondo, M., Hamada, K., Yano, K., Nishimura, M., et al., 2009. Gibberellin modulates anther development in rice via the transcriptional regulation of GAMYB. The Plant Cell 21 (5), 1453–1472.

Balafrej, H., Bogusz, D., Triqui, Z.E.A., Guedira, A., Bendaou, N., Smouni, A., et al., 2020. Zinc hyperaccumulation in plants: a review. Plants 9 (5), 562.

Banerjee, A., Tripathi, D.K., Roychoudhury, A., 2018. Hydrogen sulphide trapeze: environmental stress amelioration and phytohormone crosstalk. Plant Physiology and Biochemistry 132, 46–53.

Batista, P.F., Müller, C., Merchant, A., Fuentes, D., Silva-Filho, R.D.O., da Silva, F.B., et al., 2020. Biochemical and physiological impacts of zinc sulphate, potassium phosphite and hydrogen sulphide in mitigating stress conditions in soybean. Physiologia Plantarum 168 (2), 456–472.

Bazihizina, N., Taiti, C., Marti, L., Rodrigo-Moreno, A., Spinelli, F., Giordano, C., et al., 2014. $Zn^{2+}$-induced changes at the root level account for the increased tolerance of acclimated tobacco plants. Journal of Experimental Botany 65 (17), 4931–4942.

Bhardwaj, R., Sharma, I., Kapoor, D., Gautam, V., Kaur, R., Bali, S., et al., 2014. Brassinosteroids: improving crop productivity and abiotic stress tolerance. Physiological Mechanisms and Adaptation Strategies in Plants Under Changing Environment 161–187.

Bogamuwa, S.P., Jang, J.C., 2014. Tandem CCCH zinc finger proteins in plant growth, development and stress response. Plant and Cell Physiology 55 (8), 1367–1375.

Borker, A.R., David, K., Singhal, N., 2020. Analysis of time varying response on uptake patterns of Cu and Zn ions under application of ethylene diamine disuccinic acid and gibberellic acid in Lolium perenne. Chemosphere 260, 127541.

Brokbartold, M., Wischermann, M., Marschner, B., 2012. Plant availability and uptake of lead, zinc, and cadmium in soils contaminated with anti-corrosion paint from pylons in comparison to heavy metal contaminated urban soils. Water, Air, & Soil Pollution 223 (1), 199–213.

Cao, X., Ma, L.Q., Rhue, D.R., Appel, C.S., 2004. Mechanisms of lead, copper, and zinc retention by phosphate rock. Environmental Pollution 131 (3), 435–444.

Chandrasekhar, B., Umesha, S., Kumar, H.N., 2017. Proteomic analysis of salicylic acid enhanced disease resistance in bacterial wilt affected chilli (*Capsicum annuum*) crop. Physiological and Molecular Plant Pathology 98, 85–96.

Chattha, M.U., Hassan, M.U., Khan, I., Chattha, M.B., Mahmood, A., Nawaz, M., et al., 2017. Biofortification of wheat cultivars to combat zinc deficiency. Frontiers in Plant Science 8, 281.

Chen, S., Li, X., Yang, C., Yan, W., Liu, C., Tang, X., et al., 2021. Genome-wide identification and characterization of FCS-like zinc finger (FLZ) family genes in maize (*Zea mays*) and functional analysis of ZmFLZ25 in plant abscisic acid response. International Journal of Molecular Sciences 22 (7), 3529.

Chen, Y.T., Wang, Y., Yeh, K.C., 2017. Role of root exudates in metal acquisition and tolerance. Current Opinion in Plant Biology 39, 66–72.

Cuypers, A., Remans, T., Weyens, N., Colpaert, J., Vassilev, A., Vangronsveld, J., 2013. Soil-plant relationships of heavy metals and metalloids. Heavy Metals in Soils 161–193. Springer.

Das, J., Sarkar, P., 2018. Remediation of arsenic in mung bean (Vigna radiata) with growth enhancement by unique arsenic-resistant bacterium *Acinetobacter lwoffii*. Science of the Total Environment 624, 1106–1118.

Deenamo, N., Kuyyogsuy, A., Khompatara, K., Chanwun, T., Ekchaweng, K., Churngchow, N., 2018. Salicylic acid induces resistance in rubber tree against *Phytophthora palmivora*. International Journal of Molecular Sciences 19 (7), 1883.

Dhalaria, R., Kumar, D., Kumar, H., Nepovimova, E., Kuča, K., Torequl Islam, M., et al., 2020. Arbuscular mycorrhizal fungi as potential agents in ameliorating heavy metal stress in plants. Agronomy 10 (6), 815.

Djemal, R., Khoudi, H., 2021. The ethylene-responsive transcription factor of durum wheat, TdSHN1, confers cadmium, copper, and zinc tolerance to yeast and transgenic tobacco plants. Protoplasma 1–13.

Emamverdian, A., Ding, Y., Mokhberdoran, F., 2020. The role of salicylic acid and gibberellin signaling in plant responses to abiotic stress with an emphasis on heavy metals. Plant Signaling & Behavior 15 (7), 1777372.

Es-sbihi, F.Z., Hazzoumi, Z., Joutei, K.A., 2020. Effect of salicylic acid foliar application on growth, glandular hairs and essential oil yield in *Salvia officinalis* L. grown under zinc stress. Chemical and Biological Technologies in Agriculture 7 (1), 1–11.

Estelle, M. (Ed.), 2011. Auxin Signaling: From Synthesis to Systems Biology; A Subject Collection from Cold Spring Harbor Perspectives in Biology. Cold Spring Harbor Laboratory Press.

Faizan, M., Bhat, J.A., Noureldeen, A., Ahmad, P., Yu, F., 2021. Zinc oxide nanoparticles and 24-epibrassinolide alleviates Cu toxicity in tomato by regulating ROS scavenging, stomatal movement and photosynthesis. Ecotoxicology and Environmental Safety 218, 112293.

Falkowska, M., Pietryczuk, A., Piotrowska, A., Bajguz, A., Grygoruk, A., Czerpak, R., 2011. The effect of gibberellic acid (GA3) on growth, metal biosorption and metabolism of the green algae *Chlorella vulgaris* (Chlorophyceae) Beijerinck exposed to cadmium and lead stress. Polish Journal of Environmental Studies 20 (1), 53–59.

Feigl, G., Lehotai, N., Molnár, Á., Ördög, A., Rodríguez-Ruiz, M., Palma, J.M., et al., 2015. Zinc induces distinct changes in the metabolism of reactive oxygen and nitrogen species (ROS and RNS) in the roots of two Brassica species with different sensitivity to zinc stress. Annals of Botany 116 (4), 613–625.

Frawley, E.R., Karlinsey, J.E., Singhal, A., Libby, S.J., Doulias, P.T., Ischiropoulos, H., et al., 2018. Nitric oxide disrupts zinc homeostasis in *Salmonella enterica* serovar Typhimurium. MBio 9 (4), 01040-18.

Fukazawa, J., Teramura, H., Murakoshi, S., Nasuno, K., Nishida, N., Ito, T., et al., 2014. DELLAs function as coactivators of GAI-ASSOCIATED FACTOR1 in regulation of gibberellin homeostasis and signaling in Arabidopsis. The Plant Cell 26 (7), 2920–2938.

Gangwar, S., Singh, V.P., 2011. Indole acetic acid differently changes growth and nitrogen metabolism in *Pisum sativum* L. seedlings under chromium (VI) phytotoxicity: implication of oxidative stress. Scientia Horticulturae 129 (2), 321–328.

Gao, S., Xiao, Y., Xu, F., Gao, X., Cao, S., Zhang, F., et al., 2019. Cytokinin-dependent regulatory module underlies the maintenance of zinc nutrition in rice. New Phytologist 224 (1), 202–215.

Gautam, S., Rathoure, A.K., Chhabra, A., Pandey, S.N., 2017. Effects of nickel and zinc on biochemical parameters in plants—a review. Octa Journal of Environmental Research 5 (1).

Ghori, N.H., Ghori, T., Hayat, M.Q., Imadi, S.R., Gul, A., Altay, V., et al., 2019. Heavy metal stress and responses in plants. International Journal of Environmental Science and Technology 16 (3), 1807–1828.

Gondor, O.K., Pál, M., Darkó, É., Janda, T., Szalai, G., 2016. Salicylic acid and sodium salicylate alleviate cadmium toxicity to different extents in maize (*Zea mays* L.). PLoS One 11 (8), e0160157.

González-Alcaraz, M.N., van Gestel, C.A., 2015. Climate change effects on enchytraeid performance in metal-polluted soils explained from changes in metal bioavailability and bioaccumulation. Environmental Research 142, 177–184.

Griffiths, J., Murase, K., Rieu, I., Zentella, R., Zhang, Z.L., Powers, S.J., et al., 2006. Genetic characterization and functional analysis of the GID1 gibberellin receptors in Arabidopsis. The Plant Cell 18 (12), 3399–3414.

Han, Y.C., Fu, C.C., Kuang, J.F., Chen, J.Y., Lu, W.J., 2016. Two banana fruit ripening-related C2H2 zinc finger proteins are transcriptional repressors of ethylene biosynthetic genes. Postharvest Biology and Technology 116, 8–15.

Hedden, P., Thomas, S.G., 2012. Gibberellin biosynthesis and its regulation. Biochemical Journal 444 (1), 11–25.

Hou, Y., Liu, X., Zhang, X., Chen, X., Tao, K., 2016. Effects of key components of *S. triqueter* root exudates on fractions and bioavailability of pyrene–lead co-contaminated soils. International Journal of Environmental Science and Technology 13 (3), 887–896.

Huang, Y.T., Cai, S.Y., Ruan, X.L., Chen, S.Y., Mei, G.F., Ruan, G.H., et al., 2021. Salicylic acid enhances sunflower seed germination under $Zn^{2+}$ stress via involvement in $Zn^{2+}$ metabolic balance and phytohormone interactions. Scientia Horticulturae 275 (109702).

Huang, L., Jiang, Q., Wu, J., An, L., Zhou, Z., Wong, C., et al., 2020. Zinc finger protein 5 (ZFP5) associates with ethylene signaling to regulate the phosphate and potassium deficiency-induced root hair development in Arabidopsis. Plant Molecular Biology 102 (1), 143–158.

Hu, J., Mitchum, M.G., Barnaby, N., Ayele, B.T., Ogawa, M., Nam, E., et al., 2008. Potential sites of bioactive gibberellin production during reproductive growth in Arabidopsis. The Plant Cell 20 (2), 320–336.

Hänsch, R., Mendel, R.R., 2009. Physiological functions of mineral micronutrients (Cu, Zn, Mn, Fe, Ni, Mo, B, cl). Current Opinion in Plant Biology 12 (3), 259–266.

Jahan, B., Rasheed, F., Sehar, Z., Fatma, M., Iqbal, N., Masood, A., et al., 2021. Coordinated role of nitric oxide, ethylene, nitrogen, and sulfur in plant salt stress tolerance. Stresses 1 (3), 181–199.

Javed, T., Ali, M.M., Shabbir, R., Anwar, R., Afzal, I., Mauro, R.P., 2021. Alleviation of copper-induced stress in pea (*Pisum sativum* L.) through foliar application of gibberellic acid. Biology 10 (2), 120.

Joseph, M.P., Papdi, C., Kozma-Bognár, L., Nagy, I., López-Carbonell, M., Rigó, G., et al., 2014. The Arabidopsis ZINC FINGER PROTEIN3 interferes with abscisic acid and light signaling in seed germination and plant development. Plant Physiology 165 (3), 1203–1220.

Kai, W., Fu, Y., Wang, J., Liang, B., Li, Q., Leng, P., 2019. Functional analysis of SlNCED1 in pistil development and fruit set in tomato (*Solanum lycopersicum* L.). Scientific Reports 9 (1), 1–13.

Karle, S.B., Guru, A., Dwivedi, P., Kumar, K., 2021. Insights into the role of gasotransmitters mediating salt stress responses in plants. Journal of Plant Growth Regulation 1–17.

Kasahara, H., Hanada, A., Kuzuyama, T., Takagi, M., Kamiya, Y., Yamaguchi, S., 2002. Contribution of the mevalonate and methylerythritol phosphate pathways to the biosynthesis of gibberellins in Arabidopsis. Journal of Biological Chemistry 277 (47), 45188–45194.

Kaya, C., 2016. Nitric oxide improves high zinc tolerance in maize plants. Journal of Plant Nutrition 39 (14), 2072–2078.

Kaya, C., Ashraf, M., Akram, N.A., 2018. Hydrogen sulfide regulates the levels of key metabolites and antioxidant defense system to counteract oxidative stress in pepper (*Capsicum annuum* L.) plants exposed to high zinc regime. Environmental Science and Pollution Research 25 (13), 12612–12618.

Khan, M.I.R., Jahan, B., Alajmi, M.F., Rehman, M.T., Khan, N.A., 2019b. Exogenously-sourced ethylene modulates defense mechanisms and promotes tolerance to zinc stress in mustard (*Brassica juncea* L.). Plants 8 (12), 540.

Khan, M.I.R., Khan, N.A., 2014. Ethylene reverses photosynthetic inhibition by nickel and zinc in mustard through changes in PS II activity, photosynthetic nitrogen use efficiency, and antioxidant metabolism. Protoplasma 251 (5), 1007–1019.

Khan, A.H.A., Kiyani, A., Mirza, C.R., Butt, T.A., Barros, R., Ali, B., et al., 2021. Ornamental plants for the phytoremediation of heavy metals: present knowledge and future perspectives. Environmental Research 110780.

Khan, A.R., Wakeel, A., Muhammad, N., Liu, B., Wu, M., Liu, Y., et al., 2019. Involvement of ethylene signaling in zinc oxide nanoparticle-mediated biochemical changes in Arabidopsis thaliana leaves. Environmental Science: Nano 6 (1), 341–355.

Khare, D., Mitsuda, N., Lee, S., Song, W.Y., Hwang, D., Ohme-Takagi, M., et al., 2017. Root avoidance of toxic metals requires the GeBP-LIKE 4 transcription factor in Arabidopsis thaliana. New Phytologist 213 (3), 1257–1273.

Kodaira, K.S., Qin, F., Tran, L.S.P., Maruyama, K., Kidokoro, S., Fujita, Y., et al., 2011. Arabidopsis Cys2/His2 zinc-finger proteins AZF1 and AZF2 negatively regulate abscisic acid-repressive and auxin-inducible genes under abiotic stress conditions. Plant Physiology 157 (2), 742–756.

Kohli, S.K., Handa, N., Kaur, R., Kumar, V., Khanna, K., Bakshi, P., et al., 2017. Role of salicylic acid in heavy metal stress tolerance: insight into underlying mechanism. Salicylic Acid: A Multifaceted Hormone 123–144.

Kolbert, Z., Feigl, G., Freschi, L., Poór, P., 2019. Gasotransmitters in action: nitric oxide-ethylene crosstalk during plant growth and abiotic stress responses. Antioxidants 8 (6), 167.

Kosakivska, I.V., Vasyuk, V.A., Voytenko, L.V., 2019. Effects of endogenous abscisic acid on seed germination and morphological characteristics of two related wheats *Triticum aestivum* L. and *Triticum spelta* L. Plant Physiology And Genetics 51, 55–66.

Kramer, U., 2010. In: Merchant, S., Briggs, W.R., Ort, D. (Eds.), Metal Hyperaccumulation in Plants.

Kwon, M.J., Boyanov, M.I., Yang, J.S., Lee, S., Hwang, Y.H., Lee, J.Y., et al., 2017. Transformation of zinc-concentrate in surface and subsurface environments: implications for assessing zinc mobility/toxicity and choosing an optimal remediation strategy. Environmental Pollution 226, 346–355.

Liu, X., Chen, J., Wang, G.H., Wang, W.H., Shen, Z.J., Luo, M.R., et al., 2016. Hydrogen sulfide alleviates zinc toxicity by reducing zinc uptake and regulating genes expression of antioxidative enzymes and metallothioneins in roots of the cadmium/zinc hyperaccumulator *Solanum nigrum* L. Plant and Soil 400 (1–2), 177–192.

Liu, X., Li, R., Dai, Y., Yuan, L., Sun, Q., Zhang, S., et al., 2019. A B-box zinc finger protein, Md BBX10, enhanced salt and drought stresses tolerance in Arabidopsis. Plant Molecular Biology 99 (4), 437–447.

Liu, Y., Zhang, M., Meng, Z., Wang, B., Chen, M., 2020. Research progress on the roles of cytokinin in plant response to stress. International Journal of Molecular Sciences 21 (18), 6574.

Li, C., Lv, J., Zhao, X., Ai, X., Zhu, X., Wang, M., et al., 2010. TaCHP: a wheat zinc finger protein gene down-regulated by abscisic acid and salinity stress plays a positive role in stress tolerance. Plant Physiology 154 (1), 211–221.

Li, S., Zhao, B., Jin, M., Hu, L., Zhong, H., He, Z., et al., 2020. A comprehensive survey on the horizontal and vertical distribution of heavy metals and microorganisms in soils of a Pb/Zn smelter. Journal of Hazardous Materials 400, 123–255.

Lukatkin, A.S., Gracheva, N.V., Grishenkova, N.N., Dukhovskis, P.V., Brazaitite, A.A., 2007. Cytokinin-like growth regulators mitigate toxic action of zinc and nickel ions on maize seedlings. Russian Journal of Plant Physiology 54 (3), 381–387.

Lu, Q., Zhang, T., Zhang, W., Su, C., Yang, Y., Hu, D., et al., 2018. Alleviation of cadmium toxicity in Lemna minor by exogenous salicylic acid. Ecotoxicology and Environmental Safety 147, 500–508.

Mabrouk, B., Kaab, S.B., Rezgui, M., Majdoub, N., da Silva, J.T., Kaab, L.B.B., 2019. Salicylic acid alleviates arsenic and zinc toxicity in the process of reserve mobilization in germinating fenugreek (Trigonella foenum-graecum L.) seeds. South African Journal of Botany 124, 235–243.

Mansour, M.M., Kamel, E.A.R., 2005. Interactive effect of heavy metals and gibberellic acid on mitotic activity and some metabolic changes of *Vicia faba* L. plants. Cytologia 70 (3), 275–282.

Martínez-Ruiz, A., Cadenas, S., Lamas, S., 2011. Nitric oxide signaling: classical, less classical, and nonclassical mechanisms. Free Radical Biology and Medicine 51 (1), 17–29.

Mateos-Naranjo, E., Castellanos, E.M., Perez-Martin, A., 2014. Zinc tolerance and accumulation in the halophytic species Juncus acutus. Environmental and Experimental Botany 100, 114–121.

Mazumder, M.K., Sharma, P., Moulick, D., Tata, S.K., Choudhury, S., 2021. Salicylic acid ameliorates zinc and chromium-induced stress responses in wheat seedlings: a biochemical and computational analysis. Cereal Research Communications 1–12.

Mengel, K., Kosegarten, H., Kirkby, E.A., Appel, T., 2001. Principles of Plant Nutrition. Springer, Berlin, Germany.

Miransari, M., Abrishamchi, A., Khoshbakht, K., Niknam, V., 2014. Plant hormones as signals in arbuscular mycorrhizal symbiosis. Critical Reviews in Biotechnology 34 (2), 123–133.

Montalvo, D., Degryse, F., Da Silva, R.C., Baird, R., McLaughlin, M.J., 2016. Agronomic effectiveness of zinc sources as micronutrient fertilizer. Advances in Agronomy 139, 215–267.

Mousavi, S.R., Galavi, M., Ahmadian, G., 2007. Effect of zinc and manganese foliar application on yield, quality and enrichment on potato (Solanum tuberosum L.). Asian Journal of Plant Sciences .

Mroue, S., Simeunovic, A., Robert, H.S., 2018. Auxin production as an integrator of environmental cues for developmental growth regulation. Journal of Experimental Botany 69 (2), 201–212.

Namdjoyan, S., Kermanian, H., Soorki, A.A., Tabatabaei, S.M., Elyasi, N., 2017. Interactive effects of salicylic acid and nitric oxide in alleviating zinc toxicity of Safflower (*Carthamus tinctorius* L.). Ecotoxicology (London, England) 26 (6), 752–761.

Namdjoyan, S., Kermanian, H., Soorki, A.A., Tabatabaei, S.M., Elyasi, N., 2018. Effects of exogenous salicylic acid and sodium nitroprusside on α-tocopherol and phytochelatin biosynthesis in zinc-stressed safflower plants. Turkish Journal of Botany 42 (3), 271–279.

Nan, W., Wang, X., Yang, L., Hu, Y., Wei, Y., Liang, X., et al., 2014. Cyclic GMP is involved in auxin signalling during Arabidopsis root growth and development. Journal of Experimental Botany 65 (6), 1571–1583.

Navarro-León, E., Barrameda-Medina, Y., Lentini, M., Esposito, S., Ruiz, J.M., Blasco, B., et al., 2016. Comparative study of Zn deficiency in L. sativa and B. oleracea plants: NH4 + assimilation and nitrogen derived protective compounds. Plant Science 248, 8–16.

Palmer, C.M., Guerinot, M.L., 2009. Facing the challenges of Cu, Fe and Zn homeostasis in plants. Nature Chemical Biology 5 (5), 333–340.

Palmgren, M.G., Clemens, S., Williams, L.E., Krämer, U., Borg, S., Schjørring, J.K., et al., 2008. Zinc biofortification of cereals: problems and solutions. Trends in Plant Science 13 (9), 464–473.

Pavlíková, D., Pavlík, M., Procházková, D., Zemanová, V., Hnilička, F., Wilhelmová, N., 2014. Nitrogen metabolism and gas exchange parameters associated with zinc stress in tobacco expressing an ipt gene for cytokinin synthesis. Journal of Plant Physiology 171 (7), 559–564.

Ramakrishna, B., Rao, S.S.R., 2013. 24-Epibrassinolide maintains elevated redox state of AsA and GSH in radish (*Raphanus sativus* L.) seedlings under zinc stress. Acta Physiologiae Plantarum 35 (4), 1291–1302.

Ramireddy, E., Hosseini, S.A., Eggert, K., Gillandt, S., Gnad, H., von Wirén, N., et al., 2018. Root engineering in barley: increasing cytokinin degradation produces a larger root system, mineral enrichment in the shoot and improved drought tolerance. Plant Physiology 177 (3), 1078–1095.

Rayle, D.L., Cleland, R.E., 1992. The acid growth theory of auxin-induced cell elongation is alive and well. Plant Physiology 99 (4), 1271–1274.

Ren, F.C., Liu, T.C., Liu, H.Q., Hu, B.Y., 1993. Influence of zinc on the growth, distribution of elements, and metabolism of one-year old American ginseng plants. Journal of Plant Nutrition 16 (2), 393–405.

Ren, G., Li, L., Huang, Y., Wang, Y., Zhang, W., Zheng, R., et al., 2018. Gh WIP 2, a WIP zinc finger protein, suppresses cell expansion in Gerbera hybrida by mediating crosstalk between gibberellin, abscisic acid, and auxin. New Phytologist 219 (2), 728–742.

Rhaman, M.S., Imran, S., Rauf, F., Khatun, M., Baskin, C.C., Murata, Y., et al., 2021. Seed priming with phytohormones: an effective approach for the mitigation of abiotic stress. Plants 10 (1), 37.

Rieu, I., Ruiz-Rivero, O., Fernandez-Garcia, N., Griffiths, J., Powers, S.J., Gong, F., et al., 2008. The gibberellin biosynthetic genes AtGA20ox1 and AtGA20ox2 act, partially redundantly, to promote growth and development throughout the Arabidopsis life cycle. The Plant Journal 53 (3), 488–504.

Sadeghzadeh, B., Rengel, Z., 2011. Zinc in soils and crop nutrition. In: Hawkesford, M.J., Barraclough, P. (Eds.), The Molecular and Physiological Basis of Nutrient Use Science in Crops. Wiley-Blackwell, Hoboken, NJ, USA, pp. 335–375.

Sagardoy, R., Vázquez, S., Florez-Sarasa, I.D., Albacete, A., Ribas-Carbó, M., Flexas, J., et al., 2010. Stomatal and mesophyll conductances to CO2 are the main limitations to photosynthesis in sugar beet (Beta vulgaris) plants grown with excess zinc. New Phytologist 187 (1), 145–158.

Saidi, I., Yousfi, N., Borgi, M.A., 2017. Salicylic acid improves the antioxidant ability against arsenic-induced oxidative stress in sunflower (Helianthus annuus) seedling. Journal of Plant Nutrition 40 (16), 2326–2335.

Sakakibara, H., 2006. Cytokinins: activity, biosynthesis, and translocation. Annual Review of Plant Biology 57, 431–449.

Sakata, T., Oda, S., Tsunaga, Y., Shomura, H., Kawagishi-Kobayashi, M., Aya, K., et al., 2014. Reduction of gibberellin by low temperature disrupts pollen development in rice. Plant Physiology 164 (4), 2011–2019.

Salama, Z.A., El-Fouly, M.M., Lazova, G., Popova, L.P., 2006. Carboxylating enzymes and carbonic anhydrase functions were suppressed by zinc deficiency in maize and chickpea plants. Acta Physiologiae Plantarum 28 (5), 445–451.

Santos, A.F.D., Morais, O.M., de Mello Prado, R., Leal, A.J.F., Silva, R.P.D., 2017. Relation of toxicity in corn seeds treated with zinc and salicylic acid. Communications in Soil Science and Plant Analysis 48 (10), 1123–1131.

Schnable, P.S., Ware, D., Fulton, R.S., Stein, J.C., Wei, F., Pasternak, S., et al., 2009. The B73 maize genome: complexity, diversity, and dynamics. Science 326 (5956), 1112–1115.

Sekimoto, H., Hoshi, M., Nomura, T., Yokota, T., 1997. Zinc deficiency affects the levels of endogenous gibberellins in *Zea mays* L. Plant and Cell Physiology 38 (9), 1087–1090.

Sharma, M., Gupta, S.K., Majumder, B., Maurya, V.K., Deeba, F., Alam, A., et al., 2017. Salicylic acid mediated growth, physiological and proteomic responses in two wheat varieties under drought stress. Journal of Proteomics 163, 28–51.

Shi, W.G., Li, H., Liu, T.X., Polle, A., Peng, C.H., Luo, Z.B., 2015a. Exogenous abscisic acid alleviates zinc uptake and accumulation in P opulus × canescens exposed to excess zinc. Plant, Cell & Environment 38 (1), 207–223.

Shi, Y.F., Wang, D.L., Wang, C., Culler, A.H., Kreiser, M.A., Suresh, J., et al., 2015b. Loss of GSNOR1 function leads to compromised auxin signaling and polar auxin transport. Molecular Plant 8 (9), 1350–1365.

Sidhu, G.P.S., 2016. Physiological, biochemical and molecular mechanisms of zinc uptake, toxicity and tolerance in plants. Journal of Global Biosciences 5 (9), 4603–4633.

Singh, R., Ahirwar, N.K., Tiwari, J., Pathak, J., 2018. Review on sources and effect of heavy metal in soil: its bioremediation. International Journal of Research in Applied, Natural and Social Sciences 2018, 1–22.

De Smet, I., Tetsumura, T., De Rybel, B., Frey, N.F.D., Laplaze, L., Casimiro, I., et al., 2007. Auxin-dependent regulation of lateral root positioning in the basal meristem of Arabidopsis .

Sofo, A., Bochicchio, R., Amato, M., Rendina, N., Vitti, A., Nuzzaci, M., et al., 2017. Plant architecture, auxin homeostasis and phenol content in Arabidopsis thaliana grown in cadmium-and zinc-enriched media. Journal of Plant Physiology 216, 174–180.

Song, Y., Zhou, L., Yang, S., Wang, C., Zhang, T., Wang, J., et al., 2017. Dose-dependent sensitivity of Arabidopsis thaliana seedling root to copper is regulated by auxin homeostasis. Environmental and Experimental Botany 139, 23–30.

Sponsel, V.M., Hedden, P., 2010. Gibberellin biosynthesis and inactivation. Plant Hormones. Springer, Dordrecht, pp. 63–94.

Stella de Freitas, T.F., Stout, M.J., Sant'Ana, J., 2019. Effects of exogenous methyl jasmonate and salicylic acid on rice resistance to *Oebalus pugnax*. Pest Management Science 75 (3), 744–752.

Sturikova, H., Krystofova, O., Huska, D., Adam, V., 2018. Zinc, zinc nanoparticles and plants. Journal of Hazardous Materials 349, 101–110.

Suge, H., Takahashi, H., Arita, S., Takaki, H., 1986. Gibberellin relationships in zinc-deficient plants. Plant and Cell Physiology 27 (6), 1005–1012.

Thakur, S., Singh, L., Ab Wahid, Z., Siddiqui, M.F., Atnaw, S.M., Din, M.F.M., et al., 2016. Plant-driven removal of heavy metals from soil: uptake, translocation, tolerance mechanism, challenges, and future perspectives. Environmental Monitoring and Assessment 188 (4), 206.

Tsonev, T., Cebola Lidon, F.J., 2012. Zinc in plants—an overview. Emirates Journal of Food & Agriculture (EJFA) 24 (4).

Ullah, A., Manghwar, H., Shaban, M., Khan, A.H., Akbar, A., Ali, U., et al., 2018. Phytohormones enhanced drought tolerance in plants: a coping strategy. Environmental Science and Pollution Research 25 (33), 33103–33118.

Umair Hassan, M., Aamer, M., Umer Chattha, M., Haiying, T., Shahzad, B., Barbanti, L., et al., 2020. The critical role of zinc in plants facing the drought stress. Agriculture 10 (9), 396.

Vassilev, A., Nikolova, A., Koleva, L., Lidon, F., 2011. Effects of exccss Zn on growth and photosynthetic performance of young bean plants. Journal of Phytology 3, 6.

Wang, W., Wang, X., Huang, M., Cai, J., Zhou, Q., Dai, T., et al., 2018a. Hydrogen peroxide and abscisic acid mediate salicylic acid-induced freezing tolerance in wheat. Frontiers in Plant Science 9, 1137.

Wang, Y., Wang, Y.A., Kai, W., Zhao, B.O., Chen, P., Sun, L., et al., 2014. Transcriptional regulation of abscisic acid signal core components during cucumber seed germination and under Cu2 +, Zn2 +, NaCl and simulated acid rain stresses. Plant Physiology and Biochemistry 76, 67–76.

Wang, R., Wang, J., Zhao, L., Yang, S., Song, Y., 2015. Impact of heavy metal stresses on the growth and auxin homeostasis of Arabidopsis seedlings. Biometals: An International Journal on the Role of Metal Ions in Biology, Biochemistry, and Medicine 28 (1), 123–132.

Wang, J., Yang, S., 2021. Dose-dependent responses of Arabidopsis thaliana to zinc are mediated by auxin homeostasis and transport. Environmental and Experimental Botany 104554.

Wang, Y., Zhou, J., Zhong, J., Luo, D., Li, Z., Yang, J., et al., 2018b. Cys2His2 zinc finger transcription factor BcabaR1 positively regulates abscisic acid production in *Botrytis cinerea*. Applied and Environmental Microbiology 84 (17), e00920-18.

Wani, S.H., Kumar, V., Shriram, V., Sah, S.K., 2016. Phytohormones and their metabolic engineering for abiotic stress tolerance in crop plants. The Crop Journal 4 (3), 162–176.

Weisany, W., Sohrabi, Y., Heidari, G., Siosemardeh, A., Ghassemi-Golezani, K., 2012. Changes in antioxidant enzymes activity and plant performance by salinity stress and zinc application in soybean ('*Glycine max*' L.). Plant Omics 5 (2), 60–67.

Weng, L., Zhao, F., Li, R., Xu, C., Chen, K., Xiao, H., 2015. The zinc finger transcription factor SlZFP2 negatively regulates abscisic acid biosynthesis and fruit ripening in tomato. Plant Physiology 167 (3), 931–949.

Werner, T., Nehnevajova, E., Köllmer, I., Novák, O., Strnad, M., Krämer, U., et al., 2010. Root-specific reduction of cytokinin causes enhanced root growth, drought tolerance, and leaf mineral enrichment in Arabidopsis and tobacco. The Plant Cell 22 (12), 3905–3920.

Wintz, H., Fox, T., Wu, Y.Y., Feng, V., Chen, W., Chang, H.S., et al., 2003. Expression profiles of *Arabidopsis thaliana* in mineral deficiencies reveal novel transporters involved in metal homeostasis. Journal of Biological Chemistry 278 (48), 47644–47653.

Xu, J., Yin, H., Li, Y., Liu, X., 2010. Nitric oxide is associated with long-term zinc tolerance in *Solanum nigrum*. Plant Physiology 154 (3), 1319–1334.

Yao, X., Tian, L., Yang, J., Zhao, Y.N., Zhu, Y.X., Dai, X., et al., 2018. Auxin production in diploid microsporocytes is necessary and sufficient for early stages of pollen development. PLoS Genetics 14 (5), e1007397.

Yong, Y., Zhang, Y., Lyu, Y., 2021. Functional characterization of *Lilium lancifolium* cold-responsive Zinc Finger Homeodomain (ZFHD) gene in abscisic acid and osmotic stress tolerance. Peer J 9, e11508.

Yousif, D.Y., 2018. Effects sprayed solution of salicylic acid to prevent of wilt disease caused by *Fussarium oxysporium*May Journal of Physics: Conference Series 012001.

Yuan, H.M., Huang, X., 2016. Inhibition of root meristem growth by cadmium involves nitric oxide-mediated repression of auxin accumulation and signalling in Arabidopsis. Plant, Cell & Environment 39 (1), 120–135.

Yuan, H.M., Xu, H.H., Liu, W.C., Lu, Y.T., 2013. Copper regulates primary root elongation through PIN1-mediated auxin redistribution. Plant and Cell Physiology 54 (5), 766–778.

Zanganeh, R., Jamei, R., Rahmani, F., 2019. Role of salicylic acid and hydrogen sulfide in promoting lead stress tolerance and regulating free amino acid composition in *Zea mays* L. Acta Physiologiae Plantarum 41 (6), 1–9.

Zhabinskii, V.N., Khripach, N.B., Khripach, V.A., 2015. Steroid plant hormones: effects outside plant kingdom. Steroids 97, 87–97.

Zhang, Y., Friml, J., 2020. Auxin guides roots to avoid obstacles during gravitropic growth. The New Phytologist 225 (3), 1049.

Zhang, Y., Lan, H., Shao, Q., Wang, R., Chen, H., Tang, H., et al., 2016. An A20/AN1-type zinc finger protein modulates gibberellins and abscisic acid contents and increases sensitivity to abiotic stress in rice (Oryza sativa). Journal of Experimental Botany 67 (1), 315–326.

Zhang, H., Ni, L., Liu, Y., Wang, Y., Zhang, A., Tan, M., et al., 2012. The C2H2-type Zinc Finger Protein ZFP182 is involved in abscisic acid-induced antioxidant defense in rice F. Journal of Integrative Plant Biology 54 (7), 500–510.

Zhang, P., Sun, L., Qin, J., Wan, J., Wang, R., Li, S., et al., 2018a. cGMP is involved in Zn tolerance through the modulation of auxin redistribution in root tips. Environmental and Experimental Botany 147, 22–30.

Zhang, D., Xu, Z., Cao, S., Chen, K., Li, S., Liu, X., et al., 2018b. An uncanonical CCCH-tandem zinc-finger protein represses secondary wall synthesis and controls mechanical strength in rice. Molecular Plant 11 (1), 163–174.

Zhou, M., Han, R., Ghnaya, T., Lutts, S., 2018. Salinity influences the interactive effects of cadmium and zinc on ethylene and polyamine synthesis in the halophyte plant species *Kosteletzkya pentacarpos*. Chemosphere 209, 892–900.

Chapter 9

# Zinc, nematodes, and plant disease: role and regulation

**Kamini Devi[1], Neerja Sharma[1], Palak Bakshi[1], Mohd Ibrahim[1], Tamanna Bhardwaj[1], Kanika Khanna[1], Nitika Kapoor[1], Bilal Ahmad Mir[2], Amrit Pal Singh[3], Puja Ohri[4], Anket Sharma[5] and Renu Bhardwaj[1]**

*[1]Department of Botanical and Environmental Sciences, Guru Nanak Dev University, Amritsar, Punjab, India, [2]Department of Botany, School of Life Science, University of Kashmir, Satellite campus, Kargil, Jammu and Kashmir, India, [3]Department of Pharmaceutical Sciences, Guru Nanak Dev University, Amritsar, Punjab, India, [4]Department of Zoology, Guru Nanak Dev University, Amritsar, Punjab, India, [5]University of Maryland, College Park, MD, United States*

## 9.1 Introduction

Micronutrients including zinc (Zn), boron (B), copper (Cu), iron (Fe), chlorine (Cl), manganese (Mn), and molybdenum (Mo) play an essential role in enhancing the disease resistance which further contributes to enhancing the plants physiological and chemical mechanisms. Zinc is the most important micronutrient that diminishes the extremity of pathogen infection or any disease. Plants dependent on Zn supply for regulation of auxin, $CO_2$ level maintenance in mesophyll cell, repairing of the PS II, and other metabolic processes (Gupta et al., 2016). Zinc ions also disable the proteins either by binding with their functional sites or by dislocating the ions from the sites, thus affecting the growth system of plants. Therefore, for the proper regulation of plant metabolism, the absorption, transportation, and allocation of Zn is necessary. Also, Zn is necessary to produce starch and proteins and plays an important part in pollen tube formation during pollination. The application of nanoparticles (NPs) reduced the dependence of crop productivity on chemical fertilizers. Zinc NPs are most widely used in the nano-industry for their utilization in cosmetics, medicines, food, etc, and are 100 times more efficient than other nanoparticles. This transfer of Zn in the environment increases the accumulation of Zn nanoparticles in the ecosystem that regulates plant diseases and pathogenic infections. In addition to this, they possess high catalytic activity and environmental diffusion, hence having better

**Zinc in Plants. DOI: https://doi.org/10.1016/B978-0-323-91314-0.00002-8**

interaction with biotic factors (Dale et al., 2015; Sirelkhatim et al., 2015). The nanoparticles are synthesized from the extract of leaves, stems, fruits, roots, and seeds as it is rich in antioxidants that act as stabilizing and reducing agent. Green synthesis of zinc oxide NPs from *Rosa canina* fruit extract reduces and stabilize stress and disease in plants. Plants are used for the synthesis of nanoparticles because of their large-scale manufacturing and stability. Phytochemicals of nanoparticles such as polyphenols, vitamins, terpenoids, alkaloids, polysaccharides, and amino acids reduced the metal ions and converted them into a stable form.

The Zn is essential for the metabolism of plants. It provides nutrition, involved in various redox and enzymatic reactions. The nitrogen metabolism, enzymes responsible for energy transfer, and protein synthesis depend on zinc. It plays an important role in photosynthesis and carbohydrates metabolism. Zn also maintains the integrity of the membrane and provide resistance against pathogen. It also stimulates the activities of various enzymes of plants including ribosome, carbonic anhydrase, hydrogenase that maintain their synthesis and regulation in plant cells. Zn deficiency in plants leads to reduced plant growth, leaf size, and sterility which affects crop quality and production. Plant-sequestered excess Zn in vacuoles via HMA and MTP families' transporters that localized in the membrane of the vacuole (tonoplast) mediate the vacuolar sequestration of Zn. The accumulation of Zn in the shoots was contributed by MTP which is expressed in the whole plant whereas transporter MTP3 exclusively speeds up the hyperaccumulation in roots of plants. mRNA degradation is responsible for the regulation of zinc genes. Zinc is a cofactor of structural and catalytic proteins in several enzymes. Zinc finger protein mediates both the binding of transcription factors and their interactions within proteins. It is noteworthy that Zn plays a critical role during plant defense responses against biotic factors. It has been reported that Zn-deficient plants possess a higher susceptibility toward diseases (Helfenstein et al., 2015).

Various pests/pathogens or fungi unceasingly attack the plant body and make them susceptible to disease. But plants have refined immune systems to defend against these biotic stresses that include systemic acquired resistance (SAR), effector-triggered immunity (ETI), and pathogen-associated molecular pattern (PAMP)-triggered immunity (PTI).

Many researchers have reported the coordination between Zn and disease management caused by biotic factors such as nematodes and herbivore attacks (Khoshgoftarmanesh et al., 2010). A study conducted on several pathogens infecting soybean plants depicted that those plants fertilized with Zn effectively inhibit growth-promoting bacteria to counteract the hazardous pathogens in the surrounding zones. Therefore, Zn-sufficient genotypes are more disease-resistant and reflect positive cooperation with plant growth and development and limited pathogen susceptibility. However, the mechanisms involved behind Zn-mediated resistance are probably their action to stimulate the antioxidant responses while limiting the oxidative damages (Noman et al., 2018). In addition to this, the advanced characteristics of miRNA have been documented under the influence of Zn (Noman et al., 2017). ROS-quenching enzyme Cu/Zn-SOD is regulated at the posttranscriptional level by miRNA and miR398 targets the enzymes present in cytosol and chloroplast.

Researchers have also conducted studies by supplementing Zn or a mixture of micro-nutrients containing Zn and they observed that Zn-organic complexes inhibit nematodes, herbivores, and pathogens directly by their toxic action (Chávez-Dulanto et al., 2018). High Zn levels are toxic to pathogens and their mode of action lies behind the fact that Zn displaces the essential metal from a catalytic site in proteins. Therefore, high Zn concentration affects nematodes on their mucosal surfaces and induces starvation and sensitivity to oxidative stress markers. Also, Zn efflux mechanisms have been reported in plants that enhanced phloem Zn levels as the first defense barrier followed by high exudates in tissues requiring protection (Stolpe et al., 2017).

## 9.2 Role of zinc in plant metabolism

Trace element like zinc has an extensive role in biological systems like high affinity of protein binding in various animal and plant food sources. An adequate amount of Zn was present in legumes, grains, meat, etc. It was required for various metabolic reactions and crop nutrition, and it played a significant role in the catabolic and anabolic properties of the plants. Activation of some enzymes was carried out by Zn were involved in the maintaining of cell integrity, carbohydrate metabolism, protein synthesis, formation of pollen, and auxin synthesis. It also provides resistance and tolerance from environmental stresses by regulating the Zn-mediated gene expression of specific genes. The Zn ions were endogenously bound with ligands (mucins), but the primary pathway was mediated by a carrier protein. Numerous enzymatic functions, such as carbonic anhydrase, hydrogenase, cytochrome synthesis, and ribosomal function maintenance, are influenced by it. Additionally, it enhances the performance of enzymes engaged in protein synthesis, regulation of plant hormones, and preservation of cell membrane integrity. It also stimulated the activities of enzymes involved in protein synthesis, plant hormone regulation, maintenance of cell membrane integrity, etc.

## 9.3 Uptake and transport of zinc in plants

Depending upon zinc uptake and transport from soil to roots and shoots and its sequestration, the plants have different Zn concentrations and requirements in plant parts. Plants take up a divalent form of Zn, i.e., $Zn^{2+}$, and sometimes it may also be absorbed as ligand-zinc complexes by plant roots. There were two approaches to absorb divalent form of Zn depending on the root ligand secretions which include: (1) Zn hydroxides and phosphates solubility increased by efflux of $H^+$ions and organic acids as reductants that lead to a release of Zn divalent ions for epidermal root cells absorption. (2) The outflow of phytosiderophores made stable complexes with divalent Zn ions that fostered the entry of Zn into the epidermal root cells.

The uptake of Zn has passively involved the water molecules across the Root Cell Plasma Membrane (RCPM). The divalent Zn ions influx from hyperpolarized Root Cell Plasma Membrane by the activities of $H^+$ATPase pumps. These pumps governed the extracellular pumping of $H^+$ ions in the presence of ATP hydrolysis and the efflux of $H^+$ ions in the rhizosphere of roots resulting in hyperpolarization of RCPM which further decreased the pH of soil that elevated the uptake of the cations. The divalent Zn cations were transported through specific protein transporters present on the cell membrane.

There was a surplus of potential Zn-binding sites within the cells which have been consequently having an inflated capacity for binding Zn ions. Besides polarization of the cellular plasma membrane also generates a driving force for the uptake of cations, including $Zn^{2+}$ (Krämer, 2018). Hence, a precipitous Zn concentration gradient exists between extracellular and intracellular media, and Zn is taken in passively according to the concentration gradient by transporters of the ZIP family. Free $Zn^{2+}$ is the main plant-available form of Zn in soil solution, and ZIP transporters intercede Zn uptake from the soil solution into rhizodermal cells. In the cells of aerial parts, ZIP transporters perform Zn loading from extracellular fluids. The ZIP1 transporter is seemingly tonoplast-localized and involved in the remobilization of vacuolar Zn reserves, in case of Zn deficiency.

In the regulation of *ZIP* gene expression, the Arabidopsis transcription factors bZIP19 and bZIP23 play a key role. Through the conserved Cys/His-rich Zn sensor motif (Lilay et al., 2021) in both bZIP19 and bZIP23 proteins show direct binding to $Zn^{2+}$ ions. It was assumed that these proteins directly sense the cytosolic free $Zn^{2+}$ concentration by binding/losing Zn under conditions of Zn sufficiency/deficiency; the Zn-free form is thought to form homodimers. In the promoters of Zn-responsive genes, known as zinc deficiency response elements (ZDREs) substantially interrelate with specific sequences and, therefore, initiate the transcription of these genes and primarily regulate within-cell reaction to Zn deficiency. Plant Zn status activates bZIP19 and bZIP23 factors. Both bZIP19 and bZIP23 in supplement to the *ZIP* genes also control the activity of a small number of other genes of the primary Zn deficit reaction (Castro et al., 2017), like *NAS* genes, which encrypt enzymes engaged in the production of nicotianamine, that plays a major role in cytosolic Zn handling and Zn transport (see Clemens, 2019). With the depletion of intracellular Zn, *ZIP* and *NAS* expression is activated rapidly, and the level of this stimulation depends on the range of Zn depletion (Zlobin et al., 2020). bZIP19 and bZIP23 are functionally superfluous to a large extent, bZIP19 or bZIP23 regulates some *ZIP* members and by this means they react otherwise to changes in cellular Zn status (Inaba et al., 2015; Lilay et al., 2019). For other plant species, a comparable monitoring system for primary Zn homeostasis genes has been depicted.

The plants showed a reaction to a surplus of Zn in hyperaccumulator plants. A high level of Zn was sequestered in the vacuole to enhance tolerance and provide accumulation of Zn in the vacuole as a vital nutrient. Both the HMA and MTP members of transporters families were involved in the sequestration of Zn in a vacuole. MTP1 and MTP3 transporters specified in the vacuolar membrane (tonoplast) and mediated Zn sequestration in various types of root cells, thus providing Zn tolerance, and MTP1 also contributes to hyperaccumulation of Zn in Zn-hyperaccumulators. *MTP1* gene was expressed all over the plant, mostly in reproductive parts although *MTP3* was completely expressed in roots. MTP1 transporter was involved in the transportation of Zn directly structured by cytosolic availability of Zn ions. This transporter (MTP1) has a histidine-rich (His-rich) loop on the side of the protein cytoplasmic matrix (Tanaka et al., 2013). As soon as the level of $Zn^{2+}$ were elevated, it was bound to the histidine-rich loop, and its conformation will change. The enabling of $Zn^{2+}$ to the lumen of the vacuole through transmembrane removes the conformational restraints (Kawachi et al., 2012).

## 9.4 Role of zinc in plant under biotic stress

Zinc, the most common metal in living organisms, is associated with several proteins and enzymes in plants. As a self-defense mechanism, plants absorb an ample amount of Zn from the substrate which plays a crucial role in response to pests, pathogens, herbivores, and diseases (Stolpe et al., 2017). Although the mechanism of zinc in plant defense varies greatly as it is reported to be an important player against plant pests and diseases. The interaction between plants and pests differs with zinc response by either avoiding the invader's attack or invader's ability to bypass the defense mechanism. Plant fortification or signaling pathways are the two possible mechanisms demonstrated by several studies that activate the defense reaction by metal ions (Helfenstein et al., 2015). Decrease in severity of several diseases caused by different pathogens such as Phytophthora root rot (*Phytophthora megasperma*), early blight (*Alternaria solani*) in potato (*Solanum tuberosum*), leaf spot (*Pseudopeziza medicaginis*) in alfalfa (*Medicago sativa L.*), and peach gummosis (*Botryosphaeria dothidia*) in peach (*Prunus persica*) have been reported through Zn fertilization (Machado et al., 2018). Upregulation of zinc-binding alcohol dehydrogenase (AD) and cinnamyl AD (CAD) in pathogen-inoculated plants was reported by Kumar et al. (2016)

and Shivashankar et al. (2015). A plant containing Zn finger (Znf) family protein was found resistant to pests or pathogen infection and thus demonstrated the role of zinc in plants against pathogen attack (Wang et al., 2017). Although many studies have been conducted, the knowledge still lacks how exactly Zn affects susceptibility and resistance against pathogens. At a certain level, Zn was also found to induce susceptibility to other pathogens. It is important for organisms to tightly regulate the concentration of Zn as its excess or deficient conditions make it prooxidant (Kinraide et al., 2011). Oxidative stress and regulation of zinc finger protein are the two broad-spectrum responses involved in plant pests or pathogen interaction and are tightly regulated by zinc ions.

Zinc appears to play a key role in plant immune responses (Gupta et al., 2012). It triggers or stabilizes the metalloenzymes which affect the plant-pathogen connections (Fones and Preston, 2012). Zinc finger proteins (proteins that contain a small freely folded functional domain having one or more zinc ions to maintain its structure) regulate plant responses to biotic stress conditions (Noman et al., 2019).

Plant resistance proteins NBS-LRRs (nucleotide-binding sites-leucine rich) which contain zinc finger proteins are involved in the effector-triggered immune response (Gupta et al., 2012). It was reported that about 70 plant disease-resistance proteins were isolated from different crops and of these, 37% proteins were comprised of zinc finger binding domains. The chief role of this class of proteins was the host's resistance to pests/pathogens. Pi54 containing zinc finger domain was an R-gene that has resistance against the fungus *Magnaporthe oryzae*. Upregulation of pest/pathogen resistant genes like callose, Phenylalanine ammonia lyase (PAL), laccase, and peroxidase and some transcription factors (MAD box, bZIP, NAC6, and WRKY) was analyzed in the Pi54 gene of the transgenic line of rice. Another zinc finger protein, RAR1, possesses a defensive capacity against stripe rust pathogen of wheat crop by salicylic acid generated oxidative burst and hypersensitive response (Wang et al., 2017). $C_2H_2$ zinc finger domain contains some transcription factors responsive to herbivory in Solanaceae (Lawrence et al., 2014). Two zinc finger transcription factors StZFP1 and StZFP2 were upregulated upon insect infestation in *Solanum tuberosum* L. CaZFP1 additional $C_2H_2$ zinc finger transcription factor had been identified from *Capsicum annum* which increased resistance against infection by *Pseudomonas syringae* pv. tomato when revealed in *Arabidopsis* (Kim et al., 2004).

One transcription factor AtZAT6 belongs to a member of the family of transcription regulators cysteine2/histidine2-type zinc finger proteins (ZAT) contribute to the plant growth and positively control the response of *Arabidopsis* to biotic stress by triggering the action of salicylic acid-related genes such as PATHOGENESIS RELATED GENE 1,2 and 5 (*PR1*, *PR2*, and *PR5*) (Miller et al., 2008).

VOZs (i.e., Vascular Plant One-Zinc-Finger Proteins) are highly conserved one-zinc-finger type DNA-binding proteins in plants. Loss-of-function mutation in VOZs severely impaired the plant immune responses (Nakai et al., 2013). Nowadays, ZnO nanoparticles are used to treat various plant diseases. Table 9.1 shows the effect of ZnO nanoparticles on various plants under biotic stress.

### 9.4.1 Mechanism of regulation of zinc in diseased plants

Foliar spray of zinc oxide nanoparticles (ZnO NPs)'s helps to ameliorate the disease stress in beetroot plants infected with *Meloidogyne incognita, Pectobacterium betavasculorum*, and *Rhizoctonia solani*. ZnO NPs in concentrations of 100 mg/L and 200 mg/L were observed to inhibit the hatching and cause mortality of second-stage juveniles of *Meloidogyne incognita* and prevent the growth of *Pectobacterium betavasculorum*, and *Rhizoctonia solani*. Further ZnO NPs spray improved the plant growth by stimulating various physiological parameters (chlorophyll content and chlorophyll fluorescence) and defense-related enzymes *viz.* superoxide dismutase, catalase, phenylalanine ammonia-lyase, and polyphenol oxidase, etc. Both methods of application of ZnO NPs i.e., seed priming and foliar spray, caused the maximum reduction in disease indices and nematode population (Khan and Siddiqui, 2021).

ZnO NPs target the intestine, hypodermis, and cuticle of nematodes by affecting lipid, glycogen, and mucopolysaccharides. Dissolved $Zn^{2+}$ ions are also toxic to nematodes, and they may enhance the production of ROS which further reduces the vitality of nematodes. (Khan and Siddiqui, 2018; Gupta et al., 2015; Sávoly et al., 2016; Hou et al., 2018). ZnO NPs not only affect the nematodes that attack plants but also affect the growth and yield of crops. The application of NPs improves the growth of plants by maintaining the structural stability of cell membranes, regulating the level of auxins, promoting protein synthesis cell elongation, and providing tolerance against various stressful conditions (Cakmak 2000; Tirani et al., 2019; Siddiqui et al., 2018; Siddiqui et al., 2019a,b; Ghanepour et al., 2015). Elemental Zn being an essential micronutrient improves the nutritional status of plants and induces pathogen suppression by improving the rate of photosynthesis and the activities of antioxidative enzymes in various plant species (Faizan et al., 2018; Rizwan et al., 2019; Siddiqui et al., 2019a) (Fig. 9.1).

Recently, El-Ansary et al., (2021) reported the nematocidal properties of ZnO-NPs, ZnO-bulk, and zinc acetate in combination with oxamyl against root-knot nematode in banana plants. In this study it was observed that ZnO-NPs biosynthesized from the alga, *Ulva fasciata* was more effective to control root nematodes as compared to ZnO-bulk and oxamyl alone. Under *in-vitro* conditions, ZnO-NPs with oxamyl showed 98.91% second-stage juveniles2 (J2s) mortality of *Meloidogyne incognita* after 72 hours, while 72.86% mortality was observed without oxamyl at the same exposure time. On the other hand,

**TABLE 9.1** Effect of ZnO nanoparticles on various plants under biotic stress.

| Sr. No. | ZnO NPs as nano fertilizers | Pathogen/pest | | Effect on plant species | References |
|---|---|---|---|---|---|
| 1. | Superabsorbent hydrogel based on zinc oxide nanoparticles (ZnO-NPs) and watermelon peel waste | Fungus | *Fusarium oxysporum* | Control of *Fusarium* wilt disease reduced irrigation water for pepper plants to about one-third | Abdelaziz et al. (2021) |
| 2. | 0.20/mL ZnO NPs | Bacteria and fungi | *Pseudomonas syringae, Xanthomonas campestris, Pectobacterium carotovorum*, and *Ralstonia solanacearum* and fungal spp. are *Fusarium oxysporum* Alternaria *solani* | Increased growth, photosynthetic pigments, and proline content in tomato plants | Parveen, and Siddiqui (2021) |
| 3. | 0.25 and 0.50 mL/L $TiO_2$ and ZnO NPs | Bacteria | *X. campestris* pv. *beticola, P. betavasculorum*, and *P. syringae* pv. *aptata* | Enhanced growth, photosynthetic pigments, antioxidant enzymes, and reduced disease indices in beetroot | Siddiqui et al. (2019a,b) |
| 4. | 100 and 200 mg/L ZnO NPs | Bacteria | *Pectobacterium betavasculorum, Meloidogyne incognita*, and *Rhizoctonia solani* | Foliar spray improved dry mass, physiological and biochemical attributes in beetroot | Khan, and Siddiqui (2021) |
| 5. | 100 mg/L ZnO-NPs | Virus | Tomato mosaic virus (ToMV) | enhanced growth indices, photosynthetic attributes, and enzymatic and nonenzymatic antioxidants in *Solanum lycopersicum* L. | Sofy et al. (2021) |

| | | | | | |
|---|---|---|---|---|---|
| 6. | Graphene oxide (GO) and zinc oxide (ZnO) nanoparticles (NPs) (0.05 mg/mL and 0.10 mg/mL) | Bacteria | *Pectobacterium carotovorum, Xanthomonas campestris* pv. *carotae, Meloidogyne javanica, Alternaria dauci*, and *Fusarium sola* | Increased plant growth, pigment contents, and proline contents, reduction in galling and nematode multiplication in *Daucus carota* | Siddiqui et al. (2019a,b) |
| 7. | *Rhizobium leguminosarum* and 0.10/mL ZnO NPs | Nematode and bacteria | *Meloidogyne incognita* and *Pseudomonas syringae* pv. *pisi.* | Reduced blight disease indices, galling and nematode population in pea | Kashyap, and Siddiqui (2021) |
| 8. | ZnO NPs | Bacterium, fungus and nematode | *Ralstonia solanacearum, Phomopsis vexans*, and *Meloidogyne incognita* | Increase in plant growth parameters, chlorophyll, and carotenoid contents, greater reduction in galling and nematode population in eggplants | Khan and Siddiqui (2018) |
| 9. | 0.1 mg/ml of ZnO NPs | Bacteria and nematode | *Alternaria alternata, Fusarium oxysporum* f. sp. lentis, *Xanthomonas axonopodis* pv. *phaseoli, Pseudomonas syringae* pv. *Syringae*, and *Meloidogyne incognita.* | Increase in growth, number of pods per plant, chlorophyll, carotenoid contents, and nitrate reductase activity, reduced galling, nematode multiplication, wilt, blight, and leaf spot disease severity indices in Lens culinaris (Lentil) | Siddiqui et al. (2018) |
| 10. | *Vitex agnus-castus* essential oils and foliar spray with ZnO NPs (50 ppm) | Fungus | *Rhizoctonia solani* | Enzymatic and nonenzymatic antioxidants total flavonoids, total phenolics, tannins, proline, glycine betaine, ferric reducing antioxidant power, total antioxidant capacity, and mineral composition were increased in *Phaseolus vulgaris* (bean). | Abu-Tahon et al. (2022) |

ZnO-bulk treatment was found to be responsible for the improvement in biomass of the diseased plant. Scanning electron microscopy reported that ZnO-NPs distributed and accumulated in the body of nematodes (J2s) result in NP-mediated toxicity for nematodes.

Various disrupting multiple cellular mechanisms viz. cell wall damage due to zinc oxide-localized interaction; improved membrane permeability; ATP synthesis and response to oxidative stress in eukaryotic cells and prokaryotic cells because of mitochondria weakness and intracellular outflow; and inhibition of cell growth and ultimately cell death are major possible mode of actions of Zn against nematodes in diseased plants (Lim et al., 2012; Jeevanandam et al., 2018; Liu et al., 2020).

## 9.5 Role of zinc oxide nanoparticles

Zinc oxide nanoparticles (ZnO NPs) are annually produced between 550 and 33,400 tons globally, making them the third most used NPs (Peng et al., 2017). This is so because zinc oxide nanoparticles are considered bio-safe materials for various organisms. Many reports are stating the positive effects of ZnO NPs on plant growth and metabolism. Laurenti et al. (2015) stated that ZnO NPs usage is advantageous owing to its exclusive properties such as low energy band gap and piezoelectric effect. As well as its antifungal and antibacterial properties, high photochemical and high catalytic activities make it a valuable nanoparticle. Hence, it allows its phenomenal applications in multiple ways. Many studies have been reported stating the use of ZnO-NPs for exhilarating seed germination, plant growth, and suppressing plant diseases due to its antimicrobial activity. Singh et al. (2018) studied responses to the impact of ZnO NPs on crop plants. Application of ZnO-NPs improved enzymatic activity in plants on exposure to the pathogenic virus (Sofy et al., 2020).

## 9.6 Properties of ZnO NPs

ZnO NPs have special properties that make them undergo "quantum size effects." This is so because the dimensions of semiconductor materials lessen with time, making them reduced to a few nanometers only (Faizan et al., 2020). ZnO-NPs are antimicrobial in nature as they promote oxidative stress. ZnO-NPs affect the properties of the cell membrane and lead to ROS formation. $Zn^{2+}$ ions released from ZnO, interact with thiol groups of respiratory enzymes making them turn nonfunctional (Dwivedi et al., 2014). ZnO-NPs further result in increasing fluidity, protein leakage, and deforms cellular architecture. This ultimately leads to ROS production and permanent DNA damage. Sarwar et al. (2016) also explained the combined impact of antibiotics and ZnO-NPs in mitigating pathogenic bacterial agents. ZnO is the n-type semiconducting metal oxide (Pragati et al., 2018). Its semiconducting

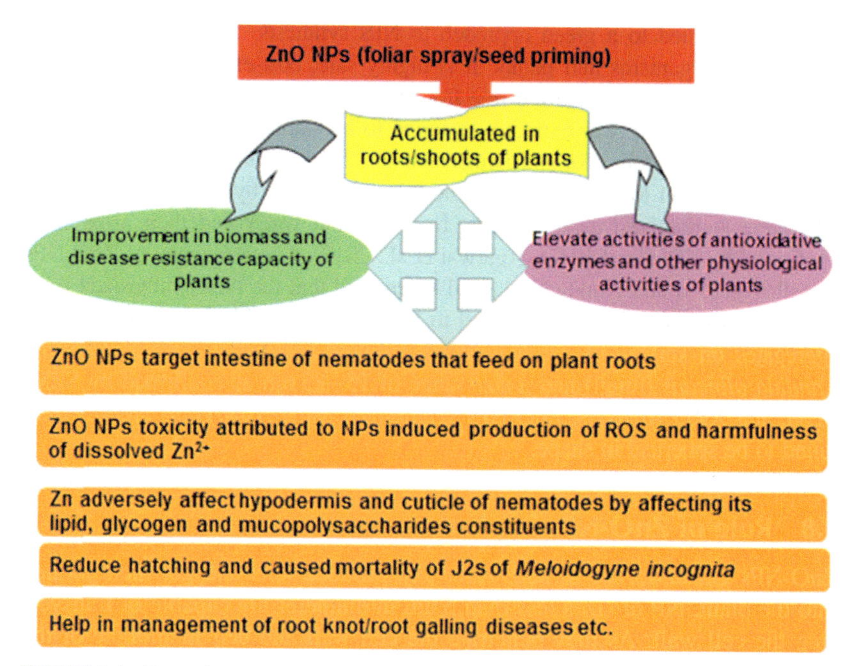


**FIGURE 9.1** Protective mode of action of ZnO NPs in nematode-stressed plants.

properties are due to a large band gap (3.37 eV) and high exciton binding energy (Stan et al., 2015). The metal oxide is enlisted under GRAS (i.e., generally recognized as safe) metal oxide list by the US FDA (Pulit-Prociak et al., 2016). ZnO NPs have piezoelectric and pyroelectric properties (Nagajyothi, et al., 2014). They have been reported to be present in different morphologies such as nanoflake, nanoflower, nanobelt, nanorod, and nanowire (Yuvakkumar et al., 2014).

## 9.7 Different methods used in nanoparticle synthesis

Nanoparticle synthesis can be achieved by physical, chemical, and green methods (Afifi et al., 2015). Physical methods comprise such methods that bring about the attraction of nanoscale particles and the formation of large, stable, well-defined nanostructures. Techniques under this type include vapor condensation, amorphous crystallization, and physical fragmentation (Elumalai et al., 2015). Physical method is expensive as it requires heavy machinery and space area for setting up of machines (Chandrasekaran et al., 2016). The second method is a chemical method that employs the use of toxic chemicals. This is not eco-friendly as it results in outlet of toxic chemicals that are detrimental to the environment. It includes chemical microemulsion, wet chemical, and spray pyrolysis (Mitra et al., 2015). The disadvantages of these methods led to a green approach to NPs synthesis. This approach makes use of microorganisms and plants. It is highly safe, cost-effective, and biocompatible. ZnO NPs so formed are free from impurities (Yuvakkumar et al., 2014). *Trifolium pratense* flower extract was found to be stable and capable of synthesizing ZnO NPs (Dobrucka and Długaszewska, 2016). Similarly, extract of *Rosa canina* was found to be adept at synthesizing ZnO NPs by FTIR studies. ZnO-NPs were synthesized using extracts from coconut water and their sizes were confirmed by TEM and XRD (Agarwal et al., 2017). The efficiency of ZnO-NPs was also studied and Kundu et al. (2014) reported that the NPs synthesized from *Pseudomonas aeruginosa* were found to be tough and stable in forming micelle aggregates on surface of carboxymethyl cellulose. ZnO-NPs synthesized using *Candida albicans* (fungi) had a size range of 15–25 nm as confirmed by SEM, TEM, and XRD Analysis. ZnO NPs synthesized using *Aspergillus* species were found to be spherical in shape.

## 9.8 Role of ZnO-NPs in plants

ZnO-NPs get dissolved in soil water and produce ions and get incorporated into the plant. NP of sizes less than 5 nm in diameter are capable of transversing the cell wall. As they penetrate the cell wall and cell membrane of the epidermis, they enter the vascular bundle (xylem) and finally stele. This finally results in the distribution and translocation of NPs to leaves. The mechanism behind this process is highly active. ZnO-NPs movement in tissues involves both apoplastic and symplastic routes. Apoplastic transport occurs via plasma membrane and through the extracellular spaces while symplastic transport involves movement between the cytoplasm of adjacent cells through plasmodesmata. Transpiration pull also aids and supports in distribution of ZnO-NPs in the above soil-plant biomass (Sun et al., 2014) (Fig. 9.2).

ZnO-NPs have been reported to foster plant growth and aid in its development. It was documented by Prasad et al. (2012) that treatment of 1000 mg/kg of ZnO NPs led to enhanced germination rate and improved seedling vigor. ZnO-NPs improved growth attributes in cucumber and green peas (Zhao et al., 2013, Mukherjee et al., 2014). Increased root growth and dry weight in onions were noticed after the application of ZnO-NP in onion (Raskar and Laware, 2014). ZnO-NPs imparted a beneficial effect on photosynthetic pigment levels in pearl millet (Tarafdar et al., 2014). ZnO-NPs play a critical role in the protection of plants against various abiotic and biotic stresses. It acts by regulating levels of osmolytes, free amino acids, and nutrients (Wang et al., 2018). According to Taran et al. (2017), foliar application of ZnO-NPs improved yield in wheat by reducing the adverse effects of drought stress. Further, Venkatachalam et al. (2017) stated that Cd and Pb toxicities were alleviated by ZnO-NPs in *Leucaena leucocephala* seedlings. The impact of Cd metal on plants and how ZnO-NPs protected the plants by modulating their redox status. Foliar application of ZnO-NPs was

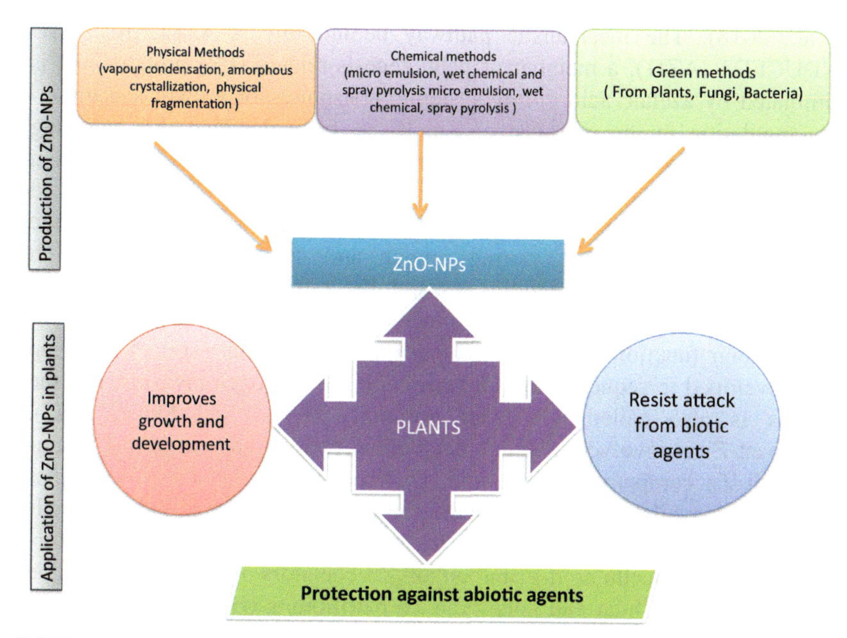


**FIGURE 9.2** Modes of production and role of ZnO-NPs in plants.

also found to be critical in alleviating the effects of salt stress on *Moringa peregrina* plants (Soliman et al., 2015) and sunflowers (Torabian et al., 2018). ZnO-NPs also aid in restoring redox balance to withstand attacks from biotic agencies. Hence, it was reported by Sofy et al. (2021) that foliar spray with ZnO-NPs can mitigate the adverse effects caused by Tomato mosaic tobamovirus (ToMV) infection. Khan and Siddiqui 2021 reported that foliar spray of ZnO NPs was more effective than seed priming in the management of root-knot, soft-rot, and root-rot disease complex of beetroot. In addition to this, ZnO NPs led to the improvement of plant dry mass and physiological and biochemical attributes. It caused a reduction in disease indices, multiplication, and the number of galls of *M. incognita*.

## 9.9 Molecular mechanism of regulation of zinc in nematode-stressed plants

Few mechanisms encompassing Zn and disease susceptibility are being evolved. Many plants' innate immune responses towards nematodes are co-linked with the synthesis of salicylic acid, ROS generation, and SA-dependent systemic resistance (Breitenbach et al., 2014). Various compounds that are involved in systemic acquired resistance namely, dicarboxylic acid and azelaic acid are present and they function during the Zn-sensing mechanism that connects the plant roots with the pathogenic responses (Bouain et al., 2018). The mechanistic pathway behind this is AZELAIC ACID INDUCED1 (AZI1), a member of pathogenesis-related (PR) proteins that is stimulated by azelaic acid during systemic resistance for growth and immunity regulation in the presence of Zn. It has been depicted that signaling upregulated the Zn and azelaic acid interactions. Lower Zn affects expression patterns of defense genes negatively while *PR1*, SA-regulated gene enhanced the expression. Strikingly, extensive studies have been highlighted in the context of plant signaling during growth and defense. Further research should be conducted to extend this research by considering Zn regulation.

Additionally, the Zn along with other micro-nutrients in pathogens have devised their functions in such a manner that their immune system has been well developed to sequestration for restricting the pathogen invasion, and the process is also called nutritional immunity (Hood and Skaar, 2012). Moreover, Zn is involved in many metabolic processes and nutritional immunity with Zn competition is feasible. Different metal chelators also impair the metal bioavailability towards pathogens. For instance, zinc induced facilitator, *ZIF1* encoding nicotianamine transport *via* tonoplast reflected disturbance in the subcellular distribution of Zn-chelators where nicotianamine functions in subcellular and interorgan Zn partitioning (Haydon et al., 2012). Furthermore, studies showed that in the biofortification of crops, Zn has been of paramount importance and contributes to ameliorating biotic stresses along with conferring plant resistance towards pathogens. Along with this when nutrient deficiencies tend to occur, the plants become more prone to pathogens. Besides, few signaling pathways are boosted under a scarcity of nutrients in order to enhance immunity. Such defense-activating pathways also modulate gene expression profiles and metabolite synthesis during nutrient deprivation along with triggering the plant defense responses (Rodríguez-Celma et al., 2016). Defensins, on the other hand, also induce innate responses and the induced expression levels of defensin-like-proteins in plants with Zn-deficient conditions have been reported (Inaba et al., 2015). Such proteins are regulated by basic region leucine-zipper transcripts *bZIP19*. It is chiefly responsible for Zn-depletion responses by upregulating the expression patterns of ZIP transporters (Assunção et al., 2010). It has been observed that *ZIP* transcription factors are active during biotic and abiotic stresses, and they are essential in maintaining plant immunity through the regulation of genes linked to PAMP-mediated immunity, effector immunity, and hormonal signaling (Noman et al., 2017). Apart from growth and development, the functional characterization of *ZIP* transcripts should be further explored in crops resistance against pathogens (Inaba et al., 2015).

In comparison to various other micronutrients, Zn has been well-known divalent cation that possesses various properties of not undergoing valency change thereby, having no redox state. Zn is regarded as a multifunctional cation as a part of numerous proteins and components of various enzymes namely, alcohol dehydrogenases, carbonic anhydrases, and superoxide dismutase. They regulate protein conformation near domains that are linked to DNA. This implies that Zn deprivation interferes with growth, metabolism, protein synthesis, and transcription under stressed conditions (Barker and Pilbeam, 2015). Zn-deficient plants contain very low superoxide dismutase, thereby, enhancing the superoxide radicals while inducing lipid peroxidation and disrupting membrane stability (Barker and Pilbeam, 2015). Additionally, the accrual of amino acids and amines takes place due to suppressed activity of protein synthase due to Zn, therefore the quantities of amino acids are substantially boosted in root exudates (Cakmak and Marschner, 1988). Nematodes have been found to be engrossed by exudates, the more the exudations in the Zn-deficient plants, the more will be these nematodes attracted in order to rapid the infection cycle (Streeter et al., 2001). The probable mechanism behind Zn-mediated resistance towards nematodes reflects that they induce the microbial activity of useful microbes encompassing biocontrol activities (Khanna et al., 2019a,b; Sharma et al., 2020; Khanna et al., 2021a,b). Moreover, they also stimulate the defensive genes encoding resistance mechanisms in plants. For instance, Zn-solubilising bacteria *Pseudomonas aeruginosa* and *Burkholderia gladioli* induced the defense genes in plants encoding phenylalanine ammonia-lyase and chalcone synthase (Khanna et al., 2019a,b). They also modulated the gene expression of genes encoding various organic acids such as malic acid, organic acid, fumaric acid, and citric acid, respectively. Henceforth, they possess antagonistic activity against nematodes due to Zn, making it a most appropriate micro-nutrient to reduce nematode infection, penetration, density, gall formation, egg hatching, etc. with utmost efficacy. It was further confirmed by de Melo Santana-Gomes et al. (2013) that the absence of application of rhizobacteria or Zn leads to a reduction in physiological, biochemical, and functional attributes of plants. The nematicidal action of Zn is mainly due to inhibition of polypeptide biosynthesis, alteration in the microbial communities within the rhizosphere, and plant physiology that mainly confers nematode resistance. According to their studies, mineral amendment of Zn inhibited the incidence of nematode infection by actively synthesizing nematicidal compounds in plants that suppressed the *Meloidogyne incognita* infection. They also reversed the adverse effects of nematodes upon root development with the enhancement of plant growth.

To understand the Zn regulation and biology of nematodes, *Caenorhabditis elegans* is a well-known model organism that possesses many advantages in terms of methods of manipulating and engineering Zn in this nematode. *C. elegans* have a minute size and life cycle, and easier cultivation practice, thereby imparting a huge potential for genetic analysis for elucidating Zn-regulation. Its genome encodes Zn transporters where their expression and function can be effectively characterized. Moreover, homeostatic mechanistic pathways have also evolved in order to understand the Zn homeostasis under limiting as well excessive conditions. The master regulator HIZR-1 is prominently involved in maintaining homeostasis. Additionally, low zinc activation promoters also function in regulatory processes. Zn has also emerged to actively play a role in cell fate determination events along with different secondary messenger signaling mechanisms in the nematodes (Earley et al., 2020).

Nematodes very effectively respond towards Zn in soil, therefore using the advanced transgenic and fluorescent approaches with *C. elegans* the studies could elope further. Like in the case of transgenic *hsp-16*-GFP-*lacZ* Zn-regulation was found to be upregulated. The fluorescent intensity of *gfp* expression also depicted fold change and these fold changes in the progeny are directly linked to Zn toxicity. Additionally, the tissue-specific *gfp* expression levels were assessed, and *hsp-16.2* gene was found to be ubiquitously expressed in nematodes. Different parameters such as LC50 endpoints, galactosidases activity, and DNA/RNA ratio are used for examining Zn exposure. The mechanism behind Zn toxicity towards nematodes lies behind the fact that nematode skin is sensitive toward Zn which induces many biological toxicities whilst, Zn when applied in liquid form is far more lethal as they drink it directly to hinder multiple pathways.

Higher quantities of Zn act toxic toward nematodes in context to undergo biological alterations that can also be transferred into their progenies. One such study conducted in *Caenorhabditis elegans* depicted multifarious toxicities of Zn in terms of life span, movement, locomotion, developmental processes, fecundity, behavior, and chemotaxis respectively. Zn causes stress within nematode tissues followed by its transfer toward progeny. According to their study, the analysis of *hsp 16–2-gfp* expression levels in embryos was found to be enhanced depicting their toxic nature towards nematode eggs. Moreover, these defects were classified based on their transferable nature. These studies also suggest that fold change of *hsp 16–2-gfp* is co-linked with Zn toxicity towards nematode eggs and their deposition. Alongside this, the evaluation of trans-generational Zn effects on nematode development also provides valuable information about maternal effects. Henceforth, the transferable activities of Zn-exposed nematodes are probably due to their deposition onto eggs.

## 9.10 Conclusion and future perspective

The mechanisms of zinc uptake and its homeostasis are not limited. The acquisition of Zn is influenced by the secretions of root exudate and root developmental system resulting in the production of bacteria (zinc-solubilizers) in the root area. The zinc provides susceptibility to plants against pathogens and other abiotic stresses. Zinc oxide nanoparticles are being used nowadays for a quantitative understanding of the regulation of expression of Zn genes in plants. The mechanism of action of ZnO nanoparticles was not specific for different host cells. The foliar spray of Zn oxide nanoparticles helps in managing plant pests/pathogens and diseases. ZnO-NPs application scavenged ROS and led to the accumulation of phenolic compounds, ascorbic acid content, and proline production in plants. Zn exposure has been observed to cause diversified defects in *C. elegans* in terms of dwindling life span, lowered generation time, locomotion, morphological defects, chemotaxis, and plasticity.

The regulation of Zn and zinc oxide (ZnO) NPs and their homeostasis will become attractive research areas. The quantitative aspect of the Zn homeostasis regulation both at the proteomics and genomics levels needs to be understood. Proper methods will be developed for measuring and analyzing the concentration of Zn ions in a plant cell. Zinc homeostasis, remobilization, and redistribution in plant cells may be an experimental approach to investigate the changes in the Zn concentration in various organs of the plant.

## References


Abdelaziz, A.M., Dacrory, S., Hashem, A.H., Attia, M.S., Hasanin, M., Fouda, H.M., et al., 2021. Protective role of zinc oxide nanoparticles based hydrogel against wilt disease of pepper plant. Biocatalysis and Agricultural Biotechnology 35, 102083.

Abu-Tahon, M.A., Mogazy, A.M., Isaac, G.S., 2022. Resistance assessment and enzymatic responses of common bean (Phaseolus vulgaris L) against Rhizoctonia solani damping-off in response to seed presoaking in Vitex agnus-castus L. oils and foliar spray with zinc oxide nanoparticles. South African Journal of Botany 146, 77–89.

Afifi, M., Almaghrabi, O.A., Kadasa, N.M., 2015. Ameliorative effect of zinc oxide nanoparticles on antioxidants and sperm characteristics in streptozotocin-induced diabetic rat testes. BioMed Research International .

Agarwal, H., Kumar, S.V., Rajeshkumar, S., 2017. A review on green synthesis of zinc oxide nanoparticles—an eco-friendly approach. Resource-Efficient Technologies 3, 406–413.

Assunção, A.G., Herrero, E., Lin, Y.F., Huettel, B., Talukdar, S., Smaczniak, C., et al., 2010. Arabidopsis thaliana transcription factors bZIP19 and bZIP23 regulate the adaptation to zinc deficiency. Proceedings of the National Academy of Sciences 107, 10296–10301.

Barker, A.V., Pilbeam, D.J. (Eds.), 2015. Handbook of Plant Nutrition. CRC press.

Bouain, N., Satbhai, S.B., Korte, A., Saenchai, C., Desbrosses, G., Berthomieu, P., et al., 2018. Natural allelic variation of the AZI1 gene controls root growth under zinc-limiting condition. PLoS Genetics 14, 1007304.

Breitenbach, H.H., Wenig, M., Wittek, F., Jordá, L., Maldonado-Alconada, A.M., Sarioglu, H., et al., 2014. Contrasting roles of the apoplastic aspartyl protease apoplastic, enhanced disease susceptibility1-dependent1 and legume lectin-like protein1 in Arabidopsis systemic acquired resistance. Plant Physiology 165, 791–809.

Cakmak, I., 2000. Tansley Review No. 111: possible roles of zinc in protecting plant cells from damage by reactive oxygen species. The New Phytologist 146, 185–205.

Cakmak, I., Marschner, H., 1988. Enhanced superoxide radical production in roots of zinc-deficient plants. Journal of Experimental Botany 39, 1449–1460.

Castro, P.H., Lilay, G.H., Muñoz-Mérida, A., Schjoerring, J.K., Azevedo, H., Assunção, A.G., 2017. Phylogenetic analysis of F-bZIP transcription factors indicates conservation of the zinc deficiency response across land plants. Scientific Reports 7, 1–14.

Chandrasekaran, R., Gnanasekar, S., Seetharaman, P., Keppanan, R., Arockiaswamy, W., Sivaperumal, S., 2016. Formulation of *Carica papaya* latex-functionalized silver

nanoparticles for its improved antibacterial and anticancer applications. Journal of Molecular Liquids. 219, 232–238.
Chávez-Dulanto, P.N., Rey, B., Ubillús, C., Rázuri, V., Bazán, R., Sarmiento, J., 2018. Foliar application of macro-and micronutrients for pest-mites control in citrus crops. Food Energy Security 7, 00132.
Clemens, S., 2019. Metal ligands in micronutrient acquisition and homeostasis. Plant, Cell and Environment 42, 2902–2912.
Dale, A.L., Lowry, G.V., Casman, E.A., 2015. Stream dynamics and chemical transformations control the environmental fate of silver and zinc oxide nanoparticles in a watershed-scale model. Environmental Science & Technology Letters 49, 7285–7293.
de Melo Santana-Gomes, S., Dias-Arieira, C.R., Roldi, M., Santo Dadazio, T., Marini, P.M., de Oliveira Barizatilde, D.A., 2013. Mineral nutrition in the control of nematodes. African Journal of Agricultural Research 8, 2413–2420.
Dobrucka, R., Długaszewska, J., 2016. Biosynthesis and antibacterial activity of ZnO nanoparticles using *Trifolium pratense* flower extract. Saudi Journal of Biological Sciences 23, 517–523.
Dwivedi, S., Wahab, R., Khan, F., Mishra, Y.K., Musarrat, J., Al-Khedhairy, A.A., 2014. Reactive oxygen species mediated bacterial biofilm inhibition via zinc oxide nanoparticles and their statistical determination. PloS one 9 (11)e111289.
Earley, B.J., Mendoza, A.D., Tan, C.H., Kornfeld, K., 2020. Zinc homeostasis and signaling in the roundworm *C. elegans*. Biochimica et Biophysica Acta (BBA)-Molecular Cell Research 118882.
El-Ansary, M.S.M., Hamouda, R.A., Elshamy, M.M., 2021. Using biosynthesized zinc oxide nanoparticles as a pesticide to alleviate the toxicity on banana infested with parasitic-nematode. Waste Biomass Valorization 1, 11.
Elumalai, K., Velmurugan, S., Ravi, S., Kathiravan, V., Ashokkumar, S., 2015. Retracted: green synthesis of zinc oxide nanoparticles using *Moringa oleifera* leaf extract and evaluation of its antimicrobial activity.
Faizan, M., Faraz, A., Yusuf, M., Khan, S.T., Hayat, S., 2018. Zinc oxide nanoparticle-mediated changes in photosynthetic efficiency and antioxidant system of tomato plants. Photosynthetica 56, 678–686.
Faizan, M., Hayat, S., Pichtel, J., 2020. Effects of zinc oxide nanoparticles on crop plants: a perspective analysis, Sustainable Agriculture Reviews, 41. Springer, Cham, pp. 83–99.
Fones, H., Preston, G.M., 2012. Reactive oxygen and oxidative stress tolerance in plant pathogenic *Pseudomonas*. FEMS Microbiology Letters 327, 1–8.
Ghanepour, S., Shakiba, M.R., Toorchi, M., Oustan, S., 2015. Role of Zn nutrition in membrane stability, leaf hydration status, and growth of common bean grown under soil moisture stress. JBES 6, 9–20.
Gupta, S., Kushwah, T., Vishwakarma, A., Yadav, S., 2015. Optimization of ZnO-NPs to investigate their safe application by assessing their effect on soil nematode *Caenorhabditis elegans*. Nanoscale Research Letters 10, 1–9.
Gupta, S.K., Rai, A.K., Kanwar, S.S., Chand, D., Singh, N.K., Sharma, T.R., 2012. The single functional blast resistance gene Pi54 activates a complex defence mechanism in rice. Journal of Experimental Botany 63, 757–772.
Gupta, S.K., Rai, A.K., Kanwar, S.S. and Sharma, T.R., 2012. Comparative analysis of zinc finger proteins involved in plant disease resistance.
Gupta, N., Ram, H., Kumar, B., 2016. Mechanism of zinc absorption in plants: uptake, transport, translocation and accumulation. Reviews in Environmental Science and Bio/Technology 15, 89–109.

Haydon, M.J., Kawachi, M., Wirtz, M., Hillmer, S., Hell, R., Krämer, U., 2012. Vacuolar nicotianamine has critical and distinct roles under iron deficiency and for zinc sequestration in Arabidopsis. The Plant Cell 24, 724–737.

Helfenstein, J., Pawlowski, M.L., Hill, C.B., Stewart, J., Lagos-Kutz, D., Bowen, et al., 2015. Zinc deficiency alters soybean susceptibility to pathogens and pests. Journal of Soil Science and Plant Nutrition 178, 896–903.

Hood, M.I., Skaar, E.P., 2012. Nutritional immunity: transition metals at the pathogen–host interface. Nature Reviews. Microbiology 10, 525–537.

Hou, J., Wu, Y., Li, X., Wei, B., Li, S., Wang, X., 2018. Toxic effects of different types of zinc oxide nanoparticles on algae, plants, invertebrates, vertebrates and microorganisms. Chemosphere 193, 852–860.

Inaba, S., Kurata, R., Kobayashi, M., Yamagishi, Y., Mori, I., Ogata, Y., et al., 2015. Identification of putative target genes of bZIP19, a transcription factor essential for Arabidopsis adaptation to Zn deficiency in roots. The Plant Journal 84, 323–334.

Jeevanandam, J., Barhoum, A., Chan, Y.S., Dufresne, A., Danquah, M.K., 2018. Review on nanoparticles and nanostructured materials: history, sources, toxicity and regulations. Beilstein Journal of Nanotechnology 9, 1050–1074.

Kashyap, D., Siddiqui, Z.A., 2021. Effect of silicon dioxide nanoparticles and Rhizobium leguminosarum alone and in combination on the growth and bacterial blight disease complex of pea caused by Meloidogyne incognita and Pseudomonas syringae pv. pisi. Archives of Phytopathology and Plant Protection 54 (9–10), 499–515.

Kawachi, M., Kobae, Y., Kogawa, S., Mimura, T., Krämer, U., Maeshima, M., 2012. Amino acid screening based on structural modeling identifies critical residues for the function, ion selectivity and structure of Arabidopsis MTP1. The FEBS journal 279, 2339–2356.

Khanna, K., Jamwal, V.L., Sharma, A., Gandhi, S.G., Ohri, P., Bhardwaj, R., et al., 2019a. Evaluation of the role of rhizobacteria in controlling root-knot nematode infection in *Lycopersicon esculentum* plants by modulation in the secondary metabolite profiles. AoB Plants 11, 069.

Khanna, K., Kohli, S.K., Ohri, P., Bhardwaj, R., 2021b. Plants-nematodes-microbes crosstalk within soil: a trade-off among friends or foes. Microbiological Research 126755.

Khanna, K., Ohri, P., Bhardwaj, R., 2021a. Genetic toolbox and regulatory circuits of plant-nematode associations. Plant Physiology and Biochemistry .

Khanna, K., Sharma, A., Ohri, P., Bhardwaj, R., Abd_Allah, E.F., Hashem, A., et al., 2019b. Impact of plant growth promoting rhizobacteria in the orchestration of *Lycopersicon esculentum* Mill. resistance to plant parasitic nematodes: a metabolomic approach to evaluate defense responses under field conditions. Biomolecules 9, 676.

Khan, M.R., Siddiqui, Z.A., 2021. Role of zinc oxide nanoparticles in the management of disease complex of beetroot (*Beta vulgaris* L.) caused by *Pectobacterium betavasculorum*, *Meloidogyne incognita* and *Rhizoctonia solani*. Horticulture, Environment, and Biotechnology 62, 225–241.

Khan, M., Siddiqui, Z.A., 2018. Zinc oxide nanoparticles for the management of *Ralstonia solanacearum*, *Phomopsis vexans* and *Meloidogyne incognita* incited disease complex of eggplant. Indian Phytopathology 71, 355–364.

Kim, S.H., Hong, J.K., Lee, S.C., Sohn, K.H., Jung, H.W., Hwang, B.K., 2004. CAZFP1, Cys 2/His 2-type zinc-finger transcription factor gene functions as a pathogen-induced early-defense gene in Capsicum annuum. Plant Molecular Biology 55, 883–904.

Kinraide, T.B., Poschenrieder, C., Kopittke, P.M., 2011. The standard electrode potential (Eq) predicts the prooxidant activity and the acute toxicity of metal ions. Journal of Inorganic Biochemistry 105, 1438–1445.

Krämer, U., 2018. Conceptualizing plant systems evolution. Current Opinion in Plant Biology 42, 66–75.

Kumar, D., Rampuria, S., Singh, N.K., Kirti, P.B., 2016. A novel zinc-binding alcohol dehydrogenase 2 from *Arachis diogoi*, expressed in resistance responses against late leaf spot pathogen, induces cell death when transexpressed in tobacco. FEBS Open Bio 6, 200–210.

Kundu, D., Hazra, C., Chatterjee, A., Chaudhari, A., Mishra, S., 2014. Extracellular biosynthesis of zinc oxide nanoparticles using *Rhodococcus pyridinivorans* NT2: multifunctional textile finishing, biosafety evaluation and in vitro drug delivery in colon carcinoma. Journal of Photochemistry and Photobiology B: Biology 140, 194–204.

Laurenti, M., Canavese, G., Sacco, A., Fontana, M., Bejtka, K., Castellino, M., et al., 2015. Nanobranched ZnO structure: p-type doping induces piezoelectric voltage generation and ferroelectric–photovoltaic effect. Advanced Materials 27, 4218–4223.

Lawrence, S.D., Novak, N.G., Jones, R.W., Farrar Jr, R.R., Blackburn, M.B., 2014. Herbivory responsive C2H2 zinc finger transcription factor protein StZFP2 from potato. Plant Physiology and Biochemistry 80, 226–233.

Lilay, G.H., Castro, P.H., Campilho, A., Assunção, A.G., 2019. The Arabidopsis bZIP19 and bZIP23 activity requires zinc deficiency–insight on regulation from complementation lines. Frontiers in Plant Science 9, 1955.

Lilay, G.H., Persson, D.P., Castro, P.H., Liao, F., Alexander, R.D., Aarts, M.G., et al., 2021. Arabidopsis bZIP19 and bZIP23 act as zinc sensors to control plant zinc status. Nature Plants 7, 137–143.

Lim, D., Roh, J.Y., Eom, H.J., Choi, J.Y., Hyun, J., Choi, J., 2012. Oxidative stress-related PMK-1 P38 MAPK activation as a mechanism for toxicity of silver nanoparticles to reproduction in the nematode *Caenorhabditis elegans*. Environmental Toxicology and Chemistry 31, 585–592.

Liu, Z., Lv, X., Xu, L., Liu, X., Zhu, X., Song, E., et al., 2020. Zinc oxide nanoparticles effectively regulate autophagic cell death by activating autophagosome formation and interfering with their maturation. Particle and Fibre Toxicology 17, 1–17.

Machado, P.P., Steiner, F., Zuffo, A.M., Machado, R.A., 2018. Could the supply of boron and zinc improve resistance of potato to early blight? Potato Research 61, 169–182.

Miller, G., Shulaev, V., Mittler, R., 2008. Reactive oxygen signaling and abiotic stress. Physiologia Plantarum 133, 481–489.

Mitra, S., Patra, P., Pradhan, S., Debnath, N., Dey, K.K., Sarkar, S., et al., 2015. Microwave synthesis of ZnO@ mSiO2 for detailed antifungal mode of action study: understanding the insights into oxidative stress. Journal of Colloid and Interface Science 444, 97–108.

Mukherjee, A., Peralta-Videa, J.R., Bandyopadhyay, S., Rico, C.M., Zhao, L., Gardea-Torresdey, J.L., 2014. Physiological effects of nanoparticulate ZnO in green peas (*Pisum sativum* L.) cultivated in soil. Metallomics. 6, 132–138.

Nagajyothi, P.C., Sreekanth, T.V.M., Tettey, C.O., Jun, Y.I., Mook, S.H., 2014. Characterization, antibacterial, antioxidant, and cytotoxic activities of ZnO nanoparticles using *Coptidis Rhizoma*. Bioorganic and Medicinal Chemistry Letters 24, 4298–4303.

Nakai, Y., Nakahira, Y., Sumida, H., Takebayashi, K., Nagasawa, Y., Yamasaki, K., et al., 2013. Vascular plant one-zinc-finger protein 1/2 transcription factors regulate abiotic and biotic stress responses in Arabidopsis. The Plant Journal 73 (5), 761–775.

Noman, A., Ali, Q., Maqsood, J., Iqbal, N., Javed, M.T., Rasool, N., et al., 2018. Deciphering physio-biochemical, yield, and nutritional quality attributes of water-stressed radish (*Raphanus sativus* L.) plants grown from Zn-Lys primed seeds. Chemosphere 195, 175–189.

Noman, A., Aqeel, M., Khalid, N., Islam, W., Sanaullah, T., Anwar, M., et al., 2019. Zinc finger protein transcription factors: integrated line of action for plant antimicrobial activity. Microbial Pathogenesis 132, 141–149.

Noman, A., Fahad, S., Aqeel, M., Ali, U., Anwar, S., Baloch, S.K., et al., 2017. miRNAs: major modulators for crop growth and development under abiotic stresses. Biotechnology Letters 39, 685–700.

Parveen, A., Siddiqui, Z.A., 2021. Zinc oxide nanoparticles affect growth, photosynthetic pigments, proline content and bacterial and fungal diseases of tomato. Arch. Phytopathol. Plant Prot. 1, 20.

Peng, C., Zhang, W., Gao, H., Li, Y., Tong, X., Li, K., et al., 2017. Behavior and potential impacts of metal-based engineered nanoparticles in aquatic environments. Nanomaterials 7, 21.

Pragati, J., Khatri, P., Rana, J.S., 2018. Green synthesis of zinc oxide nanoparticles using flower extract of *Nyctanthes arbor-tristis* and their antifungal activity. Journal of King Saud University—Science 30, 168–175.

Pulit-Prociak, J., Chwastowski, J., Kucharski, A., Banach, M., 2016. Functionalization of textiles with silver and zinc oxide nanoparticles. Applied Surface Science 385, 543–553.

Raskar, S.V., Laware, S.L., 2014. Effect of zinc oxide nanoparticles on cytology and seed germination in onion. International Journal of Current Microbiology and Applied Sciences 3, 467–473.

Rizwan, M., Ali, S., Ali, B., Adrees, M., Arshad, M., Hussain, A., et al., 2019. Zinc and iron oxide nanoparticles improved the plant growth and reduced the oxidative stress and cadmium concentration in wheat. Chemosphere 214, 269–277.

Rodríguez-Celma, J., Tsai, Y.H., Wen, T.N., Wu, Y.C., Curie, C., Schmidt, W., 2016. Systems-wide analysis of manganese deficiency-induced changes in gene activity of Arabidopsis roots. Scientific Reports 6, 1–16.

Sarwar, S., Chakraborti, S., Bera, S., Sheikh, I.A., Hoque, K.M., Chakrabarti, P., 2016. The antimicrobial activity of ZnO nanoparticles against *Vibrio cholerae*: Variation in response depends on biotype. Nanomedicine: Nanotechnology, Biology and Medicine 12, 1499–1509.

Sharma, N., Khanna, K., Manhas, R.K., Bhardwaj, R., Ohri, P., Alkahtani, J., et al., 2020. Insights into the Role of *Streptomyces hydrogenans* as the plant growth promoter, photosynthetic pigment enhancer and biocontrol agent against *Meloidogyne incognita* in *Solanum lycopersicum* seedlings. Plants 9, 1109.

Shivashankar, S., Sumathi, M., Krishnakuma, R.N.K., Ra, V.K., 2015. Role of phenolic acids and enzymes of phenylpropanoid pathway in resistance of chayote fruit (*Sechium edule*) against infestation by melon fly, *Bactrocera cucurbitae*. The Annals of Applied Biology 166, 420–433.

Siddiqui, Z.A., Khan, M.R., Abd_Allah, E.F., Parveen, A., 2019a. Titanium dioxide and zinc oxide nanoparticles affect some bacterial diseases, and growth and physiological changes of beetroot. International Journal of Vegetable Science 25, 409–430.

Siddiqui, Z.A., Khan, A., Khan, M.R., Abd-Allah, E.F., 2018. Effects of zinc oxide nanoparticles (ZnO NPs) and some plant pathogens on the growth and nodulation of lentil (Lens culinaris Medik.). Acta Phytopathologica et Entomologica Hungarica 53, 195–211.

Siddiqui, Z.A., Parveen, A., Ahmad, L., Hashem, A., 2019b. Effects of graphene oxide and zinc oxide nanoparticles on growth, chlorophyll, carotenoids, proline contents and diseases of carrot. Scientia Horticulturae 249, 374–382.

Singh, A., Singh, N.Á., Afzal, S., Singh, T., Hussain, I., 2018. Zinc oxide nanoparticles: a review of their biological synthesis, antimicrobial activity, uptake, translocation and biotransformation in plants. Journal of Materials Science 53, 185–201.

Sirelkhatim, A., Mahmud, S., Seeni, A., Kaus, N.H.M., Ann, L.C., Bakhori, S.K.M., et al., 2015. Review on zinc oxide nanoparticles: antibacterial activity and toxicity mechanism. Nano-Micro Letters 7, 219–242.

Sofy, A.R., Hmed, A.A., AbdEL-Aleem, M.A., Dawoud, R.A., Elshaarawy, R.F., Sofy, M.R., 2020. Mitigating effects of bean yellow mosaic virus infection in faba bean using new carboxymethyl chitosan-titania nanobiocomposites. International Journal of Biological Macromolecules 163, 1261–1275.

Sofy, A.R., Sofy, M.R., Hmed, A.A., Dawoud, R.A., Alnaggar, A.E.A.M., Soliman, A.M., et al., 2021. Ameliorating the adverse effects of tomato mosaic tobamovirus infecting tomato plants in Egypt by boosting immunity in tomato plants using zinc oxide nanoparticles. Molecules 26, 1337.

Soliman, A.S., El-feky, S.A., Darwish, E., 2015. Alleviation of salt stress on *Moringa peregrina* using foliar application of nanofertilizers. Journal of Horticulture and Forestry 7, 36–47.

Stan, M., Popa, A., Toloman, D., Dehelean, A., Lung, I., Katona, G., 2015. Enhanced photocatalytic degradation properties of zinc oxide nanoparticles synthesized by using plant extracts. Materials Science in Semiconductor Processing 39, 23–29.

Stolpe, C., Giehren, F., Krämer, U., Müller, C., 2017. Both heavy metal-amendment of soil and aphid-infestation increase Cd and Zn concentrations in phloem exudates of a metal-hyperaccumulating plant. Phytochemistry 139, 109–117.

Streeter, T.C., Rengel, Z., Neate, S.M., Graham, R.D., 2001. Zinc fertilisation increases tolerance to *Rhizoctonia solani* (AG 8) in *Medicago truncatula*. Plant and Soil 228, 233–242.

Sun, D., Hussain, H.I., Yi, Z., Siegele, R., Cresswell, T., Kong, L., et al., 2014. Uptake and cellular distribution, in four plant species, of fluorescently labeled mesoporous silica nanoparticles. Plant cell Reports 33, 1389–1402.

Sávoly, Z., Hrács, K., Pemmer, B., Streli, C., Záray, G., Nagy, P.I., 2016. Uptake and toxicity of nano-ZnO in the plant-feeding nematode, *Xiphinema vuittenezi*: the role of dissolved zinc and nanoparticle-specific effects. Environmental Science and Pollution Research 23, 9669–9678.

Tanaka, N., Kawachi, M., Fujiwara, T., Maeshima, M., 2013. Zinc-binding and structural properties of the histidine-rich loop of Arabidopsis thaliana vacuolar membrane zinc transporter MTP1. FEBS Open Bio 3, 218–224.

Tarafdar, J.C., Raliya, R., Mahawar, H., Rathore, I., 2014. Development of zinc nanofertilizer to enhance crop production in pearl millet (*Pennisetum americanum*). Agricultural Research 3, 257–262.

Taran, N., Storozhenko, V., Svietlova, N., Batsmanova, L., Shvartau, V., Kovalenko, M., 2017. Effect of zinc and copper nanoparticles on drought resistance of wheat seedlings. Nanoscale Research Letters 12, 1–6.

Tirani, M.M., Haghjou, M.M., Ismaili, A., 2019. Hydroponic grown tobacco plants respond to zinc oxide nanoparticles and bulk exposures by morphological, physiological and anatomical adjustments. Functional Plant Biology 46, 360–375.

Torabian, S., Zahedi, M., Khoshgoftarmanesh, A., 2018. Effect of foliar spray of zinc oxide on some antioxidant enzymes activity of sunflower under salt stress.

Venkatachalam, P., Jayaraj, M., Manikandan, R., Geetha, N., Rene, E.R., Sharma, N.C., et al., 2017. Zinc oxide nanoparticles (ZnONPs) alleviate heavy metal-induced toxicity in *Leucaena leucocephala* seedlings: a physiochemical analysis. Plant Physiology and Biochemistry 110, 59–69.

Wang, F., Jing, X., Adams, C.A., Shi, Z., Sun, Y., 2018. Decreased ZnO nanoparticle phytotoxicity to maize by arbuscular mycorrhizal fungus and organic phosphorus. Environmental Science and Pollution Research 25, 23736–23747.

Wang, X., Wang, Y., Liu, P., Ding, Y., Mu, X., Liu, X., et al., 2017. TaRar1 is involved in wheat defense against stripe rust pathogen mediated by YrSu. Frontiers in Plant Science 8, 156.

Yuvakkumar, R., Suresh, J., Nathanael, A.J., Sundrarajan, M., Hong, S.I., 2014. Novel green synthetic strategy to prepare ZnO nanocrystals using rambutan (*Nephelium lappaceum L.*) peel extract and its antibacterial applications. Materials Science and Engineering 41, 17–27.

Zhao, L., Sun, Y., Hernandez-Viezcas, J.A., Servin, A.D., Hong, J., Niu, G., et al., 2013. Influence of $CeO_2$ and ZnO nanoparticles on cucumber physiological markers and bioaccumulation of Ce and Zn: a life cycle study. Journal of Agricultural and Food Chemistry 61 (49), 11945–11951.

Zlobin, I.E., Pashkovskiy, P.P., Kartashov, A.V., Nosov, A.V., Fomenkov, A.A., Kuznetsov, V.V., 2020. The relationship between cellular Zn status and regulation of Zn homeostasis genes in plant cells. Environmental and Experimental Botany 176, 104. -104.

Chapter 10

# Zinc deficiency in plants: an insight into fortification strategies

Vishakha Sharma[1,*], Avani Maurya[1,*], Nidhi Kandhol[1], Vijay Pratap Singh[2], Shivesh Sharma[3], Jose Peralta-Videa[4] and Durgesh Kumar Tripathi[1]

*[1]Crop Nanobiology and Molecular Stress Physiology Laboratory, Amity Institute of Organic Agriculture, Amity University Uttar Pradesh, Noida, Uttar Pradesh, India, [2]Department of Botany, C.M.P. Degree College, University of Allahabad, Prayagraj, Uttar Pradesh, India, [3]Department of Biotechnology, Motilal Nehru National Institute of Technology Allahabad, Prayagraj, Uttar Pradesh, India, [4]Department of Chemistry and Biochemistry, The University of Texas at El Paso, El Paso, TX, United States*

## 10.1 Introduction

Mineral nutrients play a dominant role in the proper functioning of the plant with respect to its growth and development and its metabolic and chemical activities. Reportedly, it is observed that there are 14 mineral elements, i.e., iron (Fe), copper (Cu), boron (B), manganese (Mn), molybdenum (Mo), silicon (Si), zinc (Zn), nickel (Ni), chlorine (Cl), selenium (Si), sodium (Na), aluminum (Al), and cobalt (Co), which are needed by the plant in an adequate amount to maintain its growth and development (Vatansever et al., 2017). The absorption and transportation of these minerals are carried out by the brain of the plant, i.e., roots through xylem fibers. It is essential to facilitate these minerals to the plant as they majorly contribute to growth and development, primary and secondary metabolism of the cell defense, energy metabolism, and hormone perception (Vatansever et al., 2017). Thus, the nutritional needs of the plants should be taken into consideration so that a precise amount of nutrients can be delivered to the plant at right place, right time, and in the right amount to attain the desirable productivity, yield, and sustainability (Shah et al., 2018).

*Equal Contribution.

Zinc in Plants. DOI: https://doi.org/10.1016/B978-0-323-91314-0.00011-9

Mineral deficiency is a major problem faced in the agriculture field at present times (Sirohi et al., 2016). Plants can show visual indications of mineral deficiency. Generally, symptoms of deficiency are nutrient-specific and reflect different pattern in different crops for each essential nutrient. The deficiency of these nutrients can hamper the plant growth by alternating their growth pattern, biochemical process, antioxidant defense abilities, and their resistance to biotic and abiotic environmental stresses (Hajiboland, 2012; Das et al., 2017). Environmental factors, especially those pertaining to soil, affect the availability, uptake, and utilization of nutrients by plants (Hajiboland, 2012). There are a number of reasons leading to soil contamination which results in mineral deficiency such as industrial activities (heavy metals, effluents), land disposal (sewage water, sludge, and radionucleotides), domestic and municipal waste, agrochemicals (pesticides, fertilizers), atmospheric (acid rain, contaminated dust) (Kaur, 2021). Therefore, it is essential to either enhance the availability of nutrients to plants or to increase their strength to deal with the insufficient availabilities of essential nutrients.

Among various essential nutrients, Zn is one of the essential micronutrients, available in $Zn^{2+}$ form. It is observed that it has an important function in plant metabolism by managing the activities of hydrogenase and carbonic anhydrase enzymes as well as helps in maintaining the integrity of cellular membrane, protein synthesis, auxin regulation, plant resistance against disease, and formation of pollen (Hafeez et al., 2013). Optimum concentration of Zn is necessary for the functioning of several physiological processes in plants such as photosynthesis and sugar formation, antioxidant activity, plant fertility, production of seeds, and overall regulation of plant growth (Hussain et al., 2015; Rudani et al., 2018).

Zn deficiency is a widespread problem with most sandy, calcareous, peat soils, and soils with high phosphorus and silicon are usually deficient in Zn (Hafeez et al., 2013). Due to various environmental causes, there has been a negative impact on the supply of Zn (Ganguly et al., 2022). With a proper supply of Zn nutrients, some crops such as wheat (*Triticum aestivum*), sunflower (*Helianthus annuus*), tomato (*Solanum lycopersicum*), and red cabbage (*Brassica oleracea*) can be protected from adverse effects of Zn deficiency (Hassan et al., 2020). However, if the amount of Zn is less in plants, it results in a hindrance with respect to growth and yield (Rudani et al., 2018). It also affects the growth and development of roots and leaves. Visual symptoms of Zn deficiency are observed as chlorotic, rosetted, dwarf, and malformed leaves, as well as overall chlorosis and stunted growth of plants (Rudani et al., 2018). Zn deficiency decreases the growth and productivity of plants and its effects are also reflected in human beings. Zn deficiency is a serious problem affecting approximately half of the global population due to the low Zn content of the crops grown in Zn-deficient soils (Hafeez et al., 2013; Rudani et al., 2018). To address this challenge, a multitude of solutions, both natural and synthetic, exist to rectify Zn deficiency.

Natural approaches encompass the utilization of plant growth-promoting rhizobacteria (PGPR), renowned for their multifaceted functional activities, offering a natural, cost-effective, and environmentally friendly remedy to mitigate soil contamination while enhancing nutrient availability (Kaur et al., 2020). Synthetic ways can be the use of fertilizers, an indispensable component of modern agricultural practices, pivotal in furnishing crucial mineral elements essential for outstanding growth and blooming harvest (Khan et al., 2018). Fertilizers can help in improving the soil structure, texture, and aeration, increase the soil water retention ability, and also help in stimulating healthy root development (Assefa and Tadesse, 2019).

At present time, due to the rapid increase in population, the demand for food has also risen. Therefore, enhancing the productivity of crops while maintaining the food security is the ultimate challenge persisting in front of the agriculture sector. The value addition of Zn is the key step to increase its nutritive benefits. The concept of Zn fortification plays a significant role in Zn enrichment in crops which also favors the nutritional demand of the growing population (Jan et al., 2020). Development of new advanced methods to deal with Zn deficiency in food crops in the field, and genome-wide association studies to identify the genetic basis of Zn and its accumulation under low Zn stress environment are needed (Hacisalihoglu, 2020). In this context, this chapter covers various fortification strategies including conventional and nanotechnological techniques which help in resolving the issues of Zn deficiencies in the plant while pointing out the limitations associated with each method to suggest future research perspectives in this field.

## 10.2 Zinc deficiency in soil and impact on plant life

Zn, being one of the most important micronutrients for crops, shows a wide range of deficiency problems in plants, because it works as a regulatory cofactor of enzymes and proteins and also helps in building up of structure of plant (Alloway, 2009). It also helps in cytochrome synthesis and stabilization of ribosomal fractions (Hafeez et al., 2013). Zn deficiency in soil directly results in a decrease in plants' growth and development. The Zn concentration in soil is mainly dependent on the parent rock's chemical composition. Zn availability in soil is affected by various reasons like high calcite, high pH, high concentrations of Na, Ca, Mg, phosphate and bicarbonate, organic matter content and available Zn in soil (Alloway, 2009) (Fig. 10.1). In case of waterlogged soil, formation of small soluble Zn compounds occurs in rhizosphere, thus making it deficient in Zn content. High content of phosphorus also leads to deficiency of Zn in soil. Zn deficiency in calcareous soil with high pH is seen to a great extent due to the adsorption property of soil and due to the presence of calcium carbonate in high amounts (Nielsen, 2012). In India, approximately 50% of the soils are Zn deficient and Zn deficiency is the most commonly observed nutritional disorder affecting the majority of

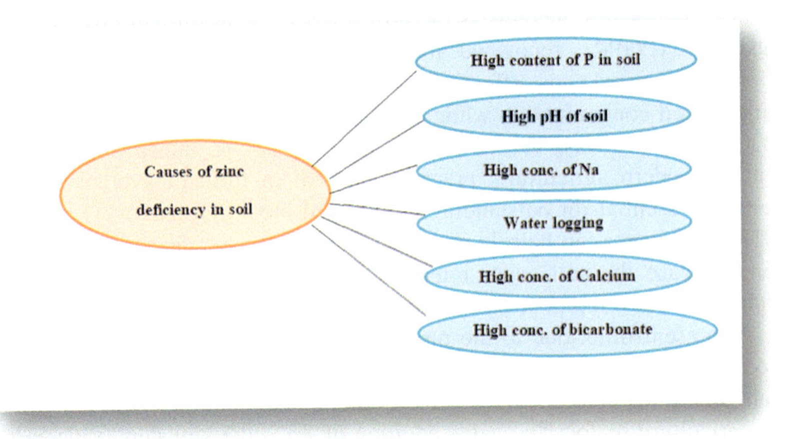


**FIGURE 10.1** Various causes responsible for induction of zinc deficiency in soil.

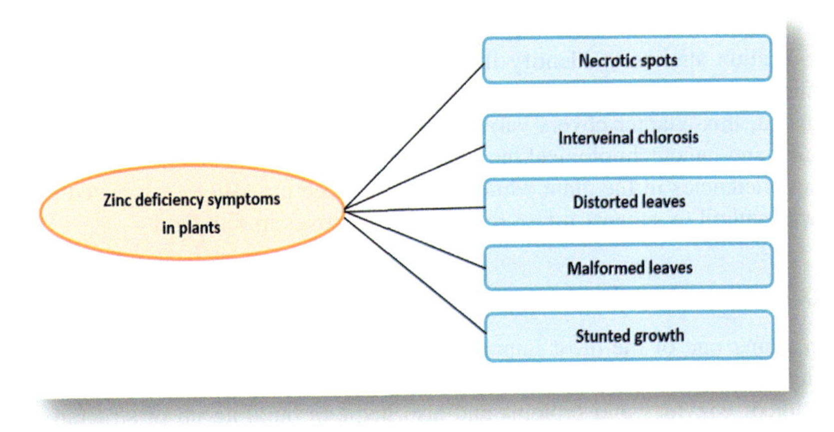


**FIGURE 10.2** Different symptoms exhibited by plants upon exposure to zinc deficiency.

the crop production (Rudani et al., 2018). Deficiency of Zn is expected to rise from 42% to 63% by 2025 due to the consistent depletion of soil fertility (Sunithakumari et al., 2016).

When crops are grown in such Zn-deficient soils, they exhibit several symptoms of deficiency. Zn deficiency results in stunting of plants, decreased number of tillers, spikelet sterility, smaller leaves, chlorosis, and lowered quality of produce (Fig. 10.2). Zn also affects the functioning of xylem resulting in difficulty for plants to uptake water. The integrity of cellular membranes and ion transport systems is mainly dependent on the optimum availability of Zn (Hafeez et al., 2013). In the maize (*Zea mays*) plant, Zn deficiency shows higher stomatal density along with a reduction in leaf and shoot size. This also shows decreased level of photosynthesis, less biomass, and imbalance in stomatal opening (Mattiello et al., 2015). Similarly, in *Oryza sativa* (rice), Zn deficiency results in brown streaks and blotches on leaves, height of plant is affected and yield is reduced due to late maturity. In extreme cases, plants may even dry and die (Wissuwa et al., 2006).

As the increasing population constantly consumes food available in the market, Zn deficient food intake results in several bad impacts on human life. Approximately, 3 billion people around the world are suffering from Zn deficiencies, with more severe conditions in areas where the population is extremely dependent on a constant diet of cereal-based foods (Hafeez et al., 2013). Zn deficiency also results in dry skin, brain disorders, decreased functional activity of the gonads in males, and memory issues (Prasad, 2012). Therefore, protecting plants against Zn deficiencies is necessary to safeguard other life forms as well. In order to deal with Zn deficiency problems in plants, conventional fortification, biofortification, and nanotechnological interventions are some potential solutions.

## 10.3 Conventional fortification strategies for mitigating zinc deficiency in plants

It is understandable that with the increasing population, food consumption is rising (Alexandratos, 2005). There are different ways through which we can increment food production to meet the increasing food demand, but the first and foremost step is to ensure that the adequate supply of nutrients reaches the plant through the application of the right techniques. There are different techniques to deal with nutrient deficiency in plants (Hajiboland, 2012). The usage of mineral fertilizers is the major driving force to increase global agricultural production required to feed the rising human population (Bindraban et al., 2015). The main challenge that fertilizer application generally faces is to deliver a small amount of fertilizer evenly over the targeted area. Proper use of machinery and technology increases its effectiveness; for example, using a Spray fertilizer technology directly on the soil surface while preparing the soil can help in increasing its efficiency (Graham, 2008).

The scarcity of Zn is very common in soils; Zn fertilizers can be used to ameliorate the Zn deficiency in plants (Khan et al., 2022). Various experiments establish the use of Zn fertilizer as an effective way to overcome Zn deficiency in plants. For example, Hidoto et al., (2017) evaluated the effect of different Zn application strategies (foliar, soil, or seed priming) on chickpea (*Cicer arietinum*) under Zn deficient soils. The results showed that Zn fertilization through all three strategies, Zn with the concentration of 0.5% $ZnSO_4.7H_2O$ improved Zn content with foliar application being the most effective method of biofortification of Zn (Hidoto et al., 2017). In another study, with a comparison of soil and foliar Zn application to increase grain Zn content of wheat (*Triticum aestivum*) grown in Zn-deficient calcareous soil, the results showed that the foliar application of Zn in wheat increased the grain Zn concentration more effectively than Zn application to soil (Zhao et al., 2014).

The technique of using foliage spray is greatly used, as it helps in the rapid utilization of the applied nutrients, provide higher resistance to disease and insect pests, helps in improving abiotic stress tolerance as well as higher resistance to physiological disorder in plants (Furness et al., 2006). In another experiment, promoting the foliar application of Zn sulfate in berries of *Vitis vinifera* on Zn deficient soil over two continuous seasons, the results displayed that foliar spraying of Zn sulfate could promote the photosynthesis as well as development of berry in vines from veraison to maturation. Meanwhile, Zn treatment also enhanced the accumulation of total soluble solid, total phenolics, total flavanols, total flavanols, tannins, and total anthocyanins in berry skin (Song et al., 2015). Scientists have also researched the agronomic traits and grain quality parameters of wheat grown in Zn-deficient soil (Esfandiari et al., 2016). They have reported that Zn sulfate application after the flowering stage increased Zn content in wheat grains effectively. Application of Zn sulfate at booting and milking stage also improved the quantity and quality of the yield of wheat plants under Zn-deficient calcareous soils (Esfandiari et al., 2016). In another experiment also, it was confirmed that Zn sulfate affects the agronomy traits and gain micronutrients in bread wheat (*Triticum aestivum*) and durum wheat (*Triticum durum*) under Zn-deficient calcareous soil (Abdoli et al., 2016). $ZnSO_4$ application through both soil and foliar methods increased the grain yield, biomass, thousand kernel weight, number of grains per spike, and number of fertile spikelets per spike (Abdoli et al., 2016). Their observations suggest that both methods of application of $ZnSO_4$ are effective in boosting wheat productivity under Zn deficiency with bread wheat providing higher grain yield and biomass than durum wheat (Abdoli et al., 2016).

Exogenous Zn application in the form of Zn sulfate to manage deficiency of Zn in plantsgets transformed into unavailable forms such as Zn (OH) and Zn ($OH_2$) depending on the soil pH (Jerlin et al., 2017). The best strategy that can be used is the application of Zn-augmented macronutrient fertilizer, to promote the application of Zn, especially within small farmers. The highly/regularly used fertilizers like urea should also be enriched with Zn. There is an advantage of using Zn fertilizer in crops as it helps in increasing the grain yield of major staple food crops such as wheat and rice. However, the usage of chemical fertilizers degrades the soil health (Doran et al., 2018). Due to the application of chemical fertilizer, the quality of soil deteriorates and regular usage of chemical fertilizer has a serious impact on soil's biochemical properties, as a result, the microbial population gets shifted (Pahalvi et al., 2021). Therefore, greener alternatives or organic farming should be adopted to maintain sustainability. Practices like green manuring help the soil in improving organic matter, water holding capacity, porosity, and carbon:nitrogen (C:N) ratio (Sajjad et al., 2019). Other practices of soil conservation, integrated nutrient management, integrated weed management, organic fertilizer, and conservation tillage can be adopted to protect the soil and maintain nutrient availability.

## 10.4 Biofortification as a greener alternative for protection against zinc deficiency

Improving food nutritional quality by conventional fortification is a method widely used for protection against Zn deficiency but it has bad impact on the environment such as leaching of chemical fertilizers, eutrophication, loss of biodiversity, and decreased soil health (Rekha and Prasad, 2006). Thus, biofortification is seen as a greener and safer alternative for improving nutritional quality of crops. Biofortification is the process of improving nutrients in food crops and is considered as a long-term strategy of providing micronutrients along with sustainability (Saltzman et al., 2013). Biofortification strategies include classical plant breeding and genetic engineering for the improvement of nutrient levels in crops, along with specialized fertilizer application by promoting the accumulation of substances that can enhance absorption of nutrients (Cakmak and Kutman, 2018) (Fig. 10.3).

Zn bio-fertilization results in better uptake of Zn, further resulting in improved plant growth, good grain size, and improved yield. Agronomic practices can assist in improving Zn availability as in the rice-wheat cropping system, where residual Zn is taken up by succeeding rice crop, and improved grain size and yield are observed (Rehman et al., 2020). Another study showed that intercropping of chickpea and wheat resulted in 2.82 times

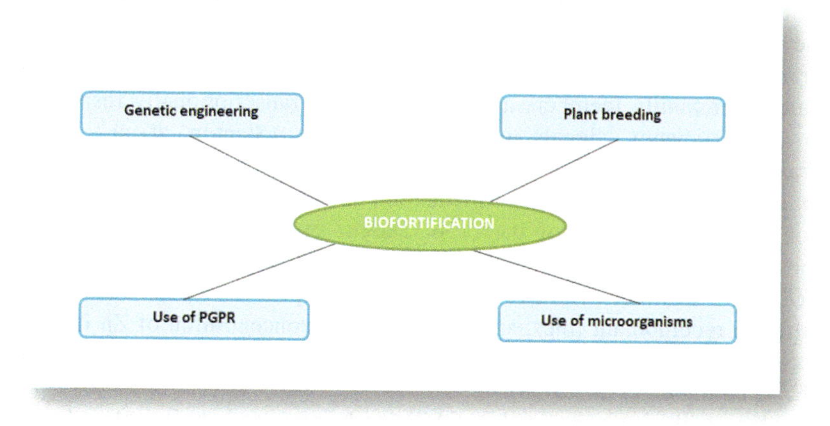


**FIGURE 10.3** Various techniques used for biofortification of plants for protection against Zn deficiency.

higher amount of Zn in chickpea seed as compared to that of Zn in monocropping (Zuo and Zhang, 2009). Furthermore, microbial strains can also help in the biofortification of plants grown under Zn deficiency. For example, an experiment was held to study the mobilization of Zn from source to sink through *Bacillus aryabhattai* strains including MDSR7, MDSR11, and MDSR14 in *Glycine max* (soybean) and wheat. Result showed that strains of MDSR7 and MDSR14 greatly worked in mobilization of Zn from its sink to source, and its concentration in endosperm of wheat and soybean suggested that these strains could be utilized as bio-inoculants for Zn biofortification (Ramesh et al., 2014).

In another study, several microbes were used for biofortification of rice that included Arbuscular mycozzihzae, *Bacillus, Enterobacter, Pseudomonas*, and *microbacterium* along with endophytic strains of *Burkholderia* sp., *Sphingomonas* sp., *Variovorax* sp., and *Sedum alfredii*. The results included a 10.4% and 20.6% increase in Zn concentration by *Sphingomonas* sp. and *Enterobacter*, respectively (Kaur et al., 2020). On showing Zn deficiency symptoms, when maize is treated with single strained Zn solubilizing rhizobacteria under axenic conditions, results in positive potential growth and can be used as a bio-inoculant to meet the nutritional requirements of food (Mumtaz et al., 2017). Similarly, when wheat is treated with *Providencia sp.* (PW5) bacteria, the accumulation and translocation of Zn from source to sink is increased resulting in a better quantity of Zn in grains (Rana et al., 2012). When wheat and soybean are treated with Zn solubilizing bacteria strains from genera *Bacillus aryabthattai*, improvement in Zn mobilization was observed in the edible portions of soybean and wheat (Hussain et al., 2018). Another study resulted in an increase in root weight, root area, and shoot weight by 74%, 75%, and 23% respectively by the inoculation of Zn mobilizing PGPR in rice plants (Shaikh and Saraf, 2017).

Studies have shown that a transgenic approach can be used to increase the Zn accumulation in edible portions of food. Zinc-Regulated Transporter–Iron-Regulated Transporter (ZRT-IRT)-like Protein (ZIP) transporter family members are reported to be transcriptionally responsive to Zn deficiency signals in different plants (Stanton et al., 2021). Transcription factors, bZIP19 and bZIP23 (regulate the adaptation to low Zn supply), were used for biofortification and showed better Zn concentration in the edible part of the crop (Hafeez et al., 2013). Another study showed that when the introduction of *Gpc-B1* (*GRAIN PROTEIN CONTENT B1*) locus from *Triticum turgidum ssp. dicoccoides* was done in different recombinant lines, and an increase in concentration of Zn in grain was observed via the enhancement of transport, uptake, and remobilization of Zn (Palmgren et al., 2008).

Along with treatments with solubilizing bacteria and strains, plant breeding also helps in dealing with micronutrient deficiency. In maize, deficiency of Zn can be treated by selections and hybridization which results in increased concentration of Zn in kernels, sequestration of Zn in endosperm, and remobilization in grains (Maqbool and Beshir, 2019). Another study showed that progenies of hybrid *Pennisetum glaucum* (pearl millet), based on *iniari* germplasm resulted in higher levels of Zn in grains (Velu et al., 2007). In Bangladesh, a Zn-biofortified wheat variety was introduced, which showed 32 and 50–55 mg/kg Zn without and with soil application of $ZnSO_4$, respectively in grains (Das et al., 2019). These studies show that genetic engineering, hybridization, and different treatments resulted in a good increase in Zn uptake from soil and Zn content in grain of crops as well, but some of these techniques can only be performed under laboratory conditions and cannot be made available on the field for large scale. Breeding and genetic biofortification methods are dependent on the interaction between genotype and environment (Joshi et al., 2010). Agronomic biofortification is also dependent on plant species, and rate, time, and place of fortification (Singh and Prasad, 2014). The application of nanoparticles to treat Zn deficiency could be an alternative for better fortification of Zn.

## 10.5 Nanotechnological interventions as a novel fortification tool

Nanotechnology is a growing sector in science equipped with several applications in almost every sector such as biomedicines, wastewater treatment, cancer therapy, cosmetic industry, and electronic and biosensors (Tripathi et al., 2017b). The manufactured nanoparticles (NPs) of size ranging from $<100$ nm have a remarkable application in various areas of the economy. In the agriculture field, nanoparticles play a crucial role as they can offer appropriate alternative sources in exchange for excessive use of chemicals, pesticides, and fungicides in agricultural crops to withstand various diseases and pests (El-Moneim et al., 2021). NPs can remarkably increase the efficiency of agricultural inputs by making use of nano-plant growth promoters, nano pesticides, nano fertilizers and nano herbicides. for the potential application in agriculture (Singh et al., 2021; Prakash et al., 2022). Nanofertilizer technology plays a key role in the agriculture sector. They help in lifting the crop growth, yield, and quality parameters along with increasing nutrient use efficiency, lowering the wastage of fertilizer application, and improving cost efficiency (Singh, 2017). It has been reported that nano fertilizers are capable of performing slow-release functions, and manage the supply of different macro and micro nutrients to the plants with precise concentration, adequate size, and additional surface area (Al-Mamun et al., 2021).

Among various micronutrients, different studies have reported the use of Zn oxide nanoparticles (ZnONPs) to treat Zn deficiency in different crops and have also observed sign of improvement in growth traits (Zulfiqar et al., 2019) (Fig. 10.4). For example, the study of the impact of application of biologically synthesized and chemically synthesized Zn oxide nanoparticle

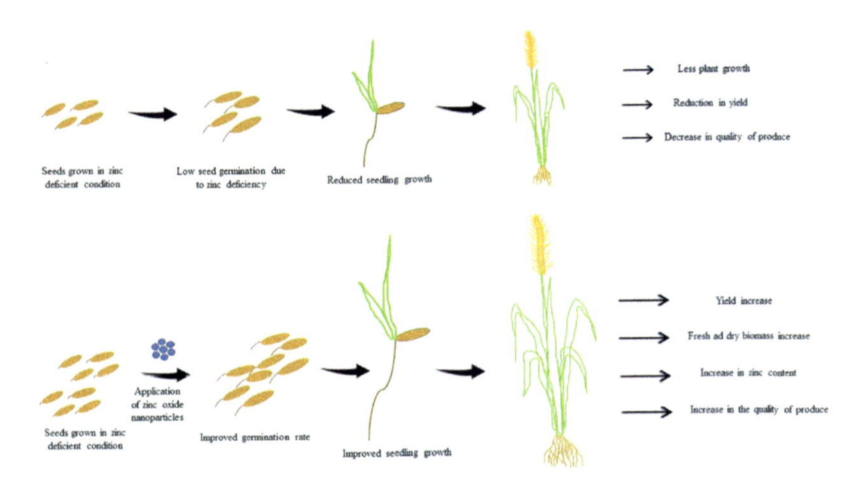


**FIGURE 10.4** Zinc oxide nanoparticles (ZnONPs)-mediated improvement of plant growth under zinc deficiency. Seed germination and seedling growth are reduced when seeds are grown in Zn-deficient soil. Overall plant growth and yield are also reduced due to Zn deficiency. However, the application of ZnONPs improved these growth traits and Zn content in the plant.

(BZnONPs and CZnONPs) on the growth of Sesame (*Sesamum indicum L.*) grown in Zn-deficient soil showed that BZnONPs can be used a nanofertilizer but CZnONPs imposed toxic impacts on the sesame plant (Narendhran et al., 2016). In another study, ZnONPs were used in rice as a nanofertilizer. The application of foliar spray of ZnONPs (5.0 g/L) on the surface of leaf in the variety PR121 grown in Zn-deficient soil resulted in a significant improvement in root characteristics and overall growth and yield parameters (Bala et al., 2019). Scientists also have studied the impact of ZnONPs (20 and 40 mg/L) on carrot (*Daucus carota*) variety cv Fire Wedge F1 Zn grown in Zn deficient soil (Awad et al., 2021). The results showed ZnONPs applied at 40 mg/L as the best treatment that provided the best plant growth in terms of shoot length, fresh and dry weights, physiological parameters such as cell membrane stability index, and nutrient uptake homeostasis (Awad et al., 2021). Yield was also measured to be highest in this case in terms of fresh and dry matter, diameter, length, and volume of carrot roots. Therefore, ZnONPs can be recommended for the sustainable production of carrot cultivation in Zn-deficient soil (Awad et al., 2021). In *Arachis hypogaea* L. grown under Zn-deficient condition, the application of 25 g of *Parthenium*-based vermicompost and ZnONPs (300 ppm) might increase the growth and yield of *A. hypogaea* (Rajiv and Vanathi, 2018).

Mahdieh et al. (2018) have experimented on the effect of seed and foliar application of ZnONPs and Zn fertilizers (Zn chelate and Zn sulfate) on rates of yield and growth of pinto bean (*Phaseolus vulgaris*) cultivar "KS21191" and "KS21193." The results depicted that ZnONPs improved vegetative parameters including plant height, root-shoot dry and fresh weight, internode length, yield (number of pods and weight of seeds), and Zn content in seed of both the cultivars suggesting that seed priming or foliar application of nanoscale nutrients can be applied to the crops to get the desired results (Mahdieh et al., 2018). Application of nanofertilizer is no doubt a promising technique that can resolve the major problem with conventional fertilizers, i.e., need for large quantities for application (Saleem et al., 2021). Therefore, the approach of introducing nanosized fertilizers, especially ZnONPs to overcome the Zn deficiency in plants seem to provide adequate results using low concentration with marvelous results (Sarkhosh et al., 2022). This would help in achieving sustainability and food security effectively and economically. Although the application of NPs is useful to the crops, there are some concerns associated with the use of NPs at large scale that need solutions. A study, on long-term foliar application, have demonstrated that the application of nanoparticle enhances the growth and physiological parameters in maize only in the first generation, while continuous application of NPs resulted in the reduced growth rate and physiological parameters such as chlorophyll content and photosynthesis (Jalali et al., 2017; Achari and Kowshik, 2018). Even ZnONPs are reported to impose toxic effects on plant growth at supraotimal concentrations (Tripathi et al., 2017a). Therefore, optimization of standard doses of NPs along with study of their implications on different components of the ecosystem including plants and other life forms must be conducted before their large-scale commercialization as nanofertilizers.

## 10.6 Conclusion

Mineral nutrients including micronutrients and macronutrients play an important role in performing physiological functions in plant metabolism. Micronutrients are essentially required for the growth of plant and help in activating enzyme reactions, but mostly these microelements remain unavailable to plants due to several reasons such as immobility, high pH of soil, greater soil salinity, high phosphorus in soil, high content of Mg, Na, Ca, bicarbonate and phosphate in the soil. Similarly, the availability of Zn is retarded due to its immobile nature and unfavorable conditions of soil. Therefore, Zn deficiency is observed to a great extent even though it is available in soil in good amounts. Zn deficiency in plants results in severe yield loss, yellowing of leaves, stunted growth, and death of plant in severe conditions. Although Zn deficiency can be treated by different methods such as conventional fortification, biofortification, and nanotechnological intervention. All these methods have a great impact in treating Zn deficiency but simultaneously they have few limitations. Conventional fortification results in the leaching of chemicals in groundwater along with problems like eutrophication. Similarly, biofortification is although better and greener alternative for uptake of Zn and protection against Zn deficiency, but these techniques are costly and cannot be adopted by small-scale farmers. In the case of fortification through NPs, the results are immensely good, the only drawback is that few nanoparticles only bind with certain ions and thus every micronutrient deficiency cannot be treated with nanoparticle technology. To overcome these problems, several new strategies are to be introduced to enrich Zn availability in edible plant parts, including biotechnological tools, nutrient management, and classical and molecular breeding practices along with future aspects. Furthermore, understanding the molecular basis of plant responses to Zn deficiency and alterations induced by different fortification methods at the gene level can help in the development of new crop varieties with improved productivity and enhanced tolerance to Zn deficiency.

## References

Abdoli, M., Esfandiari, E., Sadeghzadeh, B., Mousavi, S.B., 2016. Zinc application methods affect agronomy traits and grain micronutrients in bread and durum wheat under zinc-deficient calcareous soil. Yuzuncu Yil University Journal of Agricultural Sciences 26 (2), 202–214.

Achari, G.A., Kowshik, M., 2018. Recent developments on nanotechnology in agriculture: plant mineral nutrition, health, and interactions with soil microflora. Journal of Agricultural and Food Chemistry 66 (33), 8647–8661.

Alexandratos, N., 2005. Countries with rapid population growth and resource constraints: issues of food, agriculture, and development. Population and Development Review 31 (2), 237–258.

Alloway, B.J., 2009. Soil factors associated with zinc deficiency in crops and humans. Environmental Geochemistry and Health 31 (5), 537–548.

Al-Mamun, M.R., Hasan, M.R., Ahommed, M.S., Bacchu, M.S., Ali, M.R., Khan, M.Z.H., 2021. Nanofertilizers towards sustainable agriculture and environment. Environmental Technology & Innovation 23, 101658.

Assefa, S., Tadesse, S., 2019. The principal role of organic fertilizer on soil properties and agricultural productivity—a review. Agricultural Research & Technology Open Access Journal 22 (2), 556192.

Awad, A.A., Rady, M.M., Semida, W.M., Belal, E.E., Omran, W.M., Al-Yasi, H.M., et al., 2021. Foliar nourishment with different zinc-containing forms effectively sustains carrot performance in zinc-deficient soil. Agronomy 11 (9), 1853.

Bala, R., Kalia, A., Dhaliwal, S.S., 2019. Evaluation of efficacy of ZnO nanoparticles as remedial zinc nanofertilizer for rice. Journal of Soil Science and Plant Nutrition 19, 379–389.

Bindraban, P.S., Dimkpa, C., Nagarajan, L., Roy, A., Rabbinge, R., 2015. Revisiting fertilisers and fertilisation strategies for improved nutrient uptake by plants. Biology and Fertility of Soils 51 (8), 897–911.

Cakmak, I., Kutman, U.Á., 2018. Agronomic biofortification of cereals with zinc: a review. European Journal of Soil Science 69 (1), 172–180.

Das, S., Chaki, A.K., Hossain, A., 2019. Breeding and agronomic approaches for the biofortification of zinc in wheat (*Triticum aestivum* L.) to combat zinc deficiency in millions of a population: a Bangladesh perspective. Acta Agrobotanica 72 (2).

Das, S.K., Avasthe, R.K. and Yadav, A., 2017. Secondary and micronutrients: deficiency symptoms and management in organic farming.

Doran, J.W., Jones, A.J., Arshad, M.A., Gilley, J.E., 2018. Determinants of soil quality and health. Soil Quality and Soil Erosion. CRC Press, pp. 17–36.

El-Moneim, D.A., Dawood, M.F., Moursi, Y.S., Farghaly, A.A., Afifi, M., Sallam, A., 2021. Positive and negative effects of nanoparticles on agricultural crops. Nanotechnology for Environmental Engineering 6 (2), 21.

Esfandiari, E., Abdoli, M., Mousavi, S.B., Sadeghzadeh, B., 2016. Impact of foliar zinc application on agronomic traits and grain quality parameters of wheat grown in zinc deficient soil. Indian Journal of Plant Physiology 21, 263–270.

Furness, G.O., Thompson, A.J., Manktelow, D.W.L., 2006. A visual droplet number rating chart and fluorescent pigment sprays to estimate chemical deposition and spray coverage on plant foliage. Aspects of Applied Biology 77 (1), 171.

Ganguly, R., Sarkar, A., Dasgupta, D., Acharya, K., Keswani, C., Popova, V., et al., 2022. Unravelling the efficient applications of zinc and selenium for mitigation of abiotic stresses in plants. Agriculture 12 (10), 1551.

Graham, R.D., 2008. Micronutrient Deficiencies in Crops and Their Global Significance. Springer, Netherlands, pp. 41–61.

Hacisalihoglu, G., 2020. Zinc (Zn): the last nutrient in the alphabet and shedding light on Zn efficiency for the future of crop production under suboptimal Zn. Plants 9 (11), 1471.

Hafeez, F.Y., Abaid-Ullah, M., Hassan, M.N., 2013. Plant growth-promoting rhizobacteria as zinc mobilizers: a promising approach for cereals biofortification. Bacteria in Agrobiology: Crop Productivity 217–235.

Hajiboland, R., 2012. Effect of micronutrient deficiencies on plants stress responses. Abiotic Stress Responses in Plants: Metabolism, Productivity And Sustainability 283–329.

Hidoto, L., Worku, W., Mohammed, H., Bunyamin, T., 2017. Effects of zinc application strategy on zinc content and productivity of chickpea grown under zinc deficient soils. Journal of Soil Science and Plant Nutrition 17 (1), 112–126.

Hussain, A., Arshad, M., Zahir, Z.A., Asghar, M., 2015. Prospects of Zn solubilizing bacteria for enhancing growth of maize. Pakistan Journal of Agricultural Sciences 52 (4).

Hussain, A., Zahir, Z.A., Asghar, H.N., Ahmad, M., Jamil, M., Naveed, M. and et al., 2018. Zinc solubilizing bacteria for zinc biofortification in cereals: a step toward sustainable nutritional security. Role of Rhizospheric Microbes in Soil: Volume 2: Nutrient Management and Crop Improvement, pp. 203–227.

Jalali, M., Ghanati, F., Modarres-Sanavi, A.M., Khoshgoftarmanesh, A.H., 2017. Physiological effects of repeated foliar application of magnetite nanoparticles on maize plants. Journal of Agronomy and Crop Science 203 (6), 593–602.

Jan, B., Bhat, T.A., Sheikh, T.A., Wani, O.A., Bhat, M.A., Nazir, A., et al., 2020. Agronomic bio-fortification of rice and maize with iron and zinc: a review. International Research Journal of Pure and Applied Chemistry 21, 28–37.

Jerlin, B., Sharmila, S., Kathiresan, K., Kayalvizhi, K., 2017. Zn solubilizing bacteria from rhizospheric soil of mangroves. Journal of Microbiology and Biotechnology 2 (3), 148–155.

Joshi, A.K., Crossa, J., Arun, B., Chand, R., Trethowan, R., Vargas, M., et al., 2010. Genotype × environment interaction for zinc and iron concentration of wheat grain in eastern Gangetic plains of India. Field Crops Research 116 (3), 268–277.

Kaur, J., 2021. PGPR in management of soil toxicity. Rhizobiont in Bioremediation of Hazardous Waste 317–344.

Kaur, T., Rana, K.L., Kour, D., Sheikh, I., Yadav, N., Kumar, V., et al., 2020. Microbe-mediated biofortification for micronutrients: present status and future challenges. New and Future Developments in Microbial Biotechnology and Bioengineering. Elsevier, pp. 1–17.

Khan, M.N., Mobin, M., Abbas, Z.K., Alamri, S.A., 2018. Fertilizers and their contaminants in soils, surface and groundwater. Encyclopedia of the Anthropocene 5, 225–240.

Khan, S.T., Malik, A., Alwarthan, A., Shaik, M.R., 2022. The enormity of the zinc deficiency problem and available solutions; an overview. Arabian Journal of Chemistry 15 (3), 103668.

Mahdieh, M., Sangi, M.R., Bamdad, F., Ghanem, A., 2018. Effect of seed and foliar application of nano-zinc oxide, zinc chelate, and zinc sulphate rates on yield and growth of pinto bean (*Phaseolus vulgaris*) cultivars. Journal of Plant Nutrition 41 (18), 2401–2412.

Maqbool, M.A., Beshir, A., 2019. Zinc biofortification of maize (*Zea mays* L.): Status and challenges. Plant Breeding 138 (1), 1–28.

Mattiello, E.M., Ruiz, H.A., Neves, J.C., Ventrella, M.C., Araújo, W.L., 2015. Zinc deficiency affects physiological and anatomical characteristics in maize leaves. Journal of Plant Physiology 183, 138–143.

Mumtaz, M.Z., Ahmad, M., Jamil, M., Hussain, T., 2017. Zinc solubilizing *Bacillus* spp. potential candidates for biofortification in maize. Microbiological Research 202, 51–60.

Narendhran, S., Rajiv, P. and Sivaraj, R., 2016. Influence of zinc oxide nanoparticles on growth of *Sesamum indicum* L. in.

Nielsen, F.H., 2012. History of zinc in agriculture. Advances in Nutrition 3 (6), 783–789.

Pahalvi, H.N., Rafiya, L., Rashid, S., Nisar, B., Kamili, A.N., 2021. Chemical fertilizers and their impact on soil health. *Microbiota and BiofertilizersVol 2* Ecofriendly Tools for Reclamation of Degraded Soil Environs 1–20.

Palmgren, M.G., Clemens, S., Williams, L.E., Krämer, U., Borg, S., Schjørring, J.K., et al., 2008. Zinc biofortification of cereals: problems and solutions. Trends in Plant Science 13 (9), 464–473.

Prakash, V., Rai, P., Sharma, N.C., Singh, V.P., Tripathi, D.K., Sharma, S., et al., 2022. Application of zinc oxide nanoparticles as fertilizer boosts growth in rice plant and alleviates chromium stress by regulating genes involved in oxidative stress. Chemosphere 303, 134554.

Prasad, A.S., 2012. Discovery of human zinc deficiency: 50 years later. Journal of Trace Elements in Medicine and Biology 26 (2–3), 66–69.

Rajiv, P., Vanathi, P., 2018. Effect of parthenium based vermicompost and zinc oxide nanoparticles on growth and yield of *Arachis hypogaea* L. in zinc deficient soil. Biocatalysis and Agricultural Biotechnology 13, 251–257.

Ramesh, A., Sharma, S.K., Sharma, M.P., Yadav, N., Joshi, O.P., 2014. Inoculation of zinc solubilizing *Bacillus aryabhattai* strains for improved growth, mobilization and biofortification of zinc in soybean and wheat cultivated in Vertisols of central India. Applied Soil Ecology 73, 87–96.

Rana, A., Joshi, M., Prasanna, R., Shivay, Y.S., Nain, L., 2012. Biofortification of wheat through inoculation of plant growth promoting rhizobacteria and cyanobacteria. European Journal of Soil Biology 50, 118–126.

Rehman, A., Farooq, M., Ullah, A., Nadeem, F., Im, S.Y., Park, S.K., et al., 2020. Agronomic biofortification of zinc in Pakistan: status, benefits, and constraints. Frontiers in Sustainable Food Systems 4, 591722.

Rekha, N., S.N., Prasad, R., 2006. Pesticide residue in organic and conventional food-risk analysis. Journal of Chemical Health & Safety 13 (6), 12–19.

Rudani, L., Vishal, P., Kalavati, P., 2018. The importance of zinc in plant growth—a review. International Research Journal of Natural and Applied Sciences 5 (2), 38–48.

Sajjad, M.R., Rafique, R., Bibi, R., Umair, A., Arslan Afzal, A.A., Rafique, T., 2019. 3. Performance of green manuring for soil health and crop yield improvement. Pure and Applied Biology (PAB) 8 (2), 1543–1553.

Saleem, I., Maqsood, M.A., ur Rehman, M.Z., Aziz, T., Bhatti, I.A., Ali, S., 2021. Potassium ferrite nanoparticles on DAP to formulate slow release fertilizer with auxiliary nutrients. Ecotoxicology and Environmental Safety 215, 112148.

Saltzman, A., Birol, E., Bouis, H.E., Boy, E., De Moura, F.F., Islam, Y., et al., 2013. Biofortification: progress toward a more nourishing future. Global Food Security 2 (1), 9–17.

Sarkhosh, S., Kahrizi, D., Darvishi, E., Tourang, M., Haghighi-Mood, S., Vahedi, P., et al., 2022. Effect of zinc oxide nanoparticles (ZnO-NPs) on seed germination characteristics in two Brassicaceae family species: *Camelina sativa* and *Brassica napus* L. Journal of Nanomaterials 2022, 1–15.

Shah, A., Gupta, P. and Ajgar, Y.M., 2018, April. Macro-nutrient deficiency identification in plants using image processing and machine learning. In 2018 3rd International conference for convergence in technology (I2CT) (pp. 1–4). IEEE.

Shaikh, S., Saraf, M., 2017. Zinc biofortification: strategy to conquer zinc malnutrition through zinc solubilizing PGPR's. Biomedical Journal of Scientific & Technical Research 1 (1), 224–226.

Singh, M.D., 2017. Nano-fertilizers is a new way to increase nutrients use efficiency in crop production. International Journal of Agriculture Sciences, ISSN 9 (7), 0975–3710.

Singh, M.K., Prasad, S.K., 2014. Agronomic aspects of zinc biofortification in rice (*Oryza sativa* L.). Proceedings of the national academy of sciences, India section B: biological Sciences 84, 613–623.

Singh, R.P., Handa, R., Manchanda, G., 2021. Nanoparticles in sustainable agriculture: an emerging opportunity. Journal of Controlled Release 329, 1234–1248.

Sirohi, G., Pandey, B.K., Deveshwar, P., Giri, J., 2016. Emerging trends in epigenetic regulation of nutrient deficiency response in plants. Molecular Biotechnology 58, 159–171.

Song, C.Z., Liu, M.Y., Meng, J.F., Chi, M., Xi, Z.M., Zhang, Z.W., 2015. Promoting effect of foliage sprayed zinc sulfate on accumulation of sugar and phenolics in berries of *Vitis vinifera* cv. Merlot growing on zinc deficient soil. Molecules (Basel, Switzerland) 20 (2), 2536–2554.

Stanton, C., Sanders, D., Krämer, U., Podar, D., 2021. Zinc in plants: integrating homeostasis and biofortification. Molecular Plant .

Sunithakumari, K., Devi, S.P., Vasandha, S., 2016. Zinc solubilizing bacterial isolates from the agricultural fields of Coimbatore, Tamil Nadu, India. Current Science 196–205.

Tripathi, D.K., Mishra, R.K., Singh, S., Singh, S., Vishwakarma, K., Sharma, S., et al., 2017a. Nitric oxide ameliorates zinc oxide nanoparticles phytotoxicity in wheat seedlings: implication of the ascorbate–glutathione cycle. Frontiers in Plant Science 8, 1.

Tripathi, D.K., Singh, S., Singh, S., Pandey, R., Singh, V.P., Sharma, N.C., et al., 2017b. An overview on manufactured nanoparticles in plants: uptake, translocation, accumulation and phytotoxicity. Plant Physiology and Biochemistry 110, 2–12.

Umair Hassan, M., Aamer, M., Umer Chattha, M., Haiying, T., Shahzad, B., Barbanti, L., et al., 2020. The critical role of zinc in plants facing the drought stress. Agriculture 10 (9), 396.

Vatansever, R., Ozyigit, I.I., Filiz, E., 2017. Essential and beneficial trace elements in plants, and their transport in roots: a review. Applied Biochemistry and Biotechnology 181, 464–482.

Velu, G., Rai, K.N., Muralidharan, V., Kulkarni, V.N., Longvah, T., Raveendran, T.S., 2007. Prospects of breeding biofortified pearl millet with high grain iron and zinc content. Plant Breeding 126 (2), 182–185.

Wissuwa, M., Ismail, A.M., Yanagihara, S., 2006. Effects of zinc deficiency on rice growth and genetic factors contributing to tolerance. Plant Physiology 142 (2), 731–741.

Zhao, A.Q., Tian, X.H., Cao, Y.X., Lu, X.C., Liu, T., 2014. Comparison of soil and foliar zinc application for enhancing grain zinc content of wheat when grown on potentially zinc-deficient calcareous soils. Journal of the Science of Food and Agriculture 94 (10), 2016–2022.

Zulfiqar, F., Navarro, M., Ashraf, M., Akram, N.A., Munné-Bosch, S., 2019. Nanofertilizer use for sustainable agriculture: advantages and limitations. Plant Science 289, 110270.

Zuo, Y., Zhang, F., 2009. Iron and zinc biofortification strategies in dicot plants by intercropping with gramineous species. A review. Agronomy for Sustainable Development 29, 63–71.

Chapter 11

# Role of zinc solubilizing bacteria in sustainable agriculture

**Shikha Gupta[1], Sangeeta Pandey[1], Monika Singh[2] and Vashista Kotra[1]**
[1]*Amity Institute of Organic Agriculture, Amity University Uttar Pradesh, Noida, Uttar Pradesh, India,*
[2]*G.L. Bajaj Institute of Technology and Management, Greater Noida, Uttar Pradesh, India*

## 11.1 Introduction

The burgeoning world population which has surpassed 8 billion as per United Nations reports poses a significant threat to global food production and nutritional security (Fasusi et al., 2021). Food security is defined as "the state or condition in which all individuals have physical, economic, and social access to sufficient, safe, and nutritious food for optimal human health and well-being" (Yadav et al., 2022). The inclusion of the term nutritious entails the macro- and micronutrient content of food that are vital components of food security (Ingram, 2020). Micronutrient deficiency affects more than half of the world's population by increasing the risk of malnutrition with the highest prevalence in low and middle-income nations (Hwalla et al., 2016). Of all the micronutrients, zinc is the essential micronutrient required for the proper metabolism of living organisms including plants, animals, humans, and microbes (Ditta et al., 2022). Consequently, the deficiency of zinc known as "hidden hunger" is responsible for several metabolic abnormalities that manifest different detrimental impacts on human growth and development, primarily the impairment of the immune and cognitive development in children (de Valença et al., 2017). The production of crops and consequent consumption of zinc-enriched food-based crops is the current footstep to circumvent the zinc deficiency in developing countries in a sustainable manner.

Zinc is an important and essential micronutrient for proper plant nutrition and optimal growth and development of plants (Sharma et al., 2013). It acts as a structural, functional, and regulatory component in many physiological processes of plants such as protein synthesis, carbohydrate metabolism, maintenance of the integrity of the biological membrane, photosynthesis, chlorophyll synthesis, translation of gene into protein, catalysis of the

**Zinc in Plants. DOI: https://doi.org/10.1016/B978-0-323-91314-0.00005-3**

enzymatic reaction and provides resistance to abiotic and biotic stressed conditions (Suriyachadkun et al., 2022).

Zinc deficiency is a well-known micronutrient deficiency in crops throughout the world which hampers the growth of many crop plants such as wheat, maize, sugarcane, and rice. The deficiency of zinc is most often found in many crops resulting in stunted growth, chlorosis of leaves, and low- and poor-quality crop yield which in turn contribute to malnutrition condition in humans suffering from insufficient intake of zinc (Yadav et al., 2022). The occurrence of zinc deficiency in crops is due to the concentration of total zinc which is normally low (64 mg Zn/kg) in the soil (Alloway, 2009). The factors responsible for the low amount of plant-available forms of zinc in soil are the alkaline pH of the soil, the presence of a high concentration of bicarbonates, phosphates, $Na^{+}$ and $Mg^{+2}$ ions, prolonged irrigation with saline water or waterlogging conditions (Das and Green, 2013). The deficiency of zinc affects the health of agricultural soils and restricts the uptake of zinc by plants which in turn results in a decrease in the agronomic yield, and ultimately the nutritional quality of food degrades and adversely affects human health (Noulas et al., 2018). Moreover, plants absorb zinc as a divalent cation ($Zn^{2+}$ and $ZnOH^{+}$) and soluble organic zinc complexes, and only a small portion of total zinc becomes available to plants due to the formation of insoluble complexes in soil (Recena et al., 2021). The worldwide prevalence of Zinc deficiency in crops is due to the low solubility of Zn, rather than low Zinc availability in soil (Iqbal et al., 2010). It is predicted that the presence of a low amount of plant-available zinc in approximately 50% of agricultural soil, reduces not just output but also jeopardizes the nutritional content of grains and derivatives (Rudani et al., 2018).

In order to mitigate zinc deficiency in food-based crops and to ensure proper zinc nutrition, it is necessary to supplement the agricultural soil with zinc micronutrients through various approaches (Haroon et al., 2022). The usage of chemically based zinc-enriched fertilizers is the most practical approach for improving the zinc uptake by roots and its content in grains and thus effective in enhancing the agronomic yield. Although they play a significant role in zinc supplementation of soil, the degradation of soil nutrients and fertility, accumulation of zinc insoluble complexes, alteration of physical and chemical characteristics of soil, and cost-associated factors limit their usage in agriculture (Chakraborty et al., 2022). The rampant application of synthetic zinc fertilizers has led to economic as well as environmental pressure and could easily be converted into their insoluble form leading to poor assimilation by plants and thus exacerbating zinc deficiency in plants. In light of this, synthetic fertilizers are discouraged (Kumar et al., 2019a). Another approach widely used to alleviate zinc deficiency is conventional breeding, the creation of transgenic and molecular breeding, however, these approaches are labor-intensive, costly, and slow processes (Henriques et al., 2011).

Moving forward, it will be important not only to place greater emphasis on the enhancement of agricultural productivity of nutrient-enriched food but also to shift focus to long-term sustainable agricultural development and enhanced food security. In this regard, microbe-based zinc fertilization is gaining momentum in increasing agricultural output and improving product quality while limiting the negative footprints of chemical-based agronomic methods on natural resources and the environment. The application of zinc-solubilizing bacteria is a better alternative owing to their low cost of production, and safer and environmentally benign approach replacing all the above-mentioned approaches (Bhatt and Maheshwari, 2020). Zinc solubilizing bacteria are an important group of microorganisms that play a vital role in the solubilization of insoluble, inorganic zinc compounds through various mechanisms and to make them accessible for plant growth and development. Exploiting zinc-solubilizing bacteria as a bioinoculant could be a pragmatic strategy for replenishing soil's zinc content and synergistically improving plants' zinc content and growth (Hefferon, 2019).

Zinc solubilizing bacteria convert zinc complexes into simpler ones, making zinc available to plants through various including chelation, pH alterations in the soil, and proton extrusion. The most well-studied mechanism is the release of organic acids such as gluconic acid, butyric acid, lactic acid, and oxalic acid by rhizospheric bacteria. Organic acid synthesis increases availability by cation sequestration and lowers soil pH (Zaheer et al., 2019).

Different zinc solubilizing bacteria genera as bioinoculant have enhanced growth and zinc content in plants including *Pseudomonas*, *Rhizobium* strains (Joshi et al., 2013; Naz et al., 2016), *Bacillus aryabhattai* (Ramesh et al., 2014), and *Bacillus* sp. (Hussain et al., 2015). The bacterial-assisted biofortification provides a practical and cost-effective option to biofortify crops with an elevated content of micronutrients such as zinc in the edible parts of plants. Such biofortification strategy is extensively studied with an aim to address the global Zn malnutrition problem (Upadhayay et al., 2022a).

In this chapter, we documented the role of zinc-solubilizing bacteria in the alleviation of Zinc deficiency in plants, the zinc solubilization mechanism, and their plant growth-promoting attributes in response to abiotic and biotic stress.

## 11.2 Importance of zinc in plants

Zinc, the second most common transition metal, is an essential trace element for life as it is required for the structural integrity and catalytic activity of all the classes of enzymes including hydrolases, lyases, oxidoreductases, and transferases (Fariduddin et al., 2022). Zinc is an equally essential micronutrient required in trace amounts by plants contributing to the structural and catalytic activity of various enzymes such as DNA, RNA polymerases, and oxidoreductases, as well as in zinc finger proteins. Zinc also plays a pivotal role in growth and development, defense response against pests and diseases as well as metabolism processes of plants. It is essential for several crucial metabolic and physiological functions in plants, including glucose metabolism, photosynthesis, and biosynthesis of chlorophyll, sucrose, and starch. It is required for lipid and nucleic acid metabolism, gene expression, gene transcription, protein synthesis, auxin metabolism, antioxidant activity, and membrane integrity (Khalid et al., 2022).

The deficiency of zinc is most often found in many crops resulting in stunted growth, chlorosis of leaves, and low- and poor-quality crop yield which in turn contribute to malnutrition condition in humans, primarily affecting children suffering from insufficient intake of zinc (Pascual et al., 2016; Hafeez et al., 2013). Since plants are the primary producers and an essential component of food, their low Zn content has a direct impact on the Zn status of humans. One of the key causes of Zn insufficiency in humans is a poor diet. The occurrence of zinc deficiency in crops is associated with the low availability of plant-utilizable forms of zinc such as free ions ($Zn^{2+}$ and $ZnOH^{+}$) and soluble organic zinc complexes. Other factors responsible for zinc deficiency in crops include the alkaline pH of the soil, the presence of higher amounts of calcite, bicarbonates, sodium, and insoluble phosphate content in the soil, and excessive waterlogging conditions (Hamzah Saleem et al., 2022). There are other agronomic factors responsible for the reduced content of zinc in the soil such as extensive usage of chemical zinc-based fertilizers (Zinc oxide, Zinc carbonate, Zinc sulfate, Zinc nitrate, and Zinc chloride), inadequate irrigation system, and intensive cultivation practices (Singh et al., 2023).

Zinc is abundantly present in the earth's crust, and some of the most commonly found zinc minerals in the soil include Franklinite ($ZnFe_2O_4$), Hopeite ($Zn_3(PO_4)_2$ $4H_2O$), Smithsonite ($ZnCO_3$), Zincite ($ZnO$), and Zinkosite ($ZnSO_4$) (Khan et al., 2022).

Zinc deficiency impairs most agricultural production in approximately 50% of Indian soils particularly prevalent in states like Haryana, Maharashtra, and Karnataka. This might be due to the presence of different forms of insoluble zinc complexes such as zinc carbonate, zinc oxide, zinc sulfide, zinc ferrite, etc., which constitute more than 90% of the total zinc content present in the Indian agricultural soils and hence remains unavailable to plants (Khan et al., 2022).

## 11.3 Zinc solubilizing bacteria and mechanism of zinc solubilization

Zinc solubilizing bacteria increase the bioavailability of zinc by transforming insoluble complexes of zinc into plant-utilizable forms required for enhancing crop growth and nutritional quality. The beneficial zinc solubilizing bacteria employ different mechanisms for solubilizing zinc insoluble complexes with the production of an organic acid being the major and widely reported mechanism. Zinc solubilizing bacteria facilitate the transformation of insoluble complexes of zinc into plant-available forms by secreting organic acids (citric acid, oxalic acid, gluconic acid, 2-ketogluconic acid), protons, and hydroxyl ions which solubilize precipitated zinc by lowering the pH of the soil and thus increased the concentration of available zinc for plants. Different strains of *Pseudomonas* and *Bacillus* sp. have been reported to produce gluconic acid, hexanoic acid, pentanoic acid, and mandelic acids for zinc solubilization and consequently, improved the growth of chickpea plants under the rainfed region (Kushwaha et al., 2021). Upadhyay et al. (2021) analyzed the production of oxalic, maleic, tartaric, and fumaric acid through HPLC by zinc solubilization bacteria- *Burkholderia cepacia* and *Acinetobacter baumannii*. Similarly, the production of lactic acid, acetic acid, succinic, formic, isobutyric, and isovaleric acids has been ascribed to the possible mechanism of zinc solubilization by *Bacillus* sp. and *Dietzia maris* (Mumtaz et al., 2019; Rani et al., 2022). Additional potential mechanistic pathways for zinc solubilization include the formation of siderophores, amino acids, chelated ligands, vitamins, proton, and oxidoreductase systems on cell membranes (Upadhayay et al., 2022b).

## 11.4 Role of zinc solubilizing bacteria in abiotic and biotic stress management

Zinc, as an essential micronutrient, is vital for various biological processes within the plant system, however, its availability in soil is often limited due to a variety of abiotic and biotic factors. This is where zinc-solubilizing bacteria come in; these microorganisms are known for zinc solubilization from insoluble complexes of soil minerals, thereby making it readily available for assimilation by plants (Saleem and Khan, 2022). The mechanistic insights of zinc solubilizing bacteria-induced solubilization include the production of organic acids, siderophores, and chelating agents that break down zinc-containing minerals, releasing the bound zinc ions into the soil solution (Upadhyay et al., 2021).

In addition to zinc solubilization potential, zinc solubilizing bacteria have been shown to promote plant growth and development through various direct and indirect mechanisms such as nitrogen fixation, production of plant growth-promoting hormones, production of siderophore, mineral solubilization and induction of systemic resistance in response to abiotic stress such as drought, salinity, and heavy metal toxicity as well as pathogen attack (Verma et al., 2022).

Recent research has demonstrated that inoculation with zinc-solubilizing bacteria can significantly enhance plant growth and productivity under various environmental stresses as depicted in Fig. 11.1. For instance, Singh et al. (2022) have reported that zinc solubilizing bacteria, *Pseudomonas protegens*

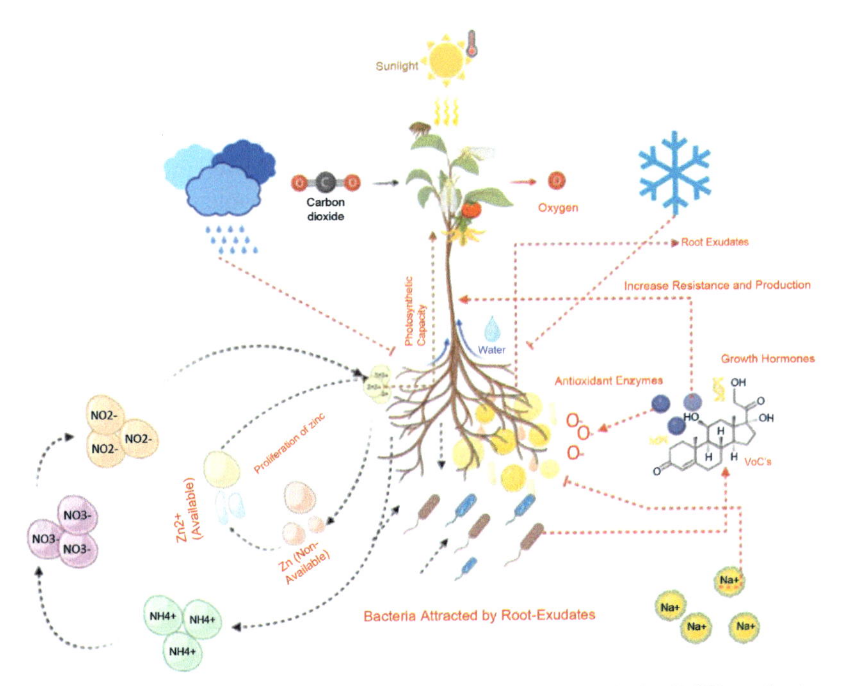


**FIGURE 11.1** A Schematic representation of zinc solubilization and other PGPR mechanisms including biotic and abiotic stress management mediated by zinc solubilizing bacteria in rhizosphere of plants.

has promoted zinc biofortification and growth promotion in wheat plants under saline stress conditions. Similarly, zinc-solubilizing bacteria *Rhizobium radiobacter* has been demonstrated to promote the growth of lettuce under saline-stress conditions (Verma et al., 2020).

In addition, zinc-solubilizing bacteria have been shown to enhance soil fertility and nutrient availability. They can solubilize phosphorus, another essential plant nutrient, from insoluble phosphate forms in soil. Phosphorus is another key plant macronutrient; however, it remains a limiting nutrient because of the less availability of orthophosphoric anionic forms for plant uptake and assimilation. They facilitate the transformation of insoluble complexes of phosphate into plant-available forms by employing various mechanisms, for instance, secretion of organic acids, protons, and hydroxyl ions which solubilize precipitated phosphate by lowering the pH of the rhizosphere and thus increase the concentration of available phosphate for plants. This makes phosphorus more available to plants and can improve their growth and productivity (Shreya et al., 2023). Another work accomplished by Martín et al. (2007) isolated *Pseudomonas fragi* strain, SAPA2 from barley rhizosphere and described its role in phosphorous (0.2% tricalcium phosphate) solubilization in strawberry plants growing in a soil amended with insoluble phosphate, thus increasing the crop yield. Several reports have documented the phosphate solubilization potential of *Pantoea agglomerans* R-42, *Pseudomonas* sp. and *Enterobacter* sp., *Burkholderia lata*, *Lysinibacillus* sp. *Bacillus* sp. is isolated from the rhizosphere of various crops and shown as an efficient plant growth-promoting rhizobacterial strain (Son et al., 2006; Suleman et al., 2018; Dinesh et al., 2018).

Zinc-solubilizing bacteria can also produce plant growth-promoting substances, such as indole-3-acetic acid (IAA), which can stimulate root growth and enhance nutrient uptake (Othman et al., 2022). Indole acetic acid (IAA) is the most important plant growth regulator that enhances many avenues of plant growth and development processes such as cell elongation, cell division, vascular tissue differentiation, fruit ripening and maturation, inhibition of bud outgrowth (apical dominance), and leaf senescence (Paque and Weijers, 2016). Several zinc solubilizing bacterial strains *Bacillus* sp., *Acinetobacter* sp., and *Paenibacillus* sp. from the rhizosphere of various crop plants such as maize and rice (Gandhi and Muralidharan 2016; Mumtaz et al., 2017) were reported to produce Indole acetic acid.

Apart from Indole acetic acid, zinc solubilizing bacteria also produce other phytohormones such as cytokinin, and gibberellins which facilitate plant growth by regulating various physiological processes (Sindhu et al., 2019). Cytokinin is another key plant growth hormone that is structurally similar to adenine and is found in low concentrations in all plant tissues, particularly root tips, shoot apex, and developing seed and fruit. They are known to promote cell division (cytokinesis) in roots and shoots, promote leaf and flower primordia creation and bud outgrowth, and prevent lateral root formation and premature leaf senescence (Bris, 2017). Likewise, Gibberellic acid is a tetracyclic diterpenoid-like phytohormone that promotes plant development and physiological processes such as seed germination, dormancy, stem and leaf elongation, pollen germination, and pollen tube growth. Gibberellins also aid in the transition from the juvenile to adult vegetative phase and the flowering stage by promoting the appropriate development of flower reproductive organs (Gupta et al., 2013). Additionally, zinc solubilizing bacteria can produce 1-aminocyclopropane-1-carboxylate (ACC) deaminase, an enzyme that helps plants tolerate environmental stresses by reducing the levels of the stress hormone, ethylene. This enzyme catalyzes the degradation of ACC, which is a precursor of ethylene, thereby reducing the levels of ethylene that accumulate in plants under stress (Othman et al., 2022).

Siderophores are low molecular weight iron-chelating secondary compounds produced by zinc solubilizing bacteria to bring the soil iron into soluble forms and provide iron nutrients to plants under iron-limited conditions. The previous investigation reported the siderophore production by zinc solubilizing bacterial strains such as *Pseudomonas japonica*, *Serratia* sp., *Burkholderia*, and *Acinetobacter* (Vaid et al., 2014; Kour et al., 2019).

Additionally, zinc solubilizing bacteria can produce chelating agents such as ethylenediaminetetraacetic acid (EDTA), which can bind to zinc ions and form stable complexes, enhancing the solubilization of zinc from soil minerals (Yadav et al., 2020).

Zinc solubilizing bacteria also have the potential to fix atmospheric nitrogen and convert it into a plant-available form, thereby enhancing plant nutrition and reducing reliance on nitrogen-enriched chemical fertilizers. This ability is mediated by the production of nitrogenase, an enzyme that catalyzes the reduction of atmospheric nitrogen to ammonia, which can be assimilated by plants (Yaghoubi Khanghahi et al., 2021).

Biocontrol is one of the potential applications of zinc-solubilizing bacteria, as it refers to the use of microorganisms to control plant diseases inflicted by pathogens. The use of zinc solubilizing bacteria as biocontrol agents involves the production of antimicrobial compounds including bacteriocins, hydrogen cyanide, and phenazine compounds, and the induction of systemic resistance in plants against pathogens. The induction of systemic resistance in plants by zinc-solubilizing bacteria involves the production of signaling molecules, such as salicylic acid and jasmonic acid, which activate the plant's defense mechanisms against pathogen attack (Jamali et al., 2020).

Overall, the multifaceted approach of zinc solubilizing bacteria makes it a promising microbial inoculant for promoting plant growth and enhancing stress tolerance in agriculture. Further research is needed to understand the mechanistic perspective involved in zinc solubilizing bacteria-mediated plant-microbe interactions and to optimize their application in different agricultural systems.

## 11.5 Biofortification of crop using zinc solubilizing bacteria

Biofortification is a process of increasing the nutritional value of food crops by enhancing the content of essential micronutrients such as iron, zinc, and vitamin A. This approach aims to address malnutrition, which affects millions of people worldwide, particularly in developing countries. Biofortification can be achieved through traditional breeding methods or genetic engineering, and it has the potential to improve the health and well-being of vulnerable populations (Nestel et al., 2006).

Biofortification has been shown to be a promising strategy to address micronutrient deficiencies, which can lead to a range of health problems, including stunted growth, impaired cognitive development, and increased susceptibility to infections. By increasing the availability of key micronutrients in staple crops such as rice, wheat, maize, and beans, biofortification can help to improve the nutritional status of populations who rely on these crops as their main source of food (Kiran et al., 2022).

The success of biofortification depends on several factors, including the genetic variability of the crop, the availability of micronutrients in the soil, and the effectiveness of the biofortification method used (Kumar et al., 2019b). In addition, biofortification should be integrated into broader efforts to address malnutrition, such as nutrition education, food fortification, and supplementation programs.

Biofortification has gained increased attention and support from governments, non-governmental organizations, and the private sector. Several international initiatives, such as the HarvestPlus program and the Biofortification Scaling Up Project, are working to promote biofortification and to develop and disseminate biofortified crops (Tekle et al., 2023).

Biofortification of zinc is essential to address zinc deficiency in populations that depend on cereal-based diets. Zinc deficiency affects more than 2 billion people worldwide and is a leading cause of stunting, impaired cognitive development, and weakened immune systems. Biofortification of staple crops with zinc can increase the intake of this essential micronutrient and reduce the prevalence of zinc deficiency in vulnerable populations (Adetola et al., 2023).

Biofortification of crops using Zinc Solubilizing Bacteria (ZSB) is an innovative and sustainable approach to improve crop productivity and reduce the use of synthetic fertilizers in agriculture (Yogi et al., 2023). Some zinc-solubilizing bacteria of genera *Pseudomonas*, *Bacillus*, *Burkholderia*, *Exiguobacterium* have been shown to play a vital role in zinc biofortification of food-based crops such as okra, chickpea, rice, wheat (Karnwal, 2021; Batool et al., 2021; Gontia-Mishra et al., 2017; Shaikh and Saraf, 2017). The application of ZSB in biofortification offers several benefits for agriculture, environmental sustainability, and human health. First, it can enhance the availability of zinc in soil, leading to improved nutrient uptake and increased plant growth and yield. Zinc deficiency can limit crop productivity and quality, and the use of ZSB can address this issue. Second, the use of ZSB can reduce the use of synthetic fertilizers, which can have negative environmental impacts, such as soil degradation, water pollution, and greenhouse gas emissions. By reducing the use of synthetic fertilizers, the application of ZSB can contribute to environmental sustainability. Third, the application of ZSB can enhance plant tolerance to biotic and abiotic stresses, such as pests, diseases, drought, and salinity. This can lead to improved crop productivity and quality, especially under adverse growing conditions. Fourth, the use of ZSB can promote soil health and sustainability by improving soil structure, increasing soil organic matter, and reducing soil erosion. Lastly, the application of ZSB can improve human health by enhancing the nutritional quality of crops, especially in regions where zinc deficiency is prevalent (Khan et al., 2023).

## 11.6 Conclusion and future perspective

Zinc is a micronutrient that plays a crucial role in the growth and development of plants, and its deficiency is a major challenge to global food security. Zinc deficiency is prevalent in soils worldwide, particularly in calcareous, alkaline, waterlogged, and sandy soils. Zinc deficiency poses a significant problem for human nutrition, particularly in developing countries where malnutrition is already a significant issue. The use of fertilizers and improved agronomic practices can lead to environmental pollution and may not be financially viable for small-scale farmers. The development of transgenic plants can be expensive and may raise concerns about the safety of genetically modified crops. In recent years, the use of zinc-solubilizing bacteria has emerged as an eco-friendly and cost-effective alternative to traditional methods for increasing the bioavailability of zinc in soil and improving crop productivity. Zinc-solubilizing bacteria are a diverse group of microorganisms that possess the ability to solubilize insoluble forms of zinc in the soil, making it available to plants. The use of zinc-solubilizing bacteria is a promising approach for replenishing zinc content in the soil, increasing crop productivity and nutritional quality of crops. The mobilization of nutrients from zinc-solubilizing bacteria is a critical function that improves the availability of zinc in soil and its uptake by plants. This, in turn, can increase the nutrient content of crops, thereby improving human nutrition. However, despite the significant potential of zinc-solubilizing bacteria in improving crop productivity and nutritional quality, several challenges need to be addressed to promote its widespread adoption. One of the significant challenges is the lack of awareness and knowledge about the use of zinc-solubilizing bacteria among farmers, extension workers, and policymakers. Therefore there is a need to raise awareness about the benefits of using zinc-solubilizing bacteria and promote its adoption among farmers, particularly in developing countries. Moreover, the development of training programs and extension services can help farmers understand the potential benefits of using zinc solubilizing bacteria and how to apply them effectively. Another challenge is the lack of standardized protocols for the isolation and characterization of zinc-solubilizing bacteria. Therefore there is a need to develop standardized protocols for the isolation and characterization of zinc-solubilizing bacteria, which can help researchers and farmers identify and select effective strains for improving crop productivity and nutritional quality. Therefore there is a need for further research to address these challenges and promote the sustainable use of zinc-solubilizing bacteria in the direction to circumvent zinc deficiency in crops and improve human health.

## Conflict of interest statement

The authors declare that the research was conducted in the absence of any commercial or financial relationships that could be construed as a potential conflict of interest.

## Acknowledgments

The authors thank DST-SERB for providing financial support with research grant ECR/2017/000080 to carry out this research work. The authors are also thankful to Amity University, Noida for providing infrastructural support.

## References

Adetola, O.Y., Taylor, J.R.N., Duodu, K.G., 2023. Can consumption of local micronutrient- and absorption enhancer-rich plant foods together with starchy staples improve bioavailable iron and zinc in diets of at-risk African populations? International Journal of Food Sciences and Nutrition 0, 1–21. Available from: https://doi.org/10.1080/09637486.2023.2182740.

Alloway, B.J., 2009. Soil factors associated with zinc deficiency in crops and humans. Environmental Geochemistry and Health 31, 537–548. Available from: https://doi.org/10.1007/s10653-009-9255-4.

Batool, S., Asghar, H.N., Shehzad, M.A., Yasin, S., Sohaib, M., Nawaz, F., et al., 2021. Zinc-solubilizing bacteria-mediated enzymatic and physiological regulations confer zinc biofortification in chickpea (Cicer arietinum L.). Journal of Soil Science and Plant Nutrition 21, 2456–2471. Available from: https://doi.org/10.1007/s42729-021-00537-6.

Bhatt, K., Maheshwari, D.K., 2020. Zinc solubilizing bacteria (Bacillus megaterium) with multifarious plant growth promoting activities alleviates growth in Capsicum annuum L. 3 Biotech 10, 36. Available from: https://doi.org/10.1007/s13205-019-2033-9.

Bris, M., 2017. Hormones in growth and development, in: Reference Module in Life Sciences. Available from: https://doi.org/10.1016/B978-0-12-809633-8.05058-5.

Chakraborty, S., Mandal, M., Chakraborty, A.P., Majumdar, S., 2022. Chapter 12 - Zinc solubilizing rhizobacteria as soil health engineer managing zinc deficiency in plants. In: Dubey, R. C., Kumar, P. (Eds.), Rhizosphere Engineering. Academic Press, pp. 215–238. Available from: https://doi.org/10.1016/B978-0-323-89973-4.00018-1.

Das, S.S., Green, A., 2013. Importance of zinc in crops and human health. Journal of SAT Agricultural Research 11, 1–7.

de Valença, A.W., Bake, A., Brouwer, I.D., Giller, K.E., 2017. Agronomic biofortification of crops to fight hidden hunger in sub-Saharan Africa. Global Food Security 12, 8–14.

Dinesh, R., Srinivasan, V., Hamza, S., Sarathambal, C., Anke Gowda, S.J., Ganeshamurthy, A. N., et al., 2018. Isolation and characterization of potential Zn solubilizing bacteria from soil and its effects on soil Zn release rates, soil available Zn and plant Zn content. Geoderma 321, 173–186. Available from: https://doi.org/10.1016/j.geoderma.2018.02.013.

Ditta, A., Ullah, N., Imtiaz, M., Li, X., Jan, A.U., Mehmood, S., et al., 2022. Zn biofortification in crops through Zn-solubilizing plant growth-promoting rhizobacteria. In: Mahmood, Q. (Ed.), Sustainable Plant Nutrition under Contaminated Environments. Springer International Publishing, Cham, pp. 115–133. Available from: https://doi.org/10.1007/978-3-030-91499-8_7.

Fariduddin, Q., Saleem, M., Khan, T.A., Hayat, S., 2022. Zinc as a versatile element in plants: an overview on its uptake, translocation, assimilatory roles, deficiency and toxicity symptoms. In: Khan, S.T., Malik, A. (Eds.), Microbial Biofertilizers and Micronutrient Availability: The Role of Zinc in Agriculture and Human Health. Springer International Publishing, Cham, pp. 137–158. Available from: https://doi.org/10.1007/978-3-030-76609-2_7.

Fasusi, O.A., Cruz, C., Babalola, O.O., 2021. Agricultural sustainability: microbial biofertilizers in rhizosphere management. Agriculture 11. Available from: https://doi.org/10.3390/agriculture11020163.

Gandhi, A., Muralidharan, G., 2016. Assessment of zinc solubilizing potentiality of Acinetobacter sp. isolated from rice rhizosphere. European Journal of Soil Biology 76, 1–8. Available from: https://doi.org/10.1016/j.ejsobi.2016.06.006.

Gontia-Mishra, I., Sapre, S., Tiwari, S., 2017. Zinc solubilizing bacteria from the rhizosphere of rice as prospective modulator of zinc biofortification in rice. Rhizosphere 3, 185–190. Available from: https://doi.org/10.1016/j.rhisph.2017.04.013.

Gupta, R., Gupta, R., Chakrabarty, S.K., 2013. Gibberellic acid in plant: still a mystery unresolved. Plant Signaling & Behavior 8, e25504. Available from: https://doi.org/10.4161/psb.25504.

Hafeez, B., Khanif, M., Saleem, M., 2013. Role of zinc in plant nutrition- a review. Journal of Experimental Agriculture International 3, 374–391. Available from: https://doi.org/10.9734/AJEA/2013/2746.

Hamzah Saleem, M., Usman, K., Rizwan, M., Al Jabri, H., Alsafran, M., 2022. Functions and strategies for enhancing zinc availability in plants for sustainable agriculture. Frontiers in Plant Science 13. Available from: https://doi.org/10.3389/fpls.2022.1033092.

Haroon, M., Khan, S.T., Malik, A., 2022. Zinc-solubilizing bacteria: an option to increase zinc uptake by plants. In: Khan, S.T., Malik, A. (Eds.), Microbial Biofertilizers and Micronutrient Availability: The Role of Zinc in Agriculture and Human Health. Springer International Publishing, Cham, pp. 207–238. Available from: https://doi.org/10.1007/978-3-030-76609-2_11.

Hefferon, K., 2019. Biotechnological approaches for generating zinc-enriched crops to combat malnutrition. Nutrients 11. Available from: https://doi.org/10.3390/nu11020253.

Henriques, A., Chalfun-Junior, A., Aarts, M., 2011. Strategies to increase zinc deficiency tolerance and homeostasis in plants. Brazilian Journal of Plant Physiology 24, 3–8. Available from: https://doi.org/10.1590/S1677-04202012000100002.

Hussain, A., Arshad, M., Zahir, Z.A., Asghar, M.A., 2015. Prospects of zinc solubilizing bacteria for enhancing growth of maize. Pakistan Journal of Agricultural Sciences 52, 915–922.

Hwalla, N., Labban, S., El, Bahn, R.A., 2016. Nutrition security is an integral component of food security. Frontiers in Life Sciences 9, 167–172. Available from: https://doi.org/10.1080/21553769.2016.1209133.

Ingram, J., 2020. Nutrition security is more than food security. Nature Food 1, 2. Available from: https://doi.org/10.1038/s43016-019-0002-4.

Iqbal, U., Jamil, N., Ali, I., Hasnain, S., 2010. Effect of zinc-phosphate-solubilizing bacterial isolates on growth of Vigna radiata. Annals of Microbiology 60, 243–248. Available from: https://doi.org/10.1007/s13213-010-0033-4.

Jamali, H., Sharma, A., Roohi, Srivastava, A.K., 2020. Biocontrol potential of Bacillus subtilis RH5 against sheath blight of rice caused by Rhizoctonia solani. Journal of Basic Microbiology 60, 268–280. Available from: https://doi.org/10.1002/jobm.201900347.

Joshi, D., Negi, G., Vaid, S., Sharma, A., 2013. Enhancement of wheat growth and Zn content in grains by zinc solubilizing bacteria. International Journal of Agriculture Environment and Biotechnology 6, 363. Available from: https://doi.org/10.5958/j.2230-732X.6.3.004.

Karnwal, A., 2021. Zinc solubilizing Pseudomonas spp. from vermicompost bestowed with multifaceted plant growth promoting properties and having prospective modulation of zinc biofortification in Abelmoschus esculentus L. Journal of Plant Nutrition 44, 1023–1038. Available from: https://doi.org/10.1080/01904167.2020.1862199.

Khalid, S., Amanullah, Ahmed, I., 2022. Enhancing zinc biofortification of wheat through integration of zinc, compost, and zinc-solubilizing bacteria. Agriculture 12. Available from: https://doi.org/10.3390/agriculture12070968.

Khan, A., Singh, A.V., Pareek, N., Arya, P., Upadhayay, V.K., Kumar Jugran, A., et al., 2023. Credibility assessment of cold adaptive Pseudomonas jesenni MP1 and P. palleroniana N26 on growth, rhizosphere dynamics, nutrient status, and yield of the kidney bean cultivated in Indian Central Himalaya. Frontiers in Plant Science 14. Available from: https://doi.org/10.3389/fpls.2023.1042053.

Khan, S.T., Malik, A., Alwarthan, A., Shaik, M.R., 2022. The enormity of the zinc deficiency problem and available solutions; an overview. Arabian Journal of Chemistry 15, 103668. Available from: https://doi.org/10.1016/j.arabjc.2021.103668.

Kiran, A., Wakeel, A., Mahmood, K., Mubaraka, R., Amin, H., Haefele, S., 2022. Biofortification of staple crops to alleviate human malnutrition: contributions and potential in developing countries. Agronomy 12, 452. Available from: https://doi.org/10.3390/agronomy12020452.

Kour, R., Jain, D., Bhojiya, A.A., Sukhwal, A., Sanadhya, S., Saheewala, H., et al., 2019. Zinc biosorption, biochemical and molecular characterization of plant growth-promoting zinc-tolerant bacteria. 3 Biotech 9, 421. Available from: https://doi.org/10.1007/s13205-019-1959-2.

Kumar, A., Dewangan, S., Lawate, P., Bahadur, I., Prajapati, S., 2019a. Zinc-solubilizing bacteria: a boon for sustainable agriculture. In: Sayyed, R.Z., Arora, N.K., Reddy, M.S. (Eds.), Plant Growth Promoting Rhizobacteria for Sustainable Stress Management, 1. Springer Singapore, Singapore, pp. 139–155. Available from: https://doi.org/10.1007/978-981-13-6536-2_8.

Kumar, S., Palve, A., Joshi, C., Srivastava, R.K., Rukhsar, 2019b. Crop biofortification for iron (Fe), zinc (Zn) and vitamin A with transgenic approaches. Heliyon 5, e01914. Available from: https://doi.org/10.1016/j.heliyon.2019.e01914.

Kushwaha, P., Srivastava, R., Pandiyan, K., Singh, A., Chakdar, H., Kashyap, P.L., et al., 2021. Enhancement in plant growth and zinc biofortification of chickpea (Cicer arietinum L.) by Bacillus altitudinis. Journal of Soil Science and Plant Nutrition 21, 922–935. Available from: https://doi.org/10.1007/s42729-021-00411-5.

Martín, L., Velázquez, E., Rivas, R., Mateos, P.F., Martínez-Molina, E., Rodríguez-Barrueco, C., et al., 2007. Effect of inoculation with a strain of Pseudomonas fragi in the growth and phosphorous content of strawberry plants. In: Velázquez, E., Rodríguez-Barrueco, C. (Eds.), First International Meeting on Microbial Phosphate Solubilization. Springer Netherlands, Dordrecht, pp. 309–315.

Mumtaz, M.Z., Ahmad, M., Jamil, M., Hussain, T., 2017. Zinc solubilizing Bacillus spp. potential candidates for biofortification in maize. Microbiological Research 202, 51–60. Available from: https://doi.org/10.1016/j.micres.2017.06.001.

Mumtaz, M.Z., Barry, K.M., Baker, A.L., Nichols, D.S., Ahmad, M., Zahir, Z.A., et al., 2019. Production of lactic and acetic acids by Bacillus sp. ZM20 and Bacillus cereus following exposure to zinc oxide: a possible mechanism for Zn solubilization. Rhizosphere 12, 100170. Available from: https://doi.org/10.1016/j.rhisph.2019.100170.

Naz, I., Ahmad, H., Khokhar, S.N., Khan, K.S., Shah, A.H., 2016. Impact of zinc solubilizing bacteria on zinc contents of wheat. American-Eurasian Journal of Agricultural & Environmental Sciences 16, 449–454.

Nestel, P., Bouis, H.E., Meenakshi, J.V., Pfeiffer, W., 2006. Biofortification of staple food crops. The Journal of Nutrition 136, 1064–1067. Available from: https://doi.org/10.1093/jn/136.4.1064.

Noulas, C., Tziouvalekas, M., Karyotis, T., 2018. Zinc in soils, water and food crops. Journal of Trace Elements in Medicine and Biology: Organ of the Society for Minerals and Trace Elements (GMS) 49, 252–260. Available from: https://doi.org/10.1016/j.jtemb.2018.02.009.

Othman, N.M.I., Othman, R., Zuan, A.T.K., Shamsuddin, A.S., Zaman, N.B.K., Sari, N.A., et al., 2022. Isolation, characterization, and identification of zinc-solubilizing bacteria (ZSB) from wetland rice fields in Peninsular Malaysia. Agriculture 12. Available from: https://doi.org/10.3390/agriculture12111823.

Paque, S., Weijers, D., 2016. Q&A: auxin: the plant molecule that influences almost anything. BMC Biology 14, 67. Available from: https://doi.org/10.1186/s12915-016-0291-0.

Pascual, M.B., Echevarria, V., Gonzalo, M.J., Hernández-Apaolaza, L., 2016. Silicon addition to soybean (Glycine max L.) plants alleviate zinc deficiency. Plant Physiology and Biochemistry: PPB / Societe Francaise de Physiologie Vegetale 108, 132–138. Available from: https://doi.org/10.1016/j.plaphy.2016.07.008.

Ramesh, A., Sharma, S.K., Sharma, M.P., Yadav, N., Joshi, O.P., 2014. Inoculation of zinc solubilizing Bacillus aryabhattai strains for improved growth, mobilization and biofortification of zinc in soybean and wheat cultivated in Vertisols of central India. Applied Soil Ecology 73, 87–96. Available from: https://doi.org/10.1016/j.apsoil.2013.08.009.

Rani, N., Kaur, G., Kaur, S., Mutreja, V., Pandey, N., 2022. Plant growth-promoting attributes of zinc solubilizing dietzia maris isolated from polyhouse rhizospheric soil of Punjab. Current Microbiology 80, 48. Available from: https://doi.org/10.1007/s00284-022-03147-2.

Recena, R., García-López, A.M., Delgado, A., 2021. Zinc Uptake by Plants as Affected by Fertilization with Zn Sulfate, Phosphorus Availability, and Soil Properties. Agronomy 11. Available from: https://doi.org/10.3390/agronomy11020390.

Rudani, K., Prajapati, K., Patel, V., 2018. The Importance of Zinc in Plant Growth -A Review, 5, 2349–4077.

Saleem, S., Khan, S.T., 2022. Development of microbes-based biofertilizer for zinc dissolution in soil. In: Khan, S.T., Malik, A. (Eds.), Microbial Biofertilizers and Micronutrient Availability: The Role of Zinc in Agriculture and Human Health. Springer International Publishing, Cham, pp. 299–329. Available from: https://doi.org/10.1007/978-3-030-76609-2_13.

Shaikh, S., Saraf, M., 2017. Biofortification of Triticum aestivum through the inoculation of zinc solubilizing plant growth promoting rhizobacteria in field experiment. Biocatalysis and Agricultural Biotechnology 9, 120–126. Available from: https://doi.org/10.1016/j.bcab.2016.12.008.

Sharma, A., Patni, B., Shankhdhar, D., Shankhdhar, S.C., 2013. Zinc – an indispensable micronutrient. Physiology and Molecular Biology of Plants 19, 11–20. Available from: https://doi.org/10.1007/s12298-012-0139-1.

Shreya, D., Amaresan, N., Supriya, N.R., 2023. Zinc solubilizing Bacillus sp (SS9) and Enterobacter sp (SS7) promote mung bean (Vigna radiata L.) growth, nutrient uptake and physiological profiles. Letters in Applied Microbiology 76, ovac063. Available from: https://doi.org/10.1093/lambio/ovac063.

Sindhu, S.S., Sharma, R., Sindhu, S., Phour, M., 2019. Plant Nutrient Management Through Inoculation of Zinc-Solubilizing Bacteria for Sustainable Agriculture. In: Giri, B., Prasad, R., Wu, Q.-S., Varma, A. (Eds.), Biofertilizers for Sustainable Agriculture and Environment. Springer International Publishing, Cham, pp. 173–201. Available from: https://doi.org/10.1007/978-3-030-18933-4_8.

Singh, A., Mishra, S., Choudhary, M., Chandra, P., Rai, A.K., Yadav, R.K., et al., 2023. Rhizobacteria improve rice zinc nutrition in deficient soils. Rhizosphere 25, 100646. Available from: https://doi.org/10.1016/j.rhisph.2022.100646.

Singh, J., Singh, A.V., Upadhayay, V.K., Khan, A., Chandra, R., 2022. Prolific contribution of Pseudomonas protegens in Zn biofortification of wheat by modulating multifaceted physiological response under saline and non-saline conditions. World Journal of Microbiology and Biotechnology 38, 227. Available from: https://doi.org/10.1007/s11274-022-03411-4.

Son, H.-J., Park, G.-T., Cha, M.-S., Heo, M.-S., 2006. Solubilization of insoluble inorganic phosphates by a novel salt- and pH-tolerant Pantoea agglomerans R-42 isolated from soybean rhizosphere. Bioresource Technology 97, 204–210. Available from: https://doi.org/10.1016/j.biortech.2005.02.021.

Suleman, M., Yasmin, S., Rasul, M., Yahya, M., Atta, B.M., Mirza, M.S., 2018. Phosphate solubilizing bacteria with glucose dehydrogenase gene for phosphorus uptake and beneficial effects on wheat. PLoS One 13, 1–28. Available from: https://doi.org/10.1371/journal.pone.0204408.

Suriyachadkun, C., Chunhachart, O., Srithaworn, M., Tangchitcharoenkhul, R., Tangjitjareonkun, J., 2022. Zinc-solubilizing *Streptomyces* spp. as bioinoculants for promoting the growth of soybean (*Glycine max* (L.) Merrill). Journal of Microbiology and Biotechnology 32, 1435–1446. Available from: https://doi.org/10.4014/jmb.2206.06058.

Tekle, D.Y., Rosewarne, E., Santos, J.A., Trieu, K., Buse, K., Palu, A., et al., 2023. Do food and nutrition policies in Ethiopia support the prevention of non-communicable diseases through population-level salt reduction measures? A policy content analysis. Nutrients 15. Available from: https://doi.org/10.3390/nu15071745.

Upadhayay, V.K., Singh, A.V., Khan, A., 2022a. Cross talk between zinc-solubilizing bacteria and plants: a short tale of bacterial-assisted zinc biofortification. Frontiers in Soil Science 1. Available from: https://doi.org/10.3389/fsoil.2021.788170.

Upadhayay, V.K., Singh, A.V., Khan, A., Sharma, A., 2022b. Contemplating the role of zinc-solubilizing bacteria in crop biofortification: an approach for sustainable bioeconomy. Frontiers in Agronomy 4. Available from: https://doi.org/10.3389/fagro.2022.903321.

Upadhyay, H., Gangola, S., Sharma, A., Singh, A., Maithani, D., Joshi, S., 2021. Contribution of zinc solubilizing bacterial isolates on enhanced zinc uptake and growth promotion of maize (Zea mays L.). Folia Microbiologica 66, 543–553. Available from: https://doi.org/10.1007/s12223-021-00863-3.

Vaid, S.K., Kumar, B.M., Sharma, A.S., Shukla, A.K., Srivastava, P.C., 2014. Effect of zn solubilizing bacteria on growth promotion and zn nutrition of rice. Journal of Soil Science and Plant Nutrition 14, 889–910.

Verma, D., Meena, R.H., Sukhwal, A., Jat, G., Meena, S.C., Upadhyay, S.K., et al., 2022. Effect of ZSB with graded levels of zinc fertilizer on yield and zinc uptake under maize cultivation. Proceedings of the National Academy of Sciences, India Section B: Biological Sciences 93. Available from: https://doi.org/10.1007/s40011-022-01433-4.

Verma, M., Singh, A., Dwivedi, D.H., Arora, N.K., 2020. Zinc and phosphate solubilizing Rhizobium radiobacter (LB2) for enhancing quality and yield of loose leaf lettuce in saline soil. Environmental Sustainability 3, 209–218. Available from: https://doi.org/10.1007/s42398-020-00110-4.

Yadav, R.C., Sharma, S.K., Ramesh, A., Sharma, K., Sharma, P.K., Varma, A., 2020. Contribution of zinc-solubilizing and -mobilizing microorganisms (ZSMM) to enhance zinc bioavailability for better soil, plant, and human health, in: Sharma, Sushil Kumar, Singh, U. B., Sahu, P.K., Singh, H.V., Sharma, Pawan Kumar (Eds.), Rhizosphere Microbes: Soil and Plant Functions. Springer Singapore, Singapore, pp. 357–386. Available from: https://doi.org/10.1007/978-981-15-9154-9_14.

Yadav, R.C., Sharma, S.K., Varma, A., Rajawat, M.V.S., Khan, M.S., Sharma, P.K., et al., 2022. Modulation in biofertilization and biofortification of wheat crop by inoculation of zinc-solubilizing rhizobacteria. Frontiers in Plant Science 13. Available from: https://doi.org/10.3389/fpls.2022.777771.

Yaghoubi Khanghahi, M., Strafella, S., Allegretta, I., Crecchio, C., 2021. Isolation of bacteria with potential plant-promoting traits and optimization of their growth conditions. Current Microbiology 78, 464–478. Available from: https://doi.org/10.1007/s00284-020-02303-w.

Yogi, A.K., Bana, R.S., Bamboriya, S.D., Choudhary, R.L., Laing, A.M., Singh, D., et al., 2023. Foliar zinc fertilization improves yield, biofortification and nutrient-use efficiency of upland rice. Nutrient Cycling in Agroecosystems 125, 453–469. Available from: https://doi.org/10.1007/s10705-023-10270-4.

Zaheer, A., Malik, A., Sher, A., Mansoor Qaisrani, M., Mehmood, A., Ullah Khan, S., et al., 2019. Isolation, characterization, and effect of phosphate-zinc-solubilizing bacterial strains on chickpea (Cicer arietinum L.) growth. Saudi Journal of Biological Sciences 26, 1061–1067. Available from: https://doi.org/10.1016/j.sjbs.2019.04.004.

Chapter 12

# Zinc and plant disease: role and regulation

Victoria J[1], Shivani Mahra[1], Kavita Tiwari[1], Sneha Tripathi[1], Samarth Sharma[1], Shivendra Sahi[2] and Shivesh Sharma[1]

[1]*Department of Biotechnology, Motilal Nehru National Institute of Technology Allahabad, Prayagraj, Uttar Pradesh, India,* [2]*Department of Biology, Saint Joseph's University, Philadelphia, PA, United States*

## 12.1 Introduction

The relevant and significant amount of nutrients is always necessary for the plants to grow and combat diseases to improve agricultural products (Hirel et al., 2011). Nutrients are highly needed from the early germination to the complete development of plants (Gioia et al., 2016). These nutrients are classified into two categories: macro- and micronutrients (Rajeswar et al., 2009; Singh et al., 2017). Calcium, phosphorous, potassium, magnesium, and nitrogen are categorized as macronutrients because they are needed in surplus amounts for plant growth and development (Bhatla et al., 2018; Hawkesford et al., 2023). Nutrients like boron, iron, silicon, zinc, nickel, copper, and molybdenum are categorized as micronutrients because they are essential in lesser amounts (St. John et al., 2013; Kirkby, 2023). Though these micronutrients are required in a lesser amount, they are significant for plants (Graham, 2008; Savarino et al., 2021). Deficiency of micronutrients affects plant growth and yield adversely because these micronutrients support plants in combating various biotic and abiotic stresses (Koç and Karayiğit., 2022; Gondal et al., 2021). Zinc plays a vital role among the micronutrients by promoting various biochemical and physiological processes for the synthesis of nucleic acids, cytochromes, activating enzymes, and an essential cofactor for several enzymes and chlorophyll formation (Tariq et al., 2023). Zn has the potential to maintain the membrane and seed stalk maturation (Gill et al., 2021). Apart from this, Zn contributed to lipid and carbohydrate metabolism, and various plant defense mechanisms through oxidation reactions in plants (Kaur and Garg, 2021). Zn also contributes to activating the Rubisco enzyme, which is distinguished for the biosynthesis of chlorophyll and thus

**Zinc in Plants. DOI: https://doi.org/10.1016/B978-0-323-91314-0.00001-6**
 

elevates the photosynthetic rate in higher plants (Poddar et al., 2020; Sharma et al., 2020). It plays a remarkable role in phytohormone formation, auxin that assists the growth of plants (Ganguly et al., 2022). According to Cakmak (2000), Zn deficiency is widespread and decreases the photosynthesis rate in crop plants like rice and wheat by inducing chlorosis (Barman et al., 2018). It is also noted that 50% of the land utilized for growing crops lacks Zn and Zn deficiency would possibly decrease the plant's survival against phytopathogens (Rani et al., 2019). Pests and plant diseases pose a threat to plant productivity. In this chapter, we will discuss zinc as a micronutrient that influences plant growth and eminent role in plant defense mechanisms against various phytopathogens.

## 12.2 Zinc bioavailability and distribution

Ablation of rocks and atmospheric contributions like dust and fire resulted in the presence of zinc in the soil naturally (Herrera et al., 2009). Zinc is present in different forms in the soil as Zn solutions, integrated and combined with the soils and organic ligands, and mostly adsorbed on the clay minerals and metallic oxides (Barrow, 1993; Sipos et al., 2008). Geographical composition in rocks contributes to the zinc profile in soils (Nowack et al., 2001). It is also reported that 78 mg of zinc is present in the earth's crust per kilogram of soil (Alloway, 2008). It is available in the sphalerite zone and in the form of ores like smithsonite, zbite, and zinkosite. Although zinc micronutrients are abundant in soil, the majority are available in insoluble form, so they are not available to plants (McBride et al., 1997). The insoluble form of zinc comprises zinc oxide, zinc chloride, zinc sulfate, zinc carbonate, and zinc phosphate. Soluble forms of Zn are the only forms of uptake by plants. It is also noticed that the content of Zn in soluble form that is available for plants ranges between 4 and 270 ppb in most regions (Solanki, 2021). Zn interacts with other micro- and macronutrients like sulfate, phosphate, chloride, and nitrate ions to form soluble complexes (Arunachalam et al., 2013). At the same time, the consumption of Zn by plants depends on the plant root activity, the microflora in the rhizosphere, and the physiochemical properties of the soil in that particular region (Khan 2005; Dotaniya and Meena, 2015). Zn phosphate and zinc sulfate are the primary ores that facilitate the existence of zinc in the soil (Aboyeji et al., 2019). By acidifying of soils, the combined or integrated form of the zinc ore can be solubilized and made available for plants (Kumar et al., 2019).

## 12.3 Physiological and biochemical functions of zinc in plants

The potential of Zn to form compounds with N, O, and S inside plant cells is attributed to its metabolic properties. By establishing a linkage with organic ligands, intracellular Zn is generally inert (Zoroddu et al., 2019; Natasha et al., 2022). The ability of plants to absorb water and the transmission in plants are impacted by the micronutrient Zn (Puga et al., 2015; Gupta et al., 2016). Moreover, it has been determined that Zn lessens the harmful effects of heat and salt intolerance (Ben Saad et al., 2010; Abdel Latef et al., 2017). It actively influences the production of the plant growth phytohormone IAA through the activation of tryptophan synthetase (Hassan et al., 2019). According to Tiwari et al. (2017), Zn actively participates in the transduction of the signals. Aldolases, dehydrogenases, isomerases, and phosphorylases are just a few of the enzymes that use Zn as a prosthetic group (Kovuri., 2016). Many proteins use zinc as a structural and catalytic component. The structural, catalytic, and catalytic sites are the three Zn ligand binding sites (Padjasek et al., 2020). Zn ensures appropriate protein folding through structural Zn sites (Zastrow and Pecoraro, 2014: Blindauer, 2015). Zn is directly involved in the catalytic activity of enzymes in catalytic sites (Li et al., 2019). Zn is found in close proximity to other elements in a catalytic site, coupled to them via glutamic acid, aspartic acid, or histidine, and bound to water (Rogowska et al., 2023). Zn is required for the synthesis of proteins and the production of energy (Umair Hassan et al., 2020). Zinc is required for the repair of the photosynthetic apparatus and the regulation of multiple biochemical activities during photosynthesis because it serves as a replacement for the photo-damaged D1 protein (Hanikenne et al., 2014; Sharma et al., 2020). Protein−protein interactions are mediated by the Zn finger protein (Laxmi, 2014). Zn interacts with membrane protein sulfhydryl groups and phospholipids to maintain the structure of macromolecules and is crucial for maintaining the integrity of cellular membranes (Rawat et al., 2021) and maintaining the movement of ions inside plant cells (Gupta et al., 2016; Caldelas and Weiss, 2017). The destabilization of plant cell membranes caused by Zn deficiency was irreversible (Sloup et al., 2017). It may cause boron and phosphorus buildup due to their nonselective entry of boron into Zn-deficient roots and leaves. Zinc is essential in the synthesis of nucleic acids and maintaining their integrity (Ward et al., 2014). Zinc is mandatory in lipid metabolism and protein synthesis as well. It is also observed that a low Zn level reflects reduced protein synthesis because a Zn shortage causes a decrease in the efficiency of RNA synthesis and ribosomes in plants (Millward, 2017; Osuna et al., 2007). Subsequently, Zn also prevents membrane injury that takes place due to the presence of reactive oxygen species (Gill and Tuteja, 2010) and Zn plays a vital role in controlling various biological processes that include inflorescence, defense response, photomorphogenesis, and seed yield (Noman et al., 2019).

## 12.4 Role of zinc in plant defense response/mechanism

Zn is involved in activating various enzymes by binding via cysteine and imidazole. The activation of enzymes regulates different plant-pathogen interactions and promotes plant defense mechanisms under both biotic and abiotic stress (Hossain et al., 2012). When Zn disentangles from the active site of the enzyme, it could cease the enzyme activity of Zn (Cloquet et al., 2008). They play a chief role with metalloenzymes in facilitating carbohydrate and lipid metabolism, protein synthesis, etc. (Roy et al., 2013). The crucial role of Zn ions can be catalytic, structural, or cocatalytic (Maret, 2013). In catalysis, it directly involves making bonds; cocatalytic Zn activity enhances the catalytic activity of one or more metals attached to it; and in structural activity, it stabilizes the tertiary structure of the enzymes (Sahoo et al., 2018). The most commonly used enzyme interaction for plant-pathogen interactions is referred to in this section. Alcohol dehydrogenases are NADP-dependent enzymes that take part in the reverse oxidation of alcohols into aldehydes and ketones (An et al., 2019). Under anaerobic conditions, microbial infestations modify the chemical reactions in plants and lead to the formation of toxic substances (Song and Ryu, 2013). Alcohols that are formed during this process are subtilized by the alcohol dehydrogenase enzyme (Iwama et al., 2015). Zn binds with this enzyme and accelerates the gene expression of different defense markers in plants to protect against stress. AD (*AdZADH2)* on binding with Zn revealed improved defense-gene expression, marker genes, and antioxidant enzymes in tobacco leaves causing cell death (Kumar et al., 2019). Zn, in combination with AD (*AdZADH2)*, upregulated defense-related gene expressions in wild peanut plants manifested with *Phaeoisariopsis personata*, a leaf spot pathogen (Kumar et al., 2019). In response to the sharp eyespot caused by *Rhizoctonia cerealis* in wheat, defense-related MBR genes (CAD, CCR, and COMT) are regulated. This regulation of monolignol-biosynthesis-related gene expression assists in the deposition of lignin in the cell wall, thereby preventing the plant from cell wall degradation (Tronchet et al., 2010). In this way, AD with Zn confers resistance against plant-pathogen interaction.

Superoxide dismutase is an antioxidant enzyme that catalyzes the dismutation of free radicals, superoxide, and hydrogen peroxide (Buettner, 2011; Ighodaro and Akinloye, 2018). Thus, it enhances the plant's defense to act against the antioxidant. Zn is pivotal to maintaining membrane integrity in plants (Souri and Hatamian, 2019). On deficiency of Zn, SOD activity suppresses, and gradually the concentration of radicals increases (Rai et al., 2012). This accumulation of reactive oxygen species causes lipid peroxidation in plants by impaired binding with phospholipid groups (Mukarram et al., 2021). Presence of Zn ion in the active site of SOD aids in significant improvement in secondary electron transfer from $Cu+$ to $HO_2$ (Ryabykh et al., 2022). Here, Zn acts as a structural factor by stabilizing the enzyme structure to enhance secondary electron transfer. Similarly, the metallothionein enzyme has the potential to act against heavy metal toxicity and osmotic stresses in plants (Rizvi et al., 2020). Zn with metallothionein plays a vital role in the pathogen defense mechanism by expressing metal regulatory transcription factor 1 (MTF-1) and suppressing the reactive oxygen and nitrogen species formed due to pathogen manifestation (Prasad and Bao, 2019). Plant defense-related markers and MTs-1 gene are more highly expressed in a plant-infected *Magnoporthe grisea* than in a control exhibiting defense properties (Lorenzo-Gutiérrez et al., 2019; Dumanović et al., 2021). MT proteins are also reported to be expressed highly against beech-scale insect and beech bark diseases (Ćalić et al., 2017). In addition, zinc protein also confers resistance against biotic stresses (Nakai et al., 2013). Zinc proteins are a well-known free-folded functional domain that consists of zinc for structure stabilization (Kluska et al., 2018). The authors reported the study of 70 disease-resistant proteins from various crops. Among these proteins, 37% contributed to the zinc finger proteins, which shows the importance of the host defense protein ZnF against pathogens (Gupta et al., 2012). Insect manifestation also regulated the expression of two Zn finger proteins in potatoes (Lawrence et al., 2014). Therefore it is evident that Zn plays a very important role in promoting plant defense mechanisms against phytopathogens via activating important enzymes. Thus the deficiency of zinc directly causes irregularities in the regulation of various defense systems in plants that provoke plants to be prone to different severe infections.

## 12.5 Zinc: a potent inhibitor of plant diseases/pathogens

The pathogens that infect plants comprise viruses, bacteria, mollicutes, protozoa, fungi, and parasitic higher plants (Harris and Pitzschke., 2020). As mentioned above, Zn is very crucial for plant growth, and in addition, it also contributes an equal amount to preventing plants from plant pathogens and insects (Nadeem et al., 2018). This plant's defense mechanism against biotrophic pathogens includes the gene for gene resistance and salicylic acid responses, while necrotrophic pathogens include jasmonic or ethylene regulation (Ben Rejeb et al., 2014). It is reported that the application of Zn in the agricultural field decreased disease manifestation in plants. The use of Zn in specific amounts against particular pathogens would gradually make plants tolerant to other types of pathogens (Dordas, 2008; Gupta et al., 2012; Al Jabri et al., 2022). Wang et al. (2016) demonstrated that Zn proteins play binary positive roles by preventing plants from developing diseases, and at the same time, promoting plant growth. The plant and pathogen responses to the applied Zn depend on the potential of pathogens against the defense mechanisms of plants and the efficiency of applying the Zn in suppressing the pathogen invasion (Cabot et al., 2019). It also relies on the environmental factors that surround the support of either plants or pathogens (Zehra et al., 2021). Zn protein-based plant defense mechanism reported that the availability of Zn enhanced gene expression of referred proteins for plant defense (Cakmak, 2000). Therefore increasing the bioavailability of Zn proteins improved the plant defense mechanism in plants to act against pathogen invasion. The results also revealed that the elevated plant defense response is not only due to the bioavailability of Zn proteins, but other mechanisms related to Zn in plants are also integrated (Lin and Aarts, 2012; Emamverdian et al., 2015). The addition of Zn also increased the antioxidant enzyme levels in plants particularly superoxide dismutase levels, under such stresses (Afzal et al., 2019). Hence, the enhanced production of superoxide enzymes brings down the generated oxygen radicals and controls the level of toxic radicals in plants (Khan et al., 2021). The use of zinc in excess of the desired amount also causes stress on plants. Plants overcome this kind of stress by regulating jasmonates and salicylic acid (Singh et al., 2016; Vankova et al., 2017). Zinc in its nanoform has been demonstrated to exhibit fungicidal and bacterial effects in plants (Singh et al., 2018). For instance, the result of applying zinc oxide nanoparticles against bacterial panicle blight disease demonstrated the elevated antibacterial response against *Burkholderia glumae* and *B. gladioli* in rice plants and concluded that this particular nanoform can be implemented as potential nanopesticides to act against plant pathogens and diseases (Ahmed et al., 2021). So, it is understood that Zn in all forms acts as a potent inhibitor in combating various plant diseases; pathogens and the role of Zn in plants against particular pathogens are elaborated in the future sections.

### 12.5.1 Regulation of zinc against bacterial pathogens/diseases

Certain studies show that zinc has the potential to combat biotic stress, specifically bacterial pathogens. Treatment of soil with zinc phosphate prevented a susceptible variety of tomato plants against *Pseudomonas syringae pv, tomato Pst.* Quaglia et al. (2021) noticed a defense response by the overexpression of salicylic-dependent PR1B1 protein. The results analysis revealed that both biochemical and morphological changes are because of antibacterial activity, which is proved by histochemical and qPCR *in vitro* assays (Quaglia et al., 2021). Zinc elevated the resistance against *Xanthomonas axonopodis* pv. *glycines* at optimal concentration in soybean. The use of Zn above the optimal concentration has reduced the efficiency of the resistance against bacterial pustules and results in phytotoxicity (Marschner, 2012). Simultaneously, it is observed that the use of less Zn than the optimal concentration enhanced the pathogen severity in soybeans (Helfenstein et al., 2015). On comparison between titanium oxide and zinc oxide nanoparticles, it was observed that Zn contributed positively to enhancing the morphological changes and resistance against bacterial diseases in beetroot (Siddiqui, et al., 2019). The nanoforms of Zn have a pivotal role in increasing both nonenzymatic and enzymatic antioxidants and have also been observed that increases chlorophyll and carotenoid content, root length, and shoot length in lentil plants *inoculated with Fusarium oxysporum* f. sp. lentis, *Alternaria alternata*, *Xanthomonas axonopodis* pv. *phaseoli*, and *Meloidogyne incognita* (Siddiqui et al., 2018). Foliar applications of Zn oxide nanoparticles controlled both bacterial and fungal diseases in tomatoes (Parveen and Siddiqui, 2021). Zn also contributed to controlling *Pseudomonas cichorii* and *Xanthomonas campestris* pv. *Vesicatoria*, which causes bacterial leaf spot diseases in ginseng and pepper, respectively (Duffy, 2007).

### 12.5.2 Regulation of zinc against viral pathogens/diseases

Micronutrients promote plant growth and development as well as being observed to confer resistance against harmful viral diseases that decrease agricultural productivity (Ajilogba et al., 2021). The studies also revealed that the application of Zn sulfate significantly reduced the severity caused by a yellow mosaic virus in mungbeans (Islam et al., 2002). The authors also noticed the recurrence of yellow mosaic disease had decreased. Under foliar spray of Zn sulfate in potatoes, the release of phenolic compounds that are essential for plant defense mechanisms against viruses was elevated. The authors also observed that the results exhibited reduced symptoms of potato viruses (Ibrahim et al., 2016). Similarly, the application of Zn sulfate in young cotton plants inhibited the disease severity by 65.34% caused by Cotton leaf curl virus and also prohibited the whitefly manifestation, which is the vector of Cotton leaf curl virus (Kalsoom et al., 2019). Zinc, along with other micronutrients like boric acid and urea, reduced the incidence of papaya ringspot virus disease in papaya, but zinc alone has no effect on the papaya ringspot virus (Deepika et al., 2021). A low level of Zn content causes the accumulation of sugars and amino acids in the plant tissues, which facilitates the sucking insects (vector) for feeding and multiplication. Hence, it increases disease severity in plants by favoring vectors (Dutta et al., 2017). At present, zinc-based nanoparticles play a crucial role in acting against viral diseases (Abdelkhalek and Al-Askar, 2020). Green synthesized Zn nanoparticles from *Mentha spicata* against tobacco mosaic virus (TMV) under greenhouse conditions showed a 96% reduction in the viral load and severity of disease incidence. Similarly, Sofy et al. (2021) observed that Zn nanoparticles incredibly improved the morphological characters and photosynthetic pigment and enhanced the enzymatic and nonenzymatic antioxidants that involve plant defense mechanisms against the TMV virus. Thus Zn in both bulk and nanoforms is active against many viral diseases in plants, increasing agricultural productivity.

### 12.5.3 Regulation of zinc against fungal pathogens

Zn plays a very crucial role in maintaining cellular integrity and acts as a defense factor against pathogenic fungal infection in plants (Zuriegat et al., 2021). Quaglia et al. (2022) demonstrated that the application of Zn in excess and at deficient levels modified the physiological properties and increased the fungal incidence in *Arabidopsis*. When the excess amount is utilized, it leads to damage to the chloroplast and causes chlorosis. When Zn is utilized in reduced amounts, it leads to a decrease in chlorophyll content and lipid peroxidation in cells (Zhang et al., 2020). When the optimal amount is utilized, the plant shows resistance to the pathogen *Botrytis cinerea* via the ROS mechanism, which is unfavorable for the pathogen. Similarly, Zn sulfate has inhibited the *Lasiodiplodia theobromae*, causing peach gummosis in peach shoots and reducing the incidence by inducing the defense-related genes PR4, CHI, PAL, PGIP, and GNS3 and altering the morphology of the fungal pathogen (Li et al., 2016). Wadhwa et al. (2014) reported the function of Zn as a cofactor in activating antioxidant enzymes in bean seedlings and the ability to decrease the incidence of root rot against *Rhizoctonia mycelia*. Zn was also demonstrated to induce different signaling pathways, like salicylic acid-dependent, ethylene-dependent (jasmonic) pathways, to be active against different pathogens' attacks by maintaining cellular membrane integrity in plants (Ferrari et al., 2007). Zn with the dose of 0.036 mM foliar spray in the chili pepper expressed enhanced physiological traits' and results displayed that Zn is effective against *Alternaria alternata* that causes leaf spot disease (Shoaib et al., 2021). Under the greenhouse solution culture experiment, Zn improved the tolerance of wheat plants against *Fusarium* invasion and foliar application of micronutrients. Zn, along with boron and manganese, was found effective against tan spot disease and early blight disease in wheat and potato, respectively (Simoglou and Dordas, 2006; Khoshgoftarmanesh et al., 2010; Machado et al., 2018). The recent development of nanotechnology for Zn nanoparticle synthesis was also reported to behave as an antifungal agent against phytopathogens. For instance, Zn oxide nanoparticles reduced the incidence of wilt, leaf spot disease, galling, and nematode phytopathogens (Siddiqui et al., 2018). These nanoparticles were reported to show resistance against replant disease in apples and decrease the copy numbers of *Fusarium solani* and *Fusarium oxysporum* (Pan et al., 2022).

### 12.5.4 Regulation of zinc against nematodes

The release and accumulation of ROS in Zn deficiency increase the root permeability, which attracts the nematode infection (Lamalakshmi Devi et al., 2017). There would be a leak of free amino acids from the roots due to the lack of Zn that inhibits protein synthesis in plants (Suganya et al., 2020). Therefore the nematode infection starts as the Zn deficiency ascends. Thus it is necessary to maintain membrane integrity under the Zn micronutrient to manage the nematode infection (Rinaldi et al., 2023). Shaukat and Siddiqui. (2003) integrated rhizobial culture with different quantities of Zn, increasing the nematicidal activity in tomatoes against root-knot nematodes. Rhizobia culture here releases nematicidal compounds that are active against root-knot development, and Zn is also observed to increase the nucleic acid stability of the culture (Shaukat and Siddiqui., 2003). A combination of soil with Zn and the P. aeruginosa IE-6S$^{+}$ strain was demonstrated to possess elevated resistance against root-knot nematodes in tomato root. The author has revealed that augmentation of Zn with the bacterial strain improved the biocontrol activity of the strain and, thereby, prevented plant pathogenic invasions (Shaukat and Siddiqui., 2003). In vitro studies reported that an increase in Zn concentration increases the nematicidal activity in tomatoes. Similarly, with other strains like PSG1 and MIG1, the nematode invasion is significantly reduced along with Zn. This study clearly states that Zn plays a very important role in inhibiting root-knot nematode damage in tomato plants. Zn at a specific dose of 60 g/L inhibited the embryonic development of *Meloidogyne incognita* in lentils and was also reported to turn down the number of galls and eggs of *M. incognita*. However, Zn alone has no significant impact on the nematode itself (Couto et al., 2016). Thus nematodes are negatively correlated with Zn in soils.

## 12.6 Future prospective and concluding remarks

Zinc as a micronutrient is phenomenal in regulating various biochemical and physiological processes against various harmful plant pathogens. It plays an important role in synthesizing various proteins like metallothionein and Zn finger and also regulates both enzymatic and nonenzymatic antioxidants essential to scavenging free radicals. In response to any pathogenic manifestation, plants release ROS that are loaded near the site of invasion. However, this accumulation of ROS should be lowered by antioxidant machinery. In Zn deficiency, the accumulation of ROS gradually increases, which becomes toxic to the plant cells, leading to disruption in membrane integrity and oxidation of cellular components that bring plant cell death. Furthermore, Zn is vital in the expression of defense-related mechanisms and sometimes causes callose deposition that stands as a barrier against infection. The application of zinc in higher concentrations may lead to extreme accumulation of ROS, which is toxic, and a reduced amount of zinc also favors sugar and amino acid deposition in the tissues of plants, which favors disease vectors. Thus maintaining the optimal amount of Zn is crucial for the plant to exhibit proper resistance against disease severity and incidence. This chapter helps in understanding plant disease management with zinc in both bulk and nanoforms, and it further paves the way for more novel applications of Zn for antipathogenic studies without any detrimental impact on plants and agricultural fields. The emerging technology in combating Zn deficiency is the implementation of plant growth-promoting rhizosphere that increase the bioavailability of Zn and other micronutrients in an economically safe and sound environment.

## Acknowledgments

Authors are thankful to Director, MNNIT Allahabad for providing necessary facilities.

## References

Abdel Latef, A.A.H., Abu Alhmad, M.F., Abdelfattah, K.E., 2017. The possible roles of priming with ZnO nanoparticles in mitigation of salinity stress in lupine (Lupinus termis) plants. Journal of Plant Growth Regulation 36, 60–70.

Abdelkhalek, A., Al-Askar, A.A., 2020. Green synthesized ZnO nanoparticles mediated by Mentha spicata extract induce plant systemic resistance against Tobacco mosaic virus. Applied Sciences 10 (15), 5054.

Aboyeji, C., Dunsin, O., Adekiya, A.O., Chinedum, C., Suleiman, K.O., Okunlola, F.O., et al., 2019. Zinc sulphate and boron-based foliar fertilizer effect on growth, yield, minerals, and heavy metal composition of groundnut (Arachis hypogaea l) grown on an alfisol. International Journal of Agronomy 2019, 1–7.

Afzal, J., Hu, C., Imtiaz, M., Elyamine, A.M., Rana, M.S., Imran, M., et al., 2019. Cadmium tolerance in rice cultivars associated with antioxidant enzymes activities and Fe/Zn concentrations. International Journal of Environmental Science and Technology 16, 4241–4252.

Ahmed, T., Wu, Z., Jiang, H., Luo, J., Noman, M., Shahid, M., et al., 2021. Bioinspired green synthesis of zinc oxide nanoparticles from a native Bacillus cereus strain RNT6: characterization and antibacterial activity against rice panicle blight pathogens Burkholderia glumae and B. gladioli. Nanomaterials 11 (4), 884.

Ajilogba, C.F., Babalola, O.O., Nikoro, D.O., 2021. Nanotechnology as vehicle for biocontrol of plant diseases in crop production. Food Security and Safety: African Perspectives 709–724.

Al Jabri, H., Saleem, M.H., Rizwan, M., Hussain, I., Usman, K., Alsafran, M., 2022. Zinc oxide nanoparticles and their biosynthesis: overview. Life 12 (4), 594.

Alloway, B.J., 2008. Zinc in Soils and Crop Nutrition.

An, J., Nie, Y., Xu, Y., 2019. Structural insights into alcohol dehydrogenases catalyzing asymmetric reductions. Critical Reviews in Biotechnology 39 (3), 366–379.

Arunachalam, P., Kannan, P., Prabukumar, G., Govindaraj, M., 2013. Zinc deficiency in Indian soils with special focus to enrich zinc in peanut. African Journal of Agricultural Research 8 (50), 6681–6688.

Barman, H., Das, S.K., Roy, A., 2018. Zinc in soil environment for plant health and management strategy. Universal Journal of Agricultural Research 6 (5), 149–154.

Barrow, N.J., 1993. Mechanisms of reaction of zinc with soil and soil components. In *Zinc in Soils and Plants*: Proceedings of the International Symposium on 'Zinc in Soils and Plants' held at The University of Western Australia, *27–28 September, 1993* (pp. 15–31). Springer Netherlands.

Ben Rejeb, I., Pastor, V., Mauch-Mani, B., 2014. Plant responses to simultaneous biotic and abiotic stress: molecular mechanisms. Plants 3 (4), 458–475.

Ben Saad, R., Zouari, N., Ben Ramdhan, W., Azaza, J., Meynard, D., Guiderdoni, E., et al., 2010. Improved drought and salt stress tolerance in transgenic tobacco overexpressing a novel A20/AN1 zinc-finger "AlSAP" gene isolated from the halophyte grass Aeluropus littoralis. Plant Molecular Biology 72, 171–190.

Bhatla, S.C., Lal, A., M., Kathpalia, R., Bhatla, S.C., 2018. Plant mineral nutrition. Plant Physiology, Development and Metabolism 37–81.

Blindauer, C.A., 2015. Advances in the molecular understanding of biological zinc transport. Chemical Communications 51 (22), 4544–4563.

Cabot, C., Martos, S., Llugany, M., Gallego, B., Tolrà, R., Poschenrieder, C., 2019. A role for zinc in plant defense against pathogens and herbivores. Frontiers in Plant Science 10, 1171.

Cakmak, I., 2000. Tansley review No. 111 possible roles of zinc in protecting plant cells from damage by reactive oxygen species. The New Phytologist 146 (2), 185–205.

Caldelas, C., Weiss, D.J., 2017. Zinc homeostasis and isotopic fractionation in plants: a review. Plant and Soil 411, 17–46.

Ćalić, I., Koch, J., Carey, D., Addo-Quaye, C., Carlson, J.E., Neale, D.B., 2017. Genome-wide association study identifies a major gene for beech bark disease resistance in American beech (Fagus grandifolia Ehrh.). BMC genomics 18 (1), 1–14.

Cloquet, C., Carignan, J., Lehmann, M.F., Vanhaecke, F., 2008. Variation in the isotopic composition of zinc in the natural environment and the use of zinc isotopes in biogeosciences: a review. Analytical and Bioanalytical Chemistry 390, 451–463.

Couto, E.A.A., Dias-Arieira, C.R., Kath, J., Homiak, J.A., Puerari, H.H., 2016. Boron and zinc inhibit embryonic development, hatching and reproduction of Meloidogyne incognita. Acta Agriculturae Scandinavica, Section B—Soil & Plant Science 66 (4), 346–352.

Deepika, S., Manoranjitham, S.K., Sendhilvel, V., Karthikeyan, G., Kavitha, C., 2021. Foliar Nutrition Enhances the Host Immunity Against Papaya Ringspot Virus.

Dordas, C., 2008. Role of nutrients in controlling plant diseases in sustainable agriculture. A review. Agronomy for Sustainable Development 28, 33–46.

Dotaniya, M.L., Meena, V.D., 2015. Rhizosphere effect on nutrient availability in soil and its uptake by plants: a review. *Proceedings of the National Academy of Sciences, India Section B: Biological Sciences*, *85*, pp.1–12.

Duffy, B., 2007. Zinc and plant disease. Mineral Nutrition and Plant Disease, pp.155–175.

Dumanović, J., Nepovimova, E., Natić, M., Kuča, K., Jaćević, V., 2021. The significance of reactive oxygen species and antioxidant defense system in plants: a concise overview. Frontiers in Plant Science 11, 552969.

Dutta, S., Ghosh, P.P., Ghorai, A.K., De Roy, M., Das, S., 2017. Micronutrients and plant disease suppression. Fertilizers and Environment News 3 (2), 2–6.

Emamverdian, A., Ding, Y., Mokhberdoran, F., Xie, Y., 2015. Heavy metal stress and some mechanisms of plant defense response. The Scientific World Journal 2015.

Ferrari, S., Galletti, R., Denoux, C., De Lorenzo, G., Ausubel, F.M., Dewdney, J., 2007. Resistance to Botrytis cinerea induced in Arabidopsis by elicitors is independent of salicylic acid, ethylene, or jasmonate signaling but requires PHYTOALEXIN DEFICIENT3. Plant Physiology 144 (1), 367–379.

Ganguly, R., Sarkar, A., Acharya, K., Keswani, C., Minkina, T., Mandzhieva, S., et al., 2022. The role of NO in the Amelioration of heavy metal stress in plants by individual application or in combination with phytohormones, especially auxin. Sustainability 14 (14), 8400.

Gill, R.A., Ahmar, S., Ali, B., Saleem, M.H., Khan, M.U., Zhou, W., et al., 2021. The role of membrane transporters in plant growth and development, and abiotic stress tolerance. International Journal of Molecular Sciences 22 (23), 12792.

Gill, S.S., Tuteja, N., 2010. Reactive oxygen species and antioxidant machinery in abiotic stress tolerance in crop plants. Plant Physiology and Biochemistry 48 (12), 909–930.

Gioia, T., Galinski, A., Lenz, H., Müller, C., Lentz, J., Heinz, K., et al., 2016. GrowScreen-PaGe, a non-invasive, high-throughput phenotyping system based on germination paper to quantify crop phenotypic diversity and plasticity of root traits under varying nutrient supply. Functional Plant Biology 44 (1), 76–93.

Gondal, A.H., Zafar, A., Zainab, D., Toor, M.D., Sohail, S., Ameen, S., et al., 2021. A detailed review study of zinc involvement in animal, plant and human nutrition. Indian Journal of Pure & Applied Biosciences 9 (2), 262–271.

Graham, R.D., 2008. Micronutrient Deficiencies in Crops and Their Global Significance. Springer Netherlands, pp. 41–61.

Gupta, N., Ram, H., Kumar, B., 2016. Mechanism of zinc absorption in plants: uptake, transport, translocation and accumulation. Reviews in Environmental Science and Bio/Technology 15, 89–109.

Gupta, S.K., Rai, A.K., Kanwar, S.S., Sharma, T.R., 2012. Comparative Analysis of Zinc Finger Proteins Involved in Plant Disease Resistance.

Hanikenne, M., Bernal, M., Urzica, E.I., 2014. Ion homeostasis in the Chloroplast. Plastid Biology 465–514.

Harris, M.O., Pitzschke, A., 2020. Plants make galls to accommodate foreigners: some are friends, most are foes. New Phytologist 225 (5), 1852–1872.

Hassan, N., Irshad, S., Saddiq, M.S., Bashir, S., Khan, S., Wahid, M.A., et al., 2019. Potential of zinc seed treatment in improving stand establishment, phenology, yield and grain biofortification of wheat. Journal of Plant Nutrition 42 (14), 1676–1692.

Hawkesford, M.J., Cakmak, I., Coskun, D., De Kok, L.J., Lambers, H., Schjoerring, J.K., et al., 2023. Functions of macronutrients. Marschner's Mineral Nutrition of Plants. Academic press, pp. 201–281.

Helfenstein, J., Pawlowski, M.L., Hill, C.B., Stewart, J., Lagos-Kutz, D., Bowen, C.R., et al., 2015. Zinc deficiency alters soybean susceptibility to pathogens and pests. Journal of Plant Nutrition and Soil Science 178 (6), 896–903.

Herrera, K.K., Tognoni, E., Omenetto, N., Smith, B.W., Winefordner, J.D., 2009. Semi-quantitative analysis of metal alloys, brass and soil samples by calibration-free laser-induced breakdown spectroscopy: recent results and considerations. Journal of Analytical Atomic Spectrometry 24 (4), 413–425.

Hirel, B., Tétu, T., Lea, P.J., Dubois, F., 2011. Improving nitrogen use efficiency in crops for sustainable agriculture. Sustainability 3 (9), 1452–1485.

Hossain, M.A., Piyatida, P., da Silva, J.A.T., Fujita, M., 2012. Molecular mechanism of heavy metal toxicity and tolerance in plants: central role of glutathione in detoxification of reactive oxygen species and methylglyoxal and in heavy metal chelation. Journal of Botany 2012.

Ibrahim, H.A., Ibrahim, M., Bondok, A., 2016. Improving growth, yield and resistance to viral diseases of potato plants through modifying some metabolites using zinc sulphate and jasmonic acid. Journal of Horticultural Science & Ornamental Plants 8, 161–172.

Ighodaro, O.M., Akinloye, O.A., 2018. First line defence antioxidants-superoxide dismutase (SOD), catalase (CAT) and glutathione peroxidase (GPX): their fundamental role in the entire antioxidant defence grid. Alexandria Journal of Medicine 54 (4), 287–293.

Islam, M.R., Ali, M.A., Islam, M.S., Golam, A.F.M., Hossain, G.F., 2002. Effect of nutrients and weeding on the incidence of mungbean mosaic. Pakistan Journal of Plant Pathology (Pakistan).

Iwama, R., Kobayashi, S., Ohta, A., Horiuchi, H., Fukuda, R., 2015. Alcohol dehydrogenases and an alcohol oxidase involved in the assimilation of exogenous fatty alcohols in Yarrowia lipolytica. FEMS Yeast Research 15 (3), fov014.

Kalsoom, H., Ali, S., Sahi, G.M., Habib, A., Zeshan, M.A., Anjum, R., et al., 2019. Differential response of micronutrients and novel insecticides to reduce cotton leaf curl virus disease and its vector in Gossypium hirsutum varieties. International Journal of Agriculture and Biology 22 (6), 1507–1512.

Kaur, H., Garg, N., 2021. Zinc toxicity in plants: a review. Planta 253 (6), 129.

Khan, A.G., 2005. Role of soil microbes in the rhizospheres of plants growing on trace metal contaminated soils in phytoremediation. Journal of trace Elements in Medicine and Biology 18 (4), 355–364.

Khan, I., Awan, S.A., Ikram, R., Rizwan, M., Akhtar, N., Yasmin, H., et al., 2021. Effects of 24-epibrassinolide on plant growth, antioxidants defense system, and endogenous hormones in two wheat varieties under drought stress. Physiologia Plantarum 172 (2), 696–706.

Khoshgoftarmanesh, A.H., Kabiri, S., Shariatmadari, H., Sharifnabi, B., Schulin, R., 2010. Zinc nutrition effect on the tolerance of wheat genotypes to Fusarium root-rot disease in a solution culture experiment. Soil Science & Plant Nutrition 56 (2), 234–243.

Kirkby, E.A., 2023. Introduction, definition, and classification of nutrients. Marschner's Mineral Nutrition of Plants. Academic press, pp. 3–9.

Kluska, K., Adamczyk, J., Krężel, A., 2018. Metal binding properties, stability and reactivity of zinc fingers. Coordination Chemistry Reviews 367, 18–64.

Koç, E., Karayiğit, B., 2022. Assessment of biofortification approaches used to improve micronutrient-dense plants that are a sustainable solution to combat hidden hunger. Journal of Soil Science and Plant Nutrition 22 (1), 475–500.

Kovuri, V.A., 2016. *A Metallomics Study on Protein Function Assignment Using ProMOL.* Rochester Institute of Technology.

Kumar, A., Dewangan, S., Lawate, P., Bahadur, I., Prajapati, S., 2019. Zinc-solubilizing bacteria: a boon for sustainable agriculture. *Plant Growth Promoting Rhizobacteria for Sustainable Stress Management: Volume 1: Rhizobacteria in Abiotic Stress Management*, pp.139–155.

Lamalakshmi Devi, E., Kumar, S., Basanta Singh, T., Sharma, S.K., Beemrote, A., Devi, C.P., et al., 2017. Adaptation strategies and defence mechanisms of plants during environmental stress. Medicinal Plants and Environmental Challenges 359–413.

Lawrence, S.D., Novak, N.G., Jones, R.W., Farrar Jr, R.R., Blackburn, M.B., 2014. Herbivory responsive C2H2 zinc finger transcription factor protein StZFP2 from potato. Plant Physiology and Biochemistry 80, 226–233.

Laxmi, A., 2014. DUF581 is plant specific FCS-like zinc finger involved in protein-protein interaction. PLoS One 9 (6), e99074.

Li, F., Bu, Y., Han, G.F., Noh, H.J., Kim, S.J., Ahmad, I., et al., 2019. Identifying the structure of Zn-N2 active sites and structural activation. Nature Communications 10 (1), 2623.

Li, Z., Fan, Y., Gao, L., Cao, X., Ye, J., Li, G., 2016. The dual roles of zinc sulfate in mitigating peach gummosis. Plant Disease 100 (2), 345–351.

Lin, Y.F., Aarts, M.G., 2012. The molecular mechanism of zinc and cadmium stress response in plants. Cellular and Molecular Life Sciences 69, 3187–3206.

Lorenzo-Gutiérrez, D., Gómez-Gil, L., Guarro, J., Roncero, M.I.G., Fernández-Bravo, A., Capilla, J., et al., 2019. Role of the Fusarium oxysporum metallothionein Mt1 in resistance to metal toxicity and virulence. Metallomics 11 (7), 1230–1240.

Machado, P.P., Steiner, F., Zuffo, A.M., Machado, R.A., 2018. Could the supply of boron and zinc improve resistance of potato to early blight? Potato Research 61, 169–182.

Maret, W., 2013. Inhibitory zinc sites in enzymes. Biometals 26, 197–204.

Marschner, H., 2012. Marschner's Mineral Nutrition of Higher Plants, Waltham, MA.

McBride, M., Sauve, S., Hendershot, W., 1997. Solubility control of Cu, Zn, Cd and Pb in contaminated soils. European Journal of Soil Science 48 (2), 337–346.

Millward, D.J., 2017. Nutrition, infection and stunting: the roles of deficiencies of individual nutrients and foods, and of inflammation, as determinants of reduced linear growth of children. Nutrition Research Reviews 30 (1), 50–72.

Mukarram, M., Choudhary, S., Kurjak, D., Petek, A., Khan, M.M.A., 2021. Drought: sensing, signalling, effects and tolerance in higher plants. Physiologia Plantarum 172 (2), 1291–1300.

Nadeem, F., Hanif, M.A., Majeed, M.I., Mushtaq, Z., 2018. Role of macronutrients and micronutrients in the growth and development of plants and prevention of deleterious plant diseases-a comprehensive review. International Journal of Chemical and Biochemical Sciences 14, 1–22.

Nakai, Y., Nakahira, Y., Sumida, H., Takebayashi, K., Nagasawa, Y., Yamasaki, K., et al., 2013. Vascular plant one-zinc-finger protein 1/2 transcription factors regulate abiotic and biotic stress responses in Arabidopsis. The Plant Journal 73 (5), 761–775.

Natasha, N., Shahid, M., Bibi, I., Iqbal, J., Khalid, S., Murtaza, B., et al., 2022. Zinc in soil-plant-human system: a data-analysis review. Science of the Total Environment 808, 152024.

Noman, A., Aqeel, M., Khalid, N., Islam, W., Sanaullah, T., Anwar, M., et al., 2019. Zinc finger protein transcription factors: integrated line of action for plant antimicrobial activity. Microbial Pathogenesis 132, 141–149.

Nowack, B., Obrecht, J.M., Schluep, M., Schulin, R., Hansmann, W., Köppel, V., 2001. Elevated lead and zinc contents in remote alpine soils of the Swiss National Park. Journal of Environmental Quality 30 (3), 919–926.

Osuna, D., Usadel, B., Morcuende, R., Gibon, Y., Bläsing, O.E., Höhne, M., et al., 2007. Temporal responses of transcripts, enzyme activities and metabolites after adding sucrose to carbon-deprived Arabidopsis seedlings. The Plant Journal 49 (3), 463–491.

Padjasek, M., Kocyła, A., Kluska, K., Kerber, O., Tran, J.B., Krężel, A., 2020. Structural zinc binding sites shaped for greater works: Structure-function relations in classical zinc finger, hook and clasp domains. Journal of Inorganic Biochemistry 204, 110955.

Pan, L., Zhao, L., Jiang, W., Wang, M., Chen, X., Shen, X., et al., 2022. Effect of zinc oxide nanoparticles on the growth of Malus hupehensis Rehd. Seedlings. Frontiers in Environmental Science 10, 82.

Parveen, A., Siddiqui, Z.A., 2021. Zinc oxide nanoparticles affect growth, photosynthetic pigments, proline content and bacterial and fungal diseases of tomato. Archives of Phytopathology and Plant Protection 54 (17–18), 1519–1538.

Poddar, K., Sarkar, D., Sarkar, A., 2020. Nanoparticles on photosynthesis of plants: effects and role. Green Nanoparticles: Synthesis and Biomedical Applications 273–287.

Prasad, A.S., Bao, B., 2019. Molecular mechanisms of zinc as a pro-antioxidant mediator: clinical therapeutic implications. Antioxidants 8 (6), 164.

Puga, A.P., Abreu, C., Melo, L.C.A., Beesley, L., 2015. Biochar application to a contaminated soil reduces the availability and plant uptake of zinc, lead and cadmium. Journal of Environmental Management 159, 86–93.

Quaglia, M., Bocchini, M., Orfei, B., D'Amato, R., Famiani, F., Moretti, C., et al., 2021. Zinc phosphate protects tomato plants against Pseudomonas syringae pv. tomato. Journal of Plant Diseases and Protection 128, 989–998.

Quaglia, M., Troni, E., D'Amato, R., Ederli, L., 2022. Effect of zinc imbalance and salicylic acid co-supply on Arabidopsis response to fungal pathogens with different lifestyles. Plant Biology 24 (1), 30–40.

R Buettner, G., 2011. Superoxide dismutase in redox biology: the roles of superoxide and hydrogen peroxide. Anti-Cancer Agents in Medicinal Chemistry (Formerly Current Medicinal Chemistry-Anti-Cancer Agents) 11 (4), 341–346.

Rai, A.C., Singh, M., Shah, K., 2012. Effect of water withdrawal on formation of free radical, proline accumulation and activities of antioxidant enzymes in ZAT12-transformed transgenic tomato plants. Plant Physiology and Biochemistry 61, 108–114.

Rajeswar, M., Rao, C.S., Balaguravaiah, D., Khan, M.A., 2009. Distribution of available macro and micronutrients in soils of Garikapadu of Krishna district of Andhra Pradesh. Journal of the Indian Society of Soil Science 57 (2), 210–213.

Rani, S., Sharma, M.K., Kumar, N., 2019. Impact of salinity and zinc application on growth, physiological and yield traits in wheat. Current Science 116 (8), 1324–1330.

Rawat, N., Singla-Pareek, S.L., Pareek, A., 2021. Membrane dynamics during individual and combined abiotic stresses in plants and tools to study the same. Physiologia Plantarum 171 (4), 653–676.

Rinaldi, L.K., Calandrelli, A., Miamoto, A., Dias-Arieira, C.R., 2023. Application of Ascophyllum nodosum extract and its nutrient components for the management of Meloidogyne javanica in soybean. Chilean Journal of Agricultural Research 83 (2), 127–136.

Rizvi, A., Zaidi, A., Ameen, F., Ahmed, B., AlKahtani, M.D., Khan, M.S., 2020. Heavy metal induced stress on wheat: phytotoxicity and microbiological management. RSC Advances 10 (63), 38379–38403.

Rogowska, A., Pryshchepa, O., Som, N.N., Śpiewak, P., Gołębiowski, A., Rafińska, K., et al., 2023. Study on the zinc ions binding to human lactoferrin. Journal of Molecular Structure 1282, 135149.

Roy, B., Baghel, R.P.S., Mohanty, T.K., Mondal, G., 2013. Zinc and male reproduction in domestic animals: a review. Indian Journal of Animal Nutrition 30 (4), 339–350.

Ryabykh, A.V., Maslova, O.A., Beznosyuk, S.A., Masalimov, A.S., 2022. The Role of Zinc Ion in the Active Site of Copper-Zinc Superoxide Dismutase.

Sahoo, P.C., Kumar, M., Puri, S.K., Ramakumar, S.S.V., 2018. Enzyme inspired complexes for industrial CO2 capture: opportunities and challenges. Journal of CO2 Utilization 24, 419–429.

Savarino, G., Corsello, A., Corsello, G., 2021. Macronutrient balance and micronutrient amounts through growth and development. Italian Journal of Pediatrics 47 (1), 1–14.

Sharma, A., Kumar, V., Shahzad, B., Ramakrishnan, M., Singh Sidhu, G.P., Bali, A.S., et al., 2020. Photosynthetic response of plants under different abiotic stresses: a review. Journal of Plant Growth Regulation 39, 509–531.

Shaukat, S.S., Siddiqui, I.A., 2003. Zinc improves biocontrol of Meloidogyne javanica by the antagonistic rhizobia. Pakistan Journal of Biological Sciences 6 (6), 575–579.

Shoaib, A., Akhtar, M., Javaid, A., Ali, H., Nisar, Z., Javed, S., 2021. Antifungal potential of zinc against leaf spot disease in chili pepper caused by Alternaria alternata. Physiology and Molecular Biology of Plants 27 (6), 1361–1376.

Siddiqui, Z.A., Khan, A., Khan, M.R., Abd-Allah, E.F., 2018. Effects of zinc oxide nanoparticles (ZnO NPs) and some plant pathogens on the growth and nodulation of lentil (Lens culinaris Medik.). Acta Phytopathologica et Entomologica Hungarica 53 (2), 195–211.

Siddiqui, Z.A., Khan, M.R., Abd_Allah, E.F., Parveen, A., 2019. Titanium dioxide and zinc oxide nanoparticles affect some bacterial diseases, and growth and physiological changes of beetroot. International Journal of Vegetable Science 25 (5), 409–430.

Simoglou, K.B., Dordas, C., 2006. Effect of foliar applied boron, manganese and zinc on tan spot in winter durum wheat. Crop Protection 25 (7), 657–663.

Singh, A., Singh, N.Á., Afzal, S., Singh, T., Hussain, I., 2018. Zinc oxide nanoparticles: a review of their biological synthesis, antimicrobial activity, uptake, translocation and biotransformation in plants. Journal of Materials Science 53 (1), 185–201.

Singh, S., Singh, A., Bashri, G., Prasad, S.M., 2016. Impact of Cd stress on cellular functioning and its amelioration by phytohormones: an overview on regulatory network. Plant Growth Regulation 80, 253–263.

Singh, V., Sarwar, A., Sharma, V., 2017. Analysis of soil and prediction of crop yield (Rice) using Machine Learning approach. International Journal of Advanced Research in Computer Science 8 (5).

Sipos, P., Németh, T., Kis, V.K., Mohai, I., 2008. Sorption of copper, zinc and lead on soil mineral phases. Chemosphere 73 (4), 461–469.

Sloup, V., Jankovská, I., Nechybová, S., Peřinková, P., Langrová, I., 2017. Zinc in the animal organism: a review. Scientia Agriculturae Bohemica 48 (1), 13–21.

Sofy, A.R., Sofy, M.R., Hmed, A.A., Dawoud, R.A., Alnaggar, A.E.A.M., Soliman, A.M., et al., 2021. Ameliorating the adverse effects of tomato mosaic tobamovirus infecting tomato plants in Egypt by boosting immunity in tomato plants using zinc oxide nanoparticles. Molecules 26 (5), 1337.

Solanki, M., 2021. The Zn as a vital micronutrient in plants. Journal of Microbiology, Biotechnology and Food Sciences 11 (3), e4026.

Song, G.C., Ryu, C.M., 2013. Two volatile organic compounds trigger plant self-defense against a bacterial pathogen and a sucking insect in cucumber under open field conditions. International Journal of Molecular Sciences 14 (5), 9803–9819.

Souri, M.K., Hatamian, M., 2019. Aminochelates in plant nutrition: a review. Journal of Plant Nutrition 42 (1), 67–78.

St. John, R.A., Christians, N.E., Liu, H., Menchyk, N.A., 2013. Secondary nutrients and micronutrient fertilization. Turfgrass: Biology, Use, and Management 56, 521–541.

Suganya, A., Saravanan, A., Manivannan, N., 2020. Role of zinc nutrition for increasing zinc availability, uptake, yield, and quality of maize (Zea mays L.) grains: an overview. Communications in Soil Science and Plant Analysis 51 (15), 2001–2021.

Tariq, A., Zeng, F., Graciano, C., Ullah, A., Sadia, S., Ahmed, Z., et al., 2023. Regulation of metabolites by nutrients in plants. Plant Ionomics: Sensing, Signaling, and Regulation 1–18.

Tiwari, S., Lata, C., Singh Chauhan, P., Prasad, V., Prasad, M., 2017. A functional genomic perspective on drought signalling and its crosstalk with phytohormone-mediated signalling pathways in plants. Current Genomics 18 (6), 469–482.

Tronchet, M., Balague, C., Kroj, T., Jouanin, L., Roby, D., 2010. Cinnamyl alcohol dehydrogenases-C and D, key enzymes in lignin biosynthesis, play an essential role in disease resistance in Arabidopsis. Molecular Plant Pathology 11 (1), 83–92.

Umair Hassan, M., Aamer, M., Umer Chattha, M., Haiying, T., Shahzad, B., Barbanti, L., et al., 2020. The critical role of zinc in plants facing the drought stress. Agriculture 10 (9), 396.

Vankova, R., Landa, P., Podlipna, R., Dobrev, P.I., Prerostova, S., Langhansova, L., et al., 2017. ZnO nanoparticle effects on hormonal pools in Arabidopsis thaliana. Science of the Total Environment 593, 535–542.

Wadhwa, N., Joshi, U.N., Mehta, N., 2014. Zinc induced enzymatic defense mechanisms in rhizoctonia root rot infected clusterbean seedlings. Journal of Botany 2014.

Wang, G., Zhang, S., Ma, X., Wang, Y., Kong, F., Meng, Q., 2016. A stress-associated NAC transcription factor (SlNAC35) from tomato plays a positive role in biotic and abiotic stresses. Physiologia Plantarum 158 (1), 45–64.

Ward, W.L., Plakos, K., DeRose, V.J., 2014. Nucleic acid catalysis: metals, nucleobases, and other cofactors. Chemical Reviews 114 (8), 4318–4342.

Zastrow, M.L., Pecoraro, V.L., 2014. Designing hydrolytic zinc metalloenzymes. Biochemistry 53 (6), 957–978.

Zehra, A., Raytekar, N.A., Meena, M., Swapnil, P., 2021. Efficiency of microbial bio-agents as elicitors in plant defense mechanism under biotic stress: a review. Current Research in Microbial Sciences 2, 100054.

Zhang, Z., Chen, Y., Li, B., Chen, T., Tian, S., 2020. Reactive oxygen species: a generalist in regulating development and pathogenicity of phytopathogenic fungi. Computational and Structural Biotechnology Journal 18, 3344–3349.

Zoroddu, M.A., Aaseth, J., Crisponi, G., Medici, S., Peana, M., Nurchi, V.M., 2019. The essential metals for humans: a brief overview. Journal of Inorganic Biochemistry 195, 120–129.

Zuriegat, Q., Zheng, Y., Liu, H., Wang, Z., Yun, Y., 2021. Current progress on pathogenicity-related transcription factors in Fusarium oxysporum. Molecular Plant Pathology 22 (7), 882–895.

Chapter 13

# Zinc and Zinc oxide nanoparticles in heavy metal/metalloids stress management in plants

Garima Balyan[1,2] and Akhilesh Kumar Pandey[1]

[1]*Department of Biotechnology, Faculty of Biosciences and Biotechnology, Invertis University, Bareilly, Uttar Pradesh, India,* [2]*Crop Nano Biology and Molecular Stress Physiology Lab, Amity Institute of Organic Agriculture, Amity University Uttar Pradesh, Noida, Uttar Pradesh, India*

## 13.1 Introduction

Throughout their life cycle, plants are subjected to different types of environmental constraint, including metal/metalloids stress, that have an adverse impact on their development and growth (Gautam et al., 2019; Pandey et al., 2019a, 2019b, 2023). Various HMs have been added to the soil by diverse human related activities such as intensive smelting, mining, application of pesticides, burning fossil fuel, and weathering and leaching of rocks (Kumar et al., 2015; Pandey et al., 2023). Heavy metals accumulation in crop plants triggered diminish yield and badly impacted the human (Peralta-Videa et al., 2009; Chakraborti et al., 2015; Kumar et al., 2015; Kumar and Trivedi, 2016). Different HMs have different biological significance and impacts on productivity of crops (Kavamura and Esposito, 2010). When the levels of heavy metals such as Nickel (Ni), Zinc (Zn), Manganese (Mn), Copper (Cu), Selenium (Se), Cobalt (Co), Molybdenum (Mo) (having some metabolic functions in plants) and Zirconium (Zr), Antimony (Sb), Arsenic (As), Mercury (Hg), Cadmium (Cd), Lead (Pb) and Chromium (Cr) (without any vital metabolic functions) reach supra-optimal levels, these significantly lower agricultural output (Kavamura and Esposito, 2010; Park et al., 2011; Rascio and Navari-Izzo, 2011; Salla et al., 2011; Foucault et al., 2013; Shahid et al., 2014; Xiong et al., 2014; Pierart et al., 2015; Gautam et al., 2020; Pandey et al., 2021). These HMs impose negative effects on physiological processes including photosynthesis and transpiration and reducing

**Zinc in Plants. DOI: https://doi.org/10.1016/B978-0-323-91314-0.00012-0**
 

seed, shoot and root growth, biomass and yield (Gautam et al., 2019, 2020, Pandey et al., 2019, 2020). They also cause chlorosis and necrosis in plants and cause oxidative damage, leading to to decrement in crop productivity (Gill, 2014; Rasheed et al., 2021). Therefore, there is an urgent need to identify a solution to decrease the uptake and accretion of metals/metalloids in crops to maintain food security.

Zn is a crucial micronutrient for humans and plants (Hafeez et al., 2013). It is needed in minor quantity by plants but is very important for several main plant physiological pathways such as protein synthesis, photosynthesis, membrane integrity, pollen development, disease resistance, and boosting the levels of antioxidant enzymes and chlorophyll in plant tissues (Yosefi et al., 2011; Hussain et al., 2015). Zn is also an integral component of several biomolecules such as lipids, proteins, and cofactors of auxins, and, therefore, it plays an important role in plant nucleic acid metabolism (Sadeghzadeh and Rengel, 2011). Plants require Zn for the regulation and maintenance of the gene expression necessary for environmental stress tolerance (Hafeez et al., 2013). Zn application has been proven to be beneficial in improving crop yield and quality (Chattha et al., 2017; Hassan et al., 2019). Different forms of Zn mitigate the negative effects imposed by adverse abiotic stress factors. Zn is also reported to regulate the uptake of HMs by plant roots and their accumulation in the plant cell and mitigate the negative effects imposed by HMs in plants (Ganguly et al., 2022).

Furthermore, compared to their bulk and molecular counterparts, nanoparticles (NPs) shows outstanding features such as larger surface area, higher reactivity, variable in particle shape and size, and a better ability to improve plant metabolism (Biswas and Wu, 2005; Giraldo et al., 2014; Siddiqui et al., 2015). Numerous NPs have been investigated for their potential to improve growth and stress tolerance of plants (Marchiol et al., 2016; Soni et al., 2023). Beneficial NPs increase rate of photosynthesis, biomass measurement, chlorophyll, sugar, osmolytes level, and antioxidant production once they have reached various plant parts (Lee et al., 2012; Abdoli et al., 2020). Moreover, NPs up regulate the gene expression pattern associated to abiotic and biotic stress (Singh et al., 2021). ZnO-NPs are also reported to show a positive impact on crops growth and to protect them against different stresses via regulating primary carbohydrate metabolism, boosting photosynthetic parameters, and activating antioxidant enzymes (Rizwan et al., 2019; Sun et al., 2019; Faizan et al., 2020; Faizan et al., 2021a; Singh et al., 2021). ZnO NPs also enhance plant nutrition by modifying the uptake of minerals and regulating the content of protein, carbohydrates, etc. to boost plant growth (Prasad et al., 2012; Zhao et al., 2014; Khan et al., 2015; Itroutwar et al., 2020; Awan et al., 2021).

Considering the protective properties exhibited by Zn and its nano-forms against enviornmental stresses in plants, the present book chapter covers the functional aspects of Zn and ZnO NPs with respect to heavy metal/metalloid stress management in plants, starting from their uptake and translocation to the mechanisms involved in Zn and ZnO NPs-induced HM stress tolerance up to the molecular level and overall conclusion.

## 13.2 Zinc transport in plants

All kinds of plants require the mineral elements known as micronutrients, of which Zn is one. Zn is essential for preserving functionality of bio-membranes and as well as crucial for the growth of generative organs and seeds. It does this by enmeshing itself into the enzymes connected to photosynthesis and energy activities. Zn is absorbed by plant roots as a divalent cation. $Zn^{2+}$ is transferred through the soil via xylem loading with the help of special transporters proteins positioned at the pericycle. The transpiration stream in the xylem is primarily responsible for long-distance Zn transfer (root to shoot) (Page and Feller, 2015). Root-specific membrane transporters actively promote $Zn^{2+}$ uptake from the soil solution. The xylem tissue's persistent H + -ATPase activity causes membrane hyper-polarization and causes outflow of positive charge ions from the cytoplasm (Pandey et al., 2019; Gautam and Pandey, 2021). Hence, the migration of $Zn^{2+}$ from the xylem to apoplastic tissue is an energy-mediated transport (Alves et al., 2004). In addition, plant roots have been observed to absorb organic-ligand-Zn complexes. Typically, Zn transport is said to occur mostly through the symplastic pathway by transporters (Gupta et al., 2016; González-Guerrero et al., 2016). Before it reaches the Casparian strip, this metal is also able to move *via* the apoplast gap in the roots (Palmer and Guerinot, 2009). Zn transport are regulated by several genes and proteins, including NRAMP (natural resistance-associated macrophage protein), ZIP (zinc regulated transporter), and HMA (heavy metal ATPase protein) (Swamy et al., 2016). Plasma membrane transporters help move material over the plasmalemma and in to the cytosol (Gautam and Pandey, 2021). Zn entry into the cytosol is facilitated by the ZIP family; outward movement of Zn to the apoplast is facilitated by the HMA family; and Zn requisition in cellular organelles like vacuole is made possible by the MTP family. Their expression changes depending on the environmental factors and the various plant tissues (Suzuki et al., 2012). The proteome of eukaryotes (9%) and bacteria (6%) are found to be build up of Zn-related proteins (Andreini et al., 2006). Aldehyde dehydrogenases, carbonate dehydratases, Zn-finger DNA-binding proteins, and (Zn/Cu) superoxide dismutase (SOD) are only a few of the enzymes that plants produce that contain Zn (Page and Feller, 2015). In eukaryotes, Zn finger proteins exists in large quantities and maintain their finger-like 3D conformation via attaching to ionic forms of Zn via peptide amino acid and have also been reported in various plants such as, maize (*Zea mays*), arabidopsis (*Arabidopsis thaliana*), wheat (*Triticum aestivum*), pepper (*Capsicum annuum*), rice (*Oryza sativa*), (Gautam et al., 2020) wild tomato (*Solanum pimpinellifolium*), potato

(*Solanum tuberosum*), and cassava (*Manihot esculenta* Crantz) (Liu et al., 2016; Yang et al., 2018; Han et al., 2020). The RING Zinc finger domain, which participates in signaling and expression of genes, binds to particular gene sequence, interacts with a variety of proteins, and mediates development, environmental adaptation and growth, is a characteristic feature of the RING Zinc finger protein (Sun et al., 2019). Plants employ many mechanisms, including the production of stress-responsive genes, the ABA system, the ubiquitin-mediated pathway, $Ca^{2+}$ and ROS mediated cascade of reaction, for zinc finger proteins to react to a range of abiotic stimuli (Gautam et al., 2020). Zn-Cu-SOD dissipates superoxide radicals, protecting membrane proteins and lipids from oxidation and thus promoting the formation of cytochrome, which is necessary for seed development (Karthika et al., 2018).

## 13.3 Zinc oxide nanoparticles transport in plants

Zinc oxide (ZnO) is characterized as an inorganic compound reflecting a broad range of performance, meaningful, encouraging, and multipurpose uses. Several environmental variables, such as drifting, photolysis, hydrolysis, microbial degradation, and leaching prevent the most of agro-chemicals used on crops from reaching their intended sites. Size, content, crystallinity, and shape are the primary factors that determine the intrinsic properties of metal NPs like ZnO. Since Zinc and Oxygen are assigned to groups II and VI, respectively, in the periodic table, it is referred to as as a type of II-VI semi-conductor (Abad et al., 2019). Its structural, mechanical, chemical, electrical, morphological, and optical features can be changed by perturbing them to the nano-dimensional level. Such altered characteristics enable the NPs to connect with cell components in a unique fashion, which facilitates the mechanical delivery of NPs inside the interior structures of cells (Rasmussen et al., 2010). Several reports have demonstrated the benefits of nanomaterials for improving soil quality, germination of seeds, plant defense mechanisms, and seedling growth. ZnO-NPs are absorbed, transferred, and stored by plants in various means that depend on both their unique features and physiology of the host plant (Kandhol et al., 2022, 2023). They enter plants by the roots, leaves, or stems and travel to the cytoplasm, epidermis, cytoplasmic space, cell walls, or vascular cells (da Cruz et al., 2019). ZnO NPs can flow in all directions through phloem channels, enabling them to enter every area of the plant (Su et al., 2019). ZnO NPs are transported to mesophyll cells after initially accumulating in the apoplast after crossing the leaf epidermis through the stomata and can localize in all cellular structures, primarily in the roots of the plant (Yuan et al., 2018).

In numerous studies on crop plants, the effects of ZnO NPs have been assessed. With elevated activities of antioxidant enzymes, nano-dimensional ZnO particles in diverse vegetable crops improve germination, pigments, sugar, and protein contents (Singh et al., 2013). ZnO NPs exhibit antibacterial action against a variety of harmful microorganisms, including *Campylobacter jejuni, Pseudomonas aeruginosa* and *Escherichia coli jejuni* (Xie et al., 2011). Moreover, these NPs also enhance $CO_2$ assimilation in plants, which produces more oxygen and stimulates root development and plant growth. Through the internalization of NPs in the endodermal and vascular tissues, ZnO NPs have a considerable impact on the cortex and root epidermis of *Lolium perenne* and offer great potential to boost agriculture productivity (Lin and Xing 2008). Due to their huge surface-to-volume ratio, ZnO NPs are significantly more effective at blocking UV rays than the bulk material (Yadav et al., 2006). Studies have shown that under salt stress, tomato plants treated with ZnO-NPs at both doses (20 and 40 mg/L), together with several plant gowth promoting bacteria (PGPBs) (*Bacillus subtilis, Lactobacillus casei, and Bacillus pumilus*), displayed enhanced morphological traits. It shows that the coexistence of ZnO-NPs and PGPB under salinity stress could modify the morphological features of tomato plants and mitigate the negative impacts of salinity stress. A small amount of ZnO-NPs improves plant metabolism by increasing the uptake of crucial nutrients like nitrogen, which then influences ion homeostasis, osmolytic biosynthesis, protein content, and harmful radical scavenging (Laware and Raskar, 2014).

## 13.4 Zinc in heavy metal/metalloid stress management in plants

Excess concentration of metals induce oxidative deterioration and cause decrement in biomass, growth, chlorophyll level, photosynthetic features and various metabolical functions in plants (Pandey et al., 2019; Gautam et al., 2020). Zn can alleviate their toxic impact by inhibiting their uptake, evoking antioxidant defense and alleviating metal toxicity to improve plant growth parameters (Hassan et al., 2017). Therefore, Zn is suggested to have a promising role in combating metals stress in several crops. Hussain et al. (2018a,b) have analyzed the impact of application of conjugated form of Zn with lysine (Zn−lys) under different level of Cr stress on physiological attributes, antioxidant enzyme level, oxidative loss, and Cr up-take in rice Hussain et al. (2018a,b). Foliar treatment of Zn-lys improved the photosynthesis, biomass, Zn and antioxiidants level and Cr uptake, thereby mitigating the adverse impact of the Cr stress on these parameters (Hussain et al., 2018a,b). Similarly, Zaheer et al. (2020b) have reported that exogenous supplementation of Zn-lys (10 mg/L) enhanced the different morpho-physiological attributes and increased antioxidative enzyme activities to reduce the oxidative losses in the leaves and roots of the Cr-stressed *Brassica napus* grown in pots irrigated with tannery wastewater containing Cr. Zaheer et al. (2020a) have also observed the alleviatory effects of Zn against Cr stress in spinach (*Spinacia oleracea* L.) in a pot experiment irrigated with tannery wastewater containing a toxic level of Cr. They reported that combined foliar spray of Fe-lys and Zn-lys decreased oxidative stress markers (EL, MDA, $H_2O_2$) by enhancing the activities of different defense enzymes like catalase (CAT), superoxide dismutase (SOD), ascorbate peroxidase (APX) peroxidase (POD) significantly, and enhanced the biomass, growth, chlorophyll status, exchangeable gaseous features and diminishes the Cr concentrations in all plant parts (Zaheer et al., 2020a). Further, Ahmad et al. (2020) have explained the importance of foliar supplementation of Zn-lys in different concentrations on maize raised under the different concentration of Cr-contaminated water. Foliar sprinkle of Zn-lys enhanced the plant performance by increasing the photosynthetic pigments (chlorophyll a, b and carotenoids), amino acids, antioxidant level and reducing the oxidative damages in maize via diminishing the enhanced production of malondialdehyde (MDA) and $H_2O_2$, indicating that exogenous supplementation of Zn can decrease Cr toxicity in crops (Ahmad et al., 2020).

Ali et al. (2022) also reported the effect of supplementation of Zn-lys as foliar sprinkle and seed primers against the toxicity of Cd stress at different concentrations on different varieties of rice and wheat. Foliar spray and seed priming of Zn-lys significantly enhanced the plant growth, improved the physiology and biochemistry, and ameliorated the toxicity caused by the uptake of Cd in wheat and rice varieties (Ali et al., 2022). Similarly, Aravind and Prasad (2003) have inspected the impact of application of Zn at various concentrations on growth of Cd-stressed *Ceratophyllum demersum*. Cadmium stress reduced the level of chlorophyll a, b, carotenoids, rate of photosynthesis, process of electron transport, and photosystem activity, while the exogenous supplementation of Zn induced the defense system of chloroplasts and associated photochemical functions and mitigated the Cd-induced toxicity of *C. demersum* (Aravind and Prasad, 2003). Wu et al. (2020) also performed a pot culture study in the greenhouse to explain the impact of foliar supplementation at different concentration of Zn on physiological and morphological parameters of wheat plant at separate growth stages. Application of Zn increased the biomass, photosynthesis and antioxidant machinery by decreasing the MDA and Cd concentrations and increasing the vigour of wheat (Wu et al., 2020). Kapur and Singh (2019) have also demonstrated that Zn elevates antioxidative defense by enhancing the levels of enzymes like APX, GR, CAT, and SOD and depleting the levels of $H_2O_2$ in Cd-stressed soybean (*Glycine max*). These studies highlight the alleviatory potential of Zn against different HM stresses and delineate that Zn can regulate plant processes at morpho-physiological as well as biochemical levels to prevent plants from the deleterious effect of toxic heavy metals (Table 13.1).

## 13.5 Zinc oxide nanoparticles in heavy metal/metalloid stress management in plants

Abiotic stressors, such as HM stress, pose significant challenges to the productivity and extension of agriculture (Pandey and Gautam, 2020;

**TABLE 13.1** Examples of growth promotion by application of different Zn forms and ZnO nanoparticles under heavy metal/ metalloid stress conditions.

| Sources of Zn alone and other forms | Effective concentration | Host plant | Stress factor | Effect on host plant | References |
|---|---|---|---|---|---|
| Lysine chelated zinc (Zn-lys) | (0, 10, 20, 30 mg/L) | *Oryza sativa* | Cr stress | Increased photosynthesis, biomass, Zn contents, and antioxidant enzyme activities and decreased Cr uptake. | Hassan et al. (2017) |
| Zn-lys | (10 mg/L) | *Brassica napus* | Cr stress | Enhanced the different morpho-physiological attributes, increased antioxidative enzyme activities, and reduced the Cr-stress. | Zaheer et al. (2020b) |
| Foliar application of Zn-lys | (0, 12.5, and 25 mM) | *Zea mays* | Cr-containing tannery wastewater | Enhanced the plant growth by increasing the photosynthetic pigments. | Ahmad et al. (2020) |
| Zn-lys as foliar spray | (0, 12.5, and 25 mM) | Wheat (Punjab-2011; Sammar) and Rice (Kisan Basmati; Chenab). | Cd stress | Improved growth attributes and antioxidative defence,reduced Cd uptake and oxidative stress markers. | Ali et al. (2022) |
| Zn | (0, 10, 20, and 40 mg/L) | *Ceratophyllum demersum* | Cd stress | Enhanced the plant growth, improved the physiology and biochemistry, and ameliorated the toxicity caused by the uptake of Cr. | Aravind and Prasad (2003) |
| Zn | (0, 10, 20, and 40 mg/L) | *Triticum aestivum* | Cd stress | Exogenous addition of Zn induced the defense system of chloroplasts and associated photochemical functions and mitigated the Cd-induced toxicity. | Wu et al. (2020) |

(*Continued*)

**TABLE 13.1** (Continued)

| Sources of Zn alone and other forms | Effective concentration | Host plant | Stress factor | Effect on host plant | References |
|---|---|---|---|---|---|
| ZnO nanoparticles with *Staphylococcus aureus* strain K1 | (0, 50, 100 mg/L) | *Triticum aestivum* | Cr stress | Application of Zn increased the photosynthesis, tissue biomass, and antioxidant enzyme activity by decreasing reactive oxygen species.Increased the levels of antioxidant enzymes, nutrient absorption, and chlorophyll content and reduced the accumulation of Cr. | Ahmad et al. (2022) |
| ZnO NPs | (0, 25, 50, 100 mg/L) | *Oryza sativa* | Cr stress | Enhanced the germination, biomass accumulation, and chlorophyll content and increased the antioxidant enzyme activities. | Basit et al. (2022) |
| Phyto-stabilized ZnO NPs | 50 mg/L | Maize (*Zea mays*) | Cr stress | Increased biochemical and physiological parameters and plant growth. | Ramzan et al. (2023) |
| ZnO NPs with *Providencia vermicola* | 10 ppm | *Luffa acutangula* | As stress | Enhanced photosynthetic pigments, water content, total sugars, and protein while reducing lipid peroxidation. | Tanveer et al. (2022) |
| ZnO NPs | 500 mg/L | *Zea mays* | Co stress | Enhanced plant growth, biomass, photosynthetic content, and antioxidant enzyme activity and reduced Co uptake | Salam et al. (2022) |
| ZnO NPs | 0, 25, 50, 75, and 100 mg/L | *Triticum aestivum* | Cd stress | Increased the dry weights of shoots, roots, plant height, and spike length and enhanced the photosynthesis process. | Rizwan et al. (2019) |

Pandey et al., 2024). To alleviate the negative impact of excess metal and metalloid, beneficial NPs such as ZnO NPs can be employed, as they can modify molecular structure, biochemistry, physiology and morphology of crops under stressful conditions (Soni et al., 2023). For instance, Ahmad et al. (2022) have researched that the application of ZnO NPs at different concentration in combination with *Staphylococcus aureus* strain K1 boosted the defense system and decreased toxic impact Cr at various concentrations in wheat seedling raised in a pot culture. Application of ZnO NPs solely or along with bacterial inoculation increased the levels of antioxidant enzymes (CAT, POD, APX, SOD), nutrient absorption, chlorophyll, carotenoids, and reduced the localization of Cr in different portions of the plant, suggesting their mitigation potential against Cr stress (Ahmad et al., 2022). Similarly, Basit et al. (2022) evaluated the implication of ZnO NPs at different concentrations to alleviate Cr stress in hydroponically grown rice (*Oryza sativa*). They observed that exogenous supplementation of ZnO NPs (100 mg/L) significantly enhanced the percentage of germination potential, germination energy, germination index, vigor index, and biomass accumulation and also improved the antioxidant enzyme activities to increase plant growth while decreasing lipid peroxidation (MDA, $H_2O_2$, EL) and Cr uptake in Cr-stressed rice seedlings (Basit et al., 2022). Furthermore, ZnO NPs (100 mg/L) treatment enhanced the level of phyto-hormone brassinosteroids, delighted chloroplast integrity, and alleviated oxidative stress that is induced by Cr accumulation in rice plants, emphasizing that ZnO NPs can retain harsh effect of Cr in plants to maintain crop growth and productivity (Basit et al., 2022). The alleviatory abilities of ZnO NPs against Cr stress in plants are also confirmed by Ramzan et al. (2023), as they have reported mitigation of the negative impact of Cr (VI) in maize through foliar supplementation of phyto-stabilized ZnO NPs by inducing an increase in length of shoot and root, weight of fresh shoot and root, chlorophyll level, total soluble sugars and proline level, antioxidant enzyme activities under varying stress levels of Cr. Prakash et al. (2022) have also inspected that when external supplement of ZnO NPs was added to the growth medium, it ameliorated the Cr effect and improved the growth of rice seedlings. It increased the photosynthetic efficiency, decreased the oxidative indicators, positively regulated the principal antioxidant genes, and optimize AsA-GSH cycle activity to mitigate Cr (VI) stress and increment in growth and productivity of rice (Prakash et al., 2022).

Tanveer et al. (2022) have established use of ZnONPs alone or in combination *Providencia vermicola* (an arsenic tolerant bacterium) and/or oxalic acid improved the growth and metabolism of *Luffa acutangula* grown on high As containing soil. The findings showed that supplementing As-stressed *L. acutangula* with ZnONPs dramatically increased level of chlorophyll, water status, proline, proteins, total sugars, and indole acetic acid despite decreasing maloandialdehyde and leakage of electrolytes in plant tissues

(Tanveer et al., 2022). Additionally, such methods raised the amount of abscisic acid, minerals, antioxidants, minimized the level of damage to leaf cell structures, and decreased As bio-accumulation in the *L. acutangula* root and shoot, highlighting that the usage of ZnO NPs separately or combined with *P. vermicola* and oxalic acid can be a feasible and environment-friendly way to alleviate As stress in *L. acutangula* (Tanveer et al., 2022). Besides As stress, ZnO NPs have also been reported to protect plants against Co stress (Salam et al., 2022). Salam et al. (2022) have demonstrated the importance of maize seed priming with ZnONPs helps in allevaition of phytotoxicity triggered by Co stress. Seed priming via ZnONPs enhanced the growth, biomass, pigment content, nutrient content, and antioxidative enzymes along with reduced Co uptake, ROS and MDA level in Co-stressed maize shoots, thereby demonstrating the potential role of ZnO NPs as a stress-mitigating tool for maintaining the yield of crops cultivated in Co-contaminated areas (Salam et al., 2022) (Table 13.1).

In the case of Cd stress, Ghouri et al. (2023) have shown that ZnO NPs induce cytological, molecular and physiological changes in Cd-stressed polyploid and diploid rice lines by improving plant's growth parameters (length, yield, dry weight, chlorophyll etc.) and regulating the extent of malondialdehyde and hydrogen peroxide through the activation of antioxidant enzymes. Overall, their study suggested that ZnO NPs application primarily maintain growth and decreases Cd accumulation in both rice lines (Ghouri et al., 2023). Hussain et al. (2018a,b) also depicted the effectiveness of ZnO-NPs in decreasing toxicity and concentration of Cd in wheat while increasing the concentration of Zn (Kandhol et al., 2023). ZnO NPs enhanced the growth, photosynthesis, and grain yield in Cd-stressed wheat plants, and overall, this study indicates that ZnO NPs might be vital player in reducing Cd and enhancing Zn bio-fortification in crops and improving their yield which can be of immense significance for handling the problem of food crisis (Hussain et al., 2018a,b). Rizwan et al. (2019) have also established a fruitful effect of seed priming with ZnO-NPs at different concentration on vigor of wheat exposed to Cd toxicity. The treatment with ZnO NPs increased the the height of plant and spike, and dry biomass of roots, shoot, grains and spikes and improved the photosynthesis process, and reducing the Cd content and EL in Cd-stressed wheat (Rizwan et al., 2019). Bashir et al. (2021) have demonstrated the efficacy of ZnO NPs in regulating Cd intake in wheat under without drought and with drought treatments raised in Cd enriched soil. ZnO NPs improved the tissue dry biomass and nullified the oxidative lossin plants provided with Cd treatment only or in mixed with drought treatment. Overall, their research suggested that ZnO NPs application to Cd-contaminated soils can promote wheat productivity in those soils under normal water as well as drought conditions, thereby paving routes towards better wheat yield against drought with metal treatment (Bashir et al., 2021). These research studies shed light on how ZnO NPs can be used as an efficient way to deal with the problem of HM stress in different plants to maintain their growth and yield during stressful conditions and also highlight that ZnO NPs can offer a sustainable nanofertilizer for crop productivity enhancement in agricultural fields (Kandhol et al., 2022; Soni et al., 2023).

## 13.6 Molecular mechanism of Zn and ZnO NPs induced regulation of metal/metalloid stress

HM pollution poses a significant threat across the world, and the intensity of HMs has recently grown dramatically, pretending a severe danger to crop development and production (Pandey et al., 2019, 2020, 2023). HMs offer substantial health risks to people when they enter human food systems. As a result, it is critical to decrease the impacts of HMs on flora and fauna by implementing proper procedures (Balyan and Pandey, 2024). In this situation, the use of micronutrients can be a vital technique for mitigating the harmful effects of HMs. Zn is an essential mineral for plant development, and Zn supplementation decreases HM-induced toxicity in plants. Zn treatment decreased HM withhold and localized in plant tissues, which is thought to be an important strategies of HM tolerance. Zn enhances membrane function, water-relationship, nutrient absorption, photosynthetic efficacy, osmolytes level, antioxidant activity, and expression of genes. Moreover, the Zn treatment boosts photosynthesis significantly via increasing the formation of plant pigments, enzymatic and photo-system activities, and preserving photosynthetic ultrastructure, resulting in superior growth against heavy metal conditions. Zinc feeding may therefore increase plant performance during HM stress by altering metaolical, biochemical and physiological processes such as anti-oxidant activity, levels of osmolytes and genes expression (Fig. 13.1).

Heat, cold, heavy metals, drought, radiation and salt stress are all important hazards to crop development and output (Pandey et al., 2020; Pandey and Gautam, 2020). HM stress is a severe form of stress that has a deleterious impact on agriculture production and human lives. Because of man-made activity, its intensity is always increasing (Alghrably et al., 2019; Umer Chattha et al., 2021; Zainab et al., 2021). HM contamination is a serious problem that requires an immediate and realistic remedy to limit the risks to various crops and soil (Gautam et al., 2019; Pandey et al., 2020, 2021; Alengebawy et al., 2021) because these HMs disrupt plant physiological and biochemical processes, impair soil health, and have a negative influence on plant performance (Poulson et al., 2021; Hasanuzzaman et al., 2019; Abdel Salam et al., 2022).

As a result, considerable steps necessity is done to limit the effects of HMs on various soils and plants, as well as their life risk. (Gautam and Pandey, 2022; Balyan and Pandey, 2024). The development of suitable technology and management methods can aid in the reduction of HM absorption through crops in polluted sites (Pandey et al., 2023).

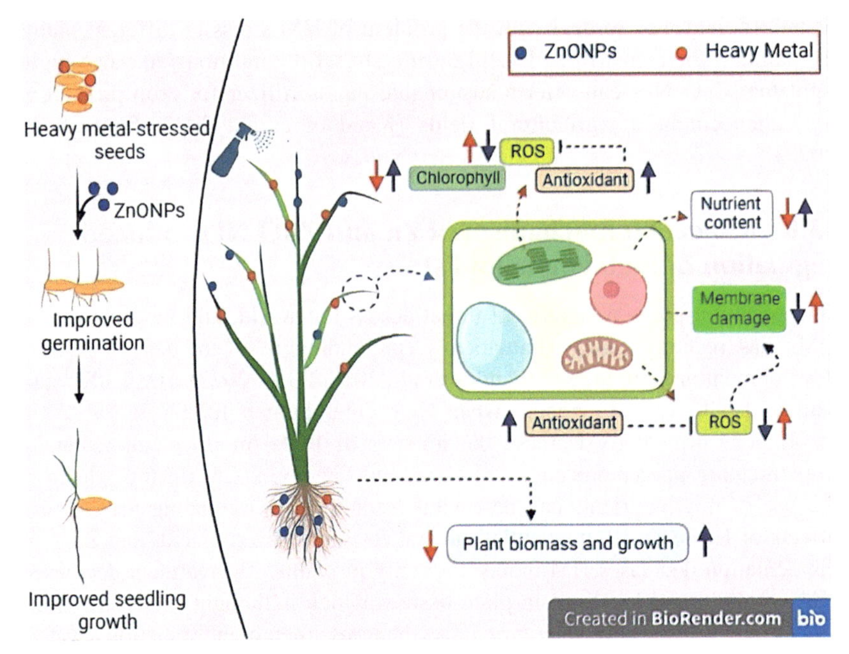


**FIGURE 13.1** Schematic representation of nanoparticles induced seed priming and management of heavy metal/metalloids stress via altering in damages in photosystem and cell membrane, production of ROS and its regulation by antioxidants in different cellular organelles, and nutrient homeoststsis.

Chemical methods, notably micronutrients, have become a viable worldwide strategy for mitigating the damaging response of HMs (Farooq et al., 2020). Zinc is a necessary component for plant growth (Chattha et al., 2017). Zn can help improve HM tolerance via enhancing the gathering of potent osmolytes, and diminishing ROS generation (Faran et al., 2019), as well as uptake and movement of HMs (Hart et al., 2005; Eckhoff, 2010; Javed et al., 2016; Qaswar et al., 2017; Sharfalddin et al., 2021; Abdelrahman et al., 2021). Zn also preserves membrane integrity and limits the creation of ROS, which causes plant cell damage (Cakmak, 2008) Moreover, Zn enhances osmolyte (Pro) accumulation and antioxidant activities (APX, CAT, POD, and SOD) while inhibiting HM absorption. Zn decreases HM translocation inside the plant body, enhancing plant tolerance to HMs (Rizwan et al., 2019; Li et al., 2020). Furthermore, Zn activates the antioxidant system and increases the production of stress-responsive genes, lowering HM-induced oxidative stress in plants (Thounaojam et al., 2014; Zhou et al., 2020).

The most frequent impact of HMs is the formation of ROS, which destroys plasma membranes and promote electrolyte disturbance (Srivastava et al., 2014; Noman et al., 2018). MDA levels rise when lipid peroxidation rises during HM stress (Medina et al., 2017). Exogenous Zn supplementation reduces MDA and $H_2O_2$ levels and safeguards membrane functions during stress condition (Zhou et al., 2020). Zn reduces metal buildup while increasing antioxidant activity (Hussain et al., 2018a; Rizwan et al., 2019). Bounded form of Zn and amino acids considerably decreased MDA, EL, and $H_2O_2$ buildup, according to Zaheer et al. (2019). Zn lowered ROS generation by increasing anti-oxidant pools. Zn treatment also increased water absorption and water level in metal-stressed plants. For example, Zaheer et al. (2022) found that combining Fe and Zn bounded with amino acids enhanced water usage efficiency against Cr stress. During As stress (10 μM and 20 μM), leaf RWC dropped by 37% and 48%, respectively. The administration of Zn (10 μM) and Si (15 μM) together considerably raised the Fe, K levels in the roots of the rice cultivated under 15 μM cadmium (Mapodzeke et al., 2021). Throughout the course of plant growth, ZnO NPs change the production of numerous hormones, chlorophyll, and glucose metabolism. It has been demonstrated that Zn may function as cofactor; vital nutrient in metabolical processes, such as provoking enzymes, osmolytes formation. ZnO NPs in conjunction with Cd and Pb metals resulted in unique genomic modifications, such as the inclusion of novel fragment of DNA and/or the lack of normal fragment and RAPD features. This work is the first to demonstrate that ZnO NPs might be crucial player in the remediation of highly metal-contaminated sites via regulating some of the critical parameters.

## 13.7 By improving photosynthetic capacity

HMs drastically lowered photosynthetic pigments and efficiency, causing diminished growth and output (Gautam et al., 2020; Gautam and Pandey, 2021). Yet, Zn can significantly increase plant photosynthetic activity when exposed to metal stress (Kolenčík et al., 2019) (Fig. 13.1). The use of Zn increases the gas exchange characteristics and restores the photosystem II (PS II). Metal stress along with Zn showed improved photosynthetic performance (Rajiv et al., 2018; Salama et al., 2019).

## 13.8 By accumulation of osmolytes

Proline (Pro), a prime player, shows an impactful link with overall defense mechanisms (Farooq et al., 2020) and aids in plasmalemma stability, the removal of excess ROS, and the control development towards stressful situations (Torabian et al., 2018). Zn treatment significantly enhanced levels of proline, activities of defense enzymes and overall performance against Cd treatment (Farooq et al., 2020). Similarly, Faizan et al. (2021b) discovered that ZnO supplementation raised proline level through boosting the expressions of proline related synthesis gene in response to Cd. Nevertheless, some scientists observed a greater increase in proline level when Zn applied (Helaly et al., 2014; Faizan et al., 2020). Metal in high concentration reduces protein level, which is related to decreased protein formation and elevated protease level, which degrades protein functionality and structure (Balestrasse et al., 2003).

Yet, the use of ZnO dramatically reduced cell survival and damages in root structures (Faizan et al., 2021b). Plants with greater photosynthetic efficacy have better stomatal characteristics. Zn inhibits HM absorption and induces HM immobilization and chelation. The application of ZnO lowers metal absorption and transportation along underground to aboveground parts, and Zn also acts as a constraint to limit process of uptake of metals by crops (Faizan et al., 2021a). Zn transport proteins are thought to be engaged in mercury absorption, and their amounts are greatly influenced by Zn levels. Higher Zn would require higher Zn transport proteins resulting in lesser transport for Hg and lowering Hg absorption in crops/ plants (Liu et al., 2022). Moreover, Zn has antioxidant properties *via* many mechanisms, including metallothionein synthesis, and assists in the creation of reduced molecular weight metallothionein-like proteins (MTLP) (Lane et al., 1987). This MTLP is high in cysteine, which is well known for binding Hg ions; as a result, Zn treatment reduces Hg toxicity by producing MTLP (Liu et al., 2022). Since amino acids generate complexes and inhibit plant absorption of chelated Zn and Fe, using them with amino acids lowered Cr uptake (Rizwan et al., 2017; Zaheer et al., 2022). Another study found that Zn treatment decreased Cd concentration in the root, stem, and leaf, except for grains, and significantly reduced Cd-induced toxicity in wheat plants (Zhou et al., 2020).

## 13.9 By soil parameters modifications

Several variables influence the level of Zn in soils, like pH, moisture, organic matter, carbonates, and clay minerals (Wang et al., 2017). Fe oxides and carbonates have an important player in maintaining soil Zn, influencing plant Zn availability (Komárek et al., 2018; Moreno-Lora and Delgado, 2020). As a result, altering level of soil Fe oxides with organic content can boost Zn levels in crops and soils. Extent of organic content in soil has a substantial impact on Zn dissolution and transport from roots to aboveground parts of plants (Cakmak, 2008). Soil pH is an essential characteristic that influences Zn availability from soil solutions. Higher level pH soils have reduced Zn bioavailability while lower pH shows greater (Sadeghzadeh, 2013). Addition of other compounds, such as sulphuric acid and sulphur, can assist lower pH of soil, greatly enhancing Zn bio-availability and assuring higher HM resistance.

## 13.10 Conclusion and future prospects

The present section outlines the developments in the study on using Zn and ZnO NPs as seed primers, soil applications, foliar sprays, and solution applications to increase plant tolerance to HM-metalloid exposure. Plants under high metal/metalloid conditions experience considerable changes in their growth attributes and diverse physiological and metabolic process alterations. Decreasing the availability of soil HM/metalloid, controlling the functioning of metal/metalloid transporters genes in crops, developing the potential of apoplastic shield to deflect metals, increasing nutrient efficiency, and strengthening the capacity of plant defense machinery are essential methods for improving plant metal-metalloid tolerance. Additionally, we provided an overview of the Zn and ZnO NPs on plants under different metal-metalloid stresses (Table 13.1) and a schematic diagram of potential ZnO NPs processes to reduce metal/metalloid in different plants (Fig. 13.1).

According to studies on HM stress management, plants respond differentially to ZnO NPs treatment levels and various administration techniques. The fact that lower dosages have negligible control consequences and that high dosages would result in additional expenses or detrimental impacts on crops is an essential issue. In addition, there are also various queries about implications of NPs in the diverse soils. Thus, it is essential to continue researching the workings of various administration techniques' beneficial impacts. Additionally, combining the use of NP types with other biological chemicals is unquestionably an exciting area of investigation. It is important to research the effects of using ZnO NPs in combination with different signaling molecules such as hydrogen sulfide, NO, CO, $Ca^{2+}$, and microbes that are resilient to metal/metalloid stress. Based on the characteristics of the soil, the types of crops, and the properties of NPs, it is thought that many processes related to growth and development and the level of metal/metalloids in crops can be managed. Further researches are needed to unveil the potential mechanisms of ZnO NPs in crops exposed to various lethal concentrations of HM/metalloids in soil.

## References


Abad, S.N.K., Moghaddam, J., Mozammel, M., Mostafaei, A., Chmielus, M., 2019. Growth mechanism and charge transport properties of hybrid Au/ZnO nanoprisms. Journal of Alloys and Compounds 777, 1386–1395.

Abdel Salam, M., Mokhtar, M., Albukhari, S.M., Baamer, D.F., Palmisano, L., Jaremko, M., et al., 2022. Synthesis and characterization of green ZnO@ polynaniline/bentonite tripartite structure (G. Zn@ PN/BE) as adsorbent for as (V) ions: Integration, steric, and energetic properties. Polymers 14 (12), 2329.

Abdelrahman, S., Alghrably, M., Campagna, M., Hauser, C.A.E., Jaremko, M., Lachowicz, J.I., 2021. Metal complex formation and anticancer activity of cu (I) and cu (ii) complexes with metformin. Molecules (Basel, Switzerland) 26 (16), 4730.

Abdoli, S., Ghassemi-Golezani, K., Alizadeh-Salteh, S., 2020. Responses of ajowan (*Trachy spermumammi* L.) to exogenous salicylic acid and iron oxide nanoparticles under salt stress. Environmental Science and Pollution Research 27, 36939–36953.

Ahmad, R., Ishaque, W., Khan, M., Ashraf, U., Riaz, M.A., Ghulam, S., et al., 2020. Relief role of lysine chelated zinc (Zn) on 6-week-old maize plants under tannery wastewater irrigation stress. International Journal of Environmental Research and Public Health 17 (14), 5161.

Ahmad, S., Mfarrej, M.F.B., El-Esawi, M.A., Waseem, M., Alatawi, A., Nafees, M., et al., 2022. Chromium-resistant *Staphylococcus aureus* alleviates chromium toxicity by developing synergistic relationships with zinc oxide nanoparticles in wheat. Ecotoxicology and Environmental Safety 230, 113142.

Alengebawy, A., Abdelkhalek, S.T., Qureshi, S.R., Wang, M.Q., 2021. Heavy metals and pesticides toxicity in agricultural soil and plants: Ecological risks and human health implications. Toxics 9 (3), 42.

Alghrably, M., Czaban, I., Jaremko, Ł., Jaremko, M., 2019. Interaction of amylin species with transition metals and membranes. Journal of Inorganic Biochemistry 191, 69–76.

Ali, S., Mfarrej, M.F.B., Hussain, A., Akram, N.A., Rizwan, M., Wang, X., et al., 2022. Zinc fortification and alleviation of cadmium stress by application of lysine chelated zinc on different varieties of wheat and rice in cadmium stressed soil. Chemosphere 295, 133829.

Alves, G., Ameglio, T., Guilliot, A., Fleurat-Lessard, P., Lacointe, A., Sakr, S., 2004. Winter variation in xylem sap pH of walnut trees: involvement of plasma membrane h + ATPase of vessel-associated cells. Tree Physiology 24 (1), 99–105. Available from: https://doi.org/10.1093/treephys/24.1.99.

Andreini, C., Banci, L., Bertini, I., Rosato, A., 2006. Zinc through the three domains of life. Journal of Proteome Research 5 (11), 3173–3178.

Aravind, P., Prasad, M.N.V., 2003. Zinc alleviates cadmium-induced oxidative stress in *Ceratophyllum demersum* L.: a free floating freshwater macrophyte. Plant Physiology and Biochemistry 41 (4), 391–397.

Awan, S., Shahzadi, K., Javad, S., Tariq, A., Ahmad, A., Ilyas, S., 2021. A preliminary study of influence of zinc oxide nanoparticles on growth parameters of Brassica oleracea var italic. Journal of the Saudi Society of Agricultural Sciences 20 (1), 18–24.

Balestrasse, K.B., Benavides, M.P., Gallego, S.M., Tomaro, M.L., 2003. Effect of cadmium stress on nitrogen metabolism in nodules and roots of soybean plants. Functional Plant Biology 30 (1), 57–64.

Balyan, G., Pandey, A.K., 2024. Root exudates, the warrior of plant life: Revolution below the ground. South African Journal of Botany 164, 280–287.

Bashir, A., Ur Rehman, M.Z., Hussaini, K.M., Adrees, M., Qayyum, M.F., Sayal, A.U., et al., 2021. Combined use of zinc nanoparticles and co-composted biochar enhanced wheat growth and decreased Cd concentration in grains under Cd and drought stress: a field study. Environmental Technology & Innovation 23, 101518.

Basit, F., Nazir, M.M., Shahid, M., Abbas, S., Javed, M.T., Naqqash, T., et al., 2022. Application of zinc oxide nanoparticles immobilizes the chromium uptake in rice plants by regulating the physiological, biochemical and cellular attributes. Physiology and Molecular Biology of Plants 28 (6), 1175–1190.

Biswas, P., Wu, C.Y., 2005. Nanoparticles and the environment. Journal of the Air & Waste Management Association 55 (6), 708–746.

Cakmak, I., 2008. Enrichment of cereal grains with zinc: agronomic or genetic biofortification? Plant and Soil 302, 1–17.

Chakraborti, D., Rahman, M.M., Mukherjee, A., Alauddin, M., Hassan, M., Dutta, R.N., et al., 2015. Groundwater arsenic contamination in Bangladesh—21 Years of research. Journal of Trace elements in Medicine and Biology 31, 237–248.

Chattha, M.U., Hassan, M.U., Khan, I., Chattha, M.B., Mahmood, A., et al., 2017. Biofortification of wheat cultivars to combat zinc deficiency. Frontiers in Plant Science 8, 281.

da Cruz, T.N., Savassa, S.M., Montanha, G.S., Ishida, J.K., de Almeida, E., Tsai, S.M., et al., 2019. A new glance on root-to-shoot in vivo zinc transport and time-dependent physiological effects of ZnSO4 and ZnO nanoparticles on plants. Scientific Reports 9 (1), 1–12.

Faizan, M., Hayat, S., Pichtel, J., 2020. Effects of zinc oxide nanoparticles on crop plants: a perspective analysis. Sustainable Agriculture Reviews 41: Nanotechnology for Plant Growth and Development 83–99.

Eckhoff, J., 2010. Using zinc to reduce cadmium in durum grain. MSU Ext. Circ 54. Available from: https://landresources.montana.edu/fertilizerfacts/documents/FF54ZnCdDurumGrain.pdf.

Faizan, M., Bhat, J.A., Chen, C., Alyemeni, M.N., Wijaya, L., Ahmad, P., et al., 2021a. Zinc oxide nanoparticles (ZnO-NPs) induce salt tolerance by improving the antioxidant system and photosynthetic machinery in tomato. Plant Physiology and Biochemistry 161, 122–130.

Faizan, M., Bhat, J.A., Hessini, K., Yu, F., Ahmad, P., 2021b. Zinc oxide nanoparticles alleviates the adverse effects of cadmium stress on *Oryza sativa* via modulation of the photosynthesis and antioxidant defense system. Ecotoxicology and Environmental Safety 220, 112401.

Faran, M., Farooq, M., Rehman, A., Nawaz, A., Saleem, M.K., Ali, N., et al., 2019. High intrinsic seed Zn concentration improves abiotic stress tolerance in wheat. Plant and Soil 437, 195–213.

Farooq, M., Ullah, A., Usman, M., Siddique, K.H., 2020. Application of zinc and biochar help to mitigate cadmium stress in bread wheat raised from seeds with high intrinsic zinc. Chemosphere 260, 127652.

Foucault, Y., Lévèque, T., Xiong, T., Schreck, E., Austruy, A., Shahid, M., Dumat, C., 2013. Green manure plants for remediation of soils polluted by metals and metalloids: Ecotoxicity and human bioavailability assessment. Chemosphere 93 (7), 1430–1435.

Ganguly, R., Sarkar, A., Dasgupta, D., Acharya, K., Keswani, C., Popova, V., et al., 2022. Unravelling the efficient applications of zinc and selenium for mitigation of abiotic stresses in plants. Agriculture 12 (10), 1551.

Gautam, A., Pandey, A.K., 2021. Aquaporins responses under challenging environmental conditions and abiotic stress tolerance in plants. The Botanical Review 1–29.

Gautam, A., Pandey, A.K., 2022. Microbial management of crop abiotic stress: Current trends and prospects. In: *Mitigation of Plant Abiotic Stress by Microorganisms*, Academic Press, pp. 53–75.

Gautam, A., Pandey, A.K., Dubey, R.S., 2019. Effect of arsenic toxicity on photosynthesis, oxidative stress and alleviation of toxicitywith herbal extracts in growing rice seedlings. Indian Journal of Agricultural Biochemistry 32 (2), 143–148.

Gautam, A., Pandey, A.K., Dubey, R.S., 2020. Unravelling molecular mechanisms for enhancing arsenic tolerance in plants: A review. Plant Gene 23, 100240.

Gautam, A., Pandey, A.K., Dubey, R.S., 2020. Azadirachta indica and Ocimum sanctum leaf extracts alleviate arsenic toxicity by reducing arsenic uptake and improving antioxidant system in rice seedlings. Physiology and Molecular Biology of Plants 26, 63–81.

Gautam, A., Pandey, P., Pandey, A.K., 2020. Proteomics in relation to abiotic stress tolerance in plants. In: *Plant Life Under Changing Environment*, Academic Press, pp. 513–541.

Ghouri, F., Shahid, M.J., Liu, J., Lai, M., Sun, L., Wu, J., et al., 2023. Polyploidy and zinc oxide nanoparticles alleviated Cd toxicity in rice by modulating oxidative stress and expression levels of sucrose and metal-transporter genes. Journal of Hazardous Materials 448, 130991.

Gill, M., 2014. Heavy metal stress in plants: a review. International Journal of Advanced Research 2 (6), 1043–1055.

Giraldo, J.P., Landry, M.P., Faltermeier, S.M., McNicholas, T.P., Iverson, N.M., Boghossian, A. A., et al., 2014. Plant nanobionics approach to augment photosynthesis and biochemical sensing. Nature Materials 13 (4), 400–408.

González-Guerrero, M., Escudero, V., Saéz, Á., Tejada-Jiménez, M., 2016. Transition metal transport in plants and associated endosymbionts: arbuscular mycorrhizal fungi and rhizobia. Frontiers in Plant Science 7, 1088.

Gupta, N., Ram, H., Kumar, B., 2016. Mechanism of zinc absorption in plants: uptake, transport, translocation and accumulation. Reviews in Environmental Science and Bio/Technology 15, 89–109.

Hafeez, B.M.K.Y., Khanif, Y.M., Saleem, M., 2013. Role of zinc in plant nutrition—a review. American Journal of Experimental Agriculture 3 (2), 374.

Han, G., Lu, C., Guo, J., Qiao, Z., Sui, N., Qiu, N., et al., 2020. C2H2 zinc finger proteins: master regulators of abiotic stress responses in plants. Frontiers in Plant Science 11, 115.

Hart, J.J., Welch, R.M., Norvell, W.A., Clarke, J.M., Kochian, L.V., 2005. Zinc effects on cadmium accumulation and partitioning in near-isogenic lines of durum wheat that differ in grain cadmium concentration. New Phytologist 167 (2), 391–401.

Hasanuzzaman, M., Alhaithloul, H.A.S., Parvin, K., Bhuyan, M.B., Tanveer, M., Mohsin, S.M., et al., 2019. Polyamine action under metal/metalloid stress: regulation of biosynthesis, metabolism, and molecular interactions. International Journal of Molecular Sciences 20 (13), 3215.

Hassan, Z., Ali, S., Rizwan, M., Hussain, A., Akbar, Z., Rasool, N. and et al., 2017. Role of zinc in alleviating heavy metal stress. *Essential Plant Nutrients: Uptake, Use Efficiency, and Management*, pp. 351–366.

Hassan, M.U., Chattha, M.U., Ullah, A., Khan, I., Qadeer, A., Aamer, M., et al., 2019. Agronomic biofortification to improve productivity and grain Zn concentration of bread wheat. International Journal of Agriculture And Biology 21, 615–620.

Helaly, M.N., El-Metwally, M.A., El-Hoseiny, H., Omar, S.A., El-Sheery, N.I., 2014. Effect of nanoparticles on biological contamination of'in vitro'cultures and organogenic regeneration of banana. Australian Journal of Crop Science 8 (4), 612–624.

Hussain, A., Arshad, M., Zahir, Z.A., Asghar, M., 2015. Prospects of zinc solubilizing bacteria for enhancing growth of maize. Pakistan Journal of Agricultural Sciences 52 (4).

Hussain, A., Ali, S., Rizwan, M., Ur Rehman, M.Z., Javed, M.R., Imran, M., et al., 2018a. Zinc oxide nanoparticles alter the wheat physiological response and reduce the cadmium uptake by plants. Environmental Pollution 242, 1518–1526.

Hussain, A., Ali, S., Rizwan, M., Zia Ur Rehman, M., Hameed, A., Hafeez, F., et al., 2018b. Role of zinc–lysine on growth and chromium uptake in rice plants under Cr stress. Journal of Plant Growth Regulation 37, 1413–1422.

Itroutwar, P.D., Govindaraju, K., Tamilselvan, S., Kannan, M., Raja, K., Subramanian, K.S., 2020. Seaweed-based biogenic ZnO nanoparticles for improving agro-morphological characteristics of rice (Oryza sativa L.). Journal of plant growth regulation 39, 717–728.

Javed, H., Naeem, A., Rengel, Z., Dahlawi, S., 2016. Timing of foliar Zn application plays a vital role in minimizing Cd accumulation in wheat. Environmental Science and Pollution Research 23, 16432–16439.

Kandhol, N., Aggarwal, B., Bansal, R., Parveen, N., Singh, V.P., Chauhan, D.K., Sonah, H., Sahi, S., Grillo, R., Peralta-Videa, J., Deshmukh, R., 2022. Nanoparticles as a potential protective agent for arsenic toxicity alleviation in plants. Environmental Pollution 300, 118887.

Kandhol, N., Rai, P., Pandey, S., Singh, S., Sharma, S., Corpas, F.J., Singh, V.P., Tripathi, D.K., 2023. Zinc induced regulation of PCR1 gene for cadmium stress resistance in rice roots. Plant Science 337, 111783.

Kapur, D., Singh, K.J., 2019. Zinc alleviates cadmium induced heavy metal stress by stimulating antioxidative defense in soybean (*Glycine max L.*) Merr crop. Journal of Applied and Natural Science 11 (2), 338–345.

Karthika, K.S., Rashmi, I., Parvathi, M.S., 2018. Biological functions, uptake and transport of essential nutrients in relation to plant growth. Plant Nutrients and Abiotic Stress Tolerance 1–49.

Kavamura, V.N., Esposito, E., 2010. Biotechnological strategies applied to the decontamination of soils polluted with heavy metals. Biotechnology Advances 28 (1), 61–69.

Khan, M., Naqvi, A.H., Ahmad, M., 2015. Comparative study of the cytotoxic and genotoxic potentials of zinc oxide and titanium dioxide nanoparticles. Toxicology Reports 2, 765–774.

Kim, H., Seomun, S., Yoon, Y., Jang, G., 2021. Jasmonic acid in plant abiotic stress tolerance and interaction with abscisic acid. Agronomy 11 (9), 1886.

Kolenčík, M., Ernst, D., Komár, M., Urík, M., Šebesta, M., Dobročka, E., et al., 2019. Effect of foliar spray application of zinc oxide nanoparticles on quantitative, nutritional, and physiological parameters of foxtail millet (*Setaria italica* L.) under field conditions. Nanomaterials 9 (11), 1559.

Komárek, M., Antelo, J., Králová, M., Veselská, V., Číhalová, S., Chrastný, V., et al., 2018. Revisiting models of Cd, Cu, Pb and Zn adsorption onto Fe (III) oxides. Chemical Geology 493, 189–198.

Kumar, S., Trivedi, P.K., 2016. Heavy metal stress signaling in plants. Plant Metal Interaction. Elsevier, pp. 585–603.

Kumar, S., Dubey, R.S., Tripathi, R.D., Chakrabarty, D., Trivedi, P.K., 2015. Omics and biotechnology of arsenic stress and detoxification in plants: current updates and prospective. Environment International 74, 221–230.

Lane, B., Kajioka, R., Kennedy, T., 1987. The wheat-germ Ec protein is a zinc-containing metallothionein. Biochemistry and Cell Biology 65 (11), 1001–1005.

Laware, S.L., Raskar, S., 2014. Influence of zinc oxide nanoparticles on growth, flowering and seed productivity in onion. International Journal of Current Microbiology Science 3 (7), 874–881.

Lee, S.H., Pie, J.-E., Kim, Y.-R., Lee, H.R., Son, S.W., Kim, M.-K., 2012. Effects of zinc oxide nanoparticles on gene expression profile in human keratinocytes. Molecular & Cellular Toxicology 8, 113–118.

Li, L., Zhang, Y., Ippolito, J.A., Xing, W., Qiu, K., Wang, Y., 2020. Cadmium foliar application affects wheat Cd, Cu, Pb and Zn accumulation. Environmental Pollution 262, 114329.

Lin, D., Xing, B., 2008. Root uptake and phytotoxicity of ZnO nanoparticles. Environmental Science & Technology 42 (15), 5580–5585.

Liu, J., Zhang, C., Wei, C., Liu, X., Wang, M., Yu, F., et al., 2016. The RING finger ubiquitin E3 ligase OsHTAS enhances heat tolerance by promoting H2O2-induced stomatal closure in rice. Plant Physiology 170 (1), 429–443.

Liu, T., Man, Y., Li, P., Zhang, H., Cheng, H., 2022. A hydroponic study on effect of zinc against mercury uptake by triticale: kinetic process and accumulation. Bulletin of Environmental Contamination and Toxicology 108 (2), 359–365.

Mapodzeke, J.M., Adil, M.F., Wei, D., Joan, H.I., Ouyang, Y., Shamsi, I.H., 2021. Modulation of key physio-biochemical and ultrastructural attributes after synergistic application of zinc and silicon on rice under cadmium stress. Plants 10 (1), 87.

Marchiol, L., Mattiello, A., Pošćić, F., Fellet, G., Zavalloni, C., Carlino, E., Musetti, R., 2016. Changes in physiological and agronomical parameters of barley (Hordeum vulgare) exposed to cerium and titanium dioxide nanoparticles. International Journal of Environmental Research and Public Health 13 (3), 332.

Medina, S., Collado-González, J., Ferreres, F., Londoño-Londoño, J., Jiménez-Cartagena, C., Guy, A., et al., 2017. Quantification of phytoprostanes–bioactive oxylipins–and phenolic compounds of *Passiflora edulis* Sims shell using UHPLC-QqQ-MS/MS and LC-IT-DAD-MS/MS. Food Chemistry 229, 1–8.

Moreno-Lora, A., Delgado, A., 2020. Factors determining Zn availability and uptake by plants in soils developed under Mediterranean climate. Geoderma 376, 114509.

Noman, A., Ali, Q., Maqsood, J., Iqbal, N., Javed, M.T., Rasool, N., et al., 2018. Deciphering physio-biochemical, yield, and nutritional quality attributes of water-stressed radish (*Raphanus sativus* L.) plants grown from Zn-Lys primed seeds. Chemosphere 195, 175–189.

Page, V., Feller, U., 2015. Heavy metals in crop plants: transport and redistribution processes on the whole plant level. Agronomy 5 (3), 447–463.

Palmer, C.M., Guerinot, M.L., 2009. Facing the challenges of Cu, Fe and Zn homeostasis in plants. Nature Chemical Biology 5 (5), 333–340.

Pandey, A.K., Borokotoky, S., Tripathi, K., Gautam, A., 2024. Interplay of hydrogen sulfide and plant metabolites under environmental stress. In: *H2S in Plants*, Academic Press, pp. 297–317.

Pandey, A.K., Gautam, A., 2020. Stress responsive gene regulation in relation to hydrogen sulfide in plants under abiotic stress. Physiologia plantarum 168 (2), 511–525.

Pandey, A.K., Gautam, A., Dubey, R.S., 2019. Transport and detoxification of metalloids in plants in relation to plant-metalloid tolerance. Plant Gene 17, 100171.

Pandey, A.K., Gautam, A., Dubey, R.S., 2021. Effect of chromium on protein oxidation, protease activity, photosynthetic parameters and alleviation of toxicity in growing rice seedlings. Indian Journal of Agricultural Biochemistry 34 (1), 39–44.

Pandey, A.K., Gautam, A., Pandey, P., Dubey, R.S., 2019. Alleviation of chromium toxicity in rice seedling using Phyllanthus emblica aqueous extract in relation to metal uptake and modulation of antioxidative defense. South African Journal of Botany 121, 306–316.

Pandey, A.K., Gautam, A., Singh, A.K., 2023. Insight to chromium homeostasis for combating chromium contamination of soil: Phytoaccumulators-based approach. Environmental Pollution 121163.

Pandey, A.K., Gedda, M.R., Verma, A.K., 2020. Effect of arsenic stress on expression pattern of a rice specific miR156j at various developmental stages and their allied co-expression target networks. Frontiers in Plant Science 11, 752.

Pandey, P., Srivastava, S., Pandey, A.K., Dubey, R.S., 2020. Abiotic-stress tolerance in plants-system biology approach. In: *Plant life under changing environment*, Academic Press, pp. 577–609.

Park, J.H., Choppala, G.K., Bolan, N.S., Chung, J.W., Chuasavathi, T., 2011. Biochar reduces the bioavailability and phytotoxicity of heavy metals. Plant and soil 348, 439–451.

Peralta-Videa, J.R., Lopez, M.L., Narayan, M., Saupe, G., Gardea-Torresdey, J., 2009. The biochemistry of environmental heavy metal uptake by plants: implications for the food chain. The International Journal of Biochemistry & Cell Biology 41 (8–9), 1665–1677.

Pierart, A., Shahid, M., Séjalon-Delmas, N., Dumat, C., 2015. Antimony bioavailability: knowledge and research perspectives for sustainable agricultures. Journal of hazardous materials 289, 219–234.

Poulson, B.G., Alsulami, Q.A., Sharfalddin, A., El Agammy, E.F., Mouffouk, F., Emwas, A.H., et al., 2021. Cyclodextrins: structural, chemical, and physical properties, and applications. Polysaccharides 3 (1), 1–31.

Prakash, V., Rai, P., Sharma, N.C., Singh, V.P., Tripathi, D.K., Sharma, S., et al., 2022. Application of zinc oxide nanoparticles as fertilizer boosts growth in rice plant and alleviates chromium stress by regulating genes involved in oxidative stress. Chemosphere 303, 134554.

Prasad, T.N.V.K.V., Sudhakar, P., Sreenivasulu, Y., Latha, P., Munaswamy, V., Reddy, K.R., Sreeprasad, T.S., Sajanlal, P.R., Pradeep, T., 2012. Effect of nanoscale zinc oxide particles on the germination, growth and yield of peanut. Journal of plant nutrition 35 (6), 905–927.

Qaswar, M., Hussain, S., Rengel, Z., 2017. Zinc fertilisation increases grain zinc and reduces grain lead and cadmium concentrations more in zinc-biofortified than standard wheat cultivar. Science of the Total Environment 605, 454–460.

Rajiv, P., Vanathi, P., Thangamani, A., 2018. An investigation of phytotoxicity using Eichhornia mediated zinc oxide nanoparticles on Helianthus annuus. Biocatalysis and Agricultural Biotechnology 16, 419–424.

Ramzan, M., Naz, G., Parveen, M., Jamil, M., Gill, S., Sharif, H.M.A., 2023. Synthesis of phytostabilized zinc oxide nanoparticles and their effects on physiological and anti-oxidative responses of *Zea mays* (L.) under chromium stress. Plant Physiology and Biochemistry.

Rascio, N., Navari-Izzo, F., 2011. Heavy metal hyperaccumulating plants: how and why do they do it? And what makes them so interesting? Plant Science 180 (2), 169–181.

Rasheed, A., Hassan, M.U., Fahad, S., Aamer, M., Batool, M., Ilyas, M., et al., 2021. Heavy metals stress and plants defense responses. Sustainable Soil and Land Management and Climate Change. CRC Press, pp. 57–82.

Rasmussen, J.W., Martinez, E., Louka, P., Wingett, D.G., 2010. Zinc oxide nanoparticles for selective destruction of tumor cells and potential for drug delivery applications. Expert Opinion on Drug Delivery 7 (9), 1063–1077.

Rizwan, M., Ali, S., Hussain, A., Ali, Q., Shakoor, M.B., Zia-ur-Rehman, M., et al., 2017. Effect of zinc-lysine on growth, yield and cadmium uptake in wheat (Triticum aestivum L.) and health risk assessment. Chemosphere 187, 35–42.

Rizwan, M., Ali, S., Ali, B., Adrees, M., Arshad, M., Hussain, A., et al., 2019. Zinc and iron oxide nanoparticles improved the plant growth and reduced the oxidative stress and cadmium concentration in wheat. Chemosphere 214, 269–277.

Sadeghzadeh, B., 2013. A review of zinc nutrition and plant breeding. Journal of Soil Science and Plant Nutrition 13 (4), 905–927.

Sadeghzadeh, B., Rengel, Z., 2011. Zinc in soils and crop nutrition. *The molecular and physiological basis of nutrient use efficiency in crops*, pp. 335–375.

Salam, A., Khan, A.R., Liu, L., Yang, S., Azhar, W., Ulhassan, Z., et al., 2022. Seed priming with zinc oxide nanoparticles downplayed ultrastructural damage and improved photosynthetic apparatus in maize under cobalt stress. Journal of Hazardous Materials 423, 127021.

Salama, D.M., Osman, S.A., Abd El-Aziz, M.E., Abd Elwahed, M.S., Shaaban, E.A., 2019. Effect of zinc oxide nanoparticles on the growth, genomic DNA, production and the quality of common dry bean (*Phaseolus vulgaris*). Biocatalysis and Agricultural Biotechnology 18, 101083.

Salla, V., Hardaway, C.J., Sneddon, J., 2011. Preliminary investigation of Spartina alterniflora for phytoextraction of selected heavy metals in soils from Southwest Louisiana. Microchemical Journal 97 (2), 207–212.

Shahid, M., Pourrut, B., Dumat, C., Nadeem, M., Aslam, M., Pinelli, E., 2014. Heavy-metal-induced reactive oxygen species: phytotoxicity and physicochemical changes in plants. Reviews of Environmental Contamination and Toxicology 232, 1–44.

Sharfalddin, A.A., Emwas, A.H., Jaremko, M., Hussien, M.A., 2021. Transition metal complexes of 6-mercaptopurine: characterization, theoretical calculation, DNA-Binding, molecular docking, and anticancer activity. Applied Organometallic Chemistry 35 (1), e6041.

Siddiqui, M.H., Al-Whaibi, M.H., Firoz, M. and Al-Khaishany, M.Y., 2015. Role of nanoparticles in plants. *Nanotechnology and plant sciences: nanoparticles and their impact on plants*, pp. 19–35.

Singh, N.B., Amist, N., Yadav, K., Singh, D., Pandey, J.K., Singh, S.C., 2013. Zinc oxide nanoparticles as fertilizer for the germination, growth and metabolism of vegetable crops. Journal of Nanoengineering and Nanomanufacturing 3 (4), 353–364.

Singh, A., Tiwari, S., Pandey, J., Lata, C., Singh, I.K., 2021. Role of nanoparticles in crop improvement and abiotic stress management. Journal of Biotechnology 337, 57–70.

Soni, S., Jha, A.B., Dubey, R.S., Sharma, P., 2023. Mitigating cadmium accumulation and toxicity in plants: The promising role of nanoparticles. Science of The Total Environment 168826.

Srivastava, R.K., Pandey, P., Rajpoot, R., Rani, A., Dubey, R.S., 2014. Cadmium and lead interactive effects on oxidative stress and antioxidative responses in rice seedlings. Protoplasma 251, 1047–1065.

Su, Y., Ashworth, V., Kim, C., Adeleye, A.S., Rolshausen, P., Roper, C., et al., 2019. Delivery, uptake, fate, and transport of engineered nanoparticles in plants: a critical review and data analysis. Environmental Science: Nano 6 (8), 2311–2331.

Sun, J., Sun, Y., Ahmed, R.I., Ren, A., Xie, M., 2019. Research progress on plant RING-finger proteins. Genes 10 (12), 973.

Suzuki, M., Bashir, K., Inoue, H., Takahashi, M., Nakanishi, H., Nishizawa, N.K., 2012. Accumulation of starch in Zn-deficient rice. Rice 5, 1–8.

Swamy, B.M., Rahman, M.A., Inabangan-Asilo, M.A., Amparado, A., Manito, C., Chadha-Mohanty, P., et al., 2016. Advances in breeding for high grain zinc in rice. Rice 9, 1–16.

Tanveer, Y., Yasmin, H., Nosheen, A., Ali, S., Ahmad, A., 2022. Ameliorative effects of plant growth promoting bacteria, zinc oxide nanoparticles and oxalic acid on Luffaacutangula grown on arsenic enriched soil. Environmental Pollution 300, 118889.

Thounaojam, T.C., Panda, P., Choudhury, S., Patra, H.K., Panda, S.K., 2014. Zinc ameliorates copper-induced oxidative stress in developing rice (*Oryza sativa* L.) seedlings. Protoplasma 251, 61–69.

Torabian, S., Zahedi, M. and Khoshgoftarmanesh, A., 2018. Effect of foliar spray of zinc oxide on some antiox idant enzymes activity of sunflower under salt stress.

Umer Chattha, M., Arif, W., Khan, I., Soufan, W., Bilal Chattha, M., Hassan, M.U., et al., 2021. Mitigation of cadmium induced oxidative stress by using organic amendments to improve the growth and yield of mash beans [*Vigna mungo* (L.)]. Agronomy 11 (11), 2152.

Wang, Q., Kong, X.P., Zhang, B.H., Wang, J., 2017. Adsorption of Zn (II) on the kaolinite (001) surfaces in aqueous environment: a combined DFT and molecular dynamics study. Applied Surface Science 414, 405–412.

Wu, C., Dun, Y., Zhang, Z., Li, M., Wu, G., 2020. Foliar application of selenium and zinc to alleviate wheat (*Triticum aestivum* L.) cadmium toxicity and uptake from cadmium-contaminated soil. Ecotoxicology and Environmental Safety 190, 110091.

Xie, Y., He, Y., Irwin, P.L., Jin, T., Shi, X., 2011. Antibacterial activity and mechanism of action of zinc oxide nanoparticles against *Campylobacter jejuni*. Applied and Environmental Microbiology 77 (7), 2325–2331.

Xiong, Y., Zhu, F., Zhao, L., Jiang, H., Zhang, Z., 2014. Heavy metal speciation in various types of fly ash from municipal solid waste incinerator. Journal of Material Cycles and Waste Management 16, 608–615.

Yadav, A., Prasad, V., Kathe, A.A., Raj, S., Yadav, D., Sundaramoorthy, C., et al., 2006. Functional finishing in cotton fabrics using zinc oxide nanoparticles. Bulletin of Materials Science 29 (6).

Yang, L., Wu, L., Chang, W., Li, Z., Miao, M., Li, Y., et al., 2018. Overexpression of the maize E3 ubiquitin ligase gene ZmAIRP4 enhances drought stress tolerance in Arabidopsis. Plant Physiology and Biochemistry 123, 34–42.

Yosefi, K., Galavi, M., Ramrodi, M., Mousavi, S.R., 2011. Effect of bio-phosphate and chemical phosphorus fertilizer accompanied with micronutrient foliar application on growth, yield and yield components of maize. Australian Journal of Crop Science 5 (2), 175–180.

Yuan, J., Chen, Y., Li, H., Lu, J., Zhao, H., Liu, M., et al., 2018. New insights into the cellular responses to iron nanoparticles in Capsicum annuum. Scientific Reports 8 (1), 3228.

Zaheer, I.E., Ali, S., Rizwan, M., Bareen, F.E., Abbas, Z., Bukhari, S.A.H., Wijaya, L., Alyemeni, M.N., Ahmad, P., 2019. Zinc-lysine prevents chromium-induced morphological, photosynthetic, and oxidative alterations in spinach irrigated with tannery wastewater. Environmental Science and Pollution Research 26, 28951–28961.

Zaheer, I.E., Ali, S., Saleem, M.H., Ali, M., Riaz, M., Javed, S., et al., 2020a. Interactive role of zinc and iron lysine on *Spinaciaoleracea* L. growth, photosynthesis and antioxidant capacity irrigated with tannery wastewater. Physiology and Molecular Biology of Plants 26, 2435–2452.

Zaheer, I.E., Ali, S., Saleem, M.H., Arslan Ashraf, M., Ali, Q., Abbas, Z., et al., 2020b. Zinc-lysine supplementation mitigates oxidative stress in rapeseed (*Brassica napus* L.) by preventing phytotoxicity of chromium, when irrigated with tannery wastewater. Plants 9 (9), 1145.

Zaheer, I.E., Ali, S., Saleem, M.H., Yousaf, H.S., Malik, A., Abbas, Z., et al., 2022. Combined application of zinc and iron-lysine and its effects on morpho-physiological traits, antioxidant capacity and chromium uptake in rapeseed (*Brassica napus* L.). PLoS One 17 (1), e0262140.

Zainab, N., Khan, A.A., Azeem, M.A., Ali, B., Wang, T., Shi, F., et al., 2021. PGPR-mediated plant growth attributes and metal extraction ability of Sesbania sesban L. in industrially contaminated soils. Agronomy 11 (9), 1820.

Zhao, C.Y., Tan, S.X., Xiao, X.Y., Qiu, X.S., Pan, J.Q., Tang, Z.X., 2014. Effects of dietary zinc oxide nanoparticles on growth performance and antioxidative status in broilers. Biological trace element research 160, 361–367.

Zhou, J., Zhang, C., Du, B., Cui, H., Fan, X., Zhou, D., et al., 2020. Effects of zinc application on cadmium (Cd) accumulation and plant growth through modulation of the antioxidant system and translocation of Cd in low-and high-Cd wheat cultivars. Environmental Pollution 265, 115045.

Chapter 14

# Zinc nutrition to plant, animals, and humans: recent updates

Aakriti Srivastava[1,2], Monika Thakur[2], Shivani Mahra[4], Vijay Pratap Singh[3], Shivesh Sharma[4] and Durgesh Kumar Tripathi[1]

*[1]Crop Nanobiology and Molecular Stress Physiology Laboratory, Amity Institute of Organic Agriculture, Amity University Uttar Pradesh, Noida, Uttar Pradesh, India, [2]Amity Institute of Food Technology, Amity University Uttar Pradesh, Sector-125, Noida, Uttar Pradesh, India, [3]Department of Botany, C.M.P. Degree College, University of Allahabad, Prayagraj, Uttar Pradesh, India, [4]Department of Biotechnology, Motilal Nehru National Institute of Technology Allahabad, Prayagraj, Uttar Pradesh, India*

## 14.1 Introduction

The global population is projected to increase by 25% by 2050, adding another 2 billion people and raising the burden of the growing food demand (Schroeder et al., 2013). The diminishing area under agriculture, development, and changes in the climate constitute the situation more serious (Malhi et al., 2021; Zabel et al., 2014; Satterthwaite et al., 2010). The accessibility of food has marginally improved due to vast agricultural practises and technological development, but the nutritional value of the goods has declined (Haddad et al., 2016). The social expense of micronutrient deficiency (MND), which is also known as "hidden hunger," is measured in millions of dollars per year and can account for as much as 11% of a country's GDP. Based on the quantities needed, macronutrients and micronutrients are the two groups into which the nutrients sustaining life are classified (Taşğın, 2017; Solanki, 2021; Kirkby, 2023). Micronutrients are essential for living beings and play serve essential roles in metabolism, growth and development, even though they are only needed in extremely small quantities. Subsequently several of the micronutrients are unable to be synthesized by living organisms, consumption of these nutrients is necessary on regular basis. without which life and optimal health will be seriously disrupted (Haddad et al., 2016). After the macronutrients i.e., Nitrogen, Phosphorus,

**Zinc in Plants. DOI: https://doi.org/10.1016/B978-0-323-91314-0.00003-X**
 

and Potassium, certain researchers believe that Zn is also the most significant mineral. Zn is a crucial mineral for plant life and animals including humans (Khan et al., 2022). Being twenty-third most common element, approximately 0.02% of the earth's surface is made up of the Zinc (i.e., likely to be bluish-white solid element having atomic weight 65.4 and its atomic number 30). Interestingly, the first report to emphasize the significance of zinc for organisms was with *Aspergillus niger* in 1869. Since several researchers studied the significance of Zn in every aspect of life globally. In the cytoplasm and organelles in various living cell systems, the proportional abundance of free Zn ions varies from 103 to 109 mol/L (Fabris, 1994). Furthermore, Zn is essential for cell growth, development and differentiation. Its significance to human immunity, neurotransmission, and proper functioning of the brain is becoming progressively more explicit (Frederickson et al., 2005; Kiouri et al., 2023). Additionally, it improves the production, preservation, and release of insulin as well as the body's ability to fight against reactive stress. Zn is similarly necessary for adequate plant development and plays a significant role in the physiology and metabolism of plants (Fig. 14.1). In addition to photosynthesis, it plays a vital role in synthesis of chlorophyll, plant reproduction, and autophagy, among other essential biochemical processes. In the study by Cabot et al. (2019), Zn impact on cereal production, seed growth, and defence against herbivores and plant diseases has been noticed. Therefore, Zn is correctly referred to as a "metal of life." Zinc has specific chemical characteristics that make it particularly valuable and significant in biological systems due to its nature as a transitional element in the periodic chart (Brown et al., 2001; Krężel and Maret, 2016). Zinc is a standard element in both natural and human environments and is crucial to many cellular processes. All higher plants and animals, including human being, require zinc for proper growth and reproduction (Fig. 14.1). Due to which Zn is referred to as an "essential trace element" or a "micronutrient." In all groups of living organisms including human immune system, zinc is essential for basic cellular functions and has a significant impact on metabolic development. (Vidyshree et al., 2016). It is essential for gene translation, stability of DNA, and the operation of more than 300 enzymes (Frassinetti et al., 2006). Zinc is an interesting metal since it doesn't have any redox properties, which allows it to move about in biological systems without causing oxidative damage—a key deficiency in iron or copper (Brown et al., 2001) (Fig 14.2).

## 14.2 Zinc nutrition to plant health

Zinc is an important micronutrient in plants as it plays an important role in many enzymes that catalyze metabolic reactions in crops (Gondal et al., 2021; Solanki, 2021). Additionally, zinc is important for protein synthesis,

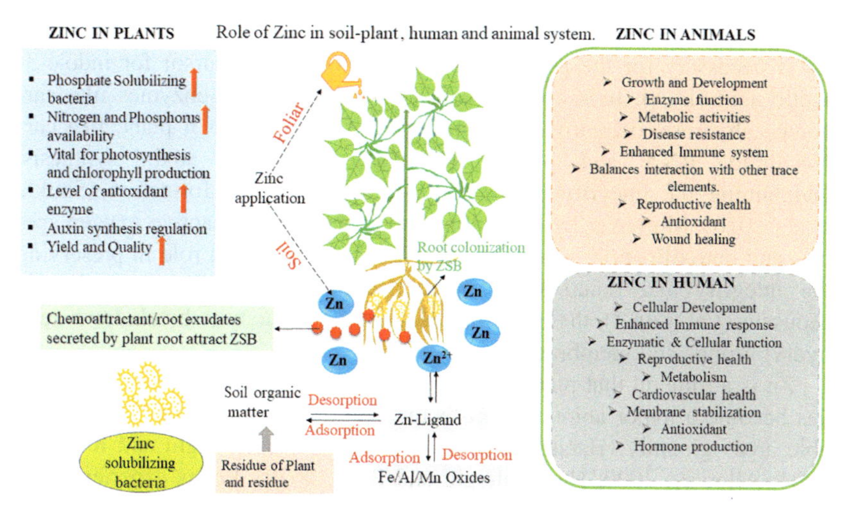


**FIGURE 14.1** Role of Zinc in soil-plant , human and animal system.

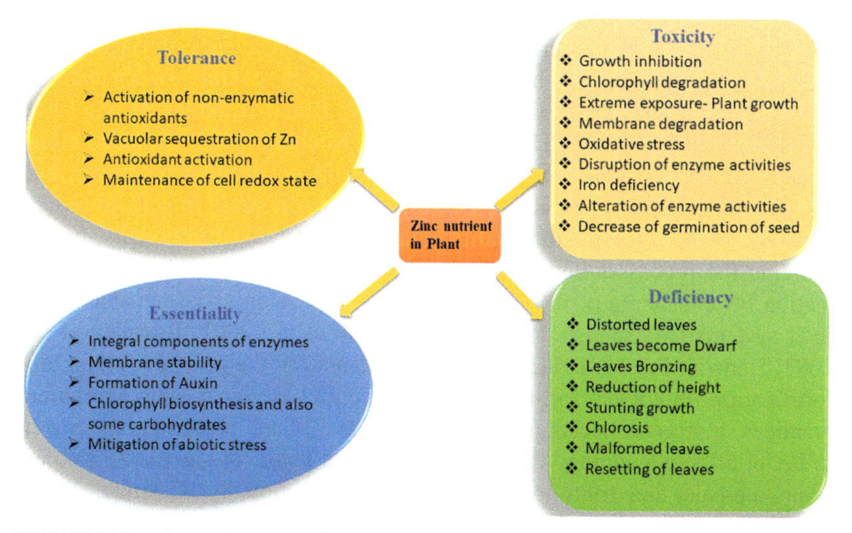


**FIGURE 14.2** Figure shows the differential responses of zinc in plants. (Adopted and modified with permission from Natasha et al., 2022).

disease resilience, photosynthesis, pollen development, cell membrane stability, pollen development, and increasing the antioxidant enzymes levels and chlorophyll in plant tissues (Hussain et al., 2015). In addition to being essential for metabolism, Zn is also necessary for plants to grow and develop healthily (Fig. 14.1). For the effective functioning of various physiological processes in plants, such as photosynthesis, production of sugar, fertility, production of seeds, growth control and disease resilience (Fig. 14.1), it is required in small but crucial concentrations (Solanki et al., 2016). In plants, Zn is necessary for the production of tryptophan, a precursor for indole-3-acetic acid (IAA). It is a component of various metallo-enzymes like carbonic anhydrase, dehydrogenases involved in metabolism of plant, like carbohydrate, RNA and protein synthesis; regulating auxin synthesis, maintaining the integrity of membranes, and in pollen formation (Shukla et al., 2016). Moreover by controlling the generation of reactive oxygen species and detoxification, zinc serves a critical physiological role in preserving the integrity and functionality of the cellular membrane. According to Mousavi et al. (2013), these substances have the capacity to break down sulfhydryl groups and membrane lipids.

Zn is a mineral that plants absorb from the soil, and Zn deficiency in soil has become a major abiotic stress factor, that affects more than 49% of aerable areas globally (Hacisalihoglu et al., 2001; Hacisalihoglu and Kochian, 2004; Alloway, 2009; Hacisalihoglu and Blair, 2020). Plants experiencing a zinc deficit commonly exhibit symptoms including short internodes and small leaves, rosetting or whirling of leaves, interveinal chlorosis of young leaves with veins remaining green in mild to moderate deficient conditions (Shukla et al., 2016). In order to decrease Zn insufficiency throughout the vulnerable areas, research has been carried out in several low-Zn nations, including Australia, India, China, Turkey and Brazil (Cakmak et al., 1998; Hacisalihoglu et al., 2001; Hacisalihoglu and Kochian, 2004; Broadley et al., 2007). So, Zn-efficient varieties of crops could be developed, grown, and identified to provide methods for controlling low-Zn stress in soils to reduce production and decreases the quality (Marschner, 1995). Additionally, understanding the processes underlying Zn effectiveness will reveal crucial information for enhancing plant nutrients as well as developing sustainable world food systems (Marschner, 1995; Hacisalihoglu et al., 2001; Broadley et al., 2007; Hamzah Saleem et al., 2022). For instance, in the past few years, biofortification (biological fortification), which enriches plants using agronomic practices, transgenic techniques, or conventional plants breeding, has proven to be an effective strategy. It provides adequate levels of Zn via vegetables, beans, cereals, and fruits to the targeted regions globally (Cakmak, 2008). Hundreds of varieties of rice (*Oryza sativa*) (Fageria, 2001; Naik et al., 2020), beans (*Phaseolus vulgaris*) (Cakmak et al., 1998), chickpeas (*Cicer arietinum*) (Khan et al., 1998: Ullah et al., 2020), and wheat (*Triticum aestivum*) (Hacisalihoglu et al., 2001) were evaluated for Zn efficiency to achieve this goal. Moreover, visual symptom evaluation methods, as well as growth and productivity under low and adequate Zn circumstances, are all included in plant Zn efficiency screening (Cakmak et al., 1998; Hacisalihoglu et al., 2004). Numerous efforts have been made over the past three decades to understand how Zn-efficient plants respond to low-Zn stress in order to develop efficient agricultural breeding techniques (Welch, 1993; Hacisalihoglu, 2020). Moreover, most of them will be grown around the globe as a result of the availability of Zn-efficient varieties (Hacisalihoglu, 2020). Root hairs that boost the amount of Zn availability in the soil can promote Zn uptake (Welch and Graham, 2005). It is widely acknowledged that soil characteristics and pH plays a significant role in determining the Zn amount that plants can utilize for the growth and development (Welch and Graham, 2005; Cakmak et al., 1998). In the process of uptake, $Zn^{2+}$ ions pass through root epidermis, endodermis, xylem, pericycle, and cortex and then it is translocated to the stem, leaves, phloem and seeds (Haslett et al., 2001). Foliar application is another most efficient method to raise grain Zn content (Zaman et al., 2018). This technology makes phloem movable and facilitates zinc's simple transfer during grain development. Numerous Zn sources, such as Zn-EDTA, Zn $(NO_3)_2$, and $ZnSO_4$, are employed for this purpose; however, co-granulating Zn with phosphorous fertilisers can reduce its efficacy due to the insoluble combination, which limits the mobility of foliar-applied Zn due to its inadequate penetration into the leaf and strong binding affinity towards leaf tissues (Montalvo et al., 2016). According to reports, Zn application through foliar $ZnSO_4$ application over 3 years at 23 experimental sites in India, Mexico, China, Zambia, Turkey, Kazakhstan, and Pakistan increased crop Zn concentration in wheat by 84% (Zou et al., 2012). Foliar application demonstrated that, total Zn absorption from wheat was shown to be equally effective as comparable with that of hydroponic systems and fortification (Signorell et al., 2019). Further, Xia et al. (2019) demonstrated that, maize crops have higher Zn concentrations and bioavailability. Intriguingly, Zn application also increased Fe concentrations in maize grains. In this manner, a single biofortification technique can increase both Zn and Fe amounts in grains. According to reports, applying $ZnSO_4$ and $FeSO_4$ together (1:2) through foliar spray increased wheat production as well as qualitative attributes of wheat (Ramzan et al., 2020). However, Niyigaba et al., 2019 reported that zinc application alone led to a considerable increase in wheat grain zinc concentration while the combined application of zinc and iron did not have similar results showing the contradictory observation. Interestingly, in addition to the foliar spray, treating seeds at the panicle and boot leaf stages can also be another efficient and alternative way of micronutrient biofortification for rice (Meena and Fathima, 2017).

## 14.3 Animal health influenced by zinc

In the 1930s and 1940s, scientists discovered that Zn was a required nutrient in rats and mice, and they later discovered that farm animals required Zn as well. The classic work of Tucker and Salmon, showed that high Ca diets fed in production of swine resulted in parakeratosis which was treated by increasing the amount of zinc in the diets, and this led to more research into the physiology and metabolism of zinc in farm animals (Tucker and Salmon, 1955). Zn is a comparatively non-toxic element, but it typically causes toxicity when taken in huge amount. Symptoms which may include dehydration, electrolyte imbalance, stomach discomfort, vertigo, fatigue, disorientation, and lack of coordination in the muscles, depending on the cause and duration of the Zn exposure (Prasad, 1976). Zinc has several roles in disease resistance and immunity. It was observed by Droke and Spears, (1993) in a study that, when lambs fed a semi-purified diet deficient in Zn showed reduced blastogenic response to a T- cell mitogen and an increased response to a T-dependent B cell mitogen. Lambs had a higher percentage of neutrophils and a lower percent of lymphocytes. The first laboratory, Beeson's Laboratory determined the Zn requirement for sheep based on semi-purified meals (Ott et al., 1965). An interesting yet not fully proven discovery indicates that sheep need double the amount of zinc in their diet for best growth, producing good quality wool, fertility in males and raising plasma zinc levels to the maximum (Ott et al., 1965). As of, cattle typically need zinc for optimal performance. Although excessive amounts of zinc (about 1.5–2.0 g per day) can be fatal if fed to calves, it is estimated that cumulative doses of 30–66 g could lead to death (Graham et al., 1987).

Zinc supplementation has observable effects, especially for enhanced weight gain in pastured calves or cows (Mayland et al., 1980). However, the level of Mn, Cu, Zn, and Co fed together in excess of permissible amounts reduce the efficiency of reproduction (Olson et al., 1999). Zn in milk increased when supplemental Zn was added to lactating cows' diets, but the amount of Zn given as a percentage of milk production decreased (Miller et al., 1965). Milk production and feed intake decreased when dairy cows were given 2000 ppm Zn, but milk and plasma Zn were increased and plasma Cu was declined whilethe feeding of 1000 ppm Zn had no adverse effects (Miller et al., 1989). In the study by Jenkins and Hidiroglou (1991) it was stated that performance was declined when pre-ruminant calves were given more than 700 ppm Zn. Calves given 706 ppm Zn showed an increase in segmented neutrophils and a reduction in prothrombin, eosinophils time and activated partial thromboplastin time (Graham et al., 1991). Moreover, Zinc has historically been added to poultry feed using more accessible and less expensive inorganic forms such sulfates, oxides, and chlorides. However, naturally occurring variations in the pH of the birds' digestive tracts might result in competitive interactions between various trace elements in addition to those that combine with phytic acids to form insoluble compounds, which are present in the majority of grain-based diets. This would prevent the trace elements from being absorbed by the body and lower the bioavailability of zinc (McDowell, 2003). Further by O'Dell et al. (1958) stated that when 15 ppm Zn was given to galvanized batteries for 4–6 weeks, chicken began to show symptoms of Zn deficiency i.e. poor development of feathers, thickening and shortening of long bones, reduced growth and rapid, labored respiration were among the symptoms in chickens recorded. More severe signs of deficiency emerged when the availability to Zn in cage coating was removed. Further studies demonstrated that, high Zn feedings in chicks resulted in MT in the gut epithelium, liver, kidney andpancreas. The specific Zn transporter plays an important role in regulating the cellular balance of zinc due to the widespread inactivation of Zn2 + ions, which are hydrophilic in nature (Bonaventura et al., 2015). This balance is maintained by two cooperative protein families: Zn transporters and metallothionenes (MTs). As birds fed diets that are high in zinc, theylost their Zn in MT when they were shifted to low Zn diets, demonstrating the importance of MT in the chick's Zn homeostasis (Oh et al., 1979). Based on Wang's study in 2002, a lack of zinc directly impacts a chicken's growth plate, causing growth plate chondrocytes to proliferate, differentiate, and die, which is why O'Dell and Savage observed stunted bones in their studies. Further, for normal offspring to be produced, hens required 65 ppm Zn and 4.0% Ca. When Zn-deficient chicks emerge, they would be feeble and unable to stand, feed, or drink without Zn supplements (Kienholz et al., 1961). Williams et al. (1989) demonstrated that, Ovarian and oviduct weights were decreased by 10 days when mature laying chickens were given 20,000 ppm Zn for 4 days. By day 18, liver and kidney concentrations were back to their pre-treatment levels, but the pancreas continued to have an elevated concentration. Additionaly, according to Dewar et al. (1982), tumors were discovered in the gizzard and pancreas of chickens after 4 days of 10,000 ppm dietary Zn. After being eliminated from the food for 28 days, the extra Zn, were vanished. Compared to controls given 20 ppm Zn, eggs from chickens fed 10,000 or 20,000 ppm Zn contained less K, Cu, and Zn (Palafox and HO-A, 1980). Zn and Cu concentrations declined weekly in a limited sample size of mare milk gathered weekly (weeks 1−8) but initial colostrum concentrations were not quantified (Schryver et al., 1986). Then deficiency of zinc in animals can easily be manifested by synthesis disorder for e.g., keratin synthesis, limited growth of limb bone and infections of eyesight, and the changes in their taste perception (as the epithelium of tongue damage) (Prasad, 2006).

## 14.4 Zinc nutrition in humans

Zinc (Zn) is also one of the essential minerals for development and the immune system. It is a crucial component of enzymes and is important for cellular development and tissue division. It is regarded as a multipurpose trace element that is engaged in numerous biological processes and has the capacity to bind over 2000 transcription factors and over 300 enzymes. It also plays an important part in cellular functioning and metabolic processes, including the oxidative stress response. Moreover, vital for immunological response, homeostasis, cell cycle progression, apoptosis, senescence, DNA replication, and repair of DNA damage (Chasapis et al., 2020). Despite being present in nature, its absorption in the human body has not been adequate. Zn shortage has a significant impact on almost 33% of the world's population, usually in non-urban areas. Prior to this study, Zn deficiency caused approximately 116,000 deaths annually throughout the globe (Galetti, 2018). Consistent dietary consumption is necessary to support these processes and keep the exchangeable Zn pool continue because the human body lacks a mechanism for long-term storage of zinc. This is true particularly during childhood, puberty, and pregnancy (Gibson and Anderson, 2009). Since Zn is important for human nourishment and growth, being the main staple food crops like rice, wheat, corn, beans, and others have higher Zn contents (Grusak and DellaPenna, 1999; Hacisalihoglu, 2020). In humans, a lack of Zinc may cause loss of appetite, anorexia, loss of taste and smell, and other symptoms, and it may affect the immune system, which cause anemia and arteriosclerosis. Zn deficiency impairs hemostasis by causing defective platelet aggregation problem, decline in T cell number, and decreased T-lymphocyte response to phytoestrogens. In reality, the only lymphocytic mitogen that occur naturally is Zinc (Keen and Gershwin, 1990; Tapiero and Tew, 2003; Gondal et al., 2021). Zn also influence hormones production development (Bjorndahl and Kvist, 2011). Zn is essential for enzyme activity as well as gene control, intra- and intercellular communication, membrane stabilization, and apoptosis (Cai et al., 2015; Chellan and Sadler, 2015; Choi et al., 2017). Health experts like the National Institutes of Health (NIH) advised adults to consume 8 mg of zinc for adult female and 11 mg of zinc for adult male everyday (NIH National Institute of Health, 2021). In the age range of 1–3 years, malnourished children have a Zn requirement of 2–4 mg/kg body weight, which is greater than that of healthy children (0.17 mg/kg), due to the higher Zn loss and lower Zn absorption in the former subgroup (Müller et al., 2001). According to reports, Zn supplementation enhances children's linear growth and weight gain and reduces cases of Zn deficiency that include acute infections of the lower respiratory tract, premature birth, and diarrheal episodes and diarrhea-induced morbidity (Liu et al., 2018). For the first 6 months, an infant receives enough Zn from exclusive breastfeeding and full nursing and nutrition (Brown et al., 2004); laterly, using the cereal-based diets however, presents a challenge to meet daily requirements and causes unfavorable deficient effects on health. The stunting rates of children under the age of five have been used as a benchmark for determining Zn deficiency, as per the new guidelines from the International Atomic Energy Agency, the World Health Organization, the International Zinc Nutrition Consultative Group, and UNICEF. According to FAO, approximately 45.4 million (6.7%) and 149.2 million (22.0%) of these children are stunted and wasted (FAO, 2021). Zn has linked up to 453,207 deaths, 4.4% of which were children, and 1.2% of the illness load (of which 3.8% were children between the ages of 0.5 and 5 years). deficit, which results in 16 million years with an impairment. (DALY) (Wessells and Brown, 2012). The daily requirement rises to approximately 0.1 and 0.7 mg in the first and fourth trimesters of pregnancy, respectively, during the gestational time (Institute of Medicine IM, 2001). Zn requirement in expectant and nursing women can be calculated by estimated through the Zn content of the tissues during pregnancy and calculating the Zn levels in the milk of lactating women (Hess and King, 2009). Zn deficiency in the woman has negative impacts on the reproductive system, such as preterm delivery. Additionally, synaptogenesis, movement, gene expression, and regular neuronal duplication are also affected (Iqbal and Ali, 2021). In pregnant women due to Zinc deficiency, risk of low birth weight and small for gestational age is also associated (Wang et al., 2015). Additionally, Zn is essential for the biosynthesis, storage and storage of male sex hormones including testosterone. Zinc deficiency is associated with dysfunctional steroidogenesis, resulting in low testosterone and progesterone levels in Leydig cells, ultimately leading to apoptosis in these cells Furthermore, zinc is essential for sperm a they are produced as a by-product for enzymes involved in replication, transcription and DNA packaging (Croxford et al., 2011). Clinically, epidermal, immunological, tissues in the gastric, skeletal, central neurological, and reproductive systems are among those affected by Zn insufficiency. Zn supplementation is therefore strongly advised for inclusion in government food, nutrition, and health programmes as well as in nutritional development instruction. To improve the effectiveness and longevity of Zn fortification, social marketing strategies should also be used.

Being a cofactor for over more than 300 enzymes, it is unique among elements. Additionally, about 3000 proteins, including signaling enzymes and transcription factors found at all phases of cellular signal transduction, are Zn-dependent (Gammoh and Rink, 2019). Given that Zn insufficiency is a present medical risk, the best markers for determining the frequency of Zn deficiency are plasma Zn concentration and a 24-hour dietary recall or dietary evaluation. Due to the scarcity of these statistics, the World Health Organization (WHO) suggests using alternative markers, such as stunting and information on dietary Zn intake from national food balance charts, to determine Zn deficiency (Wieringa et al., 2015). Zinc deficiency can disrupt the body's defense mechanisms, such as activating macrophages, natural killer, polymorphonuclear cells, the complement cascade (El-Sayed Saad et al., 2018) and adding zinc to the diet boosts the adaptive and innate immunity such as immunity against entero-toxigenic *Escherichia coli*. This improvement occurs because zinc helps to enhance the C3 complement system, improve T-cell function, and prevent phagocyte dysfunction (Sheikh et al., 2010). Likewise, Zinc is important because of its important role in malignant cell communication and metabolism and especially in apoptosis (Saravanakumar et al., 2018). nadequate levels of zinc in various endothelial cells such as fibroblasts, T cell precursors, gliomas, testes and hepatocytes can accelerate apoptosis Several zinc pathways regulate the balance between cell death and formation in the organism, and optimize zinc levels and mobility in cells (Seve et al., 2002). In addition, zinc deficiency can induce apoptosis by interfering with signaling pathways mediated by cell growth and tyrosine kinases (Clegg et al., 2005).

Since, Zn participates in numerous biochemical processes in the human body, its lacunae cause a variety of bodily symptoms that indicate a deficiency (Sangeetha et al., 2022). RDA (Recommended Dietary Allowance) of zinc and conditions linked to zinc deficiency in various age categories. The most prominent clinical symptoms of zinc deficiency can be seen in the developmental phases of growth, such as in childhood, adolescence, and also during pregnancy phase (deficiency in Foetus). These symptoms include impairment in physical growth and development. At such times, necessity of the Zn consumption becomes important (Brown et al., 2004). Lessened intake of meat especially red meat and other Zn-rich foods is said to be a factor in the elderly population's deficiencies. Additionally, there is proof that as people mature, their ability to absorb zinc decreases (Andriollo-Sanchez et al., 2005). Moreover, Zn supplementation may be useful in the management of many nutritional conditions and diseases in which Zn can be used as adjuvant therapy (Wegmüller et al., 2014) (Fig. 14.2).

## 14.5 Conclusion

In numerous ways, zinc is extremely important in growth of plant, animals as well as human being. Its functions in fundamental living processes, such as photosynthesis in plants and human brain function, are becoming increasingly apparent Zn has a significant impact on human nutrition in addition to its significance in plant growth and function. One of the main sources of these vital micronutrients for people is plants, which take up zinc from the soil. Thus, preserving human health and wellbeing depends on plants having appropriate zinc levels. This demonstrates how important it is to comprehend and control the dynamics of zinc in plant and human systems, as well as the relationship between plant health and human nutrition. With the development of Zn-efficient agricultural types that can withstand low Zn stress in soils, our knowledge of the effects of Zn on living things continues to grow and the zinc deficiency can be defeated. Enhancing human nutrition, increasing agricultural sustainability, and lowering the need for synthetic fertilizers can all be made possible with an in-depth comprehension of plant Zn efficiency techniques, cellular systems, and genes. Zn efficiency, in turn, may improve production of crop and nutritional quality for the enhancing population of the 21st century.

## References

Alloway, B.J., 2009. Soil factors associated with zinc deficiency in crops and humans. Environmental Geochemistry and Health 31, 537–548 [CrossRef].

Andriollo-Sanchez, M., Hininger-Favier, I., Meunier, N., Toti, E., Zaccaria, M., Brandolini-Bunlon, M., et al., 2005. Zinc intake and status in middle-aged and older European subjects: the ZENITH study. European Journal of Clinical Nutrition 59 (2), S37–S41.

Bjorndahl, L., Kvist, U., 2011. A model for the importance of zinc in the dynamics of human sperm chromatin stabilization after ejaculation in relation to sperm DNA vulnerability. Systems Biology in Reproductive Medicine 57, 86–92.

Bonaventura, P., Benedetti, G., Albarède, F., Miossec, P., 2015. Zinc and its role in immunity and inflammation. Autoimmunity Reviews 14 (4), 277–285.

Broadley, M.R., White, P.J., Hammond, J.P., Zelko, I., Lux, A., 2007. Zinc in plants. The New Phytologist 173, 677–702.

Brown, K.H., Rivera, J.A., Bhutta, Z., Gibson, R.S., King, J.C., Lönnerdal, B., et al., 2004. International Zinc Nutrition Consultative Group (IZiNCG) technical document #1. Assessment of the risk of zinc deficiency in populations and options for its control. Food and Nutrition Bulletin 25 (Suppl 2), S99–S203. 1.

Brown, K.H., Wuehler, S.E., Peerson, J.M., 2001. The importance of zinc in human nutrition and estimation of the global prevalence of zinc deficiency. Food and Nutrition Bulletin 22 (2), 113–125.

Cabot, C., Martos, S., Llugany, M., Gallego, B., Tolra, R., Poschenrieder, C., 2019. A Role for Zinc in Plant Defense Against Pathogens and Herbivores 10.

Cai, C., Lin, P., Zhu, H., Ko, J.K., Hwang, M., Tan, T., et al., 2015. Zinc binding to MG53 protein facilitates repair of injury to cell membranes. The Journal of Biological Chemistry 290, 13830–13839.

Cakmak, I., 2008. Enrichment of cereal grains with zinc: agronomic or genetic biofortification. Plant and Soil 302, 1–17.

Cakmak, I., Torun, B., Erenoglu, B., Oztürk, L., Marschner, H., Kalayci, M., et al., 1998. Morphological and physiological differences in the response of cereals to zinc deficiency. Euphytica 100, 349–357.

Chasapis, C.T., Ntoupa, P.S.A., Spiliopoulou, C.A., Stefanidou, M.E., 2020. Recent aspects of the effects of zinc on human health. Archives of Toxicology 94, 1443–1460.

Chellan, P., Sadler, P.J., 2015. The elements of life and medicines. Philosophical Transactions of the Royal Society A: Mathematical, Physical and Engineering Science 373, 20140182.

Choi, S.-H., Lee, K.-L., Shin, J.-H., Cho, Y.-B., Cha, S.-S., Roe, J.-H., 2017. Zinc-dependent regulation of zinc import and export genes by Zur. Nature Communications 815812.

Clegg, M.S., Hanna, L.A., Niles, B.J., Momma, T.Y., Keen, C.L., 2005. Zinc deficiency-induced cell death. IUBMB Life 57 (10), 661–669.

Croxford, T.P., McCormick, N.H., Kelleher, S.L., 2011. Moderate zinc deficiency reduces testicular Zip6 and Zip10 abundance and impairs spermatogenesis in mice. The Journal of Nutrition 141 (3), 359–365.

Dewar, W.A., Wight, P.A.L., Pearson, A., Gentle, M.J., 1982. Toxic effects of high concentrations of Zn oxide in the diet of the chick and laying hen. British Poultry Science 24, 397–404. Available from: https://www.ncbi.nlm.nihgov/pubmed/6616303.

Droke, E.A., Spears, J.W., 1993. In vitro and in vivo immunological measurements in growing lambs fed diets deficient, marginal or adequate in zinc. Journal of Nutritional Immunology 2 (1), 71–90.

El-Sayed Saad, A.M., El-Gebally, E.S.I., El-Mashad, G.M., El-Hefnawy, S.M., 2018. Effect of zinc supplementation on serum zinc and leptin levels in children on regular hemodialysis. Menoufia Medical Journal 31 (2), 664–670.

Fabris, N., 1994. Neuroendocrine-immune aging: an integrative view on the role of zinc. Annals of the New York Academy of Sciences 719, 353–368.

Fageria, N.K., 2001. Screening method of lowland rice genotypes for zinc uptake efficiency. Scientia Agricola 58, 623–626.

FAO (Food and Agriculture Organization of the United Nations), 2021. Food security and nutrition in the world security, improved nutrition and affordable healthy diets for all.

Frassinetti, S., Bronzetti, G.L., Caltavuturo, L., Cini, M., Della Croce, C., 2006. The role of zinc in life: a review. Journal of Environmental Pathology, Toxicology and Oncology 25, 3.

Frederickson, C.J., Koh, J.-Y., Bush, A.I., 2005. The neurobiology of zinc in health and disease. Nature Reviews. Neuroscience 6, 449–462.

Galetti, V., 2018. Zinc deficiency and stunting. In: Preedy, V., Patel, V. (Eds.), Handbook of Famine, Starvation, and Nutrient Deprivation. Springer. Available from: https://doi.org/10.1007/978-3-319-40007-5_93-1.

Gammoh, N.Z., Rink, L., 2019. Zinc and the immune system. Nutrition and Immunity. Springer, pp. 127–158.

Gibson, R.S., Anderson, V.P., 2009. A review of interventions based on dietary diversification or modification strategies with the potential to enhance intakes of total and absorbable zinc. Food and Nutrition Bulletin 30 (1 Suppl), S108–S143.

Gondal, A.H., Zafar, A., Zainab, D., Toor, M.D., Sohail, S., Ameen, S., et al., 2021. A detailed review study of zinc involvement in animal, plant and human nutrition. Indian Journal of Pure & Applied Biosciences 9 (2), 262–271.

Graham, T.W., Feldman, B.F., Farver, T.B., Labacitch, F., O'Neill, S.L., Thurmond, M.C., et al., 1991. Zn toxicosis of Holstein veal calves and its relationship to haematological change and associated thrombotic state. Comparative Haematology International 1, 121–128. Available from: https://doi.org/10.1007/BF00515658.

Graham, T.W., Thurmond, M.C., Clegg, M.S., Keen, C.L., Holmberg, C.A., Slanker, M.T.R., et al., 1987. An epidemiologic study Copper and Zinc Nutritional Issues for Agricultural Animal Production 157 of mortality in veal calves subsequent to an episode of Zn toxicosis on a California veal calf operation using Zn sulfate-supplemented milk replacer. JAVM 190, 1296–1301.

Grusak, M.A., DellaPenna, D., 1999. Improving the nutrient composition of plants to enhance human nutrition and health. Annual Review of Plant Biology 50, 133–161 [CrossRef].

Hacisalihoglu, G., 2020. Zinc (Zn): the last nutrient in the alphabet and shedding light on Zn efficiency for the future of crop production under suboptimal Zn. Plants 9 (11), 1471.

Hacisalihoglu, G., Blair, M., 2020. Current advances in zinc in soils and plants: implications for zinc efficiency and biofortification studies. Achieving Sustainable Crop Nutrition 76, 337–353.

Hacisalihoglu, G., Hart, J.J., Kochian, L.V., 2001. High and low affinity Zn transport systems and their possible role in Zn efficiency in bread wheat. Plant Physiology 125, 456–463.

Hacisalihoglu, G., Kochian, L.V., 2004. How do some plants tolerate low levels of soil zinc? Mechanisms of zinc efficiency in crop plants. The New Phytologist 159, 341–350 [CrossRef].

Hacisalihoglu, G., Ozturk, L., Cakmak, I., Welch, R.M., Kochian, L.V., 2004. Genotypic variation in common bean in response to Zn deficiency in calcareous soil. Plant and Soil 259, 71–83.

Haddad, L., Hawkes, C., Webb, P., Thomas, S., Beddington, J., Waage, J., et al., 2016. A new global research agenda for food. Nature 540, 30–32.

Hamzah Saleem, M., Usman, K., Rizwan, M., Al Jabri, H., Alsafran, M., 2022. Functions and strategies for enhancing zinc availability in plants for sustainable agriculture. Frontiers in Plant Science 13, 1033092.

Haslett, B.S., Reid, R.J., Rengel, Z., 2001. Zinc mobility in wheat: uptake and distribution of zinc applied to leaves or roots. Annals of Botany 87, 379–386.

Hess, S.Y., King, J.C., 2009. Effects of maternal zinc supplementation on pregnancy and lactation outcomes. Food and Nutrition Bulletin 30 (1 Suppl), S60–S78.

Hussain, A., Arshad, M., Zahir, Z.A., Asghar, M., 2015. "Prospects of zinc solubilizing bacteria for enhancing growth of maize. Pakistan Journal of Agricultural Sciences 52 (4), 915–922.

Institute of Medicine (IM), 2001. Dietary reference intakes for vitamin A, vitamin K, arsenic, boron, chromium, copper, iodine, iron, manganese, molybdenum, nickel, silicon, vanadium, and zinc. Available from: https://doi.org/10.17226/10026.

Iqbal, S., Ali, I., 2021. Effect of maternal zinc supplementation or zinc status on pregnancy complications and perinatal outcomes: an umbrella review of meta-analyses. Heliyon 7, e07540.

Jenkins, K.J., Hidiroglou, M., 1991. Tolerance of preruminant calf for excess Mn or Zn in milk replacer. Journal of Dairy Science 74, 1047–1053.

Keen, C.L., Gershwin, M.E., 1990. Zinc deficiency and immune function. Annual Review of Nutrition 10, 415–431.

Khan, H.R., McDonald, G.K., Rengel, Z., 1998. Chickpea genotypes differ in their sensitivity to Zn deficiency. Plant and Soil 198, 11–18 [CrossRef].

Khan, S.T., Malik, A., Alwarthan, A., Shaik, M.R., 2022. The enormity of the zinc deficiency problem and available solutions; an overview. Arabian Journal of Chemistry 15 (3), 103668.

Kienholz, E.W., Turk, D.E., Sunde, M.L., Hoekstra, W.G., 1961. Effects of Zn deficiency in the diets of hens. The Journal of Nutrition 75, 211–221. Available from: https://doi.org/10.1093/jn/75.2.211.

Kiouri, D.P., Tsoupra, E., Peana, M., Perlepes, S.P., Stefanidou, M.E., Chasapis, C.T., 2023. Multifunctional role of zinc in human health: an update. EXCLI Journal 22, 809.

Kirkby, E.A., 2023. Introduction, definition, and classification of nutrients. Marschner's Mineral Nutrition of Plants. Academic press, pp. 3–9.

Krężel, A., Maret, W., 2016. The biological inorganic chemistry of zinc ions. Archives of Biochemistry and Biophysics 611, 3–19.

Liu, E., Pimpin, L., Shulkin, M., Kranz, S., Duggan, C.P., Mozaffarian, D., et al., 2018. Effect of zinc supplementation on growth outcomes in children under 5 years of age. Nutrients 10 (3), 377.

Malhi, G.S., Kaur, M., Kaushik, P., 2021. Impact of climate change on agriculture and its mitigation strategies: a review. Sustainability 13 (3), 1318.

Marschner, H., 1995. Mineral Nutrition of Higher Plants. Academic Press, London, UK, p. 889.

Mayland, M.F., Rosenau, R.C., Florence, A.R., 1980. Grazing cow and calf responses to Zn supplementation. Journal of Animal Science 51, 966–974.

McDowell, L.R., 2003. Minerals in Animal and Human Nutrition, second ed. Elsevier Science BV.

Meena, N., Fathima, P., 2017. Nutrient uptake of rice as influenced by agronomic biofortification of Zn and Fe under methods of rice cultivation. Indian Journal of Pure & Applied Biosciences 5 (5), 456–459.

Miller, W.J., Amos, H.W., Gntry, R.P., Blakmon, D.M., Durrance, R.M., Crowe, C.T., et al., 1989. Long-term feeding of high zinc sulfate diets to lactating and gestating dairy cows. Journal of Dairy Science 72, 1499–1508. Available from: https://doi.org/10.3168/jds.S0022-0302(89)79260-2.

Miller, W.J., Clifton, E.M., Fowler, P.R., Perkins, H.F., 1965. Influence of high levels of dietary Zn on Zn in milk, performance and biochemistry of lactating cows. Journal of Dairy Science 48, 450–453. Available from: https://doi.org/10.3168/jds.S0022-0302(65)88251-0.

Montalvo, D., Degryse, F., da Silva, R.C., Baird, R., McLaughlin, M.J., 2016. Agronomic effectiveness of zinc sources as micronutrient fertilizer. Advances in Agronomy 139, 215–267.

Mousavi, S.R., Galavi, M., & Rezaei, M. (2013). Zinc (Zn) importance for crop production-a review.

Müller, O., Becher, H., van Zweeden, A.B., Ye, Y., Diallo, D.A., Konate, A.T., et al., 2001. Effect of zinc supplementation on malaria and other causes of morbidity in west African children: randomised double blind placebo-controlled trial. BMJ (Clinical Research ed.) 322 (7302), 1567.

Naik, S.M., Raman, A.K., Nagamallika, M., Venkateshwarlu, C., Singh, S.P., Kumar, S., et al., 2020. Genotype × environment interactions for grain iron and zinc content in rice. Journal of the Science of Food and Agriculture 100, 4150–4164 [CrossRef].

Natasha, N., Shahid, M., Bibi, I., Iqbal, J., Khalid, S., Murtaza, B., et al., 2022. Zinc in soil-plant-human system: A data-analysis review. Science of the Total Environment 808, 152024.

NIH (National Institute of Health), 2021. Zinc – Health professional fact sheet. Retrieved August 29, 2021, from https://ods.od.nih.gov/factsheets/Zinc-HealthProfessional/#en11.

Niyigaba, E., Twizerimana, A., Mugenzi, I., Ngnadong, W.A., Ye, Y.P., Wu, B.M., et al., 2019. Winter wheat grain quality, zinc and iron concentration affected by a combined foliar spray of zinc and iron fertilizers. Agronomy 9 (5), 250.

O'Dell, B.L., Newberne, P.M., Savage, J.E., 1958. Significance of dietary Zn for the growing chicken. The Journal of Nutrition 65, 503–523. Available from: https://doi.org/10.1093/jn/65.4.503.

Oh, A.H., Nakaue, H., Deagen, J.T., Whanger, P.D., Arscott, G.H., 1979. Accumulation and depletion of Zn in chick tissue metallothioneins. The Journal of Nutrition 109, 1720–1729. Available from: https://doi.org/10.1093/jn/109.10.1720.

Olson, P.A., Brink, D.R., Hickok, D.T., Carlson, M.P., Schneider, N.R., Deutscher, G.H., et al., 1999. Effects of supplementation of organic and inorganic combinations of Cu, Co, Mn, and Zn above nutrient requirement levels of postpartum two-year old cows. Journal of Animal Science 77, 522–532.

Ott, E.A., Smith, W.H., Stob, M., Parker, H.E., Harrington, R.B., Beeson, W.M., 1965. Zn requirement of the growing lamb fed a purified diet. The Journal of Nutrition 87, 459–463.

Palafox, A.L., HO-A, E., 1980. Effect of Zn toxicity in laying white leghorn pullets and hens. Poultry Science 59, 2024–2028. Available from: https://doi.org/10.3382/ps.0592024.

Prasad, A.S., 1976. Deficiency of Zn in man and its toxicity. In: Prasad, A.S. (Ed.), Trace Elements in Human Health and Disease. Vol. I Zn and Cu. Academic Press.

Prasad, R., 2006. Zinc in soils and in plant, human & animal nutrition. Indian Journal of Fertilisers 2 (9), 103.

Ramzan, Y., Hafeez, M.B., Khan, S., Nadeem, M., Rahman, S.-ur, Batool, S., et al., 2020. Biofortification with zinc and iron improves the grain quality and yield of wheat crop. International Journal of Plant Production 14, 3.

Sangeetha, V.J., Dutta, S., Moses, J.A., Anandharamakrishnan, C., 2022. Zinc nutrition and human health: overview and implications. eFood 3 (5), e17.

Saravanakumar, K., Jeevithan, E., Chelliah, R., Kathiresan, K., Wen-Hui, W., Oh, D.H., et al., 2018. Zinc-chitosan nanoparticles induced apoptosis in human acute T-lymphocyte leukemia through activation of tumor necrosis factor receptor CD95 and apoptosis-related genes. International Journal of Biological Macromolecules 119, 1144–1153.

Satterthwaite, D., McGranahan, G., Tacoli, C., 2010. Urbanization and its implications for food and farming. Philosophical Transactions of the Royal Society of London. Series B, Biological Sciences 365, 2809–2820.

Schroeder, J.I., Delhaize, E., Frommer, W.B., Guerinot, M.L., Harrison, M.J., Herrera-Estrella, L., et al., 2013. Using membrane transporters to improve crops for sustainable food production. Nature 497, 60–66.

Schryver, H.F., Oftedal, O.T., Williams, J., Soderholm, L.V., Hintz, H.F., 1986. Lactation in the horse: the mineral composition of mare milk. The Journal of Nutrition 116, 2142–2147. Available from: https://doi.org/10.1093/jn/116.11.

Seve, M., Chimienti, F., Favier, A., 2002. Role of intracellular zinc in programmed cell death. Pathologie-Biologie 50 (3), 212–221.

Sheikh, A., Shamsuzzaman, S., Ahmad, S.M., Nasrin, D., Nahar, S., Alam, M.M., et al., 2010. Zinc influences innate immune responses in children with enterotoxigenic *Escherichia coli*-induced diarrhea. The Journal of Nutrition 140 (5), 1049–1056.

Shukla, A.K., Tiwari, P.K., Pakhare, A., Prakash, C., 2016. Zinc and iron in soil, plant, animal and human health. Indian Journal of Fertilisers 12 (11), 133–149.

Signorell, C., Zimmermann, M.B., Cakmak, I., Wegmüller, R., Zeder, C., Hurrell, R., et al., 2019. Zinc absorption from agronomically biofortified wheat is similar to post-harvest fortified wheat and is a substantial source of bioavailable zinc in humans. The Journal of Nutrition 149 (5), 840–846.

Solanki, M., 2021. The Zn as a vital micronutrient in plants. Journal of Microbiology, Biotechnology and Food Sciences 11 (3), e4026.

Solanki, M., Didwania, N., Nandal, V., 2016. Potential of zinc solubilizing bacterial inoculants in fodder crops. Momentum 3, 1–4.

Tapiero, H., Tew, K.D., 2003. Trace elements in human physiology and pathology: zinc and metallothioneins. Biomedicine & Pharmacotherapy 57 (9), 399–411.

Taşğın, E., 2017. Macronutrients and micronutrients in nutrition. International Journal of Innovative Research and Reviews 1 (1), 10–15.

Tucker, H.F., Salmon, W.D., 1955. Parakeratosis or Zn deficiency in the pig, Proceedings of the Society for Experimental Biology and Medicine, 88. pp. 613–616. Available from: https://www.ncbi.nlm.nih.gov/pubmed/14371717.

Ullah, A., Farooq, M., Rehman, A., Hussain, M., Siddique, K.H.M., 2020. Zinc nutrition in chickpea (*Cicer arietinum*): a review. Crop & Pasture Science 71, 199–218.

Vidyshree, D.N., Muthuraju, R., Panneersel Vam, P., Saritha, B., Ganeshamurthy, A.N., 2016. Isolation and characterization of zinc solubilizing bacteria from stone quarry dust powder. International Journal of Agriculture Science 8, 56.

Wang, H., Hu, Y.F., Hao, J.H., Chen, Y.H., Su, P.Y., Wang, Y., et al., 2015. Maternal zinc deficiency during pregnancy elevates the risks of fetal growth restriction: a population-based birth cohort study. Scientific Reports 511262. Available from: https://doi.org/10.1038/srep11262.

Wegmüller, R., Tay, F., Zeder, C., Brnić, M., Hurrell, R.F., 2014. Zinc absorption by young adults from supplemental zinc citrate is comparable with that from zinc gluconate and higher than from zinc oxide. The Journal of Nutrition 144 (2), 132–136.

Welch, R.M., 1995. Micronutrient nutrition of plants. Critical Reviews in Plant Sciences 14, 49–82.

Welch, R.M., Graham, R.D., 2005. Agriculture: the real nexus for enhancing bioavailable micronutrients in food crops. Journal of Trace Elements in Medicine and Biology: Organ of the Society for Minerals and Trace Elements (GMS) 18, 299–307.

Wessells, K.R., Brown, K.H., 2012. Estimating the global prevalence of zinc deficiency: results based on zinc availability in national food supplies and the prevalence of stunting. PLoS One 7 (11), e50568.

Wieringa, F.T., Dijkhuizen, M.A., Fiorentino, M., Laillou, A., Berger, J., 2015. Determination of zinc status in humans: which indicator should we use? Nutrients 7 (5), 3252–3263.

Williams, S.N., Miles, R.D., Ouart, M.D., Campbell, D.R., 1989. Short-term high-level Zn feeding and tissue Zn concentration in mature laying hens. Poultry Science 68, 539–545. Available from: https://doi.org/10.3382/ps.

Xia, H., Kong, W., Wang, L., Xue, Y., Liu, W., Zhang, C., et al., 2019. Foliar Zn spraying simultaneously improved concentrations and bioavailability of Zn and Fe in maize grains irrespective of foliar sucrose supply. Agronomy 9 (7), 386.

Zabel, F., Putzenlechner, B., Mauser, W., 2014. Global agricultural land resources – a high resolution suitability evaluation and its perspectives until 2100 under climate change conditions. PLoS One 9, e107522.

Zaman, Q.U., Aslam, Z., Yaseen, M., Ihsan, M.Z., Khaliq, A., Fahad, S., Bashir, S., Ramzani, P.M., Naeem, M., 2018. Zinc biofortification in rice: leveraging agriculture to moderate hidden hunger in developing countries. Archives of Agronomy and Soil Science 64 (2), 147–161.

Zou, C.Q., Zhang, Y.Q., Rashid, A., Ram, H., Savasli, E., Arisoy, R.Z., et al., 2012. Biofortification of wheat with zinc through zinc fertilization in seven countries. Plant and Soil 361 (1), 119–130.

Chapter 15

# Zinc and plant signaling molecules special emphasis on reactive oxygen species and reactive nitrogen species

**Neeraj Kumar Dubey and Vijay Bahadur Yadav**
*Department of Botany, Rashtriya Snatkottar Mahavidyalaya, Jaunpur, Uttar Pradesh, India*

## 15.1 Introduction

Metals ions present in soil like cobalt (Co), manganese (Mn), copper (Cu), nickel (Ni), iron (Fe), molybdenum (Mo), and zinc (Zn) are very important for plant life cycle (Arif et al., 2016). Among them, micronutrient such as Fe and Zn are essential for normal plant growth and development, and required for the successful completion of the plant life cycle (Loftleidir, 2005, Arif et al., 2016). Under drought stress, Zn applications improve seed germination, and enhance plant performance in the form of productivity (Umair Hassan et al., 2020). Reactive oxygen species (ROS) and reactive nitrogen species (RNS) are important players in the dynamics of plant cell signaling during biotic and abiotic stresses (Khan et al., 2023). Biotic and abiotic stresses induce the formation of RNS production similar to ROS. RNS metabolism and signaling are correlated with ROS generation and sometimes RNS and ROS affect plants together (Corpas et al., 2013). These molecules also play an important role in the growth and development of plants, defense mechanisms, and symbiotic association during different stresses (Khan et al., 2023). Important role of ROS scavenging system in the root nodules of legumes have been also discusssed (Puppo et al., 1982) and NO in senescence and plant immunity especially bacterial infection has been mentioned earlier (Leshem and Haramaty, 1996; Noritake et al., 1996; Laxalt et al., 1997, Del Río, 2015). However, their higher concentration reduces the plant performance in the form of cell damage, and delayed plant growth (Khan et al., 2023). Biotic stress also produces an oxidative burst of RNS and ROS and it leads to changes in cell wall composition and PCD, which stop the

**Zinc in Plants. DOI: https://doi.org/10.1016/B978-0-323-91314-0.00006-5**
 

spreading of pathogens (Bleau, and Spoel, 2021). Many enzymes like catalase, ascorbate peroxidase, glutathione reductase, superoxide dismutase, and nonenzymatic antioxidants like glutathione and ascorbate regulate the homeostasis of oxidative species (Feigl et al., 2015). Furthermore, Cu and Zn with SOD of spinach leaves have been crystallized in the laboratory (Asada et al., 1973; Del Río, 2015). Excessive Zn produces stress which leads to ROS (such as superoxide, hydrogen peroxide, and hydroxyl radicals) and RNS (such as nitric oxide, nitric oxide radical, nitroxyl anion, nitrosonium cation) (Morina et al., 2010; Jain et al., 2010, Kondak et al., 2022). High ROS and RNS damage protoplasmic content like those of nucleic acid, proteins, and lipids and also break the chromosomes and produce different aberrations (Oladele et al., 2013; Sidhu, 2016, Kondal et al., 2022). High ROS oxidation of proteins and lipids produces aldehyde, glycosylation, or carbonyl groups (Madian and Regnier, 2010, Kondal et al., 2022). These products damage the membrane's natural property and thus lead to cell death (Goodarzi et al., 2020, Kondal et al., 2022). High RNS also induces the damage of protein, nucleic acid, and fatty acids through posttranslational modifications and nitration (Hess et al., 2005, Kondal et al., 2022). RNS, like NO, can help with interactions involving plants experiencing abiotic stresses from metals, symbiotic associations, biotic stress from pathogens, and plant developmental processes (Kondal et al., 2022). The role of RNS in mineral uptake, such as that of iron, has also been clearly established and it is suggested that, similar to Fe, NO may also play an important role in the uptake of other trace elements (Tewari et al., 2021). Exogenous application of NO through donor GSNO in wheat plants during hydroponic culture showed alteration in the translocation of Zn in shoots and roots, thus indicating the role of RNS in Zn homeostasis (Buet et al., 2014).

## 15.2 Zinc

Metal stress induces ROS and RNS formation in plant species (Feigl et al., 2015). Zinc is an important trace element, essential for enzymes involved in regulating cellular and biochemical processes such as lipid, protein, carbohydrate nucleic acid, and metabolism, and stored in vacuoles of the root cells (Arrivault et al., 2006). (Broadley et al., 2007; Jarosz et al. 2017; Zhang et al., 2020). Zn also played an important role in the activity of enzymes like hydrogenase and carbonic anhydrase, protein synthesis, folding of small protein domains, carbohydrate metabolism, cellular membranes integrity, phytohormone-like auxin synthesis, and pollen formation (Clemens, 2021; Kondak et al., 2022). Due to the important role of Zn in the Zn finger domain which is an important part of many membrane and nucleic acid binding proteins, Zn plays an important role in transcription and even translational processes (Broadley et al., 2007). Plants absorb $Zn^{2+}$ ions through Zn regulatory transporter proteins present in the roots from the soil solution and translocate them to the stem by xylem (Palmgren et al., 2008; Kondak et al., 2022). Thus, being a trance element, Zn concentration is very important in the completion of the plant life cycle (Broadley et al., 2007; Kondak et al., 2022). Different crops are sensitive to different levels to handle the Zn supply (Alloway, 2008; Kondak et al., 2022). Crops such as rice and maize are highly sensitive to the Zn supply while barley and tomatoes are moderately sensitive, but carrots and peas are tolerant to the Zn supply (Alloway, 2008; Kondak et al., 2022). Zn deficiency showed smaller leaves, chlorosis, stunted growth, and spikelet sterility while the excess of Zn showed similar results to different biotic and abiotic stress like drought and pathogen attack ((Ullah et al., 2019; Cabot et al., 2019; Kondak et al., 2022). Zn plays a very important role in both heat, chill stress, and waterlogged conditions and provides tolerance to plants against these stresses (Porwal et al., 2021; Kumari et al., 2022). About 145% more yield in the mustard Varuna cultivar has been observed during heat stress (Khan et al., 2014). Foliar application of Zinc oxide (ZnO) Nanoparticles (NPs) in rice, enhance chilling stress tolerance by ROS scavenging abilities and ultimately increase chlorophyll biosynthesis and plant growth (Liu et al., 2012). Zn induces seed germination in maize, maintains membrane permeability, enhances photosynthetic and water use efficiency, induces leaf area, stomatal conductance, and osmolyte accumulation, and thus improves growth and yield during water stress (Harris et al., 2005; Umair Hassan et al., 2020). Similarly, $ZnSO_4$ (1%) treatment increased grain filling and yield in maize during drought (Vazin, 2012). It is a very important factor in the generation of ROS (Zhang et al., 2020). Zn stresses induce many negative effects on plant membrane structure, inhibition of germination of seed, and reduction in the growth of plants and their roots (Feigl et al., 2015). The exact mechanism of Zn toxicity is not well explained but induction of oxidative stress and Zn deficiency might be the reason behind Zn-related abnormalities (Wintz et al., 2003; Feigl et al., 2015). Nanoparticles of zinc oxide also elevate ROS and induce induced cell death and lipid peroxidation (Zhang et al., 2020). Some plants are more tolerant to ZnSO4 than others; for example, the stress of Zn can easily be managed by *B. juncea* than *B. napus* (Feigl et al., 2015). In animal system, Zn involves in cytokine metabolism, lipid and glucose metabolism, inflammation suppressing, and activating antioxidant enzymes for ROS management (Olechnowicz et al., 2018). In plants system, Zn can induce ROS and RNS formation which reduce the plant growth and yield (Feigl et al., 2015). During oxidative stress, ROS and RNS are produced due to different environmental and pathogenic stress (Apel and Hirt, 2004; Wang et al., 2013). High levels of ROS and RNS damage biological macromolecules and thus it is very important to control their concentrations in plants (Apel and Hirt, 2004). These two reactive species ROS and RNS have overlapped signaling processes and cross-talk such as the reaction between $O_2^-$ and NO produced $ONOO^-$ (Corpas and Barroso, 2013). ZnO NPs change the homeostasis of ROS and RNS in *Brassica* species and trigger the nitration and carbonylation of protein may induce phytotoxicity (Molnár et al., 2020). Zinc chloride ($ZnCl_2$) treatment to *Solanum nigrum* accumulates NO in the root and increases programmed cell death (PCD) in primary root tips (Xu et al., 2010).

## 15.3 Reactive oxygen species and reactive nitrogen species

Plants produce ROS in cell organelles like chloroplasts, mitochondria, and peroxisomes (Tripathy and Oelmüller, 2012). Different ROS superoxide anions (O2 − ), and hydroxyl radicals (−OH) are found in plant systems (Khan et al., 2023). ROS like superoxide (O2−), singlet oxygen (1O2), hydroxyl radical (HO), and hydrogen peroxide ($H_2O_2$) are generated in plastids, mitochondria, peroxisomes, and also in cytosolic and apoplastic regions of the plant cell (Tripathy and Oelmüller, 2012). Abiotic and biotic stresses such as light intensity, temperature, salinity, nutrient deficiency, drought, and pathogen attack induce the imbalance in ROS formation and detoxification of generative oxidative stress for the plants (Tripathy and Oelmüller, 2012). With higher net ROS formation, there is photooxidative damage to DNA, proteins, and lipids and ultimately cell death. ROS act as signaling molecules for plant growth and developmental, defense processes which are later seen in the form of changes in hormone production, systemic acquired resistance, hypersensitive reaction, and programmed cell death (Tripathy and Oelmüller, 2012). Enzymes such as catalase (CAT), superoxide dismutase (SOD), ascorbate peroxidase (APX), glutathione reductase (GR), and their product like ascorbic acid (AsA), glutathione levels help in the plant to manage the upcoming stress by ROS (Rai et al., 2013). Although ROS plays an important role in the maintenance of plant growth and management during different stresses, they also cause irreversible damage to DNA and cell death (Tripathy and Oelmüller, 2012). Similar to ROS, different RNS such as NO, $NO_2$, and $HNO_2$ are found in plant systems and act as signaling molecules during plant stress management (Khan et al., 2021). Little information is available about nitrogen oxide compound formation and their role in plant systems. NO is supposed to generate in peroxisomes and play important role in life cycle of the plant from seed germination to flowering, fruiting, and also biotic−abiotic stress management (Del Río, 2015). There is a correlation between the Zn stress either in deficiency or in excess concentration with RNS like NO (Kondak et al., 2022). These RNS play an important role in plant biology during Zn stress (Kondak et al., 2022). RNS are not only involved in Zn deficiency but they also play an important role in several abiotic stress responses (Kondak et al., 2022). Computational analysis showed that there is a huge correlation between Zn transporter proteins and NO in *Arabidopsis* plant and the result indicates that NO plays an important role in Zn deficiency (Kondak et al., 2022). Exogenous NO relieves oxidative stress by inducing an antioxidant system and reduction of Zn concentration (Kondak et al., 2022). Nanoparticles of ZnO induce RNS stress, while exogenous application of NO reduces the ZnO-derived oxidative stress, as well as reducing Zn accumulation and photosynthesis enchantment (Kondak et al., 2022). Role of RNS like NO, after exogenous application of GSNO through a donor, altered the Zn translocation in the shoot and root of wheat (Buet et al., 20141 (Tewari et al., 2021). In plants like *Brassica* and *Arabidopsis*, small Zn deficiency induced NO levels in the tips of the root (Kondak et al., 2022). Exogenous application of NO on plants can increase NO content in studied tissues, and induce a number of beneficial effects such as improved nutritional status during Zn overload (Kondak et al., 2022).

## 15.4 Zinc-related protein expression and oxidative stresses mediated transgenic approaches in the management of abiotic and biotic stresses

Transgenic tomato lines having stress-inducible expression of transcription factor AtDREB1A/CBF3 by rd29A promoter showed better performance of tomato plant under water stress and having significantly high activities of antioxidant enzymes such as SOD, CAT, APX, and GR (Rai et al., 2013). Transgenic tobacco expressing cDNA clone of *Gossypium hirsutum* zinc finger protein 1 (GhZFP1) showed enhanced tolerance against salt stress and resistance to pathogen *Rhizoctonia solani* (Guo et al., 2009). Alternative oxidase (AOX) changed the ROS and RNS concentration in plants (Cvetkovska and Vanlerberghe 2012). Leaves in transgenic *Nicotiana tabacum* lacking AOX have increased concentrations of RNS and ROS (Cvetkovska and Vanlerberghe 2012). Transgenic seedlings of Arabidopsis plants overexpressing Alternative oxidase (AOX) (AOX OE) enhanced growth compared and under-expressed AOX (AOX AS) during NaCl stress (Singh et al., 2022). Both RNS and ROS concentrations decrease in transgenic lines (Singh et al., 2022). Since heavy metals such as Zn and reactive species like ROS and RNS reduce crop yield worldwide, the goal of the study should be to reduce the Zn, ROS, and RNS effect on crop yield (Feigl et al., 2015). Therefore, many transgenic plants have been generated to induce the crop yield by reducing Zn, ROS, and RNS amounts. Furthermore, in transgenic approaches to improve the nutritional quality of food crops, the authors have expressed the rice Zn transporter, OsZIP1, in tobacco and finger millet transgenic plants and found that there is increase in Zn along with Mn content in endospermic edible part of seeds (Ramegowda et al., 2013). Transgenic expression of Cu/Zn Superoxide Dismutase in transgenic tobacco also confers partial resistance to stresses like ozone-induced foliar necrosis (Pitcher and Zilinskas, 1996). Similarly, enhancement in defense against oxidative stress has been observed in transgenic potato expressing tomato Cu and Zn superoxide dismutases (Perl et al., 1993). Transgenic *Arabidopsis* having expression of putative Zinc-Transporter gene enhanced Zn resistance and accumulation in transgenic plants (Van der Zaal et al., 1999). Further considering the Zn as heavy metal, transgenic approaches to clean the Earth have become an important part of the phytoremediation concept and would provide a novel approach to getting rid of manmade pollution during mining and industrialization (Pilon-Smits, and Pilon, 2002).

## 15.5 Conclusion

It is suggested that Zn induces NO and ROS formation and if Zn stress removes ROS removing cassettes, it starts along with PCD formation to kill the Zn-affected root cells (Xu et al., 2010). Results indicate that ROS and NO play an important role in Zn accumulation and root growth. Moderate increase in NO production induces Zn-mediated inhibitory effects on root growth and high endogenous NO levels become toxic and reduce the roots length in Zn-treated plants. Studies have shown that exogenous NO can improve nutritional status during Zn stress. Further research is needed to confirm the exact relation between Zn, ROS, and RNS triangle which can help in boosting the Zn stress management along with increasing the crop yield, since Zn is not only important for plant development but it is also important for handling to other stress.

## References


Apel, K., Hirt, H., 2004. Reactive oxygen species: metabolism, oxidative stress, and signal transduction. Annual Review of Plant Biology 55, 373–399.

Arif, N., Yadav, V., Singh, S., Singh, S., Ahmad, P., Mishra, R.K., et al., 2016. Influence of high and low levels of plant-beneficial heavy metal ions on plant growth and development. Frontiers in Environmental Science 4, 69.

Arrivault, S., Senger, T., Krämer, U., 2006. The Arabidopsis metal tolerance protein AtMTP3 maintains metal homeostasis by mediating Zn exclusion from the shoot under Fe deficiency and Zn oversupply. The Plant Journal 46, 861–879.

Asada, K., Urano, M., Takahashi, M., 1973. Subcellular location of superoxide dismutase in spinach leaves and preparation and properties of crystalline spinach superoxide dismutase. European Journal of Biochemistry 36, 257–266.

Bleau, J.R., Spoel, S.H., 2021. Selective redox signaling shapes plant–pathogen interactions. Plant Physiology 186, 53–65.

Broadley, M.R., White, P.J., Hammond, J.P., Zelko, I., Lux, A., 2007. Zinc in plants. The New Phytologist 173, 677–702.

Buet, A., Moriconi, J.I., Santa-María, G.E., Simontacchi, M., 2014. An exogenous source of nitric oxide modulates zinc nutritional status in wheat plants. Plant Physiology and Biochemistry: PPB / Societe Francaise de Physiologie Vegetale 83, 337–345. Available from: https://doi.org/10.1016/j.plaphy.2014.08.020.

Cabot, C., Martos, S., Llugany, M., Gallego, B., Tolr'a, R., Poschenrieder, C., 2019. A role for zinc in plant defense against pathogens and herbivores. Frontiers in Plant Science 10. Available from: https://doi.org/10.3389/fpls.2019.011711664-462X.

Clemens, S., 2021. The cell biology of zinc. Journal of Experimental Botany. Available from: https://doi.org/10.1093/jxb/erab481.

Corpas, F.J., Barroso, J.B., 2013. Nitro-oxidative stress vs. oxidative or nitrosative stress in higher plants. The New Phytologist 199, 633–635. Available from: https://doi.org/10.1111/nph.12380.

Cvetkovska, M., Vanlerberghe, G.C., 2012. Alternative oxidase modulates leaf mitochondrial concentrations of superoxide and nitric oxide. New Phytologist 195 (1), 32–39.

Del Río, L.A., 2015. ROS and RNS in plant physiology: an overview. Journal of Experimental Botany 66 (10), 2827–2837.

Feigl, G., Lehotai, N., Molnár, A., Ördög, A., Rodríguez-Ruiz, M., Palma, J.M., et al., 2015. Zinc induces distinct changes in the metabolism of reactive oxygen and nitrogen species (ROS and RNS) in the roots of two Brassica species with different sensitivity to zinc stress. Annals of Botany 116 (4), 613–625.

Goodarzi, A., Namdjoyan, S., Soorki, A.A., 2020. Effects of exogenous melatonin and glutathione on zinc toxicity in safflower (Carthamus tinctorius L.) seedlings. Ecotoxicology and Environmental Safety 201, 110853. Available from: https://doi.org/10.1016/j.ecoenv.2020.110853.

Guo, Y.H., Yu, Y.P., Wang, D., Wu, C.A., Yang, G.D., Huang, J.G., et al., 2009. GhZFP1, a novel CCCH-type zinc finger protein from cotton, enhances salt stress tolerance and fungal disease resistance in transgenic tobacco by interacting with GZIRD21A and GZIPR5. New Phytologist 183 (1), 62–75.

Harris, D., Rashid, A., Arif, M., Yunas, M., 2005. Alleviating micronutrient deficiencies in alkaline soils of the North-West Frontier Province of Pakistan: on-farm seed priming with zinc in wheat and chickpea. In: Andersen, P., Tuladhar, J.K., Karki, K.B., Maskey, S.L. (Eds.), Micronutrients in South and South East Asia, International Centre for Integrated Mountain Development. International Centre for Integrated Mountain Development (ICIMOD), Kathmundu, Nepal, pp. 143–151.

Hess, D.T., Matsumoto, A., Kim, S.O., Marshall, H.E., Stamler, J.S., 2005. Protein Snitrosylation: purview and parameters. Nature Reviews. Molecular Cell Biology 6 (2), 150–166. Available from: https://doi.org/10.1038/nrm1569.

Jain, R., Srivastava, S., Solomon, S., Shrivastava, A.K., Chandra, A., 2010. Impact of excess zinc on growth parameters, cell division, nutrient accumulation, photosynthetic pigments and oxidative stress of sugarcane (Saccharum spp.). Acta Physiologiae Plantarum / Polish Academy of Sciences, Committee of Plant Physiology Genetics and Breeding 32, 979–986. Available from: https://doi.org/10.1007/s11738-010-0487-9.

Jarosz, M., Olbert, M., Wyszogrodzka, G., Młyniec, K., Librowski, T., 2017. Antioxidant and anti-inflammatory effects of zinc. Zinc-dependent NF-κB signaling. Inflammopharmacology 25, 11–24.

Khan, N.A., Khan, M.I.R., Asgher, M., Fatma, M., Masood, A., Syeed, S., 2014. Salinity tolerance in plants: Revisiting the role of sulfur metabolites. Journal of Plant Biochemistry & Physiology 2, 120.

Khan, M., Ali, S., Al Azzawi, T.N.I., Saqib, S., Ullah, F., Ayaz, A., et al., 2023. The key roles of ROS and RNS as a signaling molecule in plant–microbe interactions. Antioxidants 12 (2), 268.

Khan, M., Nazar, T., Pande, A., Mun, B.-G., Lee, D., Hussain, A., et al., 2021. The role of nitric oxide-induced ATILL6 in growth and disease resistance in Arabidopsis thaliana. Frontiers in Plant Science 12, 1314.

Kondak, S., Molnár, Á., Dóra, O.L.Á.H., Kolbert, Z., 2022. The role of nitric oxide (NO) in plant responses to disturbed zinc homeostasis. Plant. Stress (Amsterdam, Netherlands) 100068.

Kumari, V.V., Banerjee, P., Verma, V.C., Sukumaran, S., Chandran, M.A.S., Gopinath, K.A., et al., 2022. Plant nutrition: an effective way to alleviate abiotic stress in agricultural crops. International Journal of Molecular Sciences 23 (15), 8519.

Laxalt, A.M., Beligni, M.V., Lamattina, L., 1997. Nitric oxide preserves the level of chlorophyll in potato leaves infected by Phytophtora infestans. European Journal of Plant Pathology 103, 643–651.

Leshem, Y.Y., Haramaty, E., 1996. The characterization and contrasting effects of the nitric oxide free radical in vegetative stress and senescence of Pisum sativum Linn. foliage. Journal of Plant Physiology 148, 258–263.

Liu, C., Wu, Y., Wang, X., 2012. bZIP transcription factor OsbZIP52/RISBZ5: a potential negative regulator of cold and drought stress response in rice. Planta 235, 1157–1169. Available from: https://doi.org/10.1007/s00425-011-1564-z.

Loftleidir, H., 2005. Essential trace elements for plants. Animals and Humans. Reykjavik.

Madian, A.G., Regnier, F.E., 2010. Proteomic identification of carbonylated proteins andtheir oxidation sites. Journal of Proteome Research 9, 3766–3780. Available from: https://doi.org/10.1021/pr1002609.

Molnár, Á., Papp, M., Kovács, D.Z., Bélteky, P., Oláh, D., Feigl, G., et al., 2020. Nitro-oxidative signalling induced by chemically synthetized zinc oxide nanoparticles (ZnO NPs) in Brassica species. Chemosphere 251, 126419.

Morina, F., Jovanovic, L., Mojovic, M., Vidovic, M., Pankovic, D., Sonja Veljovic Jovanovic, S., 2010. Zinc-induced oxidative stress in Verbascum thapsus is caused by an accumulation of reactive oxygen species and quinhydrone in the cell wall. Physiologia Plantarum 140, 209–224. Available from: https://doi.org/10.1111/j.1399-3054.2010.01399.x.

Noritake, T., Kawakita, K., Doke, N., 1996. Nitric oxide induces phytoalexin accumulation in potato tuber tissues. Plant and Cell Physiology 37, 113–116.

Oladele, E.O., Odeigah, P.G.C., Taiwo, I.A., 2013. The genotoxic effect of lead and zinc on bambara groundnut (Vigna subterranean). African Journal of Environmental Science and Technology 7 (1), 9–13.

Olechnowicz, J., Tinkov, A., Skalny, A., Suliburska, J., 2018. Zinc status is associated with inflammation, oxidative stress, lipid, and glucose metabolism. The Journal of Physiological Sciences 68 (1), 19–31.

Palmgren, M.G., Clemens, S., Williams, L.E., Kraemer, U., Borg, S., Schjorring, J.K., et al., 2008. Zinc biofortification of cereals: problems and solutions. Trends in Plant Science 13, 464–473. Available from: https://doi.org/10.1016/j.tplants.2008.06.005.

Perl, A., Perl-Treves, R., Galili, S., Aviv, D., Shalgi, E., Malkin, S., et al., 1993. Enhanced oxidative-stress defense in transgenic potato expressing tomato Cu, Zn superoxide dismutases. Theoretical and Applied Genetics 85, 568–576.

Pilon-Smits, E., Pilon, M., 2002. Phytoremediation of metals using transgenic plants. Critical Reviews in Plant Sciences 21 (5), 439–456.

Pitcher, L.H., Zilinskas, B.A., 1996. Overexpression of copper/zinc superoxide dismutase in the cytosol of transgenic tobacco confers partial resistance to ozone-induced foliar necrosis. Plant Physiology 110 (2), 583–588.

Porwal, P., Sonkar, S., Singh, A.K., 2021. Nanobiotechnology. Plant Stress Enzymes Nanobiotechnology. Springer, Cham, Switzerland, pp. 327–348.

Puppo, A., Dimitrijevic, L., Rigaud, J., 1982. Possible involvement of nodule superoxide dismutase and catalase in leghemoglobin protection. Planta 156, 374–379.

Rai, G.K., Rai, N.P., Rathaur, S., Kumar, S., Singh, M., 2013. Expression of rd29A::AtDREB1A/CBF3 in tomato alleviates drought-induced oxidative stress by regulating key enzymatic and non-enzymatic antioxidants. Plant Physiology and Biochemistry 69, 90–100.

Ramegowda, Y., Venkategowda, R., Jagadish, P., Govind, G., Hanumanthareddy, R.R., Makarla, U., et al., 2013. Expression of a rice Zn transporter, OsZIP1, increases Zn concentration in tobacco and finger millet transgenic plants. Plant Biotechnology Reports 7, 309–319.

Sidhu, G.P.S., 2016. Physiological, biochemical and molecular mechanisms of zinc uptake, toxicity and tolerance in plants. Journal of Global Biosciences 5 (9), 4603–4633. Vol.

Singh, P., Kumari, A., Gupta, K.J., 2022. Alternative oxidase plays a role in minimizing ROS and RNS produced under salinity stress in Arabidopsis thaliana. Physiologia Plantarum 174 (2), e13649.

Tewari, R.K., Horemans, N., Watanabe, M., 2021. Evidence for a role of nitric oxide in iron homeostasis in plants. Journal of Experimental Botany 72 (4), 990–1006.

Tripathy, B.C., Oelmüller, R., 2012. Reactive oxygen species generation and signaling in plants. Plant Signaling & Behavior 7 (12), 1621–1633.

Ullah, A., Romdhane, L., Rehman, A., Farooq, M., 2019. Adequate zinc nutrition improves the tolerance against drought and heat stresses in chickpea. Plant Physiology and Biochemistry: PPB / Societe Francaise de Physiologie Vegetale 143, 11–18. Available from: https://doi.org/10.1016/j.plaphy.2019.08.020.

Umair Hassan, M., Aamer, M., Umer Chattha, M., Haiying, T., Shahzad, B., Barbanti, L., et al., 2020. The critical role of zinc in plants facing the drought stress. Agriculture 10 (9), 396.

Vazin, F., 2012. Effect of zinc sulfate on quantitative and qualitative characteristics of corn (Zea mays) in drought stress. Cercetări agronomice în Moldova 45, 15–24. Available from: https://doi.org/10.2478/v10298-012-0052-3.

Wang, Y., Loake, G.J., Chu, C., 2013. Cross-talk of nitric oxide and reactive oxygen species in plant programed cell death. Frontiers in Plant Science 4, 1–7.

Wintz, H., Fox, T., Wu, Y.Y., et al., 2003. Expression profiles of Arabidopsis thaliana in mineral deficiencies reveal novel transporters involved in metal homeostasis. Journal of Biological Chemistry 278, 47644–47653.

Xu, J., Yin, H., Li, Y., Liu, X., 2010. Nitric oxide is associated with long-term zinc tolerance in Solanum nigrum. Plant Physiology 154 (3), 1319–1334. Available from: https://doi.org/10.1104/pp.110.162982.

Van der Zaal, B.J., Neuteboom, L.W., Pinas, J.E., Chardonnens, A.N., Schat, H., Verkleij, J.A., et al., 1999. Overexpression of a novel Arabidopsis gene related to putative zinc-transporter genes from animals can lead to enhanced zinc resistance and accumulation. Plant Physiology 119 (3), 1047–1056.

Zhang, C., Zixuan, L., Yuhao, Z., Liang, M., Erqun, S., Yang, S., 2020. Iron free zinc oxide nanoparticles with ion-leaking properties disrupt intracellular ROS and iron shomeostasis to induce ferroptosis. Cell Death & Disease 11 (3), 183.

Chapter 16

# The contribution of rhizosphere in the supply of zinc to plants

**Ved Prakash[1], Sneha Tripathi[1], Samarth Sharma[1], Shubhangi Suri[1], Kavita Tiwari[1], Durgesh Kumar Tripathi[2] and Shivesh Sharma[1]**

*[1]Department of Biotechnology, Motilal Nehru National Institute of Technology Allahabad, Prayagraj, Uttar Pradesh, India, [2]Crop Nanobiology and Molecular Stress Physiology Laboratory, Amity Institute of Organic Agriculture, Amity University Uttar Pradesh, Noida, Uttar Pradesh, India*

## 16.1 Introduction

The rhizosphere forms a dynamic zone that expedites interaction between plants and microbes, where mutual interactive processes affect both plants and microbes (Bishnoi, 2015; Buee et al., 2009). The beneficial aspect of interaction includes the facilitation of nutrient acquisition, alleviation of stress, disease control, development and growth of plants (Adediran et al., 2016; Adele et al., 2018; Glick, 2014). Nutrient deficiency is a major hurdle to plant growth where there is a lack of nutrients or nutrients are not present in an available form which hampers the metabolic activities of plants. The reason may be low nutrient availability, poor soil microbe-plant interaction, and soil profile (Marschner et al., 2011; Rengel, 2002). Zn deficiency in soil has been widely documented. Neutral to alkaline soil often shows Zn deficiency in rice (Rehman et al., 2012).

Micronutrient has a very prominent role in plant growth and development at different stages of the life cycle (Uchida, 2000). Out of different micronutrients, Zn is essential for plant growth since it is a crucial component of several enzymes, carbohydrate metabolism, hormone synthesis, and cellular membranes; along with these, it also aids in gene expression regulation for stress tolerance in plants (Cakmak, 2000). To overcome the effect of nutrient deficiency, chemical fertilizers are often applied to the plants but they cause negative effects on soil as well as readily get changed to non-accessible forms. Apart from its breeding program for nutrient uptake, deeper insights into the molecular mechanism are needed (Rose et al., 2013).

**Zinc in Plants. DOI: https://doi.org/10.1016/B978-0-323-91314-0.00010-7**

According to the reports, global Zn inadequacy in the soil is primarily attributable to Zn insolubility rather than the lack of availability of Zn (Iqbal et al., 2010). Ramesh et al. (2014) anticipated >50% of agricultural soil in India is deficient in zinc (Ramesh et al., 2014). The utilization of Zn fertilizers is the simplest way to fulfill the Zn deficiency in soil but considering the ill effect and cost hurdles, other sustainable approaches are preferred that cause less harm to the environment (Zhao et al., 2014). Zinc solubilizing bacteria could be the apt approach to replace chemical fertilizers for improving Zn mobilization (Prathap et al., 2022). The application of beneficial soil bacteria is gaining attention for the restoration of soil and meeting Zn deficiency in agriculture (Kumar et al., 2019). Recent studies have demonstrated the use of PGPR for Zn solubilization (Hussain et al., 2015; Dhaked et al., 2017; Kamran et al., 2017; Shaikh & Saraf, 2017a,b; Zaheer et al., 2019; Ahmad et al., 2021; Zhao et al., 2024). The availability of Zn and its assimilation is readily enhanced by ZSB which in turn promotes plant growth and development. The present book chapter highlights the various factors that result in Zn mobilization along with the role of rhizospheric microbes in facilitating Zn supply to plants.

## 16.2 Factors affecting nutrient availability in the rhizosphere

The concentration of Zn in soil depends upon soil type and the influence of anthropogenic activities (Callender & Rice, 2000; Rehman et al., 2018a,b). 10 μg/g of Zn concentration in soil represents Zn deficiency while an amount more than >200 μg/g represents contamination from various sources (Alloway, 2008, 2009). However, the availability of Zn in the rhizosphere depends upon soil pH, moisture, temperature, organic matter, clay content, root exudates, and soil microorganisms.

## 16.3 Soil pH, temperature, moisture, and light intensity of soil

Zn availability is significantly influenced by soil pH. An increase in soil pH from 4.6 to 6.8 results in reduced uptake of Zn (Qadar and Azam, 2007). This is because alkaline soil favors the development of Zn precipitates containing carbonate, hydroxide, and carbonate groups (Rupa and Tomar, 1999), reducing the availability of protonated anions and free metal cations at high pH (Kiekens, 1995; Rengel, 2015). Temperature regimes alter Zn availability. In cool seasons, low temperature inhibits the decomposition of organic matter and restricts mycorrhizal colonization and root development which further restricts the uptake and translocation of Zn into shoots. In the hot season, high temperature enhances the mineralization of soil organic matter and maintains Zn supply to plants (Moraghan and Mascagni, 1991; Marschner, 1995).

Soil moisture is an important factor that dictates nutrient uptake by influencing the rate of diffusion in roots in the rhizospheric region of soil. Under drought conditions, Zn movement in the soil is hindered which results in reduced Zn uptake by plants (Bagci et al., 2007; Marschner, 2012). In waterlogged and poorly drained soils, soluble Zn concentration decreases in comparison to well-drained soils due to generation of reducing conditions which increases pH thereby enhancing the formation insoluble zinc sulfide. High pH also results in the dissolution of ferrous ($Fe^{2+}$) and manganese ($Mn^{2+}$) from hydrous oxides (inorganic complex of hydroxide, metal, and water), which may result in their competition with Zn ions for uptake in plants. (Alloway, 2008; Sadeghzadeh, 2013). Exposure to high light intensity as well as long day lengths negatively affects Zn availability by inducing impairment in ROS detoxification (Marschner and Cakmak, 1989).

## 16.4 Organic matter in soil

The uptake and availability of Zn by plant roots are determined by the amount and nature of the organic matter present. Functional groups such as SH and OH in organic materials, such as humic acids and fulvic acids, have a great affinity for metal ions like $Zn^{2+}$, increasing the bioavailability of Zn to plants via the synthesis of Zn-organic complex compounds. (Kabata-Pendias, 2010) As a result, adding organic matter to soil transfers unbound Zn into soil solutions, altering Zn's mobility and solubility. Organic matter also increases mobility by restricting Zn adsorption in the solid phase (Kiekens, 1995; Du Laing et al., 2009). Organic matter present in soil can affect the adsorption of Zn viz. soil with a higher organic content absorbs more Zn than that with a lesser organic matter content (Gurpreet-Kaur et al., 2013). In soils with low Zn content and high organic matter, Zn availability may severely hampered due to the development of stable and strong complexes that are insoluble in nature with organic matter in a solid state (Alloway, 2008).

## 16.5 Soil salinity

Saline soils found in arid and semiarid regions generally results in plants deficient in zinc. This is due to the competition among salt cations and Zn at the root zone, making Zn uptake difficult. For instance, when water used for irrigating saline–sodic soils is high in sodium ions, Zn gets leached because exchange sites are occupied by Na + (Tinker and Lauchli, 1984; Rehman et al., 2018a,b).

## 16.6 Zn interaction with other soil constituents

Zn, being a cation, interacts positively and negatively with other plant nutrients, especially anions. Zn interact positively with N, as when the rate of N application is enhanced, it improves the concentration of Zn in wheat grain

(Kutman et al., 2011), Chickpea (El Habbasha et al., 2013), and seeds (Kutman et al., 2010). This may be due to the plenitude of Zn transporters and the presence of Zn chelating nitrogenous compounds, resulting in increased root and shoot uptake (Erenoglu et al., 2011). In contrast, Zn negatively interacts with P as it is observed that during the application of P fertilizers, high P levels decreases the Zn availability (Zhang et al., 2012). This is because Zn gets immobilized at the root surface due to the formation of Zn phosphate or phytate when it comes in contact with P (Sarret et al., 2001). Cu, Cd, Mn, and Ca compete with Zn for the same adsorption sites hence, increased concentration of these cations inhibits Zn adsorption (Ullah et al., 2020). However, Zn interaction with K is positive, as Zn helps in maintaining membrane integrity, thereby reducing leakage of K and amides (Aktaş et al., 2007). Zn deficiency limits the translocation of $Fe^{2+}$ from roots to shoots, resulting in its deficiency (Rengel and Römheld, 2000) while Zn levels in wheat were found to be raised after sulfur application (Cui and Wang, 2005).

## 16.7 Zinc interaction with soil biota/mycorrhizal colonization

Zinc availability and adsorption in plants are significantly influenced by soil biota. Among the many soil- borne fungi, Arbuscular mycorrhizal fungi (AMF) are particularly remarkable as they form mutualistic associations with plants; they consume sugar from the host plant and help the plant in the uptake of water and nutrients (Diagne et al., 2020; Bhantana et al., 2021).AMF manages Zn nutrition to plants in two ways, that is, when Zn concentration is too low, AMF increases the Zn uptake through mycorrhizal hyphae, which provides more surface area for nutrient adsorption, whereas when Zn concentration is in excess, Zn translocation from root to shoot is inhibited (Bhantana et al., 2021). In addition to this, AMF increases the reach of the plant roots to pores and patches of soil that were otherwise inaccessible (Rehman et al., 2018a,b).

Plant growth-promoting rhizobacteria (PGPR) actively participate in geochemical nutrient cycles and help plants to acquire essential nutrients for growth and development (Etesami & Adl, 2020). About 90% of Zn available in soil exists in insoluble form ,making it inaccessible for plants. (Barber, 1995). However, Zn solubilizing bacteria, such as *Gluconacetobacter, Bacillus, Pseudomonas* and *Acinetobacter* (Ullah et al., 2020), improves Zn availability by making it bioavailable to plants through Zn solubilization that is bonded to $CaCO_3$ and organic complexes, and also enhances the exchangeable form of Zn in the soil (Ramesh et al., 2014).

## 16.8 Rhizospheric microflora: influencing zinc availability

Rhizospheric microflora are an important group of microorganisms inhabiting in narrow zone of the root. Root microflora contribute to plant growth promotion and nutrient recycling by an array of mechanisms. Zinc is one of the critical elements that plants require in very small amounts for proper growth, physiology, and functioning. (Jha, 2019). Zn deficiency is widespread in calcareous, high pH, and land-leveled soils of Turkey, China, Australia, and India (Noulas, et al., 2018). Zn plays an active role in photosynthesis, defense system, pollen germination, disease resistance, protein synthesis, and enzymatic activity of the plant. Plant has a self-regulated system to control soil microbial activities and mobilize these nutrients from the soil (Hakim et al., 2021). In the rhizosphere, the interaction between microbes and plants play a significant role in zinc solubilization, thus making it available to the plant (Saini et al., 2021). Well known Zn solubilizing microorganisms (ZSM) are *Bacillus aryabhattai* (Ramesh et al., 2014), *Bacillus megaterium* (Bhatt and Maheshwari, 2020), *Pseudomonas fragi* (Kamran et al., 2017), *Bacillus altitudinis* (Kushwaha et al., 2021), *Serratia marcescens, S. liquefacience*, and *Bacillus thuringiensis* (Abaid-Ullah et al., 2015), *Gluconacetobacter diazotrophicus* (Saravanan et al., 2007), and *Beauveria caledonica* (Fomina et al., 2004) ZSM solubilize Zn through acidification. ZSM produces organic acid in soil which chelate Zn ions and lowers the soil pH (Kumar et al., 2019) siderophore production, presence of proton oxido- reductive system on cell membranes and ligands chelation (Chang et al., 2005; Saravanan et al., 2011; Saini et al., 2021). The major mechanism of Zn solubilization involves the production of acid in the soil to sequester the cation of Zn under acidic pH. Inorganic acids (carbonic acid, nitric acid, and sulfuric acid) and organic acids (citric acid, oxalic acid, 5-ketogluonic acid, and gluconic acid) production by microorganisms decrease the pH of the soil and subsequently enhance zinc availability (Fasim et al., 2002). According to the reports, a 1 unit decrease in pH increases zinc availability by100 times . Siderophore production is characteristic of plant growth-promoting bacteria that helps in chelating the nutrients for plant uptake. Several studies have reported quantitatively and qualitatively on the production of siderophore by zinc solubilizing bacteria, but the class of siderophore molecule is yet unknown. Some of the fungal genera such as *Aspergillus nizer, Penicillium simplicissimum and Abisidia glauca, A. cylindrospora*, and *A. spinosa* have the capability of zinc phosphate and zinc oxide solubilization (Franz et al., 1993; Coles et al., 1999). AMF are big players in nutrient acquisition and plant growth promotion. AMF response towards Zn is negative, positive, or neutral depending upon Zn concentration in soil. Under low zinc condition, AMF enhances plant Zn nutrition whereas under high Zn concentration, AMF reduces the bioavailability of soluble zinc by converting them into an insoluble form. Acidification and complexation are the common mechanisms of Zn mobilization in mycorrhizae (Bhantana et al., 2021; Subramanian et al., 2009). The application of Zn solubilizing microbes increased the growth of plants, Zn uptake and mobilization in *Capsicum annuum* (Bhatt and Maheshwari, 2020), *Zea mays* (Hussain et al., 2015), *Triticum aestivum* (Kamran et al., 2017), and *Oryza sativa* (Bhakat et al., 2021).

## 16.9 Zinc solubilizing rhizobacteria

The rhizospheric Zn solubilizing bacteria may be employed as a solution to improve the Zn biofortification of plants. Bacterial mediated Zn solubilization study involved inoculation of Zn solubilizing rhizobacteria (ZSB) in soil containing an insoluble form of Zn such as $ZnCO_3$, ZnO, $ZnSiO_3$, ZnS, $ZnPO_4$ and it was reported that ZSB mediates the release of Zn to its available form by various mechanism (Khatoon et al., 2022; Batool et al., 2021) (Fig. 16.1). Several rhizobacteria exhibit potential zinc solubilization activities and the re-inoculation of these rhizobacteria in plant root influences the mobilization and availability of the Zn nutrient. For example, wheat inoculated with *Pseudomonas fragi* showed the highest Zn content compared to uninoculated, as *P. fragi* accelerated the bioavailability of zinc and its mobilization from root to grain (Kamran et al., 2017). A study demonstrated the high zinc solubilization ability of Bacillus spp., where treatment of rice seedlings with different *Bacillus* stains improved the growth, yield, and Zn content of rice (Shakeel et al., 2015). Recently, another study on rice plants reported higher Zn solubilization in Zn-deficient submerged soil by the application of ZSB *B. aryabhattai* (Prathap et al., 2022). A similar

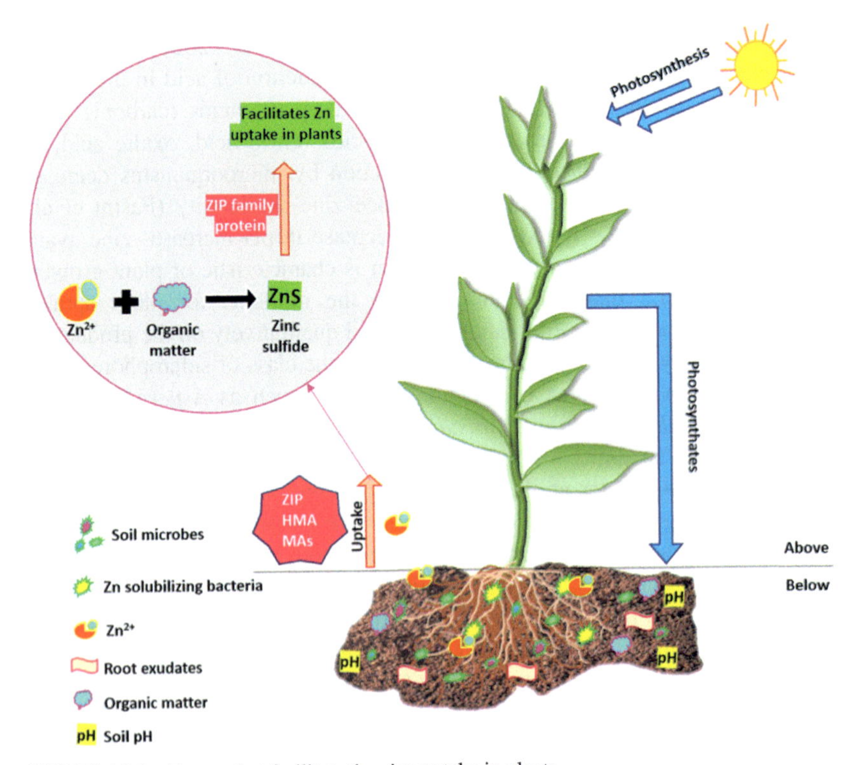


**FIGURE 16.1** Factors that facilitate the zinc uptake in plants.

observation was found in *Acinetobacter sp.* and *Serratia sp.* treated rice seedlings; both species colonized the root and influenced plant growth, root development, and zinc uptake at 0.2 mg/L of $ZnSO_4$. In rice, rhizobacteria such as *Ralstonia picketti* can benefit plants in the uptake of zinc in deficient soil through acid production especially 2-ketogluconic acids and other organic acids, including plant growth promotion and zinc fortification (Kumar et al., 2019).

## 16.10 Uptake of zinc from soil to plants

The rhizosphere provides essential nutrients to plants, which are transported through apoplastic and symplastic mechanisms (Sattelmacher, 2001; Ricachenevsky et al., 2018). The role of metal transporters ranges from the very beginning, that is, the uptake and transport of nutrients (Hall and Williams, 2003; Aibara and Miwa, 2014) (Table 16.1). Zn is mainly transported through ZIP protein family and other transporters (Ajeesh Krishna et al., 2020; Huang et al., 2020). The role of ZIP transporter family is widely identified in different species of plants such as rice, barley, maize, and barrel-clove (Lopez-Millan et al., 2004; Watts-Williams and Cavagnaro, 2018; Huang et al., 2020) for mobilization of Zn. In xylem sap, Zn gets transported either in free ions form or bound with chelators (Yokosho et al., 2009), whereas in phloem sap, it remains bound with nicotiamine (Nishiyama et al., 2012).

## 16.11 Zinc as a critical component of plant growth

Zn is a crucial micronutrient for the development and growth of plants, playing an essential role in the metabolism of plants as a cofactor of many enzymes involved in photosynthesis, protein synthesis, hormone regulation, and maintenance of membrane integrity (Tsui., 1948; Hafeez et al., 2013). Zn is a key player in enzymes such as RNA polymerases, superoxide dismutase, alcohol dehydrogenase, and carbonic anhydrase (Alloway, 2008). Zn is also required for tryptophan assimilation, a percussor of growth hormone auxin biosynthesis. Zn has a special emphasis on plant defense response because of its integrity in the copper-zinc superoxides dismutase (Cu-Zn-SOD) enzyme. Cu-Zn-SOD is involved in the detoxification of free radicles. Besides this, Zn ion is particularly important for Zn finger family proteins, acting as a transcription regulator to control cell proliferation and differentiation. Gene expression is regulated by Zn and is also present in DNA-binding proteins. Moreover, Zn is responsible for the stability of protein and chromatic structure (DNA/RNA). In soil, concentration of soluble zinc below 0.5 mg Zn/kg dry soil impaired growth and resulted in visible deficiency symptoms such as chlorosis and smaller leaves, spikelet sterility, and reduced grain yield (Cakmak., 2000). Zn concentration also influences

**TABLE 16.1** It shows different bacteria that facilitate zinc uptake in plants through different modes of action.

| Bacteria | Plant | Insoluble form of zinc | Solubilization mechanism | References |
|---|---|---|---|---|
| *Pseudomonas fluroscenes* | Sedum (Stone crop) | $Zn_3(PO_4)_2$ | Organic acid production | Upadhayay et al. (2022) |
| *Pseudomonas aeruginosa* | Sugarcane, maize | $Zn_3(PO_4)_2$, ZnO | Production of keto-gluconic acid and gluconic acid | Sunithakumari et al. (2016) |
| *Pseudomonas sp., Bacillus sp.* | Wheat, Rice, Chikpea | ZnO, ZnS and $ZnCO_3$ | Organic acid production | Rehman et al. (2018a,b), Zaheer et al. (2019) |
| *Gluconacetabacter Diazotrophicus* | Maize | ZnO, $Zn_3(PO_4)_2$ and $ZnCO_3$ | Production of gluconic acid and its derivative 5-keto-gluconic acid | Sarathambal et al. (2010) |
| *Klebsiella sp., Pseudomonas sp.* | Okra | $Zn_3(PO_4)_2$, ZnO | Organic acid production | Fasim et al. (2002) |
| *Acinetobacter sp., Burkholderia sp.* | Rice | ZnO, $ZnSO_4$ | Production of oxalic acid, gluconic acid, tartaric acid, acetic acid and formic acid | Vaid et al. (2014) |
| *Rahnella aquatillis* | Oilseed rape | ZnO | Organic acid production | He et al. (2013) |
| *Bacillus aryabhattai* | Wheat, maize | ZnO, $Zn_3(PO_4)_2$ and $ZnCO_3$ | Production of malonic acid, malic acid, succinic acid, propionic acid, citric acid, gluconic acid and keto-D-gluterate | Ramesh et al. (2014) |
| *Bacillus cereus* | Soyabean | ZnO, $Zn_3(PO_4)_2$ | Organic acid production | Sharma et al. (2012) |
| *Pseudomonas fragi, Rhizobium sp.* | Wheat, sugarcane | $Zn_3(PO_4)_2$, $ZnCO_3$ | Production of siderophores and proton | Verma et al. (2020) |

| *Pantoea agglomerans, Pantoea Dispersa* | Wheat, Soyabean | $ZnCO_3$ | Production of siderophores | Kumar et al. (2017) |
|---|---|---|---|---|
| *Serratia liquefaciens, S marcescens* | Maize, wheat | $ZnSO_4$ | Organic acid production | Othman et al. (2017) |
| *Bacillus glycinifermentans, Microbacterium oxydans, Paenarthrobacter nicotinovorans* | Wheat | $Zn_3(PO_4)_2$ | Production of keto-gluconic acid and gluconic acid | Yadav et al. (2022) |
| *Bacillus megaterium* | Sweet And chili pepper | ZnO, $Zn_3(PO_4)_2$ and $ZnCO_3$ | Production of gluconic acid | Dinesh et al. (2018), Bhatt and Maheshwari (2020) |
| *Pseudomonas taiwenensis* | Tomato | ZnO, $Zn_3(PO_4)_2$ and $ZnCO_3$ | Production of citric acid, keto-D-gluterate, propionic acid, oxalic acid and gluconic acid | Vidyashree et al. (2018) |
| *Acinetobacter calcoaceticus* | Rice | ZnO, $ZnCO_3$ | Organic acid production | Goteti et al. (2013) |
| *Agrobacterium tumefaciens* | Barley, tomato | ZnO | Exudation of organic acids and phenolic compounds | Khanghahi et al. (2018) |
| *Burkholderia cenocepacia, Pseudomonas striata* | Wheat, Maize | ZnO | Organic acid production | Pawar et al. (2015) |
| *Bacillus thuringiensis* | Soyabean, Wheat | ZnO and $ZnCO_3$ | Acidification or organic acid production | Abaid-Ullah et al. (2015) |
| *Enterobacter cloacae* | Wheat, sugarcane | $Zn_3(PO_4)_2$, $ZnCO_3$ | Siderophores and organic acid production | Kamran et al. (2017) |

environmental stress response in plants via its key role in the activation/stabilization of enzymes. Zn reduces the adverse effects of salt, drought, fungal infection, and thermal stress by influencing osmotic balance and water uptake in plants (Peck and McDonald., 2010; Wu et al., 2015) As Zn concentration is reduced, iron availability increases which initiate free radicles production and cause ROS to induce photooxidative damage (Cabot et al., 2019). Several studies provide evidence of zinc-induced stress tolerance response against many abiotic stresses in plant systems such as *Gossypium hirsutum* (Kundu et al., 2019) *Helianthus annuus* (Jan et al., 2022), *Oryza sativa* (Huang et al., 2009), and *Chrysanthemum morifolium* (Yang et al., 2014).

## 16.12 Rhizospheric engineering for enhancing Zn content

The rhizosphere is that area or region of soil immediately surrounding plant roots. This portion is influenced by various plant secretions generally known as rhizodeposits, which comprise organic acids, carbohydrates, and amino acids. This portion harbors diverse microbial communities; among them are various microbial species that perform functions that either directly or indirectly influence plant growth and development. This provides an opportunity for researchers to engineer the rhizosphere in a way to promote plant growth. This concept is known as Rhizospheric engineering. Various authors have reported such approach of exploiting plant-microbe synergy in improving overall productivity and plant growth (Ha-Tran et al., 2021; Zainab et al., 2021; Riaz et al., 2021) even under different biotic (Gupta et al., 2021; Morcillo and Manzanera, 2021) and abiotic stress (Goswami and Suresh, 2020; Sharma et al., 2021).

Rhizosphere is a function of three components viz. soil, microbes, and plants which can be manipulated or engineered to enhance yield and plant growth (Ryan et al., 2009; Hakim et al., 2021). Rhizospheric soil can be engineered either by soil amendments like fertilizers or by modulating plant traits like root architecture and root exudate secretions (Hakim et al., 2021). Former is achieved due to an increase in nutrient mobilization by modifying nutrient solubility (Tesfaye et al., 2021) while later can be achieved through plant breeding and targeted gene editing techniques which can improve Zn uptake in plants (Wei et al., 2020).

Plant uptakes the cationic form of Zinc ($Zn^{2+}$), which is present in very small amounts in the soil. The maximum fraction of Zn present in the soil is in the insoluble form generally fixed in compounds such as smithsonite ($ZnCO_3$), zinkosite ($ZnSO_4$), and phalerite (ZnS) (Shaikh and Saraf, 2017a, b). PGPR includes various zinc solubilizing bacterial strains (ZSB) like *Bacillus sp.* and *Pseudomonas fluorescens* which tend to solubilize Zn from insoluble compounds such as ZnO, ZnS, and $ZnCO_3$ (Saravanan et al., 2011). ZSB secrets various organic acids such as formic acid, oxalic acids, ketogluconic acid, and siderophore into the rhizosphere, which helps in solubilizing Zn by chelating cations bound to Zn compounds (Hakim et al., 2021). As the Zn solubility in the soil is associated with soil pH, organic acids sequester zinc cations through acidification of soil which decreases the pH in the nearby soil, hence modifying solubility of Zn present in soil (Kumar et al., 2019). Various authors have isolated ZSB strains from different crops like maize (Ahmad, 2007), rice (Yasmin, 2011), etc.

The third component of rhizospheric engineering is the plant itself which has already been mentioned and can be engineered by various plant breeding techniques. The physiological behavior of plants like Zn uptake can also be modified by phytohormones like jasmonate, brassinosteroids, and salicylic acid which are chemical messengers present in plants in very low concentrations and are involved in various developmental and growth phases of plants (Costacurta and Vanderleyden, 1995; Oleńska et al., 2020). A wide range of PGPRs are involved in phytohormone production. Auxin is a phytohormone indirectly involved in Zn uptake from soil by helping plants in root development and increasing surface area for adsorption (Kumar et al., 2019). The precursor of Indole-acetic acid (IAA) biosynthesis, Tryptophan, the most common auxin is known to be produced by an array of PGPR through four main pathways: *Erwinia, Pseudomonas, Enterobacter*, and *Klebsiella* (Tahir and Sarwar, 2013). In short, rhizospheric engineering could be one approach to deal with the Zn deficiency in plants.

## 16.13 Conclusion

Cereal crops often face a deficiency of Zn, where soil contains low organic matter, higher pH, and less nutrient supply. To overcome this, the rhizosphere may impart a significant function in improving Zn bioavailability. Out of various approaches, utilizing the Zn solubilizing rhizobacteria is the best-suited approach to mitigate zinc unavailability considering its low cost and eco-friendly aspect. The indigenous microbes interact with the plant rhizosphere and modulate the production of acids and other compounds that increase the Zn uptake in plants. The detailed understanding of these interactions and microbial role in aiding zinc availability to plants will deepen our understanding and we will be able to implement these to meet sustainable agriculture practices. The application will further help plants to withstand various abiotic stresses and will enhance crop production with better nutritional value.

## Acknowledgment

The authors are grateful to the Director, MNNIT Allahabad, Prayagraj for providing the necessary research facilities.

## References

Abaid-Ullah, M., Hassan, M.N., Jamil, M., Brader, G., Shah, M.K.N., Sessitsch, A., et al., 2015. Plant growth promoting rhizobacteria: an alternate way to improve yield and quality of wheat (*Triticum aestivum*). International Journal of Agriculture and Biology 17, 51–60.

Adediran, G.A., Ngwenya, B.T., Mosselmans, J.F.W., Heal, K.V., 2016. Bacteria–zinc co-localization implicates enhanced synthesis of cysteine-rich peptides in zinc detoxification when Brassica juncea is inoculated with Rhizobium leguminosarum. New Phytologist 209 (1), 280–293.

Adele, N.C., Ngwenya, B.T., Heal, K.V., Mosselmans, J.F.W., 2018. Soil bacteria override speciation effects on zinc phytotoxicity in zinc-contaminated soils. Environmental Science & Technology 52 (6), 3412–3421.

Ahmad M. 2007. Biochemical and Molecular Basis of Phosphate and Zinc Mobilization by PGPR in rice. M. Phil dissertation, NIBGE, Faisalabad.

Ahmad, I., Ahmad, M., Hussain, A., Jamil, M., 2021. Integrated use of phosphate-solubilizing Bacillus subtilis strain IA6 and zinc-solubilizing Bacillus sp. strain IA16: a promising approach for improving cotton growth. Folia Microbiologica 66 (1), 115–125.

Aibara, I., Miwa, K., 2014. Strategies for optimization of mineral nutrient transport in plants: multilevel regulation of nutrient-dependent dynamics of root architecture and transporter activity. Plant and Cell Physiology 55 (12), 2027–2036.

Ajeesh Krishna, T.P., Maharajan, T., Victor Roch, G., Ignacimuthu, S., Antony Ceasar, S., 2020. Structure, function, regulation and phylogenetic relationship of ZIP family transporters of plants. Frontiers in Plant Science 11, 662.

Aktaş, H., ABAK, K., Öztürk, L., Çakmak, İ., 2007. The effect of zinc on growth and shoot concentrations of sodium and potassium in pepper plants under salinity stress. Turkish Journal of Agriculture and Forestry 30 (6), 407–412.

Alloway B. 2008. 'Zinc in Soils and Crop Nutrition.' 2nd edn. (International Zinc Association: Brussels; International Fertilizer Industry Association: Paris).

Alloway, B.J., 2009. Soil factors associated with zinc deficiency in crops and humans. Environmental Geochemistry and Health 31, 537–548.

Bagci, S.A., Ekiz, H., Yilmaz, A., Cakmak, I., 2007. Effects of zinc deficiency and drought on grain yield of field-grown wheat cultivars in Central Anatolia. Journal of Agronomy and Crop Science 193, 198–206.

Barber, S.A., 1995. Soil nutrient bioavailability: a mechanistic approach. Wiley, New York.

Batool, S., Asghar, H.N., Shehzad, M.A., Yasin, S., Sohaib, M., Nawaz, F., et al., 2021. Zinc-solubilizing bacteria-mediated enzymatic and physiological regulations confer zinc biofortification in chickpea (Cicer arietinum L.). Journal of Soil Science and Plant Nutrition 21 (3), 2456–2471.

Bhakat, K., Chakraborty, A., Islam, E., 2021. Characterization of zinc solubilization potential of arsenic tolerant Burkholderia spp. isolated from rice rhizospheric soil. World Journal of Microbiology and Biotechnology 37 (3), 1–13.

Bhantana, P., Rana, M.S., Sun, X.C., Moussa, M.G., Saleem, M.H., Syaifudin, M., et al., 2021. Arbuscular mycorrhizal fungi and its major role in plant growth, zinc nutrition, phosphorous regulation and phytoremediation. Symbiosis 84 (1), 19–37.

Bhatt, K., Maheshwari, D.K., 2020. Zinc solubilizing bacteria (Bacillus megaterium) with multifarious plant growth promoting activities alleviates growth in Capsicum annuum L. 3 Biotech 10 (2), 1–10.

Bishnoi, U., 2015. PGPR interaction: an ecofriendly approach promoting the sustainable agriculture system, Advances in Botanical Research, Vol. 75. Academic Press, pp. 81–113.

Buee, M., De Boer, W., Martin, F., Van Overbeek, L., & Jurkevitch, E. 2009. The Rhizosphere Zoo: An Overview of Plant-Associated Communities of Microorganisms, Including Phages, Bacteria, Archaea, and Fungi, and of Some of Their Structuring Factors.

Cabot, C., Martos, S., Llugany, M., Gallego, B., Tolrà, R., Poschenrieder, C., 2019. A role for zinc in plant defense against pathogens and herbivores. Frontiers in Plant Science 1171.

Cakmak, I., 2000. Tansley Review No. 111: possible roles of zinc in protecting plant cells from damage by reactive oxygen species. New Phytologist 146 (2), 185–205.

Callender, E., Rice, K.C., 2000. The urban environmental gradient: anthropogenic influences on the spatial and temporal distributions of lead and zinc in sediments. Environmental Science & Technology 34 (2), 232–238.

Chang, H.B., Lin, C.W., Huang, H.J., 2005. Zinc-induced cell death in rice (Oryza sativa L.) roots. Plant Growth Regulation 46 (3), 261–266.

Coles, K.E., David, J.C., Fisher, P.J., Lappin-Scott, H.M., Macnair, M.R., 1999. Solubilisation of zinc compounds by fungi associated with the hyperaccumulator Thlaspi caerulescens. Botanical Journal of Scotland 51 (2), 237–247.

Costacurta, A., Vanderleyden, J., 1995. Synthesis of phytohormones by plant-associated bacteria. Critical Reviews in Microbiology 21 (1), 1–18.

Cui, Y., Wang, Q., 2005. Interaction effect of zinc and elemental sulfur on their uptake by spring wheat. Journal of Plant Nutrition 28, 639–649.

Dhaked, B.S., Triveni, S., Reddy, R.S., Padmaja, G., 2017. Isolation and screening of potassium and zinc solubilizing bacteria from different rhizosphere soil. International Journal of Current Microbiology and Applied Sciences 6 (8), 1271–1281.

Diagne, N., Ngom, M., Djighaly, P.I., Fall, D., Hocher, V., Svistoonoff, S., 2020. Roles of arbuscular mycorrhizal fungi on plant growth and performance: importance in biotic and abiotic stressed regulation. Diversity 12 (10), 370.

Dinesh, R., Srinivasan, V., Hamza, S., Sarathambal, C., Gowda, S.A., Ganeshamurthy, A.N., et al., 2018. Isolation and characterization of potential Zn solubilizing bacteria from soil and its effects on soil Zn release rates, soil available Zn and plant Zn content. Geoderma 321, 173–186.

Du Laing, G., Rinklebe, J., Vandecasteele, B., Meers, E., Tack, F.M., 2009. Trace metal behaviour in estuarine and riverine floodplain soils and sediments: a review. The Science of the Total Environment 407, 3972–3985.

El Habbasha, S., Mohamed, M.H., El-Lateef, E.A., Mekki, B., Ibrahim, M., 2013. Effect of combined zinc and nitrogen on yield, chemical constituents and nitrogen use efficiency of some chickpea cultivars under sandy soil conditions. World Journal of Agricultural Sciences 9 (4), 354–360.

Erenoglu, E.B., Kutman, U.B., Ceylan, Y., Yildiz, B., Cakmak, I., 2011. Improved nitrogen nutrition enhances root uptake, root-toshoot translocation and remobilization of zinc (65Zn) in wheat. The New Phytologist 189, 438–448.

Etesami, H., Adl, S.M., 2020. Plant growth-promoting rhizobacteria (PGPR) and their action mechanisms in availability of nutrients to plants. Phyto-Microbiome in Stress Regulation 11, 147–203.

Fasim, F., Ahmed, N., Parsons, R., Gadd, G.M., 2002. Solubilization of zinc salts by a bacterium isolated from the air environment of a tannery. FEMS Microbiology Letters 213 (1), 1–6.

Fomina, M., Alexander, I.J., Hillier, S., Gadd, G.M., 2004. Zinc phosphate and pyromorphite solubilization by soil plant-symbiotic fungi. Geomicrobiology Journal 21 (5), 351–366.

Franz, A., Burgstaller, W., Müller, B., Schinner, F., 1993. Influence of medium components and metabolic inhibitors on citric acid production by Penicillium simplicissimum. Microbiology (Reading, England) 139 (9), 2101–2107.

Glick, B.R., 2014. Bacteria with ACC deaminase can promote plant growth and help to feed the world. Microbiological Research 169 (1), 30–39.

Goswami, M., Suresh, D.E.K.A., 2020. Plant growth-promoting rhizobacteria—alleviators of abiotic stresses in soil: a review. Pedosphere 30 (1), 40–61.

Goteti, P.K., Emmanuel, L.D.A., Desai, S., Shaik, M.H.A., 2013. Prospective zinc solubilising bacteria for enhanced nutrient uptake and growth promotion in maize (Zea mays L.). International Journal of Microbiology 2013.

Gupta, A., Bano, A., Rai, S., Dubey, P., Khan, F., Pathak, N., et al., 2021. Plant Growth Promoting Rhizobacteria (PGPR): a sustainable agriculture to rescue the vegetation from the effect of biotic stress: a review. Letters in Applied NanoBioScience 10, 2459–2465.

Gurpreet-Kaur, Sharma, B.D., Sharma, S., 2013. Effects of organic matter and ionic strength of supporting electrolyte on zinc adsorption in benchmark soils of Punjab in Northwest India. Communications in Soil Science and Plant Analysis 44, 922–938.

Hafeez, B.M.K.Y., Khanif, Y.M., Saleem, M., 2013. Role of zinc in plant nutrition-review. American Journal of Experimental Agriculture 3 (2), 374.

Hakim, S., Naqqash, T., Nawaz, M.S., Laraib, I., Siddique, M.J., Zia, R., et al., 2021. Rhizosphere engineering with plant growth-promoting microorganisms for agriculture and ecological sustainability. Frontiers in Sustainable Food Systems 5, 16.

Hall, J.Á., Williams, L.E., 2003. Transition metal transporters in plants. Journal of Experimental Botany 54 (393), 2601–2613.

Ha-Tran, D.M., Nguyen, T.T.M., Hung, S.H., Huang, E., Huang, C.C., 2021. Roles of plant growth-promoting rhizobacteria (PGPR) in stimulating salinity stress defense in plants: a review. International Journal of Molecular Sciences 22 (6), 3154.

He, H., Ye, Z., Yang, D., Yan, J., Xiao, L., Zhong, T., et al., 2013. Characterization of endophytic Rahnella sp. JN6 from Polygonum pubescens and its potential in promoting growth and Cd, Pb, Zn uptake by Brassica napus. Chemosphere 90, 1960–1965. Available from: https://doi.org/10.1016/j.chemosphere.2012.10.057.

Huang, X.Y., Chao, D.Y., Gao, J.P., Zhu, M.Z., Shi, M., Lin, H.X., 2009. A previously unknown zinc finger protein, DST, regulates drought and salt tolerance in rice via stomatal aperture control. Genes & Development 23 (15), 1805–1817.

Huang, S., Sasaki, A., Yamaji, N., Okada, H., Mitani-Ueno, N., Ma, J.F., 2020. The ZIP transporter family member OsZIP9 contributes to root zinc uptake in rice under zinc-limited conditions. Plant Physiology 183 (3), 1224–1234.

Hussain, A., Arshad, M., Zahir, Z.A., Asghar, M., 2015. Prospects of zinc solubilizing bacteria for enhancing growth of maize. Pakistan Journal of Agricultural Sciences 52, 4.

Iqbal, U., Jamil, N., Ali, I., Hasnain, S., 2010. Effect of zinc-phosphate-solubilizing bacterial isolates on growth of Vigna radiata. Annals of Microbiology 60 (2), 243–248.

Jan, A.U., Hadi, F., Ditta, A., Suleman, M., Ullah, M., 2022. ). Zinc-induced anti-oxidative defense and osmotic adjustments to enhance drought stress tolerance in sunflower (Helianthus annuus L.). Environmental and Experimental Botany 193, 104682.

Jha, Y., 2019. The importance of zinc-mobilizing rhizosphere bacteria to the enhancement of physiology and growth parameters for paddy under salt-stress conditions. Jordan Journal of Biological Sciences 12 (2).

Kabata-Pendias, A., 2010. Trace Elements in Soils and Plants, 4th edn CRC Press, Boca Raton, FL, USA. Available from: https://doi.org/10.1017/S001447971.

Kamran, S., Shahid, I., Baig, D.N., Rizwan, M., Malik, K.A., Mehnaz, S., 2017. Contribution of zinc solubilizing bacteria in growth promotion and zinc content of wheat. Frontiers in Microbiology 8, 2593.

Khanghahi, M.Y., Ricciuti, P., Allegretta, I., Terzano, R., Crecchio, C., 2018. Solubilization of insoluble zinc compounds by zinc solubilizing bacteria (ZSB) and optimization of their growth conditions. Environmental Science and Pollution Research 25 (26), 25862–25868.

Khatoon, K., Arfi, N., Malik, A., 2022. Microbes-mediated facilitation of micronutrients uptake by plants from soil especially zinc. Microbial Biofertilizers and Micronutrient Availability. Springer, Cham, pp. 331–359.

Kiekens, L., 1995. Zinc. In: Alloway, B.J. (Ed.), Heavy Metals in Soils, 2nd edn. Blackie Academic and Professional, London, pp. 284–305.

Kumar, A., Maurya, B.R., Raghuwanshi, R., Meena, V.S., Tofazzal Islam, M., 2017. Co-inoculation with Enterobacter and rhizobacteria on yield and nutrient uptake by wheat (Triticum aestivum L.) in the alluvial soil under Indo-Gangetic plain of India. Journal of Plant Growth Regulation 36, 608–617. Available from: https://doi.org/10.1007/s00344-016-9663-5.

Kumar, A., Dewangan, S., Lawate, P., Bahadur, I., Prajapati, S., 2019. Zinc-solubilizing bacteria: a boon for sustainable agriculture. Plant Growth Promoting Rhizobacteria for Sustainable Stress Management. Springer, Singapore, pp. 139–155.

Kundu, A., Das, S., Basu, S., Kobayashi, Y., Kobayashi, Y., Koyama, H., et al., 2019. GhSTOP1, a C2H2 type zinc finger transcription factor is essential for Aluminum and proton stress tolerance and lateral root initiation in cotton. Plant Biology 21 (1), 35–44.

Kushwaha, P., Srivastava, R., Pandiyan, K., Singh, A., Chakdar, H., Kashyap, P.L., et al., 2021. Enhancement in plant growth and zinc biofortification of chickpea (Cicer arietinum L.) by Bacillus altitudinis. Journal of Soil Science and Plant Nutrition 21 (2), 922–935.

Kutman, U.B., Yildiz, B., Ozturk, L., Cakmak, I., 2010. Biofortification of durum wheat with zinc through soil and foliar application of nitrogen. Cereal Chemistry 87, 1–9.

Kutman, U.B., Yildiz, B., Cakmak, I., 2011. Improved nitrogen status enhances zinc and iron concentrations both in the whole grain and the endosperm fraction of wheat. Journal of Cereal Science 53, 118–125.

Lopez-Millan, A.F., Ellis, D.R., Grusak, M.A., 2004. Identification and characterization of several new members of the ZIP family of metal ion transporters in Medicago truncatula. Plant Molecular Biology 54 (4), 583–596.

Marschner, H., 1995. Mineral Nutrition of Higher Plants, 2nd edn Academic Press, London.

Marschner, H., Cakmak, I., 1989. High light intensity enhances chlorosis and necrosis in leaves of zinc, potassium, and magnesium deficient bean (Phaseolus vulgaris) plants. Journal of Plant Physiology 134, 308–315. Available from: https://doi.org/10.1016/S0176-1617(89)80248-2.

Marschner, P., 2012. Marschner's Mineral Nutrition of Higher Plants, 3rd edn. Academic Press, London.

Marschner, P., Crowley, D., Rengel, Z., 2011. Rhizosphere interactions between microorganisms and plants govern iron and phosphorus acquisition along the root axis: model and research methods. Soil Biology & Biochemistry 43, 883–894.

Moraghan, J., Mascagni, H., 1991. Environmental and soil factors affecting micronutrient deficiencies and toxicities. In: Mortvedt, J.J., et al., (Eds.), 'Micronutrients in Agriculture'., 2nd edn Soil Science Society of America, Madison, WI, USA, pp. 371–425.

Morcillo, R.J., Manzanera, M., 2021. The effects of plant-associated bacterial exopolysaccharides on plant abiotic stress tolerance. Metabolites 11 (6), 337.

Nishiyama, R., Kato, M., Nagata, S., Yanagisawa, S., Yoneyama, T., 2012. Identification of Zn-nicotianamine and Fe-2′-Deoxymugineic acid in the phloem sap from rice plants (Oryza sativa L.). Plant & Cell Physiology 53, 381–390. Available from: https://doi.org/10.1093/pcp/pcr188.

Noulas, C., Tziouvalekas, M., Karyotis, T., 2018. Zinc in soils, water and food crops. Journal of Trace Elements in Medicine and Biology 49, 252–260.

Oleńska, E., Małek, W., Wójcik, M., Swiecicka, I., Thijs, S., Vangronsveld, J., 2020. Beneficial features of plant growth-promoting rhizobacteria for improving plant growth and health in challenging conditions: a methodical review. The Science of the Total Environment 743, 140682. Available from: https://doi.org/10.1016/j.scitotenv.2020.140682.

Othman, N.M.I., Othman, R., Saud, H.M., Wahab, P.E.M., 2017. Effects of root colonization by zinc-solubilizing bacteria on rice plant (Oryza sativa MR219) growth. Agriculture and Natural Resources 51 (6), 532–537.

Pawar, A., Ismail, S., Mundhe, S., Patil, V.D., 2015. Solubilization of insoluble zinc compounds by different microbial isolates in vitro condition. International Journal of Tropical Agriculture 33, 865–869.

Peck, A.W., McDonald, G.K., 2010. Adequate zinc nutrition alleviates the adverse effects of heat stress in bread wheat. Plant and Soil 337 (1), 355–374.

Prathap, S., Thiyageshwari, S., Krishnamoorthy, R., Prabhaharan, J., Vimalan, B., Gopal, N.O., et al., 2022. Role of zinc solubilizing bacteria in enhancing growth and nutrient accumulation in rice plants (Oryza sativa) grown on zinc (Zn) deficient submerged soil. Journal of Soil Science and Plant Nutrition 22 (1), 971–984.

Qadar, A., Azam, Z.M., 2007. Selecting rice genotypes tolerant to zinc deficiency and sodicity: differences in concentration of major cations and sodium/potassium ratio in leaves. Journal of Plant Nutrition 30, 2061–2076. Available from: https://doi.org/10.1080/01904160701700558.

Ramesh, A., Sharma, S.K., Sharma, M.P., Yadav, N., Joshi, O.P., 2014. Inoculation of zinc solubilizing Bacillus aryabhattai strains for improved growth, mobilization and biofortification of zinc in soybean and wheat cultivated in Vertisols of central India. Applied Soil Ecology 73, 87–96. Available from: https://doi.org/10.1016/j.apsoil.2013.08.009.

Rehman, H.-u, Aziz, T., Farooq, M., Wakeel, A., Rengel, Z., 2012. Zinc nutrition in rice production systems: a review. Plant and Soil 361, 203–226.

Rehman, A., Farooq, M., Naveed, M., Nawaz, A., Shahzad, B., 2018a. Seed priming of Zn with endophytic bacteria improves the productivity and grain biofortification of bread wheat. European Journal of Agronomy 94, 98–107. Available from: https://doi.org/10.1016/j.eja.2018.01.017.

Rehman, A., Farooq, M., Ozturk, L., Asif, M., Siddique, K.H., 2018b. Zinc nutrition in wheat-based cropping systems. Plant and Soil 422 (1), 283–315.

Rengel, Z., 2002. Agronomic approaches to increasing zinc concentration in staple food crops. In: Cakmak, I., Welch, R.M. (Eds.), Impacts of Agriculture on Human Health and Nutrition. UNESCO, EOLSS Publishers, Oxford, UK.

Rengel, Z., 2015. Availability of Mn, Zn and Fe in the rhizosphere. Journal of Soil Science and Plant Nutrition 15, 397–409. Available from: https://doi.org/10.4067/S0718-95162015005000036.

Rengel, Z., Römheld, V., 2000. Root exudation and Fe uptake and transport in wheat genotypes differing in tolerance to Zn deficiency. Plant and Soil 222, 25–34. Available from: https://doi.org/10.1023/A:1004799027861.

Riaz, U., Murtaza, G., Anum, W., Samreen, T., Sarfraz, M., Nazir, M.Z., 2021. Plant Growth-Promoting Rhizobacteria (PGPR) as biofertilizers and biopesticides. Microbiota and Biofertilizers. Springer, Cham, pp. 181–196.

Ricachenevsky, F.K., de Araújo Junior, A.T., Fett, J.P., Sperotto, R.A., 2018. You shall not pass: root vacuoles as a symplastic checkpoint for metal translocation to shoots and possible application to grain nutritional quality. Frontiers in Plant Science 9, 412.

Rose, T.J., Impa, S.M., Rose, M.T., Pariasca-Tanaka, J., Mori, A., Heuer, S., et al., 2013. Enhancing phosphorus and zinc acquisition efficiency in rice: a critical review of root traits and their potential utility in rice breeding. Annals of Botany 112 (2), 331–345.

Rupa, T.R., Tomar, K.P., 1999. Zinc desorption kinetics as influenced by pH and phosphorus in soils. Communications in Soil Science and Plant Analysis 30, 1951–1962.

Ryan, P.R., Dessaux, Y., Thomashow, L.S., Weller, D.M., 2009. Rhizosphere engineering and management for sustainable agriculture. Plant and Soil 321 (1), 363–383.

Sadeghzadeh, B., 2013. A review of zinc nutrition and plant breeding. Journal of Soil Science and Plant Nutrition 13, 905–927. Available from: https://doi.org/10.4067/S0718-95162013005000072.

Saini, P., Nagpal, S., Saini, P., Kumar, A., Gani, M., 2021. Microbial mediated zinc solubilization in legumes for sustainable agriculture. Phytomicrobiome Interactions and Sustainable Agriculture. Wiley, pp. 254–276.

Sarathambal, C., Thangaraju, M., Paulraj, C., Gomathy, M., 2010. Assessing the Zinc solubilization ability of Gluconacetobacter diazotrophicus in maize rhizosphere using labelled (65) Zn compounds. Indian Journal of Microbiology 50, 103–109. Available from: https://doi.org/10.1007/s12088-010-0066-1.

Saravanan, V.S., Madhaiyan, M., Thangaraju, M., 2007. Solubilization of zinc compounds by the diazotrophic, plant growth promoting bacterium Gluconacetobacter diazotrophicus. Chemosphere 66 (9), 1794–1798.

Saravanan, V.S., Kumar, M.R., Sa, T.M., 2011. Microbial zinc solubilization and their role on plants. In: Maheshwari, D. (Ed.), Bacteria in Agrobiology: Plant Nutrient Management. Springer, Berlin, Heidelberg, pp. 47–63. Available from: https://doi.org/10.1007/978-3-642-21061-7_3.

Sarret, G., Vangronsveld, J., Manceau, A., Musso, M., D'Haen, J., Menthonnex, J.-J., et al., 2001. Accumulation forms of Zn and Pb in Phaseolus vulgarisin the presence and absence of EDTA. Environmental Science & Technology 35, 2854–2859. Available from: https://doi.org/10.1021/es000219d.

Sattelmacher, B., 2001. The apoplast and its significance for plant mineral nutrition. New Phytologist 149 (2), 167–192.

Shaikh, S., Saraf, M., 2017a. Biofortification of Triticum aestivum through the inoculation of zinc solubilizing plant growth promoting rhizobacteria in field experiment. Biocatalysis and Agricultural Biotechnology 9, 120–126.

Shaikh, S., Saraf, M., 2017b. Zinc biofortification: strategy to conquer zinc malnutrition through zinc solubilizing PGPR's. Biomedical Journal of Scientific & Technical Research 1, 224–226. Available from: https://doi.org/10.26717/BJSTR.2017.01.000158.

Shakeel, M., Rais, A., Hassan, M.N., Hafeez, F.Y., 2015. Root associated Bacillus sp. improves growth, yield and zinc translocation for basmati rice (Oryza sativa) varieties. Frontiers in Microbiology 6, 1286.

Sharma, S.K., Sharma, M.P., Ramesh, A., Joshi, O.P., 2012. Characterization of zinc-solubilizing Bacillus isolates and their potential to influence zinc assimilation in soybean seeds. Journal of Microbiology and Biotechnology 22, 352–359. Available from: https://doi.org/10.4014/jmb.1106.05063.

Sharma, S., Chandra, D., Sharma, A.K., 2021. Rhizosphere plant–microbe interactions under abiotic stress. Rhizosphere Biology: Interactions Between Microbes and Plants. Springer, Singapore, pp. 195–216.

Subramanian, K.S., Tenshia, V., Jayalakshmi, K., Ramach, V., 2009. Role of arbuscular mycorrhizal fungus (Glomus intraradices)(fungus aided) in zinc nutrition of maize. Journal of Agricultural Biotechnology and Sustainable Development 1 (1), 029–038.

Sunithakumari, K., Devi, S.P., Vasandha, S., 2016. Zinc solubilizing bacterial isolates from the agricultural fields of Coimbatore, Tamil Nadu, India. Current Science 196–205.

Tahir, M., Sarwar, M.A., 2013. Plant growth promoting rhizobacteria (PGPR): a budding complement of synthetic fertilizers for improving crop production. Pakistan Journal of Life and Social Sciences 11, 1–7.

Tesfaye, F., Liu, X., Zheng, J., Cheng, K., Bian, R., Zhang, X., et al., 2021. Could biochar amendment be a tool to improve soil availability and plant uptake of phosphorus? A meta-analysis of published experiments. Environmental Science and Pollution Research 28 (26), 34108–34120.

Tinker, P.B., Lauchli, A., 1984. Advances in Plant Nutrition. Academic. Publishers, San Diego.

Tsui, C., 1948. The role of zinc in auxin synthesis in the tomato plan. American Journal of Botany 35, 172–179.

Uchida, R., 2000. Essential nutrients for plant growth: nutrient functions and deficiency symptoms. Plant Nutrient Management in Hawaii's Soils 4, 31–55.

Ullah, A., Farooq, M., Rehman, A., Hussain, M., Siddique, K.H., 2020. Zinc nutrition in chickpea (Cicer arietinum): a review. Crop and Pasture Science 71 (3), 199–218.

Upadhayay, V.K., Singh, A.V., Khan, A., 2022. Cross talk between zinc-solubilizing bacteria and plants: a short tale of bacterial-assisted zinc biofortification. Frontiers in Soil Science 1, 788170.

Vaid, S.K., Kumar, B., Sharma, A., Shukla, A.K., Srivastava, P.C., 2014. Effect of zinc solubilizing bacteria on growth promotion and zinc nutrition of rice. Journal of Soil Science and Plant Nutrition 14, 889–910. Available from: https://doi.org/10.4067/S0718-95162014005000007.

Verma, M., Singh, A., Dwivedi, D.H., Arora, N.K., 2020. Zinc and phosphate solubilizing Rhizobium radiobacter (LB2) for enhancing quality and yield of loose leaf lettuce in saline soil. Environmental sustainability 3, 209–218. Available from: https://doi.org/10.1007/s42398-020-00110-4.

Vidyashree, D.N., Muthuraju, N., Panneerselvam, P., Mitra, D., 2018. Organic acids production by zinc solubilizing bacterial isolates. International Journal of Current Microbiology and Applied Sciences 7 (10), 626–633.

Watts-Williams, S.J., Cavagnaro, T.R., 2018. Arbuscular mycorrhizal fungi increase grain zinc concentration and modify the expression of root ZIP transporter genes in a modern barley (Hordeum vulgare) cultivar. Plant Science 274, 163–170.

Wei, R., Wang, X., Zhang, W., Shen, J., Zhang, H., Gao, Y., et al., 2020. The improved phosphorus utilization and reduced phosphorus consumption of ppk-expressing transgenic rice. Field Crops Research 248, 107715. Available from: https://doi.org/10.1016/j.fcr.2020.107715.

Wu, S., Hu, C., Tan, Q., Li, L., Shi, K., Zheng, Y., et al., 2015. Drought stress tolerance mediated by zinc-induced antioxidative defense and osmotic adjustment in cotton (Gossypium hirsutum). Acta Physiologiae Plantarum 37 (8), 1–9.

Yadav, R.C., Sharma, S.K., Varma, A., Rajawat, M.V.S., Khan, M.S., Sharma, P.K., et al., 2022. Modulation in biofertilization and biofortification of wheat crop by inoculation of zinc-solubilizing rhizobacteria. Frontiers in Plant Science 13, 777771-777771.

Yang, Y., Ma, C., Xu, Y., Wei, Q., Imtiaz, M., Lan, H., et al., 2014. A zinc finger protein regulates flowering time and abiotic stress tolerance in chrysanthemum by modulating gibberellin biosynthesis. The Plant Cell 26 (5), 2038–2054.

Yasmin, S., 2011. Characterization of Growth Promoting and Antagonistic Bacteria Associated with Rhizosphere of Cotton and Rice. NIBGE, Faisalabad.

Yokosho, K., Yamaji, N., Ueno, D., Mitani, N., Ma, J.F., 2009. OsFRDL1 is a citrate transporter required for efficient translocation of iron in rice. Plant Physiology 149, 297–305. Available from: https://doi.org/10.1104/pp.108.128132.

Zaheer, A., Malik, A., Sher, A., Qaisrani, M.M., Mehmood, A., Khan, S.U., et al., 2019. Isolation, characterization, and effect of phosphate-zinc-solubilizing bacterial strains on chickpea (Cicer arietinum L.) growth. Saudi Journal of Biological Sciences 26 (5), 1061–1067.

Zainab, N., Khan, A.A., Azeem, M.A., Ali, B., Wang, T., Shi, F., et al., 2021. PGPR-mediated plant growth attributes and metal extraction ability of Sesbania sesban l. in industrially contaminated soils. Agronomy 11 (9), 1820.

Zhang, Y.Q., Deng, Y., Chen, R.Y., Cui, Z.L., Chen, X.P., Yost, R., et al., 2012. The reduction in zinc concentration of wheat grain upon increased phosphorus-fertilization and its mitigation by foliar Zn application. Plant and Soil 361, 143–152.

Zhao, A.Q., Tian, X.H., Cao, Y.X., Lu, X.C., Liu, T., 2014. Comparison of soil and foliar zinc application for enhancing grain zinc content of wheat when grown on potentially zinc-deficient calcareous soils. Journal of the Science of Food and Agriculture 94 (10), 2016–2022.

Zhao, Y., Yao, J., Li, H., Sunahara, G., Li, M., Tang, C., Duran, R., Ma, B., Liu, H., Feng, L., Zhu, J., 2024. Effects of three plant growth-promoting bacterial symbiosis with ryegrass for remediation of Cd, Pb, and Zn soil in a mining area. Journal of Environmental Management 353, 120167.

# Index

*Note*: Page numbers followed by "*f*" and "*t*" refer to figures and tables, respectively.

## A

## B

## G



## H

## N

# O



# P

## R

## S

9780323913140